VOLUME SIX HUNDRED SEVENTY-EIGHT

METHODS IN ENZYMOLOGY

Small Angle Scattering Part B: Methods for Structural Interpretation

METHODS IN ENZYMOLOGY

VOLUME SIX HUNDRED SEVENTY-EIGHT

METHODS IN ENZYMOLOGY

Small Angle Scattering Part B: Methods for Structural Interpretation

Edited by

JOHN A. TAINER
Robert A. Welch Chair in Chemistry
Departments of Molecular & Cellular Oncology
and of Cancer Biology Director Structural Biology
University of Texas MD Anderson Cancer Center
Houston, TX

ELSEVIER

ACADEMIC PRESS
An imprint of Elsevier

Academic Press is an imprint of Elsevier
50 Hampshire Street, 5th Floor, Cambridge, MA 02139, United States
525 B Street, Suite 1650, San Diego, CA 92101, United States
The Boulevard, Langford Lane, Kidlington, Oxford OX5 1GB, United Kingdom
125 London Wall, London, EC2Y 5AS, United Kingdom

First edition 2023

ISBN: 978-0-323-99181-0
ISSN: 0076-6879

For information on all Academic Press publications
visit our website at https://www.elsevier.com/books-and-journals

Publisher: Zoe Kruze
Developmental Editor: Federico Paulo S. Mendoza
Production Project Manager: Sudharshini Renganathan
Cover Designer: Greg Harris

Typeset by STRAIVE, India

Contents

Contributors

Chris A. Brosey
Department of Molecular and Cellular Oncology, The University of Texas MD Anderson Cancer Center, Houston, TX, United States

Kathryn Burnett
Molecular Biophysics and Integrated Bioimaging Division, Lawrence Berkeley National Laboratory, Berkeley, CA, United States

Lester Carter
Stanford Synchrotron Radiation Lightsource, SLAC National Accelerator Laboratory, Menlo Park, CA, United States

Leonie Chatzimagas
Theoretical Physics and Center for Biophysics, Saarland University, Saarbrücken, Germany

Naga Babu Chinnam
Department of Molecular and Cellular Oncology, The University of Texas MD Anderson Cancer Center, Houston, TX, United States

Scott Classen
Molecular Biophysics and Integrated Bioimaging Division, Lawrence Berkeley National Laboratory, Berkeley, CA, United States

Tomer Cohen
The Rachel and Selim Benin School of Computer Science and Engineering, The Hebrew University of Jerusalem, Jerusalem, Israel

Joshua T. Del Mundo
Department of Chemical Engineering, The Pennsylvania State University, University Park, PA, United States

Florent Delhommel
Institute of Structural Biology, Helmholtz Zentrum München, Neuherberg; Bavarian NMR Center, Department of Chemistry, Technical University of Munich, Garching, Germany

Eric Fagerberg
Division of Theoretical Chemistry, Lund University, Lund, Sweden

Frank Gabel
Université Grenoble Alpes; Commissariat à l'Energie Atomique et aux Energies Alternatives, Direction de la Recherche Fondamentale; Centre National de la Recherche Scientifique, Institut de Biologie Structurale, Grenoble, France

Enrique D. Gomez
Department of Chemical Engineering; Department of Materials Science and Engineering and Materials Research Institute, The Pennsylvania State University, University Park, PA, United States

Esther W. Gomez
Department of Chemical Engineering; Department of Biomedical Engineering, The Pennsylvania State University, University Park, PA, United States

Thomas D. Grant
Department of Structural Biology, Jacobs School of Medicine and Biomedical Sciences, SUNY University at Buffalo, Buffalo, NY, United States

Matan Halfon
The Rachel and Selim Benin School of Computer Science and Engineering, The Hebrew University of Jerusalem, Jerusalem, Israel

Michal Hammel
Molecular Biophysics and Integrated Bioimaging Division, Lawrence Berkeley National Laboratory, Berkeley, CA, United States

Jochen S. Hub
Theoretical Physics and Center for Biophysics, Saarland University, Saarbrücken, Germany

Greg L. Hura
Molecular Biophysics and Integrated Bioimaging Division, Lawrence Berkeley National Laboratory, Berkeley; Department of Chemistry and Biochemistry, University of California Santa Cruz, Santa Cruz, CA, United States

Tushar Jain
Computational Biology, Adimab LLC, Palo Alto, CA, United States

Cy M. Jeffries
European Molecular Biology Laboratory (EMBL), Hamburg Unit, c/o Deutsches Elektronen-Synchrotron, Hamburg, Germany

Darin E. Jones
Department of Pharmaceutical Sciences, University of Arkansas for Medical Sciences, Little Rock, AR, United States

Annette Eva Langkilde
Department of Drug Design and Pharmacology, University of Copenhagen, Copenhagen, Denmark

Samuel Lenton
Drug Delivery and Biophysics of Biopharmaceuticals, Department of Pharmacy, University of Copenhagen, Copenhagen, Denmark; Division of Theoretical Chemistry, Lund University, Lund, Sweden

Santiago Martínez-Lumbreras
Institute of Structural Biology, Helmholtz Zentrum München, Neuherberg; Bavarian NMR Center, Department of Chemistry, Technical University of Munich, Garching, Germany

Haydyn D.T. Mertens
European Molecular Biology Laboratory, Hamburg Unit, Hamburg, Germany

Davide Moiani
Department of Molecular and Cellular Oncology, The University of Texas MD Anderson Cancer Center, Houston, TX, United States

Daniel T. Murray
Molecular Biophysics and Integrated Bioimaging Division, Lawrence Berkeley National Laboratory, Berkeley, CA, United States

Sintu Rongpipi
Advanced Light Source and Materials Science Division, Lawrence Berkeley National Laboratory, Berkeley, CA, United States

Michael Sattler
Institute of Structural Biology, Helmholtz Zentrum München, Neuherberg; Bavarian NMR Center, Department of Chemistry, Technical University of Munich, Garching, Germany

Dina Schneidman-Duhovny
The Rachel and Selim Benin School of Computer Science and Engineering, The Hebrew University of Jerusalem, Jerusalem, Israel

Beth Sharkey
High-Throughput Expression, Adimab LLC, Lebanon, NH, United States

Runze Shen
Department of Molecular and Cellular Oncology, The University of Texas MD Anderson Cancer Center, Houston, TX, United States

David S. Shin
Molecular Biophysics and Integrated Bioimaging Division, Lawrence Berkeley National Laboratory, Berkeley, CA, United States

Arvind Sivasubramanian
Computational Biology, Adimab LLC, Palo Alto, CA, United States

Marie Skepö
Division of Theoretical Chemistry, Lund University; LINXS—Institute of Advanced Neutron and X-ray Science, Lund, Sweden

Aleem Syed
Department of Molecular and Cellular Oncology, The University of Texas MD Anderson Cancer Center, Houston, TX, United States

John A. Tainer
Department of Molecular and Cellular Oncology, The University of Texas MD Anderson Cancer Center; Department of Cancer Biology, University of Texas MD Anderson Cancer Center, Houston, TX; Molecular Biophysics and Integrated Bioimaging Division, Lawrence Berkeley National Laboratory, Berkeley, CA, United States

Jill Trewhella
School of Life and Environmental Sciences, The University of Sydney, Sydney, NSW, Australia

Susan E. Tsutakawa
Molecular Biophysics and Integrated Bioimaging Division, Lawrence Berkeley National Laboratory, Berkeley, CA, United States

Mark Tully
BioSAXS beamline, BM29, European Synchrotron Radiation Facility, ESRF, Grenoble, France

Bente Vestergaard
Department of Drug Design and Pharmacology, University of Copenhagen, Copenhagen, Denmark

Andrew E. Whitten
Australian Nuclear Science and Technology Organisation, Lucas Heights, NSW, Australia

Preface

"*And yet it moves*"—This phrase, attributed to Galileo Galilei, emphasizes the critical importance of defining matter with its movement over time for understanding a system.

Advances in experimental and computational structural methods are increasingly making transformative improvements in defining and predicting biological mechanisms. At the heart of these improvements is the critical ability to visualize both structural detail at the level of their molecular building blocks along with their changing conformation and assemblies as a function of time. Prior *Methods in Enzymology* volumes on diffraction methods and cryo-electron microscopy have presented detailed techniques for accurate and often high-resolution structures for crystallized and cryo-cooled biological macromolecules and assemblies. Here the contributing authors for *Scattering Methods for Biological Macromolecules* focus on procedures employing the scattering of X-rays (SAXS) and neutrons (SANS) by biological macromolecules, as divided into Volume 678, *Part A, Methods for Structural Investigation* and Volume 679, *Part B, Methods for Structural Interpretation*. These scattering methods can comprehensively measure and model molecular structures in solution, including their flexibility and dynamic functional conformations and assemblies. The analyses not only can approach near physiological conditions but also can broadly evaluate solution states and conditions so as to better understand the relationships between macromolecular structures, their solution environments, and their functionally important flexibility and dynamics. In essence, defining the time-dependent conformation and assembly state of macromolecules in their biological environment is foundational and predictive knowledge for their activities and regulation. In this context, these volumes encompass the capabilities that make scattering methods a premier approach for measuring molecular conformations and assemblies in both space and time, as well as for efficiently leveraging information on atomic models from all other techniques.

Each volume is organized into sections and topics pertaining to the spectrum of scattering methods and the logical sequence of steps involved in first investigating (Part A) and then interpreting structure (Part B). Part A is primarily devoted to the experimental aspects of scattering experiments and data, starting from sample preparation and handling, followed by methods of data collection and data assessment. Part B primarily discusses data and structural interpretation, covering various aspects of scattering measurements and methods for their analysis as well as the resulting structural insights. Armed by the information in these articles, the reader can proceed knowledgeably and productively toward their structural goals with an appreciation of the practical and fundamental aspects of each technique, including its potential to substantially augment combined and integrated structural approaches.

In providing these two volumes on scattering methods, the authors and I collectively strived to attain a threefold overall goal: (1) to give biochemists and biologists an introduction to the field of macromolecular structure analysis by scattering, offering them practical guidance to available techniques and pathways to gain functionally important insights into macromolecular structures; (2) to give scattering practitioners a comprehensive spectrum of current techniques available to them from robust to state-of-the-art and into pushing the technology's frontiers; and (3) to provide sufficient knowledge and guidance for interested researchers to effectively apply scattering approaches to combined methods and integrated structural biology. The authors of these chapters include both those upon whose shoulders we stand in developing new methods as well as established leaders and new pioneers in extending the reach and impact of scattering methods for structural biology.

In Volume 677 Graewert and Svergun provide the strong foundation for *Part A, Methods for Structural Investigation* in Chapter 1, by comprehensively describing sample environments and requirements for biological SAXS. In Chapters 2–5, Pollack, Duff, Krueger, and Jeffries and colleagues provide expert information on equipment and procedures to harness the power of neutrons for contrast and other experiments. In Chapters 6–7, Rosenberg and Vachette and their colleagues detail considerations and methods for coupling size-exclusion chromatography with scattering (SEC-SAXS). Tidow and Gabel and their colleagues then describe powerful approaches to time-resolved SAXS (Chapter 8) and SANS (Chapter 9). Vestergaard and Langkilde (Chapter 10) provide an essential guide for the application of SAXS to the biologically and medically important process of protein fibrillation. Pioneering methods for high-pressure SAXS experiments and for soft X-ray scattering are described by Gillilan (Chapter 11) and Gomez and colleagues (Chapter 12), respectively. In Chapter 13, Yang discusses SAXS on lipid membranes and in Chapter 14 Todow and colleagues present SANS on membrane protein by stealth constitution systems. Predicting scattering with explicit solvent is detailed by Hub and Chatzimagas, who describe the WAXSiS web server for calculation of SAXS/WAXS curves based on explicit-solvent molecular dynamics, as a useful tool to extract maximum information from the SAXS profile (Chapter 15). In Chapter 16, Skepö and colleagues describe sample preparation and data collection considerations for applying SAXS to highly flexible protein samples. Wang and colleagues (Chapter 17) consider SAXS analyses of RNA, including an approach to simplifying exploration of conformational space and tying this in with SAXS constraints. In Chapter 18, Bernadó and Sagar provide unique information for enhanced biological insights from SAS data, including conformational rearrangements and transient biomolecular complexes, to anchor this volume.

Trewhella provides the foundation for Volume 676, *Part B, Methods for Structural Interpretation* in Chapter 1, with insights into the method and its development into a mainstream technique at the frontier of integrative structural biology, including key standards for data quality, model validation, publication, and data sharing. In Chapter 2, Hub and Chatzimagas describe a rigorous protocol for Bayesian interpretation of SAXS-driven MD simulations, models, and methods that will aid understanding of biological function by enabling understanding of conformational dynamics. Whitten and Jeffries describe SANS data analysis and modeling (Chapter 3), including a clear theoretical overview plus a step-by-step procedure for available software. Gabel describes time-resolved small-angle neutron scattering (Chapter 4) with an exemplary macromolecular system including the practical use of temperature and thermophilic systems to control kinetics. In Chapter 5, Gomez and colleagues describe informative procedures for predicting scattering contrast, identifying energies plus methods for interpretation suitable for measurements using resonant soft X-ray scattering (RSoXS) and near-edge X-ray absorption fine structure (NEXAFS) spectroscopy to gain structural insights. Grant (Chapter 6) describes his ab initio modeling algorithm DENSS, which allows examination of low-resolution density fluctuations inside the particle envelope, which would seem to have many biological applications. In Chapter 7, Mertens considers the components and developments in the ATSAS package for its user-driven applications informing data reduction, analysis, and modeling strategies. Schneidman and colleagues describe a protocol for accurate integrative modeling of antibody-antigen complexes in Chapter 8, including SAXS profiles, protein-protein docking, deep-learning models, and statistical potentials. In Chapter 9, Sattler and colleagues present the integration of

SAS and NMR as complementary solution structural techniques to better describe structure with conformational dynamics with application guidelines depending upon the nature of the system. Skepö and coauthors present SAXS spectra analyses for highly flexible protein sample solutions in Chapter 10 with the definable parameters, model-free analysis methods, plus advanced methods to examine ensembles and crowded solutions. In Chapter 11, Brosey and colleagues present an efficient protocol for applying high-throughput SAXS to screening the binding of chemical ligands. In Chapter 12, Tsutakawa and colleagues consider SAXS applications to improved structure prediction. Vestergaard and Langkilde (Chapter 13) describe analysis of protein fiber structures by SAXS. In Chapter 14, Hura and Murray reveal elegant web-accessible approaches to analyses of similarity and correlated SAXS data sets to complete the volume.

In practice it is nearly impossible to navigate through scattering experiments, controls, data validation and analysis, and data interpretation, which can be conceptually simple but are usually technically sophisticated, without more methods detail than is available in published papers, which typically are space limited and focused upon new results. In their sharing of techniques and methods that they have painstakingly developed, validated, and optimized by years of tests and efforts, the authors herein made a remarkable contribution and investment toward the reproducibility and advanced impact of scattering for structural biology. In this regard, I acknowledge the generosity of all the contributing authors who strived to reveal their insights and explicit instructions on how to utilize their valuable methods to our readers. We furthermore thank our many reviewers, whose time and attention contributed valuable suggestions to improve the chapters. I am grateful to Anna Pyle and David Christianson (Co-Editors-in-Chief of *Methods in Enzymology*) for the invitation and strong support. Most credit for the quality of the two volumes should go to the authors and reviewers, along with David Christianson, who was a force of nature in implementing reviewing and editing. I thank Paulo Mendoza and the Elsevier staff for their patient and expert efforts at all stages to strengthen the volumes for readers. I acknowledge my National Cancer Institute R35 CA220430 support for my efforts on the project "*Mesocale and Nanoscale Technologies Integrated by Structures for DNA Repair Complexes* (MANTIS-DRC)." I admit that I am responsible for any mistakes and omissions. Yet, due to the combined efforts of authors, reviewers, and editors, the chapters in these two volumes provide a depth of coverage and clarity of presentation suitable for readers spanning from students to experts.

Furthermore, the combined chapters show that scattering methods have advanced to the point that these articles are likely to become classic "go to" references that stand the test of time. So we expect that many laboratories will find these collected methods helpful as a starting point to adapt for their own systems and to extend these methods for in-depth understanding of mechanism, regulation, and functions. Reflecting on these chapters, it is an honor to be the editor of both volumes, where leaders and innovators I have admired, along with early investigators who are pushing methods forward into new frontiers, provide inspiration and insights to me and others working on hybrid methods to unveil biology at the nanoscale. In my view, the collected articles in these two volumes provide a critical enabling path to defining macromolecular structures and mechanisms in solution, including biologically relevant flexibility in both functionally important complexes and conformations.

John A. Tainer
University of Texas MD Anderson Cancer Center

CHAPTER ONE

Data quality assurance, model validation, and data sharing for biomolecular structures from small-angle scattering

Jill Trewhella*
School of Life and Environmental Sciences, The University of Sydney, Sydney, NSW, Australia
*Corresponding author: e-mail address: jill.trewhella@sydney.edu.au

Contents

Abstract

Key to small-angle scattering (SAS) maturing and becoming a mainstream structural biology technique was the work done by the SAS community to establish standards for data quality, model validation and data sharing. Through a consultative process spanning more than a decade and a half, guidelines for publication have been established that include criteria for evaluating data quality and for model validation. In this process gaps were identified that stimulated innovation and development of new tools, for example new measures of model ambiguity and of the goodness-of-fit of a model to SAS data that complement the traditional global fit parameter χ^2. The need for a global repository for biomolecular SAS data and models was identified and the SASBDB was established as a searchable, curated, freely accessible, downloadable database of experimental data, experimental conditions, sample details, derived models, and their fit to the data. Importantly, the SASBDB uses a common dictionary format that supports archiving of structures solved using integrative methods to support seamless data exchange with a federated system of public databanks that includes

Methods in Enzymology, Volume 678
ISSN 0076-6879
https://doi.org/10.1016/bs.mie.2022.11.002

the world-wide Protein Data Bank (wwPDB) as the major repository for structural biology. Thus, biomolecular SAS is now well-positioned to achieve its full potential as a mainstream structural biology technique contributing at the frontier of integrative structural biology and meeting "best practice" standards for data quality assurance and data sharing.

1. Becoming a mainstream structural biology technique

In their classic monograph on small-angle scattering (SAS) (Guinier & Fournet, 1955) predicted that the technique would have its greatest impact in structural biology. Their reasoning was that biomolecules fold into defined structures according to the laws of physics and chemistry and thus can be prepared as solutions of monodispersed particles having uniform size and shape; the fundamental requirement for interpretating SAS data in terms of a structure. Despite this apparent advantage, however, it was not until the 1990s that the numbers of biomolecular SAS citations began a seemingly exponential rise from a few 10s per year to >17,600 per year for the 5-year period 2017–2021 (Web of Science Core Collection search using topics "protein" and "small-angle scattering"). Significant to the timing of this rise was the publication of the first program to rapidly calculate the scattering profile of a biomolecular from atomic coordinates (Svergun, Barberato, & Koch, 1995). The emergence of this critical missing link connecting high-resolution structures and low-resolution solution SAS data dramatically increased the potential impact of SAS for understanding the relationship between biomolecular structure and function. The subsequent 30-year sustained growth in citations for biomolecular SAS was also fuelled by continuing advances in capabilities for data analysis and interpretation accompanied by innovation in sample preparation and instrumentation. Additionally, there was substantial investment in the development of commercial benchtop laboratory SAXS instrumentation as well as increasing numbers of dedicated biomolecular SAS facilities at international synchrotron and neutron facilities that opened the field to SAS experts and non-experts alike (for recent reviews, see Brosey & Tainer, 2019; Chen & Hennig, 2018; Da Vela & Svergun, 2020; Hub, 2018; Kursula, 2021; Mahieu & Gabel, 2018; Schroer & Svergun, 2018; Trewhella, 2016, 2022; Tuukkanen, Spilotros, & Svergun, 2017).

By 1989 virtually all journals required deposition of atomic coordinates for reported structures to the Protein Data Bank (PDB) (Berman, Kleywegt, Nakamura, & Markley, 2012). With coordinates thus broadly available, structural interpretation of biomolecular SAS routinely took advantage of knowledge of high-resolution structures to link solution structural information to function. An early example is the small-angle X-ray scattering (SAXS) study that showed that the helix connecting the globular Ca^{2+}-binding domains of calmodulin had to be flexible in solution rather than adopting the extended conformation seen in the crystal structure (Heidorn & Trewhella, 1988). Further SAXS and small-angle neutron scattering (SANS) studies showed how this flexibility facilitated the dramatic conformational rearrangements calmodulin underwent to accommodate binding to structurally distinct regulatory targets (Heidorn et al., 1989; Trewhella, Blumenthal, Rokop, & Seeger, 1990). With the emergence of methods to optimize the fit of atomistic models to SAS data, applications have increasingly sought to characterize solution structures using atomic coordinates from high-resolution techniques (for example, Comoletti et al., 2010; Custodio et al., 2020; Groothuizen et al., 2013; Hammel et al., 2021; Hammel & Tainer, 2021; Hennig, Wang, Sonntag, Gabel, & Sattler, 2013; Kelly et al., 2020; Kim, Bothe, Frederick, Holder, & Markley, 2014; Lin et al., 2020; Lycksell et al., 2021; Muller et al., 2013; Rambo & Tainer, 2013; Sauve et al., 2015; Takala et al., 2014; Tidow et al., 2012; Whitten, Jeffries, Harris, & Trewhella, 2008; Wilamowski et al., 2021).

As the structural biology frontier has moved steadily to larger and more complex structures that cannot be solved using a single technique, integrative methods that combine disparate experimental techniques with computational methods (Rout & Sali, 2019; Sali, 2021) have grown to the extent that the world-wide Protein Data Bank (wwPDB) announced (Burley et al., 2017) and now is implementing (Vallat et al., 2021; Vallat, Webb, Westbrook, Sali, & Berman, 2018) a prototype system, PDB-Dev, for depositing integrative (or "hybrid) structural models. As of September 2022, SAS data are included in 16 of the 104 integrative structures deposited in PDB-Dev. Thus, we are witnessing a natural confluence for the paths of biomolecular SAS and integrative methods (Grishaev, 2017; Schneidman-Duhovny, Kim, & Sali, 2012; Schroer & Svergun, 2018).

Biomolecular SAS has evolved from the early days where its application was often to monitor the radius of gyration for biomolecules in solution, providing insights, for example, into oligomerization states or large-scale conformational changes upon ligand binding, to today where it is contributing to solving increasingly detailed structures of nature's most complex biomolecular assemblies. In the process, the field has moved from being largely a domain for the physical scientist to a mainstream technique for the generalist structural biologist. For biomolecular SAS in the context of three-dimensional (3D) structure analysis to achieve its full potential, as was the case for crystallography, the community had to take up the challenge to develop and agree standards for data quality, model validation, publication, and data sharing.

2. The path to standards and data sharing

The International Union of Crystallography (IUCr) has been a leader for standardization, data quality and data sharing, with a long history of highly productive collaborations in support of structural biology. Thus, it was the IUCr Commission for Small-Angle Scattering (CSAS) (https://www.iucr.org/resources/commissions/small-angle-scattering) that provided a natural platform to begin community discussions regarding data quality and reporting standards for biomolecular SAS and 3D modelling.

Historical context for the IUCr CSAS starts with the formation of the IUCr in the post-WWII period, motivated by recognition in the scientific community of the need for a body to "promote international cooperation, publication of research, standardization of units and methods and to form a focus for the relation of crystallography to other sciences" (https://www.iucr.org/iucr/history/early-history). These goals were championed by many of the most recognizable names in the history of scattering and diffraction, such as Ewald, von Laue, and Bragg. Notably, André Guinier, commonly referred to as the father of SAS (Trewhella, 2021), was a member of the international committee that authored the March 1947 report recommending the objectives of what would become the IUCr, which was formally admitted to the International Council of Scientific Unions (now the International Council for Science) on 7 April 1947. In 1995, the SAS Special Interest Group of the American Crystallographic Association (ACA) met in Montreal and discussed the desirability of setting up a broader international group to further the interests of researchers in SAS. The links between crystallography and SAS long having been recognized, the group determined that forming a Commission within the IUCr would be the most appropriate path. Further discussions ensued at SAS-96 (Campinas, Brazil, 1996), the 10th meeting in the triennial SAS conference series that brings together the international SAS community. A report and draft terms of reference ensued that were forwarded to the IUCr at the XVII Congress in Seattle where the IUCr General Assembly voted to form a permanent Commission for SAS (CSAS) at its meeting on 14 August 1996 (https://www.iucr.org/resources/commissions/small-angle-scattering/history).

At a meeting of the IUCr CSAS during SAS2006 (Kyoto, Japan, 2006) there was a discussion regarding establishing standards for publishing the results of biomolecular SAS experiments, with an emphasis on experiments aimed at structural modelling. The discussion was lively, with concerns raised regarding the potential for standards to restrict opportunities for publishing legitimate results, or for unreasonable or misguided standards to evolve. Nevertheless, there was support for the idea that for SAS to achieve its full potential in the rapidly evolving field of structural biology an agreed framework for reporting results was needed. The discussion made it clear that community uptake of such a framework would benefit from focusing the discussion on "guidelines" rather than "standards." Further, success would require a structured, yet flexible process and ample opportunity for community input. The fora that would provide opportunity for this input would include CSAS, the IUCr Journals Commission, open meetings at the triennial SAS meetings, as well as SAS special interest groups at national meetings. With a commitment to broad consultation, the process began with leadership from experts across the international community (including SAS experimenters, providers/developers of analysis tools and instrumentation) there were ongoing discussions at future national and international meetings, and reporting of progress in open access articles. Critical to progress in reaching each of the major milestones on this path (Table 1) was allowing time to embed new practices and obtain the required resources.

Table 1 Timeline of milestones leading to publication guidelines and requirements for data deposition for biomolecular SAS.

Year	Activity	Documentation
2006	Discussion of the need for standards for biomolecular SAS publication at the IUCr CSAS meeting at the Kyoto triennial SAS meeting (SAS2006).	
2012	IUCr Journals Commission accepts CSAS endorsed draft publication guidelines for biomolecular SAS that are published, along with their rationale, in an open access paper.	Jacques, Guss, Svergun, and Trewhella (2012)
	Commentary on the importance of quality control and a standard reporting framework for biomolecular SAS.	Jacques, Guss, and Trewhella (2012)

Continued

Table 1 Timeline of milestones leading to publication guidelines and requirements for data deposition for biomolecular SAS.—cont'd

Year	Activity	Documentation
2013	wwPDB SASvtf, established to consider if PDB should accept models based on SAS data, makes 6 recommendations, for: 1. a global SAS data repository 2. a standard dictionary to manage SAS data 3. options for different kinds of models (atomistic and shape), 4. criteria for data quality and the extent to which a model fits the data 5. the need to consider archiving of a broader set of data types, and 6. bringing thought leaders together to define what to archive and how	Trewhella et al. (2013)
2014	SASBDB is developed in response to SASvtf recommendations 1 and 2.	Valentini, Kikhney, Previtali, Jeffries, and Svergun (2015)
2015	New methods developed for assessing model ambiguity and for complementing the global goodness-of-fit parameters χ^2 to identify and assess significance of local deviations developed in response to SASvtf recommendation 4.	Petoukhov and Svergun (2015) Franke, Jeffries, and Svergun (2015)
	Per SASvtf recommendations 5 and 6, IHMvtf meeting and report makes 5 recommendations for structures solved using integrative methods: 1. models plus all experimental data, metadata, and protocols should be archived 2. a flexible model representation is required 3. procedures for estimating the uncertainty to be developed, validated, and adopted 4. a federated system of model and data archives should be created 5. publication standards for integrative models should be established	Sali et al. (2015)
2016	The sasCIF extension of the PDBx/mmCIF is extended to include experimental data and models with relevant meta data for data deposition and management in support of IHMvtf recommendation 4.	Kachala, Westbrook, and Svergun (2016)

Table 1 Timeline of milestones leading to publication guidelines and requirements for data deposition for biomolecular SAS.—cont'd

Year	Activity	Documentation
2017	SASvtf and NMRvtf charged with detailing what data and meta-data are required for SAS and SAS/NMR model deposition and validation in PDB.	
	Preparation of updated guidelines for publication of biomolecular SAS through the work of an expanded SASvtf plus CSAS members and consultants.	
	Streamlining between PDB OneDep and SASBDB implemented for structures solved by jointly refining NMR and SAS data.	Young et al. (2017) and Kikhney, Borges, Molodenskiy, Jeffries, and Svergun (2020)
	Updated guidelines presented in platform talk and discussed at CSAS open meeting at the IUCr 24th General Assembly and Congress (Hyderabad, August 2017) for comment prior to publication as an open access paper.	Trewhella et al. (2017)
	PDB-Dev is announced for archiving integrative structures and SAS was the first technique after crystallography, NMR and EM to be included with data quality and model fit validation.	Burley et al. (2017) Vallat et al. (2021)
2019	SASvtf initiates round-robin project to assess limits of reproducibility for SAS data and provide a benchmark data set for modeling.	
2022	*IUCrJ* announces the IUCr biology journals will require deposition of biomolecular SAS data into SASBDB for review purposes.	Baker et al. (2022)
2022	Publication of "A Round Robin Approach Provides a Detailed Assessment of Biomolecular Small-Angle Scattering Data Reproducibility and Yields Consensus Curves for Benchmarking."	Trewhella et al. (2022)

3. Draft publication guidelines and plans for data archiving

In 2012, one outcome of the prior 6 years of community consultation was that the IUCr Commission on Journals, with the support of CSAS, accepted draft publication guidelines for biomolecular SAS. These guidelines were accompanied by the explicit statement that "It is not the intention of this document to define a quality requirement for SAS experiments that would be acceptable for publication. Rather, the purpose is to outline the way in which SAS experiments should be presented in order to enable the reader to independently assess the quality of any interpretations made by the authors." The draft guidelines provided essential and preferred reporting requirements for sample quality, data acquisition, experimental validation, and modelling. The rationale for draft guidelines was published as an open access paper (Jacques, Guss, Svergun, & Trewhella, 2012) and there was a related commentary piece on the importance of quality control and a standard reporting framework for biomolecular SAS (Jacques, Guss, & Trewhella, 2012).

In July of 2012, the first meeting of the newly formed SAS validation task force (SASvtf) of wwPDB was held at the Rutgers Center for Structural Biology (RSCB). From the establishment of the PDB in 1971 with just 7 deposited X-ray crystallography structures, the wwPDB has become the major resource for all structural biology methods (Berman, Vallat, & Lawson, 2020) with 188,000 structures as of March 2022. As the structural biology field grew and diversified, the PDB embraced techniques beyond crystallography (first NMR, then 3D electron microscopy). In the process, deposition guidelines were developed with standard formats designed to be extensible so that not only was the structure represented, but also the supporting metadata such as information on source organisms, samples, workflows, authors, citations, software, and model quality metrics (Bourne et al., 1997). Depositions in the wwPDB that co-refined NMR and SAS were a growing population, and yet there was no place to hold the SAS data supporting these structures. There were also some SAS-only deposited structures, but without SAS-specific supporting information. The initial charge of the SASvtf therefore was to answer the question:

- Should the PDB accept (some types of) models based on SAS studies (an alternative could be, for instance, to capture such models in a separate database)?

- If yes:
  - o Which types of models should be included (and which should not)?
  - o What are the minimum requirements for these models?
  - o What are the requirements regarding the supporting experimental data that need to be deposited?
  - o What validation procedures should be applied in the deposition and annotation process (pertaining to the quality of the model, the quality of the data, and the quality of the fit of the model to the data)?

Part answer to these questions lay in the draft guidelines now endorsed by the IUCr Commissions, but with further detail required. The report from the first meeting of the SASvtf made 6 recommendations (Trewhella et al., 2013):

1. A global data repository is needed that holds standard format X-ray and neutron SAS data that is searchable and freely accessible for download
2. A standard dictionary is required for definitions of terms for data collection and for managing the SAS data repository
3. Options should be provided for including in a repository SAS-derived shape and atomistic models based on rigid-body refinement against SAS data along with specific information regarding the uniqueness and uncertainty of the model and the protocol used to obtain it
4. Criteria need to be agreed on for assessment of the quality of SAS data and accuracy of SAS-derived models and the extent to which a given model fits SAS data
5. With the increasing diversity of structural biology data and models being generated, archiving options for models derived from diverse data will be required
6. Thought leaders from the various structural biology disciplines should jointly define what to archive in the PDB and what complementary archives might be needed, taking into account both scientific needs and funding

The SASvtf report emphasized that a prerequisite for archiving any model should be the availability of the data so that the model can be critically evaluated against the original data and any subsequent data. Two high priority issues for solution SAS were extensively discussed in terms a need for better tools for quantitative assessment: (1) the inherent limits of 1D solution scattering profiles for defining unique solutions and (2) the limits of χ^2 as a measure of the quality of a model fit to the data. The reduced χ^2, while the standard goodness-of-fit measure used historically in SAS, has two problems. First it is highly dependent on error amplitudes and accurate error propagation. Second, as a global fit parameter it fails to identify significant local deviations in model fits that reflect significant structural features not accurately accounted for in the model (see 2.4 Structure Modelling in Trewhella et al., 2017). Stimulated in part by the SASvtf report there were two developments: (1) a method to provide a quantitative assessment of the ambiguity associated with spherically averaged single particle scattering and the uniqueness of a shape reconstructed from the given data set (Petoukhov & Svergun, 2015), and (2) a new correlation map, goodness of fit approach to complement and overcome the inherent limitations of χ^2 (Franke et al., 2015).

Also, in part response to SASvtf recommendation 6, the wwPDB Integrative/Hybrid Methods validation task force (IHMvtf) was formed and hosted a meeting at the European Bioinformatics Institute in Hinxton, UK (October 2014). The meeting brought together 38 leading experts in a variety of experimental structural biology techniques, computational structural biology, integrative modeling and visualization, and data archiving to address a series of questions central to the future of structural biology: "How should integrative models be represented? How should the data and integrative models be validated? What data should be archived? How should the data and models be archived? What information should accompany the publication of integrative models?" The two-day workshop led to a report with 5 recommendations (Sali et al., 2015):

1. In addition to archiving the models themselves, all relevant experimental data, and metadata as well as experimental and computational protocols should be archived
2. A flexible model representation needs to be developed, allowing for multi-scale models, multi-state models, ensembles of models, and models related by time or other order
3. Procedures for estimating the uncertainty of integrative models should be developed, validated, and adopted
4. A federated system of model and data archives should be created
5. Publication standards for integrative models should be established

With momentum clearly building in the broader structural biology community for data quality standards and model validation, along with data and model archiving, the ongoing work on these objectives within the SAS community meant we were well-positioned for the envisioned future for integrative structural biology.

4. A data archive for SAS as part of a federated system for integrative structural biology

The SAS Biological Data Bank (SASBDB) (Valentini et al., 2015) was established in 2014 as a "curated repository of freely accessible and fully searchable SAS experimental data, which are deposited together with the relevant experimental conditions, sample details, instrument characteristic and derived models" (https://wwwsasbdb.org). SASBDB was developed with recommendations of the wwPDB SASvtf and IHMvtf in view. Importantly, to facilitate seamless interchange of data in a federated system of data bases, SASBDB was designed to import and export data using the sasCIF, an extension of the PDBx/mmCIF (macro-molecular Crystallographic Information Format) data representation that is mandatory for submission to the PDB (Adams et al., 2019). The sasCIF was first established in 2000 (Malfois & Svergun, 2000) and further developed and extended in 2016 (Kachala et al., 2016) as a standard dictionary of terms including experimental information, results and models, and relevant metadata regarding SAS data acquisition and analysis for deposition and management in an archive. Importantly, the CIF format is infinitely extensible and as such the sasCIF can be continuously updated to include new terms and definitions.

An NSF sponsored workshop for Federating Structural Models and Data held during the 63rd Annual Meeting of the Biophysical Society (Baltimore, MD, 2019) has since made recommendations for standards and data exchange and for validation of integrative models. These recommendations include continuing to develop the dictionary for integrative structures using the PDBx/mmCIF framework and to promote common data standards that enable efficient data exchange in support of a validation pipeline for integrative structures that would include measures of data quality, criteria to assess models and their fit to the data and the uncertainty of the model (Berman et al., 2019).

5. The current publication guidelines and data archiving requirements

In 2017, the SASvtf received a new charge from the wwPDB leadership, which was to work in conjunction with the NMR validation task force (NMRvtf) to advise on data and meta-data required for SAS and SAS/NMR model deposition and validation in the PDB. To respond to this tasking, an expanded SASvtf membership was established to include broader biomolecular SAS expertise and wider geographical coverage. The expanded SASvtf then reviewed the 2012 draft guidelines and 2013 SASvtf report and considered in detail the recommendation that "criteria need to be agreed upon for the assessment of the quality of deposited data and the accuracy of SAS-derived models, and the extent to which a given model fits the SAS data." The work proceeded with exchanges of correspondence, a virtual (5 May) meeting and a physical meeting (at the 67th Annual Meeting of the American Crystallographic Association, New Orleans, 26–30 May 2017). A further expanded group of 22 leading SAS experimenters, instrument scientists, analysis program developers, as well as experts in crystallography and NMR, came together to support a consensus set of "updated" publication guidelines that was presented to the NMRvtf and wwPDB leadership group.

On 28 July, streamlining between the PDB's OneDep deposition system (Young et al., 2017) and SASBDB via an API (Application Programming Interface) was implemented. For the first time, SAXS data supporting NMR/SAXS structures would be available for reviewers and once published to the broader community (Kikhney et al., 2020). Moreover, SAS was the first technique beyond crystallography, NMR, and electron microscopy to be formally included in the prototype archive for integrative/hybrid methods PDB-Dev (Vallat et al., 2021).

A paper describing the updated guidelines was prepared for submission to the IUCr journal *Acta Crystallographica D Structural Biology* and a talk was scheduled for the IUCr 24th General Assembly and Congress (Hyderabad, August 2017) to present them and invite comment. The guidelines were also discussed in detail at the open meeting of CSAS and the Hyderabad meeting.

The 2017 guidelines (Trewhella et al., 2017) is an open access paper providing a detailed reporting framework that, when used, enables readers of any paper that includes biomolecular SAXS or SANS data to "independently assess the quality of the data and the basis for any interpretations presented." The guidelines include recommendations for: sample details; data acquisition and reduction; data presentation, analysis, and validation; and structure modeling (Fig. 1). A detailed rationale is provided for each recommendation, including recommendations to ensure adequate treatment of structural ambiguity, and complementing χ^2 as a goodness-of-fit parameter with measures that clearly identify local regions of misfit. The latter is in the form of an

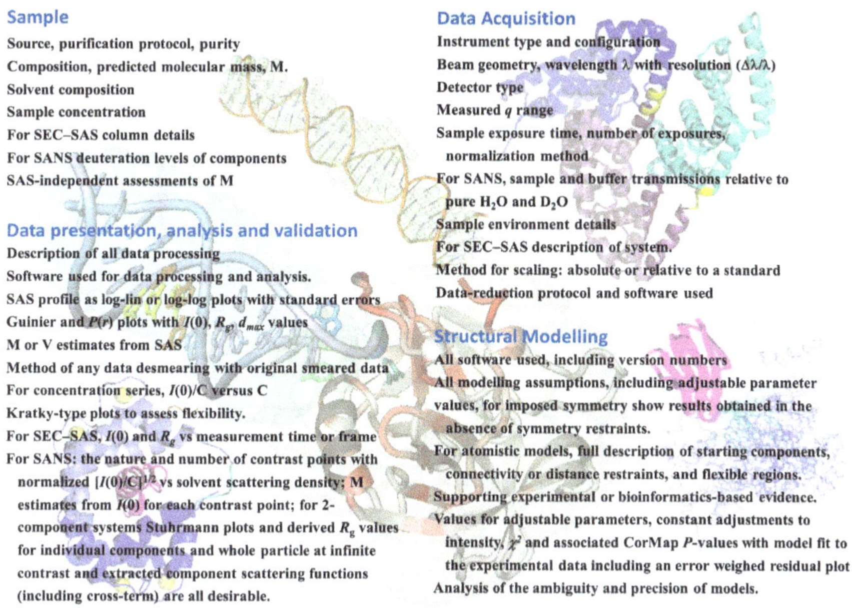

Fig. 1 Brief summary of the main elements of the publication guidelines from Trewhella et al. (2017). In addition to the four main areas for best practice reporting, the guidelines also recommend SAS data deposition in a public archive. The figure background is made up of example structures modelled against SAS data. (Abbreviations: λ is the wavelength of the radiation, C biomolecular concentration, R_g radius of gyration, M and V molecular weight, and volume, respectively, $I(0)$ forward scattered intensity, $P(r)$ distance distribution function of the scattering particle.)

error weighted difference plot of a model with respect to the data combined with results from the correlation map analysis (Franke et al., 2015) that assigns an error-independent statistical significance to local deviations. The guidelines also recommend deposition of the SAS data in a public archive. As a demonstration, the reporting guidelines were applied to SAXS data for three example proteins that demonstrate different challenges for SAS: calmodulin as a 2-domain protein with a flexible interconnecting helix; bovine serum albumin that has a preponderance for aggregation and some flexibility; the highly charged homo-tetramer glucose isomerase (a.k.a. xylose isomerase) that shows interparticle interference effects at relatively low-concentrations (1 mg/mL). The reporting framework is used to illustrate essential steps required before choosing a modeling strategy and then approaches to model validation. Since going online the 2017 guidelines paper continues to be downloaded at high rates (>14,000 downloads as of 24 November 2022 since going online). To facilitate easy reporting, the article is accompanied by downloadable templates for the recommended tables.

In 2022, the biology and medicine editors of the high-profile, open-access, peer review journal of the IUCr (*IUCrJ*) published an editorial regarding their policies in support of FAIR and FACT principles of publishing (Baker et al., 2022). FAIR refers to the recommendation that scientific data should be Findable, Accessible, Interoperable, and Reusable (Wilkinson et al., 2016), while FACT stands for Fair, Accurate, Confidential and Transparent (van der Aalst, Bichler, & Heinzl, 2017). As proponents of FAIR and FACT, the IUCr has supported a variety of standards for publication and the now universally supported mandatory submission of macromolecular crystallographic data (in the form of structure coordinates and reduced diffraction data). Given the progress in the development of standards and data archiving for NMR, cryoEM and SAS, IUCr biology journals (*Acta Crystallographica D* and *F* (*Structural Biology* and *Structural Biology Communications*, respectively), and *IUCrJ*) will now require deposition of these data in a public archive, as is already the case for crystallography (Baker et al., 2022). The requirement for SAS is deposition in the SASBDB, to include buffer subtracted scattering curves along with validation reports. Detailed requirements for SAS reporting for these journals have been updated to include SAS data deposition and adherence to the 2107 publication guidelines (with updated template tables informed by community consultations, including discussions at the triennial SAS meeting SAS2022 in Campinas Brazil).

As noted already, a growing number of structures solved using integrative methods include SAS data; examples include the five-column, tholos-like architecture of a supramolecular structure formed by a proteasome-assembling chaperone homolog using native mass spectrometry, SAS, high-speed atomic force microscopy, and electron microscopy (Yagi-Utsumi et al., 2020), and the 552 protein, 50 MDa nuclear pore complex using a wide range of experimental data including SAXS data for 18 of the nuclear core proteins that provide information on the structures of components in solution and validation of their shapes in the assembly (Kim et al., 2018). Structures deposited in PDB-Dev with SAS data have links to the SASBDB for the contributing SAS data and a data and model validation report based on the 2017 publication guidelines. The SASBDB entry similarly reports the essential information for data and model validation, accompanied by a sliding colour scale that is akin to those used by the wwPDB in the validation reports (see SASBDB example for SAS data for an Ig-domain of nucleoporin pom152 https://www.sasbdb.org/data/SASDBV9/ and the NMR structure of the Pom152 Ig-like repeat 5TVZ in the wwwPDB https://www.rcsb.org/structure/5TVZ).

6. Quantifying data reproducibility and establishing a consensus experimental benchmark data set

Since the first calculations of SAS profiles in solution from atomic coordinates, considerable progress has been achieved in understanding the strengths and limitations of different approaches (reviewed in Brosey & Tainer, 2019; Hub, 2018; Manalastas-Cantos et al., 2021) that were tested against various data sets. However, until now there had been no quantitative evaluation of SAS data reproducibility and there was no benchmark data set against which to evaluate different methods. Further, there remain outstanding questions regarding SAS profile calculation, perhaps most significantly how to model the hydration layer. The hydration layer, and for that matter the ion cloud in the case of highly charged biomolecules, has significant impact of the measured SAS profile and remains an active area of research. Most methods for SAS profile calculation account for the hydration layer by simply optimizing the fit to an experimental scattering profile by adding free parameters (e.g. for modelling SAXS profiles CRYSOL defines a shell of dummy atoms whose volume and contrast is optimized (Svergun et al., 1995); FoXS uses form factors of individual atoms with scale factors to optimize (Schneidman-Duhovny, Hammel, Tainer, & Sali, 2013); AQUASAXS defines an explicit water shell with varying density and scale factors to optimize (Poitevin, Orland, Doniach, Koehl, & Delarue, 2011); AXES uses precomputed water boxes to model the water distribution around an inserted protein with multiple scale factors to optimize (Grishaev, Guo, Irving, & Bax, 2010). As an alternative, WAXSiS uses explicit-solvent all-atom molecular dynamics (MD) simulation to calculate the SAS profile, then comparison to experimental data generally requires a scale factor and a constant background adjustment to account for solvent subtraction errors (Chen & Hub, 2014).

With the increasing uptake of requirements on publication of SAS data and models, it seemed opportune to use a round-robin approach to obtain high-quality SAS data on a set of proteins that would allow for the assessment of the limits of SAS data reproducibility and provide a consensus benchmark data set to compare different approaches to the calculation of SAS profiles from atomic coordinates. Such a project began in 2019, with the support of the SASvtf and IUCr CSAS members and consultants and is now completed (Trewhella et al., 2022). Five proteins were measured (RNase A, lysozyme, xylanase, urate oxidase and xylose isomerase) at 12 SAXS and four SANS centres. In total, 171 SAXS data sets and 76 SANS data sets were submitted for evaluation. Measurements in batch mode as well as using in-line size-exclusion chromatography (SEC) were made using SAXS and SANS, in H_2O and D_2O buffers. The data confirmed the value of in-line SEC for assuring mono-dispersed samples. Merging of data collected in SEC-SAS and batch-SAS modes was successful in eliminating small amounts of aggregate or inter-particle interference while optimizing the statistical quality of the data over the widest q-range possible. Consensus SAXS and SANS profiles with associated errors over extended q-ranges (0–1 Å^{-1} and in some instances to 2.2 Å^{-1}) for each of the candidate proteins are now deposited in SASBDB along with the original measurement data. With these data, we can now answer quantitatively questions regarding reproducibility of biomolecular SAS data and begin to account for residual differences between the consensus profiles and calculated profiles. This work engaged 45 co-authors from the Americas, Europe and Asia drawn from the growing community of biomolecular SAS researchers.

7. Future opportunities

The five consensus SAS profiles from the round robin experiments provide a core set for improving any one approach to theoretical SAS profile prediction. It would by highly advantageous to expand and improve upon this core set. With measurement of just one protein per year using the protocols developed in the round robin, the core set could be doubled in just 5 years. All that is needed is sufficient sample to distribute for ~10 independent measurements, and just one protein a year would double the consensus set in 5 years. Additionally, as new methods for combining data are developed, the deposited original measurements for the core set and any additions could be used to produce an improved set of consensus data. Indeed, these data provide an incentive for innovation in data combining methods and the resolution of outstanding questions of how to ensure accurate error propagation. Finally, newly collected data from new instrumentation with new procedures on high quality samples will improve reproducibility and ultimately the quality achievable for a consensus profile.

8. Conclusions

As was the case in the early days of protein crystallography, there was hesitancy in the biomolecular SAS community to embrace standards for data quality and model validation. Nevertheless, with open communication among continuously increasing constituencies, and allowing for time to embed new practices, the community worked together to address the fundamental requirements necessary to ensure quality in the results from biomolecular SAS and to support data sharing. The process was useful in articulating long standing issues that stimulated innovation in the development of new capabilities. The consistent philosophy driving the work has been to facilitate independent assessment of the quality of the data and the basis for its interpretation in terms of 3D structures. While many of the data quality standards apply equally to broader applications of biomolecular SAS, for example large-scale screening or time-resolved experiments, these important studies will have distinct requirements that need to be defined by their respective communities, perhaps using similar approaches to the structured process described here. In this way, biomolecular SAS can sustain its position as a modern, mainstream structural biology technique.

Acknowledgments

Thank you to J. Mitchell Guss for his critical reading of this manuscript and helpful comments, and for his continued encouragement and support of the work described here. Thanks also to the wwPDB leadership team, and to Helen Berman in particular, for the encouragement and support provided with the establishment of the SASvtf. Finally, many thanks to the members of the SAS community broadly, the SASvtf, and CSAS members and consultants who contributed to the discussions, papers, methods, and resource development necessary to achieve the goals for biomolecular SAS that eventually were achieved and had their begining with those embryonic discussions at the CSAS meeting in Kyoto during SAS2006.

References

Adams, P. D., Afonine, P. V., Baskaran, K., Berman, H. M., Berrisford, J., Bricogne, G., et al. (2019). Announcing mandatory submission of PDBx/mmCIF format files for crystallographic depositions to the protein data bank (PDB). *Acta Crystallographica. Section D, Biological Crystallography*, *75*(Pt. 4), 451–454. https://doi.org/10.1107/S2059798319004522.

Baker, E. N., Bond, C. S., Garman, E. F., Newman, J., Read, R. J., & van Raaij, M. J. (2022). Submission of structural biology data for review purposes. *IUCrJ*, *9*(Pt. 1), 1–2. https://doi.org/10.1107/S2052252521012999.

Berman, H. M., Adams, P. D., Bonvin, A. A., Burley, S. K., Carragher, B., Chiu, W., et al. (2019). Federating structural models and data: Outcomes from a workshop on archiving integrative structures. *Structure*, *27*(12), 1745–1759. https://doi.org/10.1016/j.str.2019.11.002.

Berman, H. M., Kleywegt, G. J., Nakamura, H., & Markley, J. L. (2012). The protein data bank at 40: Reflecting on the past to prepare for the future. *Structure, 20*(3), 391–396. https://doi.org/10.1016/j.str.2012.01.010.

Berman, H. M., Vallat, B., & Lawson, C. L. (2020). The data universe of structural biology. *IUCrJ*, 7(Pt. 4), 630–638. https://doi.org/10.1107/S205225252000562X.

Bourne, P. E., Berman, H. M., McMahon, B., Watenpaugh, K. D., Westbrook, J. D., & Fitzgerald, P. M. (1997). Macromolecular crystallographic information file. *Methods in Enzymology, 277*, 571–590. https://doi.org/10.1016/s0076-6879(97)77032-0.

Brosey, C. A., & Tainer, J. A. (2019). Evolving SAXS versatility: Solution X-ray scattering for macromolecular architecture, functional landscapes, and integrative structural biology. *Current Opinion in Structural Biology, 58*, 197–213. https://doi.org/10.1016/j.sbi.2019.04.004.

Burley, S. K., Kurisu, G., Markley, J. L., Nakamura, H., Velankar, S., Berman, H. M., et al. (2017). PDB-Dev: A prototype system for depositing integrative/hybrid structural models. *Structure, 25*(9), 1317–1318. https://doi.org/10.1016/j.str.2017.08.001.

Chen, P. C., & Hennig, J. (2018). The role of small-angle scattering in structure-based screening applications. *Biophysical Reviews, 10*(5), 1295–1310. https://doi.org/10.1007/s12551-018-0464-x.

Chen, P. C., & Hub, J. S. (2014). Validating solution ensembles from molecular dynamics simulation by wide-angle X-ray scattering data. *Biophysical Journal, 107*(2), 435–447. https://doi.org/10.1016/j.bpj.2014.06.006.

Comoletti, D., Miller, M. T., Jeffries, C. M., Wilson, J., Demeler, B., Taylor, P., et al. (2010). The macromolecular architecture of extracellular domain of alphaNRXN1: Domain organization, flexibility, and insights into trans-synaptic disposition. *Structure, 18*(8), 1044–1053. https://doi.org/10.1016/j.str.2010.06.005.

Custodio, T. F., Das, H., Sheward, D. J., Hanke, L., Pazicky, S., Pieprzyk, J., et al. (2020). Selection, biophysical and structural analysis of synthetic nanobodies that effectively neutralize SARS-CoV-2. *Nature Communications, 11*(1), 5588. https://doi.org/10.1038/s41467-020-19204-y.

Da Vela, S., & Svergun, D. I. (2020). Methods, development and applications of small-angle X-ray scattering to characterize biological macromolecules in solution. *Current Research in Structural Biology, 2*, 164–170. https://doi.org/10.1016/j.crstbi.2020.08.004.

Franke, D., Jeffries, C. M., & Svergun, D. I. (2015). Correlation map, a goodness-of-fit test for one-dimensional X-ray scattering spectra. *Nature Methods, 12*(5), 419–422. https://doi.org/10.1038/nmeth.3358.

Grishaev, A. V. (2017). Hybrid applications of solution scattering to aid structural biology. *Advances in Experimental Medicine and Biology, 1009*, 215–227. https://doi.org/10.1007/978-981-10-6038-0_13.

Grishaev, A., Guo, L., Irving, T., & Bax, A. (2010). Improved fitting of solution X-ray scattering data to macromolecular structures and structural ensembles by explicit water modeling. *Journal of the American Chemical Society, 132*(44), 15484–15486. https://doi.org/10.1021/ja106173n.

Groothuizen, F. S., Fish, A., Petoukhov, M. V., Reumer, A., Manelyte, L., Winterwerp, H. H., et al. (2013). Using stable MutS dimers and tetramers to quantitatively analyze DNA mismatch recognition and sliding clamp formation. *Nucleic Acids Research, 41*(17), 8166–8181. https://doi.org/10.1093/nar/gkt582.

Guinier, A., & Fournet, G. (1955). In C. B. Walker, & M. G. Mayer (Eds.), *Small-angle scattering of X-rays*. New York: John Wiley & Sons. Trans.

Hammel, M., Rosenberg, D. J., Bierma, J., Hura, G. L., Thapar, R., Lees-Miller, S. P., et al. (2021). Visualizing functional dynamicity in the DNA-dependent protein kinase holoenzyme DNA-PK complex by integrating SAXS with cryo-EM. *Progress in Biophysics and Molecular Biology, 163*, 74–86. https://doi.org/10.1016/j.pbiomolbio.2020.09.003.

Hammel, M., & Tainer, J. A. (2021). X-ray scattering reveals disordered linkers and dynamic interfaces in complexes and mechanisms for DNA double-strand break repair impacting cell and cancer biology. *Protein Science*, *30*(9), 1735–1756. https://doi.org/10.1002/pro.4133.

Heidorn, D. B., Seeger, P. A., Rokop, S. E., Blumenthal, D. K., Means, A. R., Crespi, H., et al. (1989). Changes in the structure of calmodulin induced by a peptide based on the calmodulin-binding domain of myosin light chain kinase. *Biochemistry*, *28*(16), 6757–6764. https://doi.org/10.1021/bi00442a032.

Heidorn, D. B., & Trewhella, J. (1988). Comparison of the crystal and solution structures of calmodulin and troponin C. *Biochemistry*, *27*(3), 909–915. https://doi.org/10.1021/bi00403a011.

Hennig, J., Wang, I., Sonntag, M., Gabel, F., & Sattler, M. (2013). Combining NMR and small angle X-ray and neutron scattering in the structural analysis of a ternary protein-RNA complex. *Journal of Biomolecular NMR*, *56*(1), 17–30. https://doi.org/10.1007/s10858-013-9719-9.

Hub, J. S. (2018). Interpreting solution X-ray scattering data using molecular simulations. *Current Opinion in Structural Biology*, *49*, 18–26. https://doi.org/10.1016/j.sbi.2017.11.002.

Jacques, D. A., Guss, J. M., Svergun, D. I., & Trewhella, J. (2012). Publication guidelines for structural modelling of small-angle scattering data from biomolecules in solution. *Acta Crystallographica. Section D, Biological Crystallography*, *68*(Pt 6), 620–626. https://doi.org/10.1107/S0907444912012073.

Jacques, D. A., Guss, J. M., & Trewhella, J. (2012). Reliable structural interpretation of small-angle scattering data from bio-molecules in solution—The importance of quality control and a standard reporting framework. *BMC Structural Biology*, *12*, 9. https://doi.org/10.1186/1472-6807-12-9.

Kachala, M., Westbrook, J., & Svergun, D. (2016). Extension of the sasCIF format and its applications for data processing and deposition. *Journal of Applied Crystallography*, *49*(Pt. 1), 302–310. https://doi.org/10.1107/S1600576715024942.

Kelly, J. A., Olson, A. N., Neupane, K., Munshi, S., San Emeterio, J., Pollack, L., et al. (2020). Structural and functional conservation of the programmed-1 ribosomal frameshift signal of SARS coronavirus 2 (SARS-CoV-2). *The Journal of Biological Chemistry*, *295*(31), 10741–10748. https://doi.org/10.1074/jbc.AC120.013449.

Kikhney, A. G., Borges, C. R., Molodenskiy, D. S., Jeffries, C. M., & Svergun, D. I. (2020). SASBDB: Towards an automatically curated and validated repository for biological scattering data. *Protein Science*, *29*(1), 66–75. https://doi.org/10.1002/pro.3731.

Kim, J. H., Bothe, J. R., Frederick, R. O., Holder, J. C., & Markley, J. L. (2014). Role of IscX in iron-sulfur cluster biogenesis in Escherichia coli. *Journal of the American Chemical Society*, *136*(22), 7933–7942. https://doi.org/10.1021/ja501260h.

Kim, S. J., Fernandez-Martinez, J., Nudelman, I., Shi, Y., Zhang, W., Raveh, B., et al. (2018). Integrative structure and functional anatomy of a nuclear pore complex. *Nature*, *555*(7697), 475–482. https://doi.org/10.1038/nature26003.

Kursula, P. (2021). Small-angle X-ray scattering for the proteomics community: Current overview and future potential. *Expert Review of Proteomics*, *1-8*. https://doi.org/10.1080/14789450.2021.1951242.

Lin, S. M., Lin, S. C., Hsu, J. N., Chang, C. K., Chien, C. M., Wang, Y. S., et al. (2020). Structure-based stabilization of non-native protein-protein interactions of coronavirus nucleocapsid proteins in antiviral drug design. *Journal of Medicinal Chemistry*, *63*(6), 3131–3141. https://doi.org/10.1021/acs.jmedchem.9b01913.

Lycksell, M., Rovsnik, U., Bergh, C., Johansen, N. T., Martel, A., Porcar, L., et al. (2021). Probing solution structure of the pentameric ligand-gated ion channel GLIC by small-angle neutron scattering. *Proceedings of the National Academy of Sciences of the United States of America*, *118*(37). https://doi.org/10.1073/pnas.2108006118.

Mahieu, E., & Gabel, F. (2018). Biological small-angle neutron scattering: Recent results and development. *Acta Crystallographica. Section D, Biological Crystallography*, *74*(Pt 8), 715–726. https://doi.org/10.1107/S2059798318005016.

Malfois, M., & Svergun, D. I. (2000). sasCIF: An extension of core crystallographic information file for SAS. *Journal of Applied Crystallography*, *33*, 812–816. https://doi.org/10.1107/S0021889800001357.

Manalastas-Cantos, K., Konarev, P. V., Hajizadeh, N. R., Kikhney, A. G., Petoukhov, M. V., Molodenskiy, D. S., et al. (2021). ATSAS 3.0: Expanded functionality and new tools for small-angle scattering data analysis. *Journal of Applied Crystallography*, *54*(Pt 1), 343–355. https://doi.org/10.1107/S1600576720013412.

Muller, R., Grawert, M. A., Kern, T., Madl, T., Peschek, J., Sattler, M., et al. (2013). High-resolution structures of the IgM fc domains reveal principles of its hexamer formation. *Proceedings of the National Academy of Sciences of the United States of America*, *110*(25), 10183–10188. https://doi.org/10.1073/pnas.1300547110.

Petoukhov, M. V., & Svergun, D. I. (2015). Ambiguity assessment of small-angle scattering curves from monodisperse systems. *Acta Crystallographica. Section D, Biological Crystallography*, *71*(Pt. 5), 1051–1058. https://doi.org/10.1107/S1399004715002576.

Poitevin, F., Orland, H., Doniach, S., Koehl, P., & Delarue, M. (2011). AquaSAXS: A web server for computation and fitting of SAXS profiles with non-uniformally hydrated atomic models. *Nucleic Acids Research*, *39*(Web Server issue), W184–W189. https://doi.org/10.1093/nar/gkr430.

Rambo, R. P., & Tainer, J. A. (2013). Super-resolution in solution X-ray scattering and its applications to structural systems biology. *Annual Review of Biophysics*, *42*, 415–441. https://doi.org/10.1146/annurev-biophys-083012-130301.

Rout, M. P., & Sali, A. (2019). Principles for integrative structural biology studies. *Cell*, *177*(6), 1384–1403. https://doi.org/10.1016/j.cell.2019.05.016.

Sali, A. (2021). From integrative structural biology to cell biology. *The Journal of Biological Chemistry*, *296*, 100743. https://doi.org/10.1016/j.jbc.2021.100743.

Sali, A., Berman, H. M., Schwede, T., Trewhella, J., Kleywegt, G., Burley, S. K., et al. (2015). Outcome of the first wwPDB hybrid/integrative methods task force workshop. *Structure*, *23*(7), 1156–1167. https://doi.org/10.1016/j.str.2015.05.013.

Sauve, V., Lilov, A., Seirafi, M., Vranas, M., Rasool, S., Kozlov, G., et al. (2015). A Ubl/ubiquitin switch in the activation of Parkin. *The EMBO Journal*, *34*(20), 2492–2505. https://doi.org/10.15252/embj.201592237.

Schneidman-Duhovny, D., Hammel, M., Tainer, J. A., & Sali, A. (2013). Accurate SAXS profile computation and its assessment by contrast variation experiments. *Biophysical Journal*, *105*(4), 962–974. https://doi.org/10.1016/j.bpj.2013.07.020.

Schneidman-Duhovny, D., Kim, S. J., & Sali, A. (2012). Integrative structural modeling with small angle X-ray scattering profiles. *BMC Structural Biology*, *12*, 17. https://doi.org/10.1186/1472-6807-12-17.

Schroer, M. A., & Svergun, D. I. (2018). Recent developments in small-angle X-ray scattering and hybrid method approaches for biomacromolecular solutions. *Emerging Topics in Life Sciences*, *2*(1), 69–79. https://doi.org/10.1042/ETLS20170138.

Svergun, D., Barberato, C., & Koch, M. H. J. (1995). CRYSOL—A program to evaluate x-ray solution scattering of biological macromolecules from atomic coordinates. *Journal of Applied Crystallography*, *28*, 768–773. https://doi.org/10.1107/S0021889895007047.

Takala, H., Bjorling, A., Berntsson, O., Lehtivuori, H., Niebling, S., Hoernke, M., et al. (2014). Signal amplification and transduction in phytochrome photosensors. *Nature*, *509*(7499), 245–248. https://doi.org/10.1038/nature13310.

Tidow, H., Poulsen, L. R., Andreeva, A., Knudsen, M., Hein, K. L., Wiuf, C., et al. (2012). A bimodular mechanism of calcium control in eukaryotes. *Nature*, *491*(7424), 468–472. https://doi.org/10.1038/nature11539.

Trewhella, J. (2016). Small-angle scattering and 3D structure interpretation. *Current Opinion in Structural Biology*, *40*, 1–7. https://doi.org/10.1016/j.sbi.2016.05.003.

Trewhella, J. (2021). Growing a thriving international community for small-angle scattering through collaboration. *Journal of Applied Crystallography*, *54*(Pt 4), 1029–1033. https://doi.org/10.1107/S1600576721007561.

Trewhella, J. (2022). Recent advances in small-angle scattering and its expanding impact in structural biology. *Structure*, *30*(1), 15–23. https://doi.org/10.1016/j.str.2021.09.008.

Trewhella, J., Blumenthal, D. K., Rokop, S. E., & Seeger, P. A. (1990). Small-angle scattering studies show distinct conformations of calmodulin in its complexes with two peptides based on the regulatory domain of the catalytic subunit of phosphorylase kinase. *Biochemistry*, *29*(40), 9316–9324. https://doi.org/10.1021/bi00492a003.

Trewhella, J., Duff, A. P., Durand, D., Gabel, F., Guss, J. M., Hendrickson, W. A., et al. (2017). 2017 publication guidelines for structural modelling of small-angle scattering data from biomolecules in solution: An update. *Acta Crystallographica. Section D, Biological Crystallography*, *73*(Pt 9), 710–728. https://doi.org/10.1107/S2059798317011597.

Trewhella, J., Hendrickson, W. A., Kleywegt, G. J., Sali, A., Sato, M., Schwede, T., et al. (2013). Report of the wwPDB small-angle scattering task force: Data requirements for biomolecular modeling and the PDB. *Structure*, *21*(6), 875–881. https://doi.org/10.1016/j.str.2013.04.020.

Trewhella, J., Vachette, P., Bierma, J., Blanchet, C., Brookes, E., Chakravarthy, S., et al. (2022). A round Robin approach provides a detailed assessment of biomolecular small-angle scattering data reproducibility and yields consensus curves for benchmarking. *Acta Crystallographica. Section D, Biological Crystallography*, *78*(11), 1315–1336. https://doi.org/10.1107/S2059798322009184.

Tuukkanen, A. T., Spilotros, A., & Svergun, D. I. (2017). Progress in small-angle scattering from biological solutions at high-brilliance synchrotrons. *IUCrJ*, *4*(Pt. 5), 518–528. https://doi.org/10.1107/S2052252517008740.

Valentini, E., Kikhney, A. G., Previtali, G., Jeffries, C. M., & Svergun, D. I. (2015). SASBDB, a repository for biological small-angle scattering data. *Nucleic Acids Research*, *43*(Database issue), D357–D363. https://doi.org/10.1093/nar/gku1047.

Vallat, B., Webb, B., Fayazi, M., Voinea, S., Tangmunarunkit, H., Ganesan, S. J., et al. (2021). New system for archiving integrative structures. *Acta Crystallographica. Section D, Biological Crystallography*, 77(Pt. 12), 1486–1496. https://doi.org/10.1107/S2059798321010871.

Vallat, B., Webb, B., Westbrook, J. D., Sali, A., & Berman, H. M. (2018). Development of a prototype system for archiving integrative/hybrid structure models of biological macromolecules. *Structure*, *26*(6), 894–904 e892. https://doi.org/10.1016/j.str.2018.03.011.

van der Aalst, W. M. P., Bichler, M., & Heinzl, A. (2017). Responsible data science. *Business and Information Systems Engineering*, *59*(5), 311–313. https://doi.org/10.1007/s12599-017-0487-z.

Whitten, A. E., Jeffries, C. M., Harris, S. P., & Trewhella, J. (2008). Cardiac myosin-binding protein C decorates F-actin: Implications for cardiac function. *Proceedings of the National Academy of Sciences of the United States of America*, *105*(47), 18360–18365. https://doi.org/10.1073/pnas.0808903105.

Wilamowski, M., Hammel, M., Leite, W., Zhang, Q., Kim, Y., Weiss, K. L., et al. (2021). Transient and stabilized complexes of Nsp7, Nsp8, and Nsp12 in SARS-CoV-2 replication. *Biophysical Journal*, *120*(15), 3152–3165. https://doi.org/10.1016/j.bpj.2021.06.006.

Wilkinson, M. D., Dumontier, M., Aalbersberg, I. J., Appleton, G., Axton, M., Baak, A., et al. (2016). The FAIR guiding principles for scientific data management and stewardship. *Scientific Data*, *3*, 160018. https://doi.org/10.1038/sdata.2016.18.

Yagi-Utsumi, M., Sikdar, A., Song, C., Park, J., Inoue, R., Watanabe, H., et al. (2020). Supramolecular tholos-like architecture constituted by archaeal proteins without functional annotation. *Scientific Reports*, *10*(1), 1540. https://doi.org/10.1038/s41598-020-58371-2.

Young, J. Y., Westbrook, J. D., Feng, Z., Sala, R., Peisach, E., Oldfield, T. J., et al. (2017). OneDep: Unified wwPDB system for deposition, biocuration, and validation of macromolecular structures in the PDB archive. *Structure*, *25*(3), 536–545. https://doi.org/10.1016/j.str.2017.01.004.

CHAPTER TWO

Structure and ensemble refinement against SAXS data: Combining MD simulations with Bayesian inference or with the maximum entropy principle

Leonie Chatzimagas and Jochen S. Hub*
Theoretical Physics and Center for Biophysics, Saarland University, Saarbrücken, Germany
*Corresponding author: e-mail address: jochen.hub@uni-saarland.de

Contents

Methods in Enzymology, Volume 678
ISSN 0076-6879
https://doi.org/10.1016/bs.mie.2022.09.014

Abstract

Small-angle X-ray scattering (SAXS) is a powerful method for tracking conformational transitions of proteins or soft-matter complexes in solution. However, the interpretation of the experimental data is challenged by the low spatial resolution and the low information content of the data, which lead to a high risk of overinterpreting the data. Here, we illustrate how SAXS data can be integrated into all-atom molecular dynamics (MD) simulation to derive atomic structures or heterogeneous ensembles that are compatible with the data. Besides providing atomistic insight, the MD simulation adds physicochemical information, as encoded in the MD force fields, which greatly reduces the risk of overinterpretation. We present an introduction into the theory of SAXS-driven MD simulations as implemented in GROMACS-SWAXS, a modified version of the GROMACS simulation software. We discuss SAXS-driven parallel-replica ensemble refinement with commitment to the maximum entropy principle as well as a Bayesian formulation of SAXS-driven structure refinement. Practical considerations for running and interpreting the simulations are presented. The methods are freely available via GitLab at https://gitlab.com/cbjh/gromacs-swaxs.

1. Introduction

Understanding the function of biomolecules requires understanding of their conformational dynamics. An increasingly popular method for tracking conformational transition of biomolecules is small-angle X-ray scattering (SAXS), which provides structural information that is not accessible by other techniques. Unlike NMR spectroscopy, which probes local distances and angles of smaller biomolecules, SAXS provides information on the overall shape and is applicable to both small and large biomolecules. Unlike crystallography or cryoelectron microscopy, SAXS probes molecules at ambient temperatures in solution, enabling experiments that track conformational transition after application of external stimuli. The accuracy of SAXS data has greatly improved over recent years thanks to sample delivery coupled with size exclusion chromatography (SEC-SAXS), thereby reducing sample aggregation artifacts, single-photon counting detractors, and standards for sample preparation (Berthaud, Manzi, Pérez, & Mangenot, 2012; Jeffries et al., 2016). Software and algorithms for data analysis and for SAXS-based structural modeling has greatly developed (Gräwert & Svergun, 2020). These properties and developments establish SAXS as an indispensable tool for integrative structural biology (Brosey & Tainer, 2019; Rout & Sali, 2019).

The interpretation of the SAXS data is challenged by the low information content of the experimental signals. Because the biomolecules are randomly oriented during solution scattering, SAXS curves $I(q)$ represent an orientational average and, consequently, are only a one-dimensional smooth function of momentum transfer $q = 4\pi \sin(\theta)/\lambda$. Here, λ is the X-ray wavelength and 2θ is the scattering angle. The number of data points in $I(q)$ that provide independent structural information is estimated by the number of Shannon–Nyquist channels (Moore, 1980; Rambo & Tainer, 2013)

$$N_{\text{Shan}} = (q_{\text{max}} - q_{\text{min}})D/\pi, \quad (1)$$

where q_{max} and q_{min} denote the maximum and minimum momentum transfer in $I(q)$, respectively, and D is the maximum diameter of the solute. N_{Shan} is in the range of 5–30 for many SAXS experiments. For comparison, even a small protein with 100 residues contains approximately 200 flexible backbone angles, demonstrating that SAXS data is by far insufficient for defining all degrees of freedom of a biomolecule. Consequently, structure refinement against SAXS data is highly ambiguous, that is, many different structures fit the data equally well. Challenges due to the low information content are enhanced by the presence of heterogeneous ensembles and by uncertainties in the experimental and predicted SAXS curves. The low information content of the data together with larger number of degrees of freedom leads to a significant risk of overinterpretation upon fitting structural models against experimental data (Hub, 2018; Putnam, Hammel, Hura, & Tainer, 2007).

To mitigate the risk of overinterpretation during structure refinement, two main strategies have been devised. First, nearly all degrees of freedom of the biomolecule have been constrained, leading to methods such as rigid-body or normal mode refinement (Gorba, Miyashita, & Tama, 2008; Pelikan, Hura, & Hammel, 2009; Petoukhov & Svergun, 2005). Such methods use only simple energy terms to discriminate plausible from prohibited conformations, such as volume exclusion terms between protein domains or a multidimensional harmonic potential along normal modes. Second, physicochemical information has been added to the SAXS data with the help of atomistic molecular dynamics (MD) simulations (Björling, Niebling, Marcellini, van der Spoel, & Westenhoff, 2015; Chen & Hub, 2015; Kimanius, Pettersson, Schluckebier, Lindahl, & Andersson, 2015; Paissoni, Jussupow, & Camilloni, 2020). During MD simulations, all degrees of freedom are kept flexible, but the force field restrains the biomolecule to realistic conformations with acceptable free energy. Simulations with a coupling to experimental SAXS curves have been coined SAXS-driven or SAXS-guided MD simulations.

Several previous studies described SAXS-driven simulations, in which the MD simulations were coupled to experimental SAXS data with an energetic restraint. Björling et al. (2015) presented a method with a focus on the interpretation of SAXS curve *differences* between two conformational states. This study neglected effects from the hydration layer on the SAXS curve, rationalized by the fact that hydration layer effects may approximately cancel upon taking SAXS curve differences. Our group developed SAXS-driven MD focused on absolute SAXS curves and implemented the method into an extension of the GROMACS software, coined GROMACS-SWAXS (https://gitlab.com/cbjh/gromacs-swaxs). The implementation uses explicit-solvent SAXS curve calculations using atomistic representations for the hydration layer and excluded solvent (Chen & Hub, 2014, 2015). Kimanius et al. (2015) suggested refining protein structures against SAXS data using metadynamics. However, since the SAXS curve calculations neglected the buffer subtraction, the implementation was not yet ready for refinement against experimental data. Paissoni et al. (2020) provided a method for SAXS-driven simulations implemented in PLUMED, which was primarily motivated with the aim of accelerating SAXS-driven simulations. This method maps the atomistic model to a coarse-grained representation for computing a SAXS curve. The method neglects effects from the hydration layer, and the coarse-grained approximation is limited to smaller scattering angles (Bernetti & Bussi, 2021). However, as discussed in this chapter and previously, the SAXS-driven MD simulations implemented in GROMACS-SWAXS are subject to only a small computational overhead between 5% and 20%, suggesting that SAXS-driven MD simulations based on all-atom SAXS predictions are likewise computationally feasible. Hsu, Leshchev, Kosheleva, Kohlstedt, and Chen (2020) presented SAXS-driven MD simulations utilizing the Debye equation, similar to Björling et al. (2015); however, these authors included effects from the hydration layer by increasing atomic form factors of solvent-exposed atoms, similar to the FoXS method (Schneidman-Duhovny, Hammel, & Sali, 2010).

This chapter describes the refinement of protein and soft-matter complexes by coupling all-atom MD simulations to experimental SAXS data, as implemented in GROMACS-SWAXS. We first describe SAXS-driven MD simulations coupled with a harmonic restraint to the data (Chen & Hub, 2015). The method was extended for simultaneous refinement against SAXS and several small-angle neutron scattering (SANS) data sets collected at different D_2O concentrations (Chen et al., 2019). We discuss how SAXS-guided structure refinement is embedded into rigorous probability theory, as the obtained MD ensemble may be interpreted as the posterior distribution of a Bayesian inference problem (Shevchuk & Hub, 2017). Such viewpoint enables statements of confidence intervals, which are still underdeveloped in SAXS-guided modeling. Finally, we describe the refinement of heterogeneous ensembles against SAXS data using a minimal bias, that is, with commitment to Jaynes' principle of maximum entropy (Hermann & Hub, 2019; Ivanović, Hermann, Wójcik, Pérez, & Hub, 2020).

The methods described here are freely available via an extension of the GROMACS simulation software (Abraham et al., 2015) developed at https://gitlab.com/cbjh/gromacs-swaxs. We present the theoretical basis of SAXS-driven MD simulations. We discuss conceptual considerations as well as practical guidelines for running and interpreting the calculations. Additional tutorials and documentation are available at https://cbjh.gitlab.io/gromacs-swaxs-docs. Although we here refer mostly to SAXS, the methods described here are likewise applicable for refinement against small-angle neutron scattering (SANS) data or for refinement against combined SAXS/SANS data.

SAXS-driven MD simulations require a forward model for predicting SAXS curves from given MD simulations. The methods implemented in GROMACS-SWAXS utilize explicit-solvent calculations, thereby taking explicit representations of the hydration layer and excluded solvent into account (Chen & Hub, 2014; Merzel & Smith, 2002; Park, Bardhan, Roux, & Makowski, 2009). For more details on explicit-solvent SAXS predictions, we refer to chapter "Predicting solution scattering patterns with explicit-solvent molecular simulations" by Chatzimagas and Hub in Part A of this monograph.

2. SAXS-driven molecular dynamics simulations

2.1 Experiment-supported energetic bias

Similar to refinement against other experimental data (Jack & Levitt, 1978), SAXS-guided refinement is implemented by augmenting the MD force field energy $V_{FF}(R)$ with an experiment-derived energy $E_{exp}(R, D)$ that drives the simulation into conformations R that are compatible with the data D:

$$E_{hybrid} = V_{FF}(R) + E_{exp}(R,D) \tag{2}$$

Here, the data is given by the experimental SAXS intensities $I_{exp}(q_i)$ and errors $\sigma(q_i)$, where $i = 1,\ldots,N_{exp}$ and N_{exp} is the number of experimental data points. The MD simulation is coupled to the data using a harmonic restraint on the SAXS intensities following

$$E_{\text{exp}}(R(t), D) = \frac{1}{2} f_c k_B T \sum_{i=1}^{N_{\text{exp}}} \frac{[I_c(q_i; R(t)) - (f I_{\text{exp}}(q_i) + c)]^2}{[f\sigma(q_i)]^2}. \quad (3)$$

Here, I_c is the SAXS curve calculated on-the-fly from the MD simulation and f_c is an overall force constant, whereas k_B and T denote the Boltzmann constant and the temperature. The factor 1/2 is introduced to enable a Bayesian interpretation of the refined ensemble in the special case $f_c = 1$ as described below. The symbols f and c denote fitting parameter that adjust the overall scale and an offset of the experimental curve relative to the calculated curve; f and c are adjusted throughout the simulation by minimizing E_{exp}. Hence, only those differences between I_c and I_{exp} that cannot be explained by f and c contribute to E_{exp} and, thereby, drive the simulation. The offset c absorbs uncertainties due to the buffer subtraction and may be omitted when coupling to high-precision SAXS data.

As an example, Fig. 1 presents structure refinement of the nuclear exportin CRM1, which transports protein cargos across the nuclear pore (Chen & Hub, 2015). Free simulations of CRM1 are highly force field dependent, evident from the fact that simulations with the Amber99sb or with the Charmm22* force fields lead to overly open or overly closed conformations, respectively (Fig. 1A, B, D top). Simulations with each force field exhibit poor agreement with experimental SAXS data (Fig. 1C, black and solid yellow or green curves). Upon restraining the simulations to SAXS data, excellent agreement with the data is obtained (Fig. 1C, dashed yellow or green curves), and both two force fields lead to a consensus partly open conformation (Fig. 1D bottom). Hence, the SAXS data overrules imperfections in the two force fields. Fig. 1C shows large residuals at low q between the experimental and the computed SAXS curves. These large residuals are a consequence of very small experimental errors, while the overall error is likely dominated by systematic errors. If systematic errors would be ignored, such large residuals would lead to excessively large SAXS-derived forces, hence asking for methods for modeling systematic errors as described below (see Sections 2.3 and 3.2).

2.2 Increasing the computational efficiency by smoothing and re-binning the experimental curve

Owing to the large number of pixels on modern X-ray detectors, $I_{\text{exp}}(q_i)$ is heavily oversampled. $I_{\text{exp}}(q_i)$ contains typically 800–2500 noisy estimates of

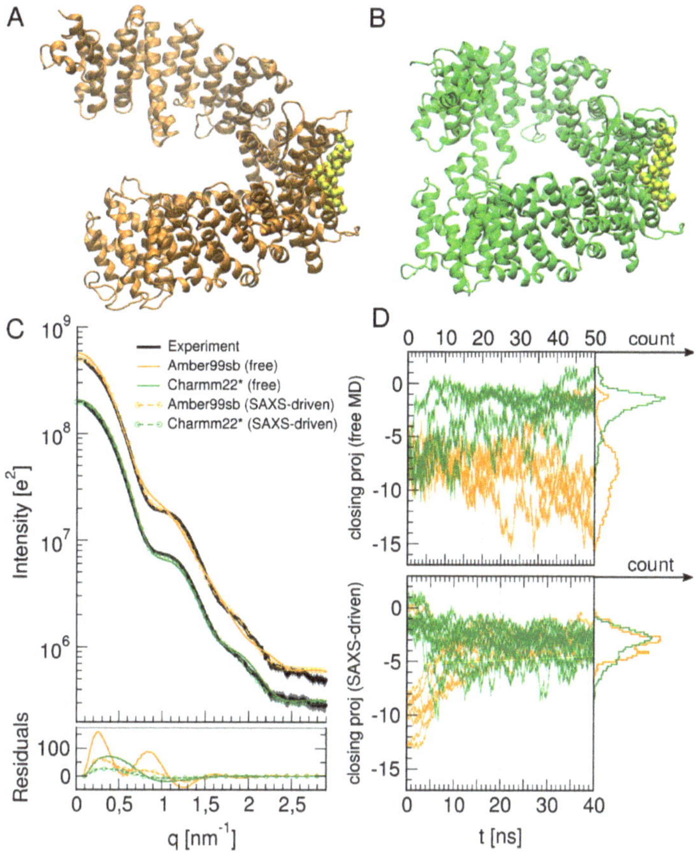

Fig. 1 Structure refinement of the nuclear exportin CRM1 against SAXS data. Free MD simulations with the Amber99sb (*yellow*) or Charmm22* (*green*) force fields lead to different conformations (A/B; D, *top*), which both disagree with experimental SAXS data ($\chi^2 = 1397$ and $\chi^2 = 429$ for Amber99sb and Charmm22*, respectively) (C, see legend). During SAXS-driven simulations, the different force fields lead to similar conformations (D, *bottom*), in excellent agreement with the data ($\chi^2 = 263$ and $\chi^2 = 57$ for Amber99sb and Charmm22*, respectively). *Adapted and reused with permission from Chen, P.-c., & Hub, J. S. (2015). Interpretation of solution X-ray scattering by explicit-solvent molecular dynamics. Biophysical Journal, 108, 2573–2584.*

the true underlying smooth SAXS curve, which contains only few independent data points ($N_{\text{exp}} \gg N_{\text{Shan}}$). Consequently, the formulation in Eq. (3) is inefficient, since it requires the calculation of $I_c(q_i; R)$ for a large number N_{exp} of q-points.

The computational cost is reduced by smoothing the raw experimental data, thereby merging neighboring $I_{\text{exp}}(q_i)$ q-points within the same

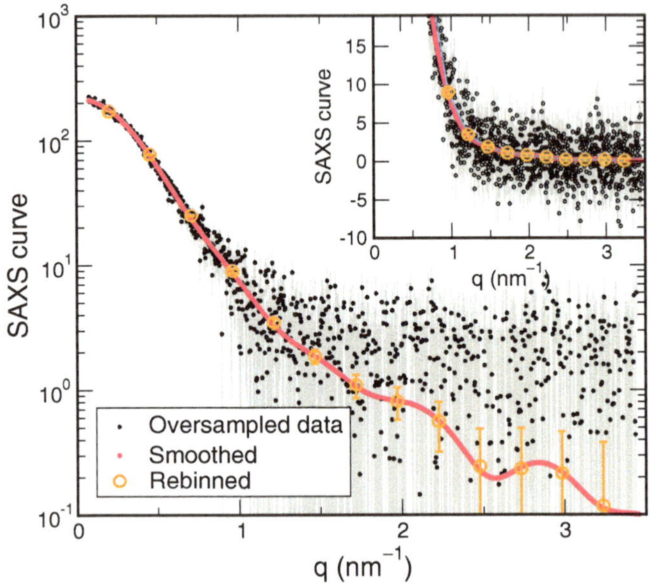

Fig. 2 Smoothing and rebinning the experimental curve to increase the computational efficiency of SAXS-driven MD simulations. *Black dots*: Oversampled experimental SAXS data for bovine serum albumin dimers taken from Jeffries et al. (2016). Curve smoothed with DATGNOM (*pink*) and resampled with DATREGRID, here using 1.5 *q*-points per Shannon bin (Manalastas-Cantos et al., 2021).

Shannon bin into a smoothed curve $\bar{I}_{\text{exp}}(q)$ (Fig. 2). A shell script for smoothing the curve based on the ATSAS module DATGNOM (Manalastas-Cantos et al., 2021) is available at https://cbjh.gitlab.io/gromacs-swaxs-docs. Upon smoothing the curve, N_{exp} raw experimental data points are merged into a smooth curve with only N_{Shan} independent features, suggesting that the uncertainties of the smoothed curve follow $\bar{\sigma}^2(q) \approx \sigma^2(q)/n_s$, where $n_s = N_{\text{exp}}/N_{\text{Shan}}$ is the number of experimental points per Shannon bin. Let $\Delta I(q_i) := I_c(q_i; R) - (fI_{\text{exp}}(q_i) + c)$ denote the residuals between experimental and calculated curve and

$$\chi^2 = \sum_{i=1}^{N_{\text{exp}}} \left(\frac{\Delta I(q_i)}{f\sigma(q_i)} \right)^2, \tag{4}$$

such that $E_{\text{exp}} = f_c k_B T \chi^2/2$. Let us further decompose the residuals $\Delta I_i(q_i) = \Delta \bar{I}_i + \delta I_i$ into contributions (i) relative to the smoothed curve $\Delta \bar{I}_i$, which can be fitted by adjusting the biomolecular structure R, and

(ii) owing to statistical noise in the data δI_i, which cannot be fitted. Then, χ^2 can be rewritten as

$$\chi^2 = \sum_{i=1}^{N_{\text{exp}}} \left(\frac{\Delta \bar{I}(q_i)}{f\sigma(q_i)} \right)^2 + \sum_{i=1}^{N_{\text{exp}}} \left(\frac{\delta I(q_i)}{f\sigma(q_i)} \right)^2 + 2 \sum_{i=1}^{N_{\text{exp}}} \frac{\Delta \bar{I}(q_i)\, \delta I(q_i)}{[f\sigma(q_i)]^2} \quad (5)$$

The third term in Eq. (5) vanishes approximately because the noise $\delta I(q_i)$ is symmetrically distributed around zero. The second term adds a constant offset of ≈ 1 per experimental data point, i.e., a total offset of N_{exp} to χ^2 that cannot be fitted. The first term quantifies the deviation with respect to the smoothed curve and contains all the relevant structural information. Using that $\Delta \bar{I}(q_i)$ and $\sigma(q_i)$ are approximately constant within each Shannon bin, the first term of Eq. (5) simplifies to a sum over the Shannon bins,

$$\chi^2 \approx \sum_{b=1}^{N_{\text{Shan}}} \frac{\Delta \bar{I}(q_b)^2}{f^2 \bar{\sigma}(q_b)^2} + N_{\text{exp}}, \quad (6)$$

where we used $\bar{\sigma}^2(q) \approx \sigma^2(q)/n_s$. Hence, up to a structurally irrelevant constant offset N_{exp} owing to experimental noise, χ^2 and E_{exp} can be evaluated using only the N_{Shan} intensities and errors of the smoothed curve. In case that more than N_{Shan} q-points are used to evaluate E_{exp}, a correcting prefactor is needed and we arrive at

$$E'_{\text{exp}} \approx \frac{1}{2} f_c\, k_B T \frac{N_{\text{Shan}}}{N_{\text{used}}} \sum_{i=1}^{N_{\text{used}}} \frac{[I_c(q_i; R) - (f \bar{I}_{\text{exp}}(q_i) + c)]^2}{[f \bar{\sigma}(q_i)]^2}, \quad (7)$$

where N_{used} may be chosen in the range of 1–2 N_{Shan}. Evidently, using the formulation in Eq. (7) greatly improves the computational efficiency of SAXS-driven MD simulations as it requires ~30 to 300 times fewer SAXS intensity calculations as compared to the formulation in Eq. (3).

2.3 Accounting for systematic and calculated errors

SAXS data collected with single-photon counting detectors may be subject to tiny statistical errors $\sigma_{\text{exp}}(q)$ at small scattering angles. Consequently, the overall uncertainty of the data may be dominated by unknown systematic errors, for instance owing to imprecise buffer subtraction or minor undetected radiation damage. When applying only the tiny statistical errors, the SAXS-derived energies take spuriously large values at small angles due to the $1/\bar{\sigma}(q_i)^2$ term, which would further propagate into large SAXS-derived forces.

Such problems are avoided by modeling of systematic errors. We previously modeled systematic errors via an uncertainty $\delta\rho_{\rm buf}$ of 0.1% to 1% of the buffer density (Chen & Hub, 2015). The uncertainty $\delta\rho_{\rm buf}$ propagates into an uncertainty $\sigma_{\rm buf}(q)$ of the SAXS curve, which is large at low angles and small at wide angles. Consequently, $\sigma_{\rm buf}(q)$ dominates the overall uncertainty at low angles and avoids spuriously large SAXS-derived forces, whereas $\bar{\sigma}(q)$ dominates at wide angles. Apart from the systematic errors, GROMACS-SWAXS takes statistical errors of the calculated curve $\sigma_{\rm c}(q_i)$ into account (Fig. 3). This is implemented by replacing the errors $\bar{\sigma}$ in Eq. (7) with $\sigma_{\rm tot}^2 = f^2\bar{\sigma}^2 + \sigma_{\rm buf}^2 + \sigma_{\rm c}^2$, leading to the final SAXS-derived energy applied by GROMACS-SWAXS:

$$E_{\rm exp}^{\rm f} \approx \frac{1}{2} f_c\, k_B T \frac{N_{\rm Shan}}{N_{\rm used}} \sum_{i=1}^{N_{\rm used}} \frac{[I_{\rm c}(q_i;R) - (f\bar{I}_{\rm exp}(q_i)+c)]^2}{f^2\bar{\sigma}^2(q_i)+\sigma_{\rm buf}^2(q_i)+\sigma_{\rm c}^2(q_i)}, \tag{8}$$

2.4 On-the-fly averaging of the calculated SAXS curve

Explicit-solvent SAXS curve predictions require averaging over multiple MD frames for computing a SAXS curve. The number of frames required for obtaining a converged SAXS curve depends on the contrast between the biomolecule and the pure buffer; the smaller the electron density contrast,

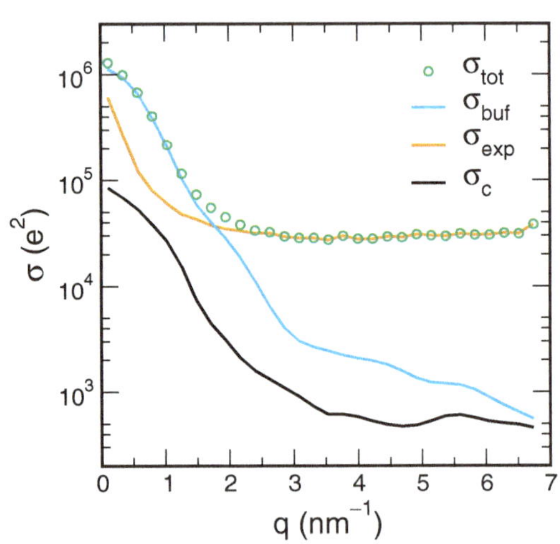

Fig. 3 Typical errors applied during SAXS-driven MD. At small angles, systematic errors $\sigma_{\rm buf}$ (*blue*) dominate the total uncertainty $\sigma_{\rm tot}$ (*green circles*), here modeled by a buffer density uncertainty of 1%. At wide angles, experimental errors dominate (*orange*). Calculated statistical errors are typically insignificant (*black*).

the more frames are required to converge the SAXS curve. During SAXS-driven MD simulations discussed here, the SAXS curve is therefore computed as a running temporal average with a memory kernel that decays exponentially into the past. The on-the-fly average of a quantity X at simulation time t is then given by

$$\langle X\rangle(t) = \mathcal{N}^{-1}\int_0^t X(t')\, \mathrm{e}^{-(t-t')/\tau} dt', \tag{9}$$

where τ is the memory time, typically chosen between 50 and 200 ps, and $\mathcal{N}$ is a normalization constant. Likewise, on-the-fly averaging of the scattering amplitudes of the biomolecular and the pure-solvent system using Eq. (9) provide an on-the-fly averaged SAXS curve (Chen & Hub, 2015).

Implementing the coupling to the experimental SAXS curve with an on-the-fly averaged calculated curve has two key advantages: first, thermal fluctuations on the time scale 1–2 τ such as solvent, side chain, or loop fluctuations are taken into account before comparing the calculated with the experimental curve in Eq. (8). Such thermal fluctuations may significantly influence the SAXS curve of proteins at moderate scattering angles ($q > 0.25$ Å^{-1}) (Chen & Hub, 2014; Moore, 2014; Tiede, Zhang, & Seifert, 2002). Second, there is no need to compute SAXS intensities every MD step, but instead one SAXS update every ~0.5 ps is sufficient. The update interval together with the memory time must be chosen such that the SAXS curves converge within τ. The longer update interval renders SAXS-driven MD simulation computationally efficient with an overhead of only 5–20% relative to unbiased simulations. However, owing to the on-the-fly average, the dynamics are not conservative because the energy $E^{\mathrm{f}}_{\mathrm{exp}}$ depends not only on the current but also on previous conformations. Hence, a reasonably tight temperature coupling scheme is required to avoid energy drifts.

2.5 SAXS-derived forces applied during MD

SAXS-derived forces applied in the MD simulations are given by the negative gradients of the SAXS-derived energy,

$$\mathbf{F}_k = -\nabla_k E^{\mathrm{f}}_{\mathrm{exp}}(R, D) \tag{10}$$

$$= -f_c\, k_B T \frac{N_{\mathrm{Shan}}}{N_{\mathrm{used}}} \sum_{i=1}^{N_{\mathrm{used}}} \frac{I_c(q_i; R(t)) - \left(f \bar{I}_{\mathrm{exp}}(q_i) + c\right)}{\sigma_{\mathrm{tot}}(q_i)^2} \nabla_k I_c(q_i; R(t)), \tag{11}$$

where $\nabla_k I_c(q_i; R(t))$ denotes the on-the-fly averaged gradients of the SAXS intensities with respect to the coordinates of atom k. For large biomolecular systems, storing these gradients may require several Gigabytes of memory (Chen & Hub, 2015).

GROMACS-SWAXS computes the forces $\mathbf{F}_k$ and the gradients $\nabla_k I_c(q_i;R(t))$ only for the solute atoms but not for the solvent atoms. Therefore, in addition to the derivatives of the I_c with respect to the solute coordinates, a correction factor

$$f_{\text{contrast}} = \frac{\rho_{\text{solu}} - \rho_{\text{solv}}}{\rho_{\text{solu}}} \tag{12}$$

is applied, where ρ_{solu} and ρ_{solv} denote the average solute and solvent densities, respectively. Here, density refers to the scattering length density, that is, to the electron density in SAXS or to the neutron scattering length density in SANS. As a numerical example, the factor f_{contrast} accounts for the fact that, upon moving a protein domain with density $\rho_{\text{solu}} = 440$ e nm^{-3} inside solvent with density $\rho_{\text{solv}} = 334$ e nm^{-3}, only the contrast of 106 e nm^{-3} is moved. Consequently, change of $I_c(q_i)$ upon a domain movement in mobile water is reduced by the factor $f_{\text{contrast}} = 0.24$ as compared to a domain movement at fixed water positions.

When coupling to SANS data measured at a large D_2O concentration, the contrast and the factor f_{contrast} may even become negative because a D_2O buffer exhibits a larger neutron scattering length density than proteins. Consequently, upon moving a protein domain to the left, the contrast may move to the right; such effect is correctly taken into account by a negative factor f_{contrast}.

2.6 Protocol A

To carry out SWAXS-driven MD simulations with the GROMACS-SWAXS implementation published at https://gitlab.com/cbjh/gromacs-swaxs, the following protocol is recommended:

1. Download and compile GROMACS-SWAXS, following the installation instructions of official GROMACS. Alternatively, GROMACS-SWAXS can be easily installed using Spack, which is available at many high-performance computing centers.
2. Setup the MD simulation system following freely available GROMACS tutorials. Make sure to use an approximately 1 nm larger simulation box compared to common simulations via the GROMACS module `gmx editconf -d`.

3. First compute a SAXS curve from an equilibrium simulation using the rerun functionality of the GROMACS-SWAXS `gmx mdrun` module, following tutorials at https://cbjh.gitlab.io/gromacs-swaxs-docs and as described in chapter "Predicting solution scattering patterns with explicit-solvent molecular simulations" by Chatzimagas and Hub in Part A of this monograph. Computing a SAXS curve involves the setup of a pure-solvent simulation systems, definition of the atomic form factors with the `gmx genscatt` module, and building of the envelope with `gmx genenv`, which defines the solvent region contributing to the SAXS curve. From visual inspection of the SAXS curve, is it plausible that a conformational transition of the biomolecule explains the deviation between the calculated and the experimental curve?
4. Generate the run-input (tpr) file with the `gmx grompp` module. For a typical SAXS-driven MD simulation, the MD parameter (mdp) file may look as follows:

```
; read scattering info from topology
define              = -DSCATTER
; turn on SAXS calc. and, optionally, multiple SANS calc.
scatt-coupl         = xray [neutron neutron ...]
; solute group
waxs-solute         = Protein ; or Protein-Masses
; solvent group
waxs-solvent        = Water_and_ions
; rotational fit group as defined with gmx genenv
waxs-rotfit         = C-alpha
; memory time for on-the-fly average
waxs-tau            = 200
; turn on SAXS energy E_exp slowly, e.g. over 5ns
waxs-t-target       = 5000
; define coupling potential type. Alternative: log
waxs-potential      = linear
; use experiental, calculated and systematic errors
waxs-weights        = exp+calc+solvdens
; force constant, use 1-5. fc=1 enables Bayesian interpretation
waxs-fc             = 1
; fit scale f and offset c of experimental curve
waxs-Iexp-fit       = scale-and-offset
; update on-the-fly averaged SAXS curve every 0.5ps with
dt=0.002
waxs-nstcalc        = 250
```

```
; nr of q-values, N_used. Use approx. 1.5 N_Shan
waxs-nq             = 15
; qmin and qmax in nm^(-1)
waxs-startq         = 0
waxs-endq           = 5
; nr of q-vectors for orientational average, use ~0.2*(D*qmax)^2
waxs-nsphere        = 500
; density of solvent (e/nm3), used for a precise buffer subtraction
waxs-solvdens       = 334
; relative uncertainty of the solvent density, used to
; estimate systematic errors. Use 0.1% to 0.5%
waxs-solvdens-uncert = 0.005
; D2O concentration (e.g. 0% and 100%) for each SANS calc.
waxs-deuter-conc = [0 1]
```

5. Smooth the experimental target curve. Upon smoothing the curve, the errors are reduced by a factor of $n_s^{1/2}$, where n_s is the number of experimental data points per Shannon bin. A Shell script for this purpose is available at https://cbjh.gitlab.io/gromacs-swaxs-docs. The shell script requires installation of the ATSAS software (Manalastas-Cantos et al., 2021).
6. Finally, specify the envelope files with environment variables and run the SAXS-driven MD:

```
export GMX_ENVELOPE_FILE=envelope.dat
export GMX_WAXS_FIT_REFFILE=envelope-ref.gro
sw=/path/to/pure-solvent.tpr
fw=/path/to/pure-solvent.xtc
gmx mdrun -s topol.tpr -fw $fw -sw $sw -is target.xvg ...
```

Here, `pure-solvent.tpr` and `pure-solvent.xtc` are the run-input and trajectory files of the pure-solvent systems used for computing the buffer subtraction. These have already been set up for computing the SAXS curve from an equilibrium simulation in step 3.
7. To analyze the simulation, visualize it in a molecular viewer and validate that the conformation is reasonable. Inspect in `waxs_final.xvg` whether the simulation was capable of finding a conformation that is compatible with the target SAXS curve. Inspect the SAXS-derived energy $E_{\mathrm{exp}}^{\mathrm{f}}$, which is stored in the energy (edr) file and available via the `gmx energy` command. Validate that the energy is in the range of several k_BT after the structure has been refined. The contributions of individual q-points to $E_{\mathrm{exp}}^{\mathrm{f}}$, available in `waxs_pot.xvg`, may reveal q-regions that could not be explained.

8. Since `waxs_final.xvg` represents the final on-the-fly average of the calculated curve, it represents only the final 1–2 τ ($\sim$200 ps) of the simulation. Use the rerun functionality of mdrun (`gmx mdrun -rerun traj.xtc`) to compute the SAXS curve that is uniformly averaged over a longer part of the SAXS-driven simulation, for instance over the final 10 ns. A uniform average is enabled with the mdp option `waxs-tau = 0`. Use more q-points, such as `waxs-nq = 100`.

3. SAXS-driven MD as a tool for Bayesian inference of molecular structures

3.1 Posterior, likelihood, and prior distributions

Following pioneering work by Rieping, Habeck, and coworkers on refinement against NMR data (Habeck, Rieping, & Nilges, 2006; Rieping, Habeck, & Nilges, 2005), SAXS-driven MD simulations have been reformulated as tool for Bayesian inference of biomolecular structures from experimental data. Accordingly, the goal of structure refinement is to find the conditional probability $P(R, \theta|D, K)$ that quantifies the plausibility for the biomolecular structure R in the light of the experimental data D and prior physical knowledge K. In the context of SAXS-guided modeling, the data D is given by the experimental curve $I_{\text{exp}}(q)$ and its errors, whereas the prior knowledge K represents the MD force field, the starting conformation of the simulation, and other prior experience. The symbol θ summarizes so-called nuisance parameters such as the unknown fitting parameters or unknown systematic errors, which must be estimated simultaneously with the structure.

In Bayesian inference, $P(R, \theta|D, K)$ should *not* be interpreted as probabilities of random events, as common in the "frequentist interpretation" of probability theory, but instead as a quantification of our state of knowledge and ignorance. A wide distribution $P(R, \theta|D, K)$ implies that many different conformations R are compatible with D and K, implying a high degree of ignorance and large confidence intervals on R. By the same token, a narrow $P(R, \theta|D, K)$ implies that only few structures are plausible in the light of D and K, implying precise knowledge of R and small confidence intervals on R. Hence, computing $P(R, \theta|D, K)$ provides rigorous confidence intervals for the refined structure founded in probability theory.

The conditional probability is evaluated using Bayes' theorem,

$$P(R, \theta|D, K) \propto L(D|R, \theta, K)\ \pi(R|K)\ \pi(\theta|K), \tag{13}$$

where $\pi(R|K)$ and $\pi(\theta|K)$ denote the prior distributions of conformations and of the nuisance parameters, respectively. $L(D|R, \theta, K)$ is the likelihood that the data D was measured given that the structure R was present in the experiment and given a specific set of nuisance parameters θ. The distribution $P(R, \theta|D, K)$ is called *posterior distribution*. When running MD simulations, the prior $\pi(R|K)$ is naturally given by the Boltzmann distribution

$$\pi(R|K) \propto \mathrm{e}^{-V_{\mathrm{FF}}(R)/k_B T} \tag{14}$$

with the MD force field $V_{\mathrm{FF}}(R)$, i.e., by the ensemble obtained before incorporating the data D. A reasonable choice for $\pi(\theta|K)$ is less obvious; if no prior information on the nuisance parameters is available, a noninformative prior such as a flat or a Jeffreys prior may be used. It is advisable to test multiple priors in order to exclude that the conclusions depend on the choice of the priors. For instance, testing multiple choices of $\pi(R|K)$ corresponds to running simulations with difference force fields V_{FF}.

Assuming Gaussian independent errors σ_{tot}, the likelihood function is

$$\begin{aligned} L(D|R,\theta,K) &\propto \prod_{i=1}^{N_{\exp}} \exp\left(-\frac{[I_{\mathrm{c}}(q_i;R) - I'_{\exp}(q_i;\theta)]^2}{2[f\sigma(q_i;\theta)]^2}\right) \\ &= \exp\left(-\frac{1}{2}\sum_{i=1}^{N_{\exp}} \frac{[I_{\mathrm{c}}(q_i;R,\theta) - I'_{\exp}(q_i;\theta)]^2}{[f\sigma(q_i;\theta)]^2}\right), \end{aligned} \tag{15}$$

Here, the $I'_{\exp}(q_i;\theta) = f\ I_{\exp}(q_i) + c$ is the experimental data adjusted by the scale f and the offset c, which appear as nuisance parameters (cf. Eq. 3). In GROMACS-SWAXS, the likelihood function is modified twofold relative to Eq. (15). First, as described above, the raw experimental intensities $(I_{\exp}, \sigma)$ and errors are replaced with the smoothed curve $(\bar{I}_{\exp}, \bar{\sigma})$, thereby replacing the sum over $N_{\exp}$ values with a sum over N_{Shan} values. Second, the errors are augmented with the calculated and systematic errors, leading to the final likelihood function:

$$L^{\mathrm{f}}(D|R,\theta,K) \propto \exp\left(-\frac{1}{2}\frac{N_{\mathrm{Shan}}}{N_{\mathrm{used}}}\sum_{j=1}^{N_{\mathrm{used}}} \frac{\left(I_{\mathrm{c}}(q_j;R) - \bar{I}'_{\exp}(q_j;\theta)\right)^2}{f^2\bar{\sigma}^2(q_i) + \sigma^2_{\mathrm{buf}}(q_i;\theta) + \sigma^2_{\mathrm{c}}(q_i)}\right) \tag{16}$$

Hence, L^{f} contains three nuisance parameters $\theta = \{f, c, \delta\rho_{\mathrm{buf}}\}$.

The posterior distribution $P(R, \theta|D, K)$ cannot be computed analytically but is instead obtained by importance sampling. Here, Newtonian dynamics as implemented by MD simulations are used for sampling the conformations.

By taking the negative logarithm of the posterior, we turn the probability functions into energy terms,

$$\begin{aligned} E &= -k_B T \ln P(R,\theta|D,K) \\ &= V_{\mathrm{FF}}(R) + E^{\mathrm{f}}_{\mathrm{exp}}(R,\, D,\, \theta)\big|_{f_c=1} - k_B T \ln \pi(\theta|K), \end{aligned} \tag{17}$$

where we used Eqs. (8), (13), (14), and (16). Several aspects of this results are notable:

- The original SAXS-derived energy Eq. (8) required the choice of a force constant f_c that weights the experimental data relative to the force field V_{FF}. Using Bayesian inference, in contrast, no force constant is needed because the weight of the experimental data is fully determined by probability theory. In practice, it may be useful to first drive a conformational transition with a larger f_c and, in a follow-up simulation, sample the posterior with $f_c = 1$.
- Sampling the posterior $P(R, \theta|D, K)$ has been implemented using Gibbs sampling. Accordingly, nuisance parameters θ are sampled with Metropolis Monte-Carlo at fixed conformations R, followed by Newtonian dynamics of R at fixed θ, and so on.
- Sampling the fitting parameters f and c is not required because they can be marginalized out analytically at the level of the likelihood when assuming flat priors for f and c, i.e., $\pi(f|K) = \pi(c|K) = \text{const}$. Then, the likelihood in Eq. (16) is replaced with $\widetilde{L}(D|R,\delta\rho_{\mathrm{buf}},K) = \int L(D|R,\theta,K)\,\mathrm{d}f\,\mathrm{d}c$, thereby taking "all possible" values of f and c into account. Evaluating this integral shows that $\widetilde{L}$ takes the same form as L, except that f and c are replaced with their maximum-likelihood estimates f_{ml} and c_{ml} (Shevchuk & Hub, 2017).

3.2 Bayesian treatment of systematic errors at small angles

GROMACS-SWAXS models systematic errors at small angles via an uncertainty of the buffer density $\delta\rho_{\mathrm{buf}}$ (see above and Fig. 3). In Bayesian SAXS-driven MD, $\delta\rho_{\mathrm{buf}}$ can be treated as one of the nuisance parameters θ (Shevchuk & Hub, 2017). Accordingly, the relative uncertainty $\delta\rho_{\mathrm{buf}}$ is sampled simultaneously with the structure R to obtain a joint posterior distribution over structures and buffer density uncertainties $P(R, \delta\rho_{\mathrm{buf}}|D, K)$. GROMACS-SWAXS applies a Gaussian prior distribution for $\delta\rho_{\mathrm{buf}}$. Let $r_{\mathrm{buf}} := \delta\rho_{\mathrm{buf}}/\rho_{\mathrm{buf}}$, then the prior is

$$\pi(r_{\mathrm{buf}}|K) \propto \exp\left(-r^2_{\mathrm{buf}}/2\epsilon^2_{\mathrm{buf}}\right) \tag{18}$$

where ρ_{buf} is the solvent density and ϵ_{buf} is given with the mdp parameter `waxs-solvdens-uncert`, typically set between 0.1% and 1%. This algorithm detects automatically whether the experimental data is biased by systematic errors at small angles. Namely, if the MD force field permits a conformational transition that explains the experimental data $I_{\mathrm{exp}}(q)$, the posterior of $\delta\rho_{\mathrm{buf}}$ would peak at small values, indicating that systematic errors are not required to explain the experimental data (Fig. 4, blue solid). In contrast, if the MD simulations does not find a conformation that explains I_{exp} at small angles, the posterior of $\delta\rho_{\mathrm{buf}}$ would peak at larger values, indicating that significant systematic errors are plausible in the light of the data and the force field (Fig. 4, orange dashed).

Clearly, treatment of systematic errors in SAXS-based modeling is still underdeveloped. To harvest increasingly finer details in SAXS data, it will

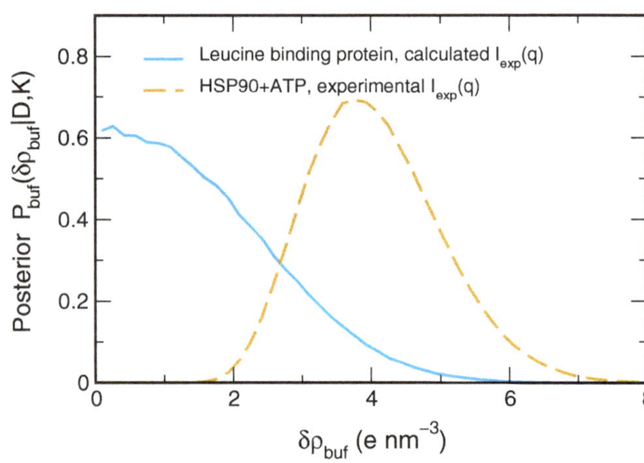

Fig. 4 Example of posterior distributions over the uncertainty of the buffer density $\delta\rho_{\mathrm{buf}}$ taken from SAXS-driven MD simulations of leucine binding protein (LBP, *solid curve*) or HSP90 bound to ATP (*dashed curve*). For LBP, the posterior peaks at $\delta\rho_{\mathrm{buf}} = 0$; hence, systematic errors are not required to explain the data, but they cannot be excluded either. This finding is expected since the SAXS-driven MD simulation were carried out against a calculated target curve without any uncertainties. In contrast, the posterior for HSP90 peaks at larger $\delta\rho_{\mathrm{buf}} = 4$ e nm^{-3}, suggesting that significant systematic errors are strictly required to explain the data. This finding reflects that the SAXS-driven MD was carried out against experimental data with substantial systematic errors at small angles. *Data taken from Shevchuk, R., & Hub, J. S. (2017). Bayesian refinement of protein structures and ensembles against SAXS data using molecular dynamics. PLOS Computational Biology, 13, e1005800.*

be useful to explicitly model other sources of errors, such as a small fraction of aggregated samples. We believe that Bayesian inference will provide a rigorous framework to learn systematic errors simultaneously with the biomolecular structures.

3.3 Protocol B

The Bayesian interpretation of SAXS-driven MD simulations are enabled by using a force constant of unity, and optionally, by sampling $\delta\rho_{buf}$. The mdp file for a Bayesian SAXS-driven MD simulation should contain, apart from the options described above:

```
waxs-fc= 1
waxs-solvdens-uncert= 0.005
waxs-solvdens-uncert-bayesian= yes
```

The simulation frames in the trajectory of the SAXS-driven MD may be interpreted as samples from a high-dimensional posterior distribution over conformations $P(R|D, K)$. Obtaining posteriors over intuitive properties, such as the center-of-mass distance d_{com} between two domains, is straightforward: d_{com} values may be extracted from the trajectory frames and plotted as a histogram. The histograms are the posterior $P_{com}(d_{com}|D, K)$ over the center-of-mass distances, suggesting that the peak position and the width of P_{com} provide the most plausible d_{com} and its uncertainty in the light of the data and the force field. Mathematically, obtaining $P_{com}(d_{com}|D, K)$ from $P(R|D, K)$ would involve a marginalization, that is the integration over all other degrees of freedom except d_{com}; however, since the trajectory of the Bayesian SAXS-driven MD contains samples of d_{com}, there is no need carry out a marginalization in practice.

Samples from the posterior over the relative uncertainty of the buffer density, $\delta\rho_{buf}/\rho_{buf}$, are written into separate output file `waxs_solvDensUncert.xvg` allowing the calculation of a histogram and, hence, the posterior over the uncertainty of the buffer density.

4. Maximum-entropy ensemble refinement against SAXS data

4.1 Theoretical background

Since SAXS is a solution method, experimental SAXS intensities represent the average over a structural ensemble. For structurally stable proteins, the ensemble may be approximated by a single, most prominent conformation, enabling the use of structure refinement methods described above. In contrast, for solutes that adopt a heterogeneous ensemble, the ensemble is adequately represented by a large number of conformations or by conformational distributions. Typical examples are intrinsically disordered proteins (IDPs) and proteins with disordered regions, domains connected with flexible linkers, or dynamic soft-matter complexes. Upon refining such heterogeneous ensembles against experimental SAXS data $I_{\mathrm{exp}}(q)$, the SAXS curves computed from the individual conformations may differ from $I_{\mathrm{exp}}(q)$, and only the ensemble-averaged computed curve should agree with the data.

However, ensemble refinement against SAXS data is an ill-posed problem because many different ensembles would agree with $I_{\mathrm{exp}}(q)$, even if the conformational space is restrained with an all-atom force field. Two strategies have been put forward for choosing a justified refined ensemble from all the ensembles that satisfies the data (Ravera, Sgheri, Parigi, & Luchinat, 2016). Following the strategy of maximum parsimony, the aim is to explain the data with as few conformations as possible. Such approach is most justified if the biomolecule adopts a few well-defined conformational states. The second approach is founded in statistical physics and is based on Jaynes' maximum-entropy principle (Boomsma, Ferkinghoff-Borg, & Lindorff-Larsen, 2014; Cesari, Reißer, & Bussi, 2018; Hermann & Hub, 2019; Jaynes, 1957). According to Jaynes, we should choose the ensemble distribution with the greatest uncertainty (or with the least information) that satisfies a given set of constraints. In the context of structure refinement, the constraints are our requested agreement with the experimental data. The principle is satisfied by finding an ensemble distribution that maximizes the Shannon entropy

$$S(p) = -\sum_i p(R_i) \ln p(R_i) \tag{19}$$

under the given constrains.

In ensemble refinement, however, we are typically interested in refining a prior ensemble from a free MD simulation against experimental data, suggesting that it is useful to maximize the *relative entropy* between the unbiased and the refined ensemble (Caticha, 2004). Because the relative entropy is the negative of the Kullback–Leibler divergence $D_{\mathrm{KL}}(p_1|p_0)$ (Kullback & Leibler, 1951), maximizing the relative entropy implies that we should find a refined ensemble distribution $p_1(R_i)$ that minimizes

$$D_{\mathrm{KL}}(p_1|p_0) = \sum_i p_1(R_i) \ln \frac{p_1(R_i)}{p_0(R_i)} \tag{20}$$

under the given constrains, where $p_0(R_i)$ is the prior ensemble distribution. Taking the prior from an unbiased MD ensemble, the aim is to find an updated ensemble that is (i) in agreement with the data and (ii) updated as minimally as possible with respect to the unbiased prior ensemble. In other words, the ensemble is only updated as strictly needed to explain the data, while any bias that is not supported by the data is avoided. Specifically, the formulation assures that the ensemble is not updated if the prior ensemble is already in agreement with the data.

The minimization problem can be solved with the help of Lagrangian multipliers, where one multiplier is required for each experimental constraint (Pitera & Chodera, 2012). However, the Lagrangian multipliers must be optimized in an iterative manner, which may be tedious in presence of a larger number of experimental constraints (Boomsma et al., 2014). An alternative for implementing a minimal bias is the parallel-replica approach. Here, several copies of the system (replicas) are simulated in parallel, and only the calculated data averaged over the replicas is restrained to the data with a harmonic restraint (Fig. 5). Roux and Weare (2013) and Cavalli, Camilloni, and Vendruscolo (2013) showed that the replica-averaged harmonic restraint imposes a minimal bias in the limit of a large number of replicas.

4.2 Parallel-replica ensemble refinement against SAXS data

Parallel-replica refinement against SAXS data is illustrated for an IDP in Fig. 5. First, the SAXS intensity is averaged over the replicas intensities $I_c(q_i, R_\alpha)$,

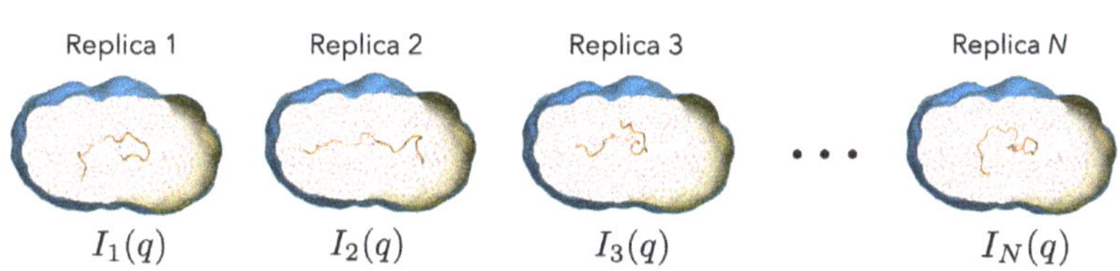

Fig. 5 Illustration of parallel-replica ensemble refinement. N replicas are simulated simultaneously, each providing a calculated curve $I_1,\ldots,I_N$. Coupling the replica-averaged SAXS curve to the experiment with a harmonic restraint leads to the maximum entropy ensemble.

$$\bar{I}_c(q_i; R_1,\ldots,R_{N_{\mathrm{rep}}}) = \frac{1}{N_{\mathrm{rep}}} \sum_{\alpha=1}^{N_{\mathrm{rep}}} I_c(q_i, R_\alpha), \tag{21}$$

where α is the replica index and N_{rep} is the number of replicas. Then, the systems are coupled to the data with a harmonic restraint similar to Eq. (8), except that the SAXS curve from a single simulation is substituted by the replica-averaged curve:

$$E\mathrm{exp}\left(R_1,\cdots,R_{N_{\mathrm{rep}}};I\mathrm{exp}\right) = \frac{1}{2} N_{\mathrm{rep}} f_c k_B T \frac{N_{\mathrm{Shan}}}{N_{\mathrm{used}}} \sum_{j}^{N_{\mathrm{used}}} \frac{\left[\bar{I}_c\left(q_j;R_1,\cdots R_N\right) - \bar{I}'_{\mathrm{exp}}\left(q_j\right)\right]^2}{f^2\,\bar{\sigma}^2\left(q_j\right)} \tag{22}$$

Here, $\bar{I}'_{\mathrm{exp}}$ denotes the smoothed experimental curve adjusted by the fitting parameters f and c, as used above. The biasing energy is multiplied with N_{rep}, such that the factor cancels with $1/N_{\mathrm{rep}}$ in Eq. (21) when taking the derivatives with respect to atomic coordinates, as done for computing the SAXS-derived forces (Hummer & Köfinger, 2015).

4.3 Choosing the number of replicas

The number of replicas that are required to follow the maximum entropy principle depends on the system (Boomsma et al., 2014; Hermann & Hub, 2019). A possible strategy for finding a good value for N_{rep} is to investigate distributions $h(\xi)$ of a few important degrees of freedom ξ, such as distribution of the radius of gyration of an IDP or of the moments of inertia of a soft-matter complex. Accordingly, the Kullback–Leibler divergence $D_{\mathrm{KL}}(h_1|h_0)$ between the biased distribution $h_1(\xi)$ and unbiased distribution $h_0(\xi)$ may be plotted vs the number of replicas. A sufficient value of N_{rep} would be indicated by a plateau region of such plot.

A disadvantage of D_{KL} is its numerical instability; namely, since p_0 appears in the denominator of Eq. (20), D_{KL} is unstable if some conformations of the biased ensemble p_1 were hardly sampled in the unbiased distribution p_0. A numerically more stable alternative is given by the Jensen–Shannon divergence, which may be considered as a smoothed and symmetrized version of D_{KL},

$$D_{\mathrm{JS}}(h_1|h_0) = [D_{\mathrm{KL}}(h_0|h_M) + D_{\mathrm{KL}}(h_1|h_M)]/2, \tag{23}$$

where $h_M = (h_0 + h_1)/2$ is the average of h_0 and h_1 (Hermann & Hub, 2019). Another useful measure is given by the entropy of biased distributions h_1, $S[h_1] = \int h_1(\xi) \ln h_1(\xi) d\xi$. If the number of replicas is too low, the ensemble is still overly biased, which may lead to overly narrow ensemble and overly narrow distributions $h_1(\xi)$. Hence, it is useful to plot the entropy (or even simply the width) of $h_1(\xi)$ versus the number of replicas. For the refinement of an IDP ensemble we found previously that using only 4–8 replicas were sufficient. This finding is likely explained by the fact that, when using explicit-solvent SAXS predictions, $I_c(q_i, R_\alpha)$ represent on-the-fly averages with a memory time of 50–200 ps. Hence, even the SAXS curves from individual replicas represent some conformation heterogeneity, explaining why only few replicas are sufficient to represent the heterogeneous overall shape of the IDP ensemble that is encoded by the data.

4.4 Protocol C

Multireplica SAXS-driven MD simulations are set up similar to the single-replica refinement described above.

1. Compile GROMACS-SWAXS with MPI support using
```
cmake -DGMX_MPI=ON ...
```
2. When using four replicas, for instance, use the following mdp options:
```
waxs-ensemble-type     = maxent-ensemble
waxs-ensemble-nstates  = 4
waxs-scale-i0          = no ; yes with small contrast, e.g. with IDP
gen-vel                = yes
```
With the `gen-vel = yes` option, the replicas start with different initial velocities, providing independent simulations. For solutes with a small contrast, such as an IDP, the forward scattering intensity $I_c(q = 0)$ may not converge within the memory time τ. To greatly accelerate the convergence, $I_c(q = 0)$ may be fixed to the forward intensity of the target curve by adding a small constant density to the solvent, turned on with the mdp option `waxs-scale-i0 = yes`. Prepare one tpr file for each replica, and place the tpr files into different subdirectories 000, 001, etc:
```
gmx grompp -f maxent.mdp -o 000/topol.tpr
gmx grompp -f maxent.mdp -o 001/topol.tpr etc.
```
3. Run the multireplica simulation with the `-multidir` functionality of mdrun:

```
sw=/path/to/pure-solvent.tpr
fw=/path/to/pure-solvent.xtc
target=$(realpath Itarget_trans.xvg)
export GMX_ENVELOPE_FILE=$(realpath envelope.dat)
export GMX_WAXS_FIT_REFFILE=$(realpath envelope-ref.gro)
mpiexec -np 4 gmx_mpi mdrun \
-s topol.tpr -sw $sw -fw $fw -is $target -multidir 000 001
```

4. Carry out the same analysis as described above for regular SAXS-driven MD. Now, the file `waxs_final.xvg` contains the final on-the-fly average of the replica-averaged SAXS curve. Carry out a rerun with the trajectories of the four replicas to compute a uniformly averaged SAXS curve for each replica, and henceforth average the calculated curve. Inspect whether the replica-averaged curve agrees with the experimental target curve.
5. Compute distributions $h_1(\xi)$ of interesting observables ξ, combined from all replica trajectories, such as the distribution of the radius of gyration or the secondary structure content of an IDP. Such distributions quantify the heterogeneity of the ensemble.
6. With a new simulation system, redo the simulation with increasing numbers of replicas N_{rep}. Recompute distributions over observables using the same aggregated simulation time (e.g., 1×400 ns, 2×200 ns, 4×100 ns, etc.). Inspect convergence of the entropy (or the width) of the distributions $h_1(\xi)$ as function of N_{rep}.

4.5 Example: Ensemble refinement of a detergent micelle

Fig. 6 presents a multireplica-average ensemble refinement of a DDM detergent micelle using an increasing number of 1–10 replicas. The distributions of moments of inertia as computed from the aggregated simulations strongly depend on the number of replicas. Namely, the distributions from single-copy refinement are overly narrow, indicative of an overly restrained ensemble in violation of the maximum entropy principle (Fig. 6C, black). Using a larger number of replicas, the distributions are wider, reflecting a larger degree of heterogeneity. Critically, all simulations reveal quantitative agreement with the data, even when using only a single replica (Fig. 6B). This demonstrates that agreement with the data does by far not guarantee that the ensemble is correct.

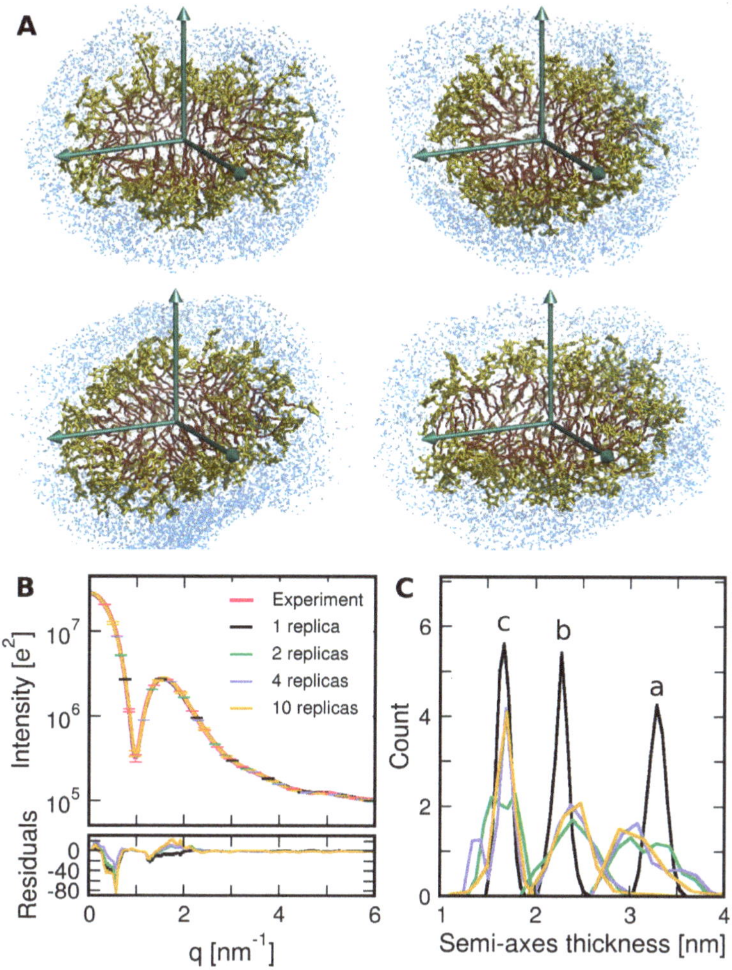

Fig. 6 (A) Parallel-replica ensemble refinement of a DDM detergent micelle. (B) Excellent agreement is found with the data, irrespective of the number of replicas between 1 and 10 ($\chi^2 = 129, \chi^2 = 80, \chi^2 = 37$, and $\chi^2 = 161$ for 1, 2, 4, and 10 replicas, respectively). (C) However, different numbers of replicas lead to different conformation ensembles, here quantified by the distributions of the micelle core semiaxes a, b, and c, demonstrating that agreement with the data does by far not guarantee that the ensemble is correct. *Reprinted with permission from Ivanović, M. T., Hermann, M. R., Wójcik, M., Pérez, J., & Hub, J. S. (2020). Small-angle X-ray scattering curves of detergent micelles: Effects of asymmetry, shape fluctuations, disorder, and atomic details. The Journal of Physical Chemistry Letters, 11(3), 945–951. https://doi.org/10.1021/acs.jpclett.9b03154, Copyright 2020 American Chemical Society.*

5. Discussion: Conceptual considerations and recommendations

5.1 SAXS-driven MD simulations (should) feel only a weak bias by the SAXS data

Structural data is typically insufficient for defining all degrees of freedom of a biomolecule. This is true not only for SAXS data, but also for data from X-ray crystallography, NMR spectroscopy, or cryoelectron microscopy. To avoid overfitting during structure refinement, structural data is complemented with additional physicochemical information that restraints the biomolecule into realistic conformations. The required amount of additional information critically depends on the information content of the data; the lower information content of the data, the more predictive additional information is needed. For instance, excluding atomic overlaps and restraining chemical bond geometries is often sufficient for the refinement of atomic models against crystallographic data but would be by far insufficient for refinement against SAXS data.

During SAXS-driven MD, the structure is largely imposed by the all-atom force field, which restraints not only chemical geometries but also maintains a proper hydrogen bond network and favorable electrostatic interactions. Overfitting is avoided by using only a small force constant f_c for the SAXS-derived restraints in the order of unity, such that biomolecular dynamics are largely controlled by the force field $V_{FF}(R)$ whereas the energy E_{exp} only mildly pushes the biomolecule into agreement with the data (see Eqs. 8 and 17). Indeed, E_{exp}^{f} takes values of few up to tens of kilojoules per mole, whereas the potential energy contributions from Lennard-Jones or Coulomb interactions are typically in the range of hundred thousands to millions of kilojoules per mole.

Inspection of the SAXS-derived potentials is also advised to prevent overinterpretation of the structural ensemble. The potential should converge over simulation time to values in the range of several k_BT (thermal energies). Caution is advised if individual q-points give rise to high potentials, indicating that parts of the SAXS curve cannot be explained by the simulation ensemble. Generally, it is highly desirable to cross-validate the derived ensemble against independent structural or biochemical data that has not been used during the refinement.

5.2 Accelerating transitions with SAXS data and sampling limitations

SAXS-driven MD simulations have been used to accelerate large-scale conformational transitions of biomolecules, which would require prohibitively long simulation times during unbiased simulations. For instance, we showed that SAXS data may be used to drive large-scale opening transition of the large proteins Hsp90 or ATCase in MD simulations (Chen & Hub, 2015; Shevchuk & Hub, 2017). However, SAXS-driven MD works as an enhanced sampling technique only if the SAXS-derived forces point into the direction of the sought-after conformation transition. This is typically true in the case of large-scale domain motions, in particular if these motions modulate the radius of gyration. In contrast, achieving a complex, nonlinear rearrangements in the simulation such as the folding of an unstructured tail into an α-helix is far more challenging. In such cases, the correct final state might be detectable via a low χ^2 in Eq. (5), but it is unlikely that SAXS-derived forces would accelerate the folding transition. Hence, for guiding complex transitions with SAXS data, SAXS-driven MD may be combined with established enhanced sampling techniques such as simulated tempering or Hamiltonian replica exchange.

5.3 Further analysis

The trajectories obtained from SAXS-driven MD are typically post-processed to provide insight that is not directly accessible from the SAXS data. This includes, for instance, analysis of (i) the heterogeneity of the refined ensembles, (ii) interactions at the atomic level such as secondary structure content or hydrogen bond patterns, (iii) transition pathways, or (iv) structural parameters such as the persistence lengths or end-to-end distances of IDPs. In addition, refined ensembles may provide insight into thermodynamic driving forces by means of free energy calculations, which have so far been underexposed in the context of SAXS-driven MD simulations.

6. Applications

Although SAXS-driven MD simulations have emerged recently, they have provided insight into a range of (bio)molecular systems. To mention a few, parallel-replica ensemble refinement of detergent micelles was used to identify capabilities and limitations of simple geometric models in explaining

SAXS data (Ivanović et al., 2020). WAXS-driven MD solved the salt-dependent solution structures of RNA triplex motifs (Chen, He, Kirmizialtin, & Pollack, 2022), the dynamics of RNA upon ligand or ion binding (He, Henning-Knechtel, & Kirmizialtin, 2022), or a large-scale opening transition of a phytochrome (Björling et al., 2015). In a related approach, combined with metainference and metadynamics, simulations restrained to SAXS data characterized a flexible polyubiquitin chain (Jussupow et al., 2020). Disordered intermediate states of unfolding bovine α-lactalbumin were obtained by restraining MD simulations to time-resolved SAXS data (Hsu, Leshchev, Kosheleva, Kohlstedt, & Chen, 2021). Ongoing efforts in our group involve heterogeneous ensembles of RNA–protein complexes involved in translational control. We expect that more applications will emerge as the implementations are now freely available and widely documented.

7. Summary

SAXS is an increasingly valuable tool of integrative structural biology thanks to the accuracy of data collected at modern SEC-SAXS beamlines and thanks to its structural information content, which is not accessible by other techniques. However, the interpretation of the data is challenged by the low information content of the experimental signals, leading to the risk of overinterpreting the data. This risk is mitigated by using MD simulations that add physicochemical information to the data.

In this chapter, we presented three approaches for refining structural ensembles against SAXS data by integrating the experimental data on-the-fly into all-atom MD simulations:

(i) During SAXS-driven MD, an MD simulation is coupled to experimental SAXS data using harmonic restrains on the data, thereby refining an ensemble that may be approximated by a single, most prominent conformation. The SAXS-driven simulations promote conformational transitions compatible with the data. They are capable of overcoming force field imperfections and sampling limitations of unbiased simulations, given that the transitions are geometrically simple.

(ii) When using SAXS-driven MD as a tool for Bayesian inference of biomolecular structures from a given SAXS curve, the MD simulation samples the sought-after posterior distribution. The posterior quantifies the plausibility of a biomolecular structure in the light of the experimental data and prior physicochemical information, as encoded in the MD force fields. This Bayesian framework may estimate systematic errors at small angles simultaneously with the structure, which enables one to assess whether systematic errors are required to explain the experimental data.

(iii) The parallel-replica approach allows one to refine heterogeneous ensembles with commitment to the maximum-entropy principle. Here, the SAXS curve averaged over several replicas is coupled to the experimental data using harmonic restrains, such that the updated ensemble is compatible with the data but biased as minimally as possible with respect to the unbiased ensemble.

All approaches require a forward model for calculating the SAXS intensities from the atomistic structures. We discussed structure and ensemble refinement based on explicit-solvent SAXS curve calculations, as used in the WAXSiS method, thereby accurately representing the hydration layer and the excluded solvent (Chen & Hub, 2014; Knight & Hub, 2015). An implementation of the methods described here employing the explicit-solvent SAXS curve calculations is freely available in GROMACS-SWAXS (https://gitlab.com/cbjh/gromacs-swaxs).

Acknowledgments

We thank Milos Ivanovic for preparing Fig. 6A. This study was supported by the Deutsche Forschungsgemeinschaft (HU 1971/3-1).

References

Abraham, M. J., Murtola, T., Schulz, R., Páll, S., Smith, J. C., Hess, B., & Lindahl, E. (2015). GROMACS: High performance molecular simulations through multi-level parallelism from laptops to supercomputers. *SoftwareX*, *1*, 19–25.

Bernetti, M., & Bussi, G. (2021). Comparing state-of-the-art approaches to back-calculate SAXS spectra from atomistic molecular dynamics simulations. *The European Physical Journal B*, *94*(9), 180. https://doi.org/10.1140/epjb/s10051-021-00186-9.

Berthaud, A., Manzi, J., Pérez, J., & Mangenot, S. (2012). Modeling detergent organization around aquaporin-0 using small-angle X-ray scattering. *Journal of the American Chemical Society*, *134*(24), 10080–10088.

Björling, A., Niebling, S., Marcellini, M., van der Spoel, D., & Westenhoff, S. (2015). Deciphering solution scattering data with experimentally guided molecular dynamics simulations. *Journal of Chemical Theory and Computation*, *11*, 780–787.

Boomsma, W., Ferkinghoff-Borg, J., & Lindorff-Larsen, K. (2014). Combining experiments and simulations using the maximum entropy principle. *PLOS Computational Biology*, *10*(2), e1003406.

Brosey, C. A., & Tainer, J. A. (2019). Evolving SAXS versatility: Solution X-ray scattering for macromolecular architecture, functional landscapes, and integrative structural biology. *Current Opinion in Structural Biology*, *58*, 197–213. https://doi.org/10.1016/j.sbi.2019.04.004.

Caticha, A. (2004). Relative entropy and inductive inference. In *AIP conference proceedings: Vol. 707* (pp. 75–96). American Institute of Physics.

Cavalli, A., Camilloni, C., & Vendruscolo, M. (2013). Molecular dynamics simulations with replica-averaged structural restraints generate structural ensembles according to the maximum entropy principle. *The Journal of Chemical Physics*, *138*, 03B603.

Cesari, A., Reißer, S., & Bussi, G. (2018). Using the maximum entropy principle to combine simulations and solution experiments. *Computation*, *6*, 15.

Chen, P.-c., & Hub, J. S. (2014). Validating solution ensembles from molecular dynamics simulation by wide-angle X-ray scattering data. *Biophysical Journal*, *107*, 435–447.

Chen, P.-c., & Hub, J. S. (2015). Interpretation of solution X-ray scattering by explicit-solvent molecular dynamics. *Biophysical Journal*, *108*, 2573–2584.

Chen, P.-c., Shevchuk, R., Strnad, F. M., Lorenz, C., Karge, L., Gilles, R., et al. (2019). Combined small-angle X-ray and neutron scattering restraints in molecular dynamics simulations. *Journal of Chemical Theory and Computation*, *15*(8), 4687–4698. https://doi.org/10.1021/acs.jctc.9b00292.

Chen, Y.-L., He, W., Kirmizialtin, S., & Pollack, L. (2022). Insights into the structural stability of major groove RNA triplexes by WAXS-guided MD simulations. *Cell Reports Physical Science*, *3*(7), 100971.

Gorba, C., Miyashita, O., & Tama, F. (2008). Normal-mode flexible fitting of high-resolution structure of biological molecules toward one-dimensional low-resolution data. *Biophysical Journal*, *94*(5), 1589–1599.

Gräwert, T. W., & Svergun, D. I. (2020). Structural modeling using solution small-angle X-ray scattering (SAXS). *Journal of Molecular Biology*, *432*(9), 3078–3092. https://doi.org/10.1016/j.jmb.2020.01.030.

Habeck, M., Rieping, W., & Nilges, M. (2006). Weighting of experimental evidence in macromolecular structure determination. *Proceedings of the National Academy of Sciences of the United States of America*, *103*(6), 1756–1761.

He, W., Henning-Knechtel, A., & Kirmizialtin, S. (2022). Visualizing RNA structures by SAXS-driven MD simulations. *Frontiers in Bioinformatics*, *2*, 781949. https://doi.org/10.3389/fbinf.

Hermann, M. R., & Hub, J. S. (2019). SAXS-restrained ensemble simulations of intrinsically disordered proteins with commitment to the principle of maximum entropy. *Journal of Chemical Theory and Computation*, *15*(9), 5103–5115. https://doi.org/10.1021/acs.jctc.9b00338.

Hsu, D. J., Leshchev, D., Kosheleva, I., Kohlstedt, K. L., & Chen, L. X. (2020). Integrating solvation shell structure in experimentally driven molecular dynamics using X-ray solution scattering data. *The Journal of Chemical Physics*, *152*(20), 204115. https://doi.org/10.1063/5.0007158.

Hsu, D. J., Leshchev, D., Kosheleva, I., Kohlstedt, K. L., & Chen, L. X. (2021). Unfolding bovine α-lactalbumin with T-jump: Characterizing disordered intermediates via time-resolved x-ray solution scattering and molecular dynamics simulations. *The Journal of Chemical Physics*, *154*(10), 105101.

Hub, J. S. (2018). Interpreting solution X-ray scattering data using molecular simulations. *Current Opinion in Structural Biology*, *49*, 18–26.

Hummer, G., & Köfinger, J. (2015). Bayesian ensemble refinement by replica simulations and reweighting. *The Journal of Chemical Physics*, *143*(24), 12B634_1.

Ivanović, M. T., Hermann, M. R., Wójcik, M., Pérez, J., & Hub, J. S. (2020). Small-angle X-ray scattering curves of detergent micelles: Effects of asymmetry, shape fluctuations, disorder, and atomic details. *The Journal of Physical Chemistry Letters*, *11*(3), 945–951. https://doi.org/10.1021/acs.jpclett.9b03154.

Jack, A., & Levitt, M. (1978). Refinement of large structures by simultaneous minimization of energy and R factor. *Acta Crystallographica. Section A, Foundations of Crystallography*, *34*(6), 931–935.

Jaynes, E. T. (1957). Information theory and statistical mechanics. *Physical Review, 106*(4), 620.

Jeffries, C. M., Graewert, M. A., Blanchet, C. E., Langley, D. B., Whitten, A. E., & Svergun, D. I. (2016). Preparing monodisperse macromolecular samples for successful biological small-angle X-ray and neutron-scattering experiments. *Nature Protocols, 11*(11), 2122–2153.

Jussupow, A., Messias, A. C., Stehle, R., Geerlof, A., Solbak, S. M., Paissoni, C., et al. (2020). The dynamics of linear polyubiquitin. *Science Advances, 6*(42), eabc3786.

Kimanius, D., Pettersson, I., Schluckebier, G., Lindahl, E., & Andersson, M. (2015). SAXS-guided metadynamics. *Journal of Chemical Theory and Computation, 11*(7), 3491–3498.

Knight, C. J., & Hub, J. S. (2015). WAXSiS: A web server for the calculation of SAXS/WAXS curves based on explicit-solvent molecular dynamics. *Nucleic Acids Research, 43*, W225–W230.

Kullback, S., & Leibler, R. A. (1951). On information and sufficiency. *The Annals of Mathematical Statistics, 22*, 79–86.

Manalastas-Cantos, K., Konarev, P. V., Hajizadeh, N. R., Kikhney, A. G., Petoukhov, M. V., Molodenskiy, D. S., et al. (2021). *ATSAS 3.0*: Expanded functionality and new tools for small-angle scattering data analysis. *Journal of Applied Crystallography, 54*(1), 343–355. https://doi.org/10.1107/S1600576720013412.

Merzel, F., & Smith, J. C. (2002). Is the first hydration shell of lysozyme of higher density than bulk water? *Proceedings of the National Academy of Sciences of the United States of America, 99*, 5378–5383.

Moore, P. B. (1980). Small-angle scattering. Information content and error analysis. *Journal of Applied Crystallography, 13*(2), 168–175.

Moore, P. B. (2014). The effects of thermal disorder on the solution-scattering profiles of macromolecules. *Biophysical Journal, 106*, 1489–1496.

Paissoni, C., Jussupow, A., & Camilloni, C. (2020). Determination of protein structural ensembles by hybrid-resolution SAXS restrained molecular dynamics. *Journal of Chemical Theory and Computation, 16*(4), 2825–2834. https://doi.org/10.1021/acs.jctc.9b01181.

Park, S., Bardhan, J. P., Roux, B., & Makowski, L. (2009). Simulated X-ray scattering of protein solutions using explicit-solvent models. *The Journal of Chemical Physics, 130*, 134114. https://doi.org/10.1063/1.3099611.

Pelikan, M., Hura, G. L., & Hammel, M. (2009). Structure and flexibility within proteins as identified through small angle X-ray scattering. *General Physiology and Biophysics, 28*(2), 174.

Petoukhov, M. V., & Svergun, D. I. (2005). Global rigid body modeling of macromolecular complexes against small-angle scattering data. *Biophysical Journal, 89*(2), 1237–1250.

Pitera, J. W., & Chodera, J. D. (2012). On the use of experimental observations to bias simulated ensembles. *Journal of Chemical Theory and Computation, 8*(10), 3445–3451. https://doi.org/10.1021/ct300112v.

Putnam, C. D., Hammel, M., Hura, G. L., & Tainer, J. A. (2007). X-ray solution scattering (SAXS) combined with crystallography and computation: Defining accurate macromolecular structures, conformations and assemblies in solution. *Quarterly Reviews of Biophysics, 40*(3), 191–285. https://doi.org/10.1017/S0033583507004635.

Rambo, R. P., & Tainer, J. A. (2013). Accurate assessment of mass, models and resolution by small-angle scattering. *Nature, 496*(7446), 477.

Ravera, E., Sgheri, L., Parigi, G., & Luchinat, C. (2016). A critical assessment of methods to recover information from averaged data. *Physical Chemistry Chemical Physics, 18*(8), 5686–5701.

Rieping, W., Habeck, M., & Nilges, M. (2005). Inferential structure determination. *Science*, *309*(5732), 303–306.

Rout, M. P., & Sali, A. (2019). Principles for integrative structural biology studies. *Cell*, *177*(6), 1384–1403. https://doi.org/10.1016/j.cell.2019.05.016.

Roux, B., & Weare, J. (2013). On the statistical equivalence of restrained-ensemble simulations with the maximum entropy method. *The Journal of Chemical Physics*, *138*(8), 084107.

Schneidman-Duhovny, D., Hammel, M., & Sali, A. (2010). FoXS: A web server for rapid computation and fitting of SAXS profiles. *Nucleic Acids Research*, *38*, W540–W544.

Shevchuk, R., & Hub, J. S. (2017). Bayesian refinement of protein structures and ensembles against SAXS data using molecular dynamics. *PLOS Computational Biology*, *13*, e1005800.

Tiede, D. M., Zhang, R., & Seifert, S. (2002). Protein conformations explored by difference high-angle solution x-ray scattering: Oxidation state and temperature dependent changes in cytochrome c. *Biochemistry*, *41*(21), 6605–6614.

CHAPTER THREE

Data analysis and modeling of small-angle neutron scattering data with contrast variation from bio-macromolecular complexes

Andrew E. Whitten[a,*] and Cy M. Jeffries[b,*]
[a]Australian Nuclear Science and Technology Organisation, Lucas Heights, NSW, Australia
[b]European Molecular Biology Laboratory (EMBL), Hamburg Unit, c/o Deutsches Elektronen-Synchrotron, Hamburg, Germany
*Corresponding authors: e-mail address: andrew.whitten@ansto.gov.au; cy.jeffries@embl-hamburg.de

Contents

Methods in Enzymology, Volume 678
ISSN 0076-6879
https://doi.org/10.1016/bs.mie.2022.09.015

Abstract

Small-angle neutron scattering (SANS) with contrast variation (CV) is a valuable technique in the structural biology toolchest. Accurate structural parameters—e.g., radii of gyration, volumes, dimensions, and distance distribution(s)—can be derived from the SANS-CV data to yield the shape and disposition of the individual components within stable complexes. Contrast variation is achieved through the substitution of hydrogen isotopes (^{1}H for ^{2}H) in molecules and solvents to alter the neutron scattering properties of each component of a complex. While SANS-CV can be used a stand-alone technique for interrogating the overall structure of biomacromolecules in solution, it also complements other methods such as small-angle X-ray scattering, crystallography, nuclear magnetic resonance, and cryo-electron microscopy. Undertaking a SANS-CV experiment is challenging, due in part to the preparation of significant quantities of monodisperse samples that may require deuterium (^{2}H) labeling. Nevertheless, SANS-CV can be used to study a diverse range biomacromolecular complexes including protein-protein and protein-nucleic acid systems, membrane proteins, and flexible systems resistant to crystallization. This chapter describes how to approach the data analysis and modeling of SANS data, including: (1) Analysis of the forward scattering ($I(0)$) and calculation of theoretical estimates of contrast; (2) Analysis of the contrast dependence of the radius of gyration using the Stuhrmann plot and parallel axis theorem; (3) Calculation of composite scattering functions to evaluate the size, shape, and dispositions of individual components within a complex, and; (4) Development of real-space models to fit the SANS-CV data using volume-element bead modeling or atomistic rigid body modeling.

Abbreviations

CV contrast variation
FWHM full width at half maximum height
IFT indirect Fourier transformation
RI refractive index
SANS small-angle neutron scattering
SAXS small-angle X-ray scattering
UV ultraviolet

1. Introduction

Small-angle scattering is an established technique in structural biology used to structurally characterize proteins and other biomolecules in solution (Brosey & Tainer, 2019; Da Vela & Svergun, 2020; Mahieu & Gabel, 2018; Trewhella, 2022). For the study of biomacromolecular complexes or assemblies, small-angle X-ray Scattering (SAXS) can provide low-resolution structural data (shape information) for the entire complex, but without information from complementary techniques it becomes increasingly difficult to uniquely define the internal arrangement of different components within a complex. To determine this arrangement, a different perspective is required. One approach is to systematically vary the scattering signal from each component, which will provide information on the relationship between the different components.

The average scattering-length density of a molecule ($\overline{\rho}$), is defined as

$$\overline{\rho} = \frac{1}{V}\sum\nolimits_i b_i, \tag{1}$$

where V is the volume of the molecule, and b_i is the coherent scattering length of each atom (i) in the molecule. The scattering length density of any system, be it a macromolecule or solvent, is dependent on the types of atoms present, and the nature of the radiation used. Electromagnetic X-ray radiation is scattered by electrons, hence, $\overline{\rho}$ (X-ray) is dependent on the number of electrons in the molecule, which is a function of its elemental/chemical composition. Neutrons, on the other hand, are neutral molecules scattered primarily by nuclei, and as different isotopes of the same element scatter neutrons differently, $\overline{\rho}$ (Neutron) is dependent on the isotopic composition of the sample. A related quantity of central importance to the scattering experiment is the average contrast ($\Delta\overline{\rho}$), which is simply the difference between the average scattering-length density of a molecule $\overline{\rho}$ and the scattering length density of the solvent (ρ_s)

$$\Delta\overline{\rho} = \overline{\rho} - \rho_s. \tag{2}$$

Contrast is a measure of the excess scattering length density of a macromolecule with respect to the solvent, and thus its propensity to scatter radiation. From Eq. (2) it can be seen that $\Delta\overline{\rho}$ can be altered by changing $\overline{\rho}$ and/or ρ_s. For X-rays, changing $\overline{\rho}$ requires chemical modification of the molecule (Grishaev, Anthis, & Clore, 2012), but ρ_s can be changed through the addition of salts containing high atomic number elements (Gabel, Engilberge, Perez, & Girard, 2019; Tardieu et al., 1976), or through the incorporation of high concentrations of sugars into the solvent (Kirste & Stuhrman, 1967). For X-ray scattering from molecules with non-uniform contrast such as lipoproteins (Tardieu et al., 1976) or protein-nucleic acid complexes (Pabit et al., 2020; Tokuda, Pabit, & Pollack, 2016), this approach can provide additional information regarding the internal structure of the complex as changing ρ_s will differentially change the contrast of each component of the complex. This is because $\overline{\rho}$ (X-ray) for lipids or the phosphate rich nucleic acid component differs from that of the protein. However, this approach is not useful for the study of protein-protein complexes, as changing the scattering-length density of the solvent leaves the contrast of both components of the complex approximately equal.

Neutron scattering opens a range of possibilities in relation to manipulation of the contrast of biomolecules (Breyton et al., 2013; Jacques & Trewhella, 2010; Lapinaite, Carlomagno, & Gabel, 2020; Mahieu & Gabel, 2018; Trewhella, 2022). Through isotopic labeling of a molecule, and/or varying the isotopic ratio of a given element in the solvent it is possible to change both $\overline{\rho}$ and ρ_s, and consequently $\Delta\overline{\rho}$, without altering the elemental composition of the molecule or solvent. The most relevant isotopic substitution for neutron scattering from biomolecules is between the two stable isotopes of hydrogen, ^{1}H (protium or H), and ^{2}H (deuterium or D). Hydrogen is ubiquitous in biological systems and H and D have very different scattering lengths. While D has a positive scattering length similar in magnitude to most of the significant isotopes found in biomolecules (^{12}C, ^{14}N, ^{16}O, ^{31}P, and ^{32}S), H has a scattering length that is about half that magnitude and opposite in sign to that of D. The negative scattering length indicates the neutron experiences an attractive potential as it scattered by the H nucleus, which means the phase of the scattered neutron is preserved (Bacon, 1975). Conversely, a positive scattering length indicates the neutron experiences a repulsive potential as it is scattered, resulting in a 180° phase change relative to the incident neutron. The phase difference between neutrons scattered by H nuclei and neutrons scattered by other common nuclei results in destructive interference.

At a conceptual level, understanding how the substitution of H for D changes the neutron contrast of biomolecules is reasonably straightforward, and without difficulty it can be recognized that if the solution conditions are adjusted such that $\Delta\overline{\rho} = 0$ for one component of the complex, it will be rendered practically "invisible." This means that any scattering data collected will be dominated by the visible component of the complex. This approach, known as contrast matching, is a special case of contrast variation (CV), where CV exploits the scattering length differences of H and D to not only provide structural information for each component within a complex, but also the disposition of each component within a complex. While the interpretation of contrast matching data is relatively intuitive, understanding CV data is often complicated.

This chapter provide a theoretical overview of the different approaches to analyze SANS-CV data, followed by a step-by-step procedure for each analysis or modeling technique. The procedures outlined assume the reader is familiar with Guinier analysis (Guinier & Fournet, 1955), and indirect

Fourier transformation (IFT) of scattering data (Glatter, 1977; Svergun, Semenyuk, & Feigin, 1988). Further, the advice and procedures described herein assume that the design of the small-angle scattering experiment is appropriate, the samples have been well prepared, the data were collected for adequate lengths of time, and all required corrections have been applied during the reduction of the 2-dimensional scattering images to obtain a number of 1-dimensional scattering profiles (preferably) on an absolute scale. Practical advice regarding these aspects of SANS experiments is covered elsewhere (Jeffries et al., 2016; Krueger, 2017; Trewhella et al., 2017; Whitten & Trewhella, 2009) (see Chapter "Planning, executing and assessing the validity of SANS contrast variation experiments" by Krueger, Vol. 1 and Chapter "Technical considerations for small-angle neutron scattering from biological macromolecules in solution: Cross sections, contrasts, instrument setup and measurement" by Pietras et al., Vol. 1).

2. Analysis of the forward scattering intensity, *I*(0), and calculation of contrast

Generally, the first analysis technique applied to small-angle scattering data is the Guinier plot (Guinier & Fournet, 1955), which is a linear fit to the natural logarithm of the measured scattering intensities, $I(q)$, vs the square of the scattering vector (or momentum transfer) q, for small values of q (where $q=(4\pi\sin\theta)/\lambda$, 2θ is the scattering angle, and λ is the wavelength). The Guinier plot provides estimates of two parameters: the forward scattering intensity ($I(0)$) by extrapolation to $q=0$, which provides information on macromolecular volume; and the radius of gyration (R_g) from the slope of the linear fit, which is equal to $-q^2R_g{}^2/3$ and provides information regarding the distribution of scattering centers within the molecule. This section will focus on the analysis of $I(0)$ from a CV series, while analysis of R_g will be covered in Section 3.

For a solution of non-interacting, identical macromolecules, after background scattering contributions from the solvent have been subtracted, $\Delta\overline{\rho}$ is directly related to the measured scattering intensities

$$I(q) = N(\Delta\overline{\rho}V)^2P(q), \tag{3}$$

where N is the number density of molecules in solution (the concentration of molecules per unit volume), and $P(q)$ is the form factor, which encodes information regarding the shape of the molecule. If $P(q)$ is normalized such that $P(0)=1$, it follows that

$$I(0) = N(\Delta\bar{\rho}V)^2. \quad (4)$$

In other words, $I(q)$ extrapolated to $q=0$ depends linearly on the concentration, quadratically on both $\Delta\bar{\rho}$ and V, and does not depend on the molecular shape. Thus, V can be estimated from the measured $I(q)$ vs q data with knowledge of N and $\Delta\bar{\rho}$. For series of measurements where the contrast of the molecule of interest is varied, there is more information available than just molecular size, and it is informative to look at the dependence of $I(0)$ on the fraction of D_2O (f_{D_2O}) in an aqueous solvent. For a homogeneous solution of molecules, $\Delta\bar{\rho}$ is linearly dependent on f_{D_2O} (Jacrot, 1976; Whitten, Cai, & Trewhella, 2008) (see Chapter "Deuteration for biological SANS: Case studies, success and challenges in chemistry and biology" by Duff, Vol. 1). Thus, from Eq. (4) it follows that

$$\frac{I(0)}{N} \propto \left(f_{D_2O} - f_{MP}\right)^2, \quad (5)$$

where f_{MP} corresponds to the match-point of the scattering molecule, which is the f_{D_2O} value where $\Delta\bar{\rho} = 0$. Consequently, for an ideal contrast variation series from homogeneous solutions, a plot of $I(0)/N$ vs f_{D_2O} will be quadratic with a single root of f_{MP}. Fitting a function of the form of Eq. (5) is the most natural approach to estimate f_{MP}, and given the versatility of modern statistical software, fitting a function of this form is straightforward. However, as Eq. (5) is not in a form compatible with regular polynomial regression (i.e., it is not of the form $ax^2 + bx + c$), it represents a practical issue for basic statistical software. To solve this issue, the square root of both sides of Eq. (5) can be taken to obtain:

$$\sqrt{\frac{I(0)}{N}} \propto \left|f_{D_2O} - f_{MP}\right|. \quad (6)$$

Eq. (6) describes a V-shaped function centered on the match-point of the scattering molecule. To facilitate fitting using linear regression to solve for f_{MP}, the $\sqrt{I(0)/N}$ terms where $f_{D_2O} > f_{MP}$ are assigned negative values, thus converting the V-shaped function to a straight line that intercepts the horizontal axis at the match point for the scattering molecule. For ideal samples, analysis using Eqs. (5) or (6) will give reliable estimates of the match-point of the molecule. Fig. 1 outlines the nature of potential issues that can be identified using the $I(0)$ analysis in terms of Eq. (5), but ultimately personal preference will dictate which approach is used.

a. Homogeneous Solution	b. Dissociating Complex	c. Oligomerisation/ Aggregation
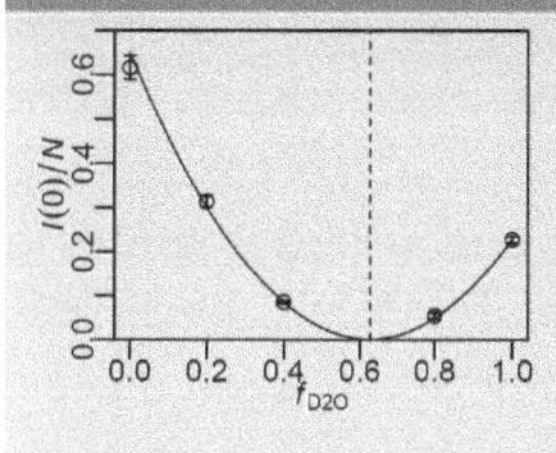	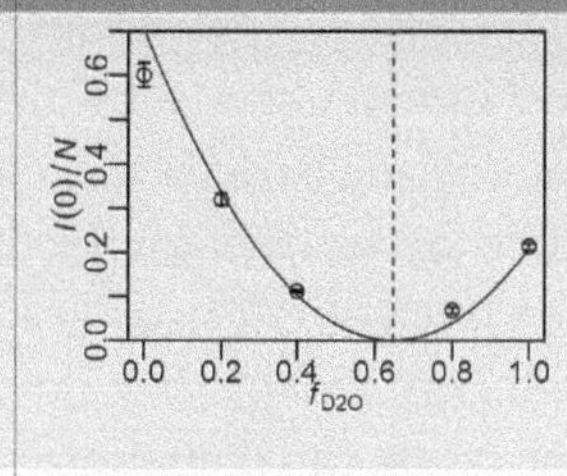	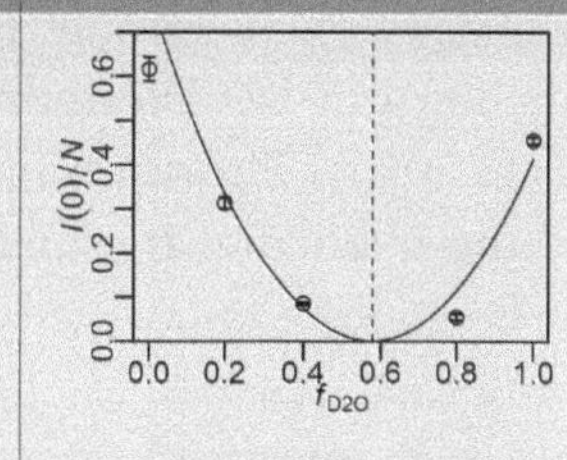
The dependency of $I(0)$ with f_{D20} should be represented accurately by a parabolic curve that intersects the f_{D20} axis at a single point, corresponding to the match-point of the complex.	The dependency of $I(0)$ with f_{D20} should be reasonably well represented by a parabolic curve that intersects the f_{D20} axis at a single point. The parabola will typically miss data points at the extremities of the plot, but the deviation of the estimated match-point from the real value is usually small. Of note, $I(0)$ values from a dissociating complex can be accurately represented by general quadratic function, but that function will not intersect the f_{D20} axis.	The dependency of $I(0)$ with f_{D20} will not be reasonably well represented by a parabolic curve that intersects the f_{D20} axis at a single point. The parabola will typically miss many data points, and the deviation of the estimated match-point from the real value is usually significant.
No corrective actions required.	Corrective action will be sample dependent. Where the size of the complex is large, the signal from the complex may dwarf the signal from the monomeric material at many contrast points, allowing normal treatment of the data. However, if dissociation is significant, data will need to be analysed as a mixture, precluding any of the further analysis presented in this chapter.	Corrective action will involve removing one or more contrast points and re-running the analysis. The contrast points identified as problematic would also be excluded from all further analysis and modelling. In the example above it is reasonably clear that the contrast point at $f_{D20} = 1.0$ is the problematic contrast point, but it is some cases, multiple contrast points may be affected.

Fig. 1 Typical trends observed in the analysis of $I(0)$ from contrast variation data, with corresponding corrective actions for solutions of (A) Homogeneous complex, (B) Dissociating complex, (C) Oligomerized or aggregated complex.

If a plot of either $I(0)/N$ or $\sqrt{I(0)/N}$ vs f_{D_2O} behaves as expected, the experimental estimate of the match-point can be compared to theoretical estimates. In principle, the calculation of $\Delta\overline{\rho}$ is straightforward. From the primary sequence of a protein or nucleic acid, $\overline{\rho}$ can be calculated by summing the scattering lengths of each atom in the molecule and dividing this by an estimate of the volume of the molecule. This calculation can then be repeated for each component of the solvent, and the contrast can be calculated using Eq. (2). These calculations can be done by hand using tabulated values (Jacrot, 1976), or a software tool (Clifton & Myatt, n.d.; Sarachan, Curtis, & Krueger, 2013; Whitten, Cai, & Trewhella, 2008). However, there are a number of factors that need to be taken into consideration that are not immediately obvious. Assuming the primary sequence of each component of the complex is known, as is the stoichiometry of the complex, the properties of a biomolecular complex that influence $\Delta\overline{\rho}$ are: the extent of (non-exchangeable) deuterium labeling; the volume of each component; and the proportion of labile H that exchanges with the solvent (solvent accessible exchangeable H). Of these properties, the extent of deuterium labeling can be determined independently by mass-spectrometry (Jeffries et al., 2016). The molecular volume (strictly the volume of solvent excluded by the molecule) is estimated by summing effective tabulated volumes for each residue (amino acid or nucleotide) in a molecule. Volumes estimated using this approach are generally reliable but do deviate for small proteins with high surface-to-volume ratios, or when there are significant unstructured regions, as solvated residues occupy less space than amino acids in a folded structure (Murphy, Matubayasi, Payne, & Levy, 1998). In the case of small proteins, one can turn to experimentally determined partial specific volumes (Trewhella et al., 2017). The volumes of each component of the complex are not uniquely determined by the $I(0)$ analysis, but small tweaks to the volume of one, or both, of the components may yield better agreement between the theoretical and experimental determined values of f_{MP}. Estimates of the number of solvent accessible exchangeable H are difficult to make without information from NMR or mass spectrometry, however, there is evidence from protein neutron crystallography (Mason, Bentley, & McIntyre, 1984; Schoenborn, 1972) and NMR (Krauss & Cowburn, 1981; Persson & Halle, 2015) that backbone amide hydrogens (especially toward the core of a protein) are not fully exchanged in deuterated solvents. In practice, the total fraction of labile hydrogen atoms accessible by the solvent will be between 0.95 and 1.0. Given that the number of exchangeable H positions is small relative to the total number of H positions within the bound volume of macromolecules, the effect of incomplete labile H exchange with the solvent on $\Delta\overline{\rho}$ is expected to be small. Nonetheless, it is a parameter that can be varied when calculating the $\Delta\overline{\rho}$ values for a macromolecular complex.

2.1 Programs

- The Contrast tool in the MULCh package (Whitten et al., 2008)
- R (R Development Core Team, 2013), running in RStudio (RStudio Team, 2020)

2.2 Procedure

1. If not performed prior to data collection, measure the concentration of biomacromolecule in solution for each contrast point using ultraviolet absorbance (UV) spectroscopy, refractive index (RI) measurements, or colorimetric assays (see note 1).

2. Check the D content of any labeled macromolecules and each solvent. The (non-exchangeable) D content of a labeled macromolecule can be determined using mass-spectrometry. The deuterium content of solvents can be inferred from densitometry measurements (the density of an aqueous solution will increase by ~0.00107 g/cm^3 for every 1% v/v increase in the D_2O concentration), infrared spectroscopy, or more conveniently using neutron transmission measurements (see note 2).
3. Determine $I(0)$ together with an estimate of the uncertainty in $I(0)$ for each contrast point via Guinier analysis (see notes 3 and 4). These parameters and their uncertainties can also be estimated using IFT to obtain the pair-distance distribution function, $p(r)$. However, at least in the first instance, Guinier analysis should be performed, and you may choose to include these plots in a publication or as supplementary information (see note 5).
4. Analyze the dependence of $I(0)$ upon contrast by fitting either Eq. (6) (option a) or Eq. (5) (option b) to the data.
   a. Using the contrast module of the MULCh web interface, select the number of contrast points, and enter f_{D_2O} (a value between 0 and 1), $I(0)$, $\sigma(I(0))$, and N (see note 6). At the same time MULCh will attempt to calculate the contrast of the molecule and its components (go to step 4).
   b. Instead of MULCh, a statistical package such as R can be used to do the fitting. Load a file containing f_{D_2O}, $I(0)$, $\sigma(I(0))$, and N. Then fit a function of the form $I(0)/N = a(f_{D_2O} - f_{MP})^2$ (see note 7).
5. Identify $I(0)$ values that do not have the expected dependence on f_{D_2O} and consider if they are diagnostic of dissociation ($I(0)$ values too small) or perhaps aggregation ($I(0)$ values too high), the latter sometimes induced with increasing D_2O concentrations (Fig. 1). If the dataset contains significant dissociated material, alternative analysis techniques may be required to model the data. If aggregation is indicated, $I(0)$ analysis should be repeated with datasets displaying signs of aggregation removed.
6. Using a program such as MULCh, (see note 8) small variations of the subunit volume and the fraction of accessible labile hydrogen atoms may be needed to obtain agreement between the theoretical and experimental match-points (within the estimated uncertainty). Once this agreement is achieved, you will have estimates of contrast for the components of the complex consistent with the experimental data, which will be used in further analysis.

2.3 Notes

1. An accurate concentration determination for your complex is important for later analysis (i.e., for determination of molecular mass). For most protein complexes, a UV absorbance measurement at 280 nm (which is sensitive to tyrosine and tryptophan content) using a NanoDrop or similar instrument can be sufficiently accurate when corrected by an absorption coefficient (determined either experimentally or theoretically using the Protparam server (Gasteiger et al., 2005)). It is also possible to obtain protein concentrations using shorter UV wavelengths that are not dependent on the presence of tyrosine and tryptophan in a protein (Anthis & Clore, 2013). Refractive Index (RI) provides an alternative to UV spectroscopy for determining the concentration of proteins, nucleic acids, and protein-nucleic acid complexes. It is especially convenient in cases where proteins lack aromatic amino acids, such as many intrinsically disordered proteins, or where nucleic acids saturate spectrophotometric systems. The RI increments (dn/dc) are quite similar for proteins (0.16–0.2 mL/g), nucleic acids (0.17–0.19 mL/g), and for proteins, dn/dc maybe calculated based on the amino acid sequence (Schuck, Perugini, Gonzales, Howlett, & Schubert, 2002). In the absence of these preferred methods, a well-calibrated colorimetric assay can be used (e.g., a Bradford assay).
2. H incoherently scatters a large proportion of neutrons, which effectively attenuates the incident beam. A plot of $(\ln T(f_{D_2O}))/l(f_{D_2O})$ vs f_{D_2O} should be linear, where $T(f_{D_2O})$ is the transmission, and $l(f_{D_2O})$ is the pathlength of a sample (or buffer) at a given f_{D_2O}. Deviations from linearity indicate the H:D ratio is not at the expected value.
3. The number of points used in the Guinier analysis will likely differ between contrast points as the "apparent" size of the scatterer will change. A range should be chosen such that there are no obvious systematic differences between the experimental data and the fit, but the chosen range of the plot should be $q.R_g < 1.3$, However, this value is an upper estimate; for elongated molecules a more conservative approach would be to choose a range such that $q.R_g < 0.8–1.0$. Additionally, at low-q deviations may be observed that are due to parasitic scattering, or attractive/repulsive intermolecular interactions. The effects of parasitic scattering may be resolved by carefully re-subtracting data with an appropriate background or excluding the first few $I(q)$ points from the fit. The effect of intermolecular interactions usually extends further into the Guinier region and represents a significant problem. It would be

anticipated that measurements are made in the dilute regime so that repulsive interactions are minimal. However, attractive interactions and/or aggregation may manifest in some solutions due to sample handling issues or may be induced by D_2O in the solvent. In either case, SANS data collected from samples exhibiting the presence of attractive interactions (oligomers formed by specific intermolecular interactions) and/or aggregates (large molecules formed by non-specific intermolecular interactions) must be excluded from all further analysis.

4. Guinier analysis of contrast points collected close to the whole molecule match-point can be difficult due to the significant reduction in $\Delta\overline{\rho}$ resulting in very low signal-to-noise. Further, the internal scattering density fluctuations in the molecule can give rise to a signal in the mid-q region that is larger than the signal at low-q. As will be discussed in detail in Section 3, near the whole-molecule match-point, it is possible to have negative R_g values, where in combination with low-single-to-noise data, it may be difficult to identify the correct Guinier region. Thus, in such cases, it is common that both $I(0)$ and R_g are often overestimated by Guinier analysis. This overestimate does not usually significantly influence the analysis of $I(0)$ as the values are small and the relative uncertainty is high.
5. Using a series expansion, a first order approximation to the uncertainty in a function $f(x)$, given x and the associated uncertainty $\sigma(x)$, is

$$\sigma(f(x)) \approx \left| \frac{df(x)}{dx} \sigma(x) \right|. \tag{7}$$

Thus, to generate a Guinier plot for publication with error bars,

$$\sigma(\ln I(q)) \approx \frac{\sigma(I(q))}{I(q)}. \tag{8}$$

6. MULCh takes f_{D_2O}, $I(0)$, $\sigma(I(0))$ and N (protein concentration) as input. It performs a preliminary 2nd order polynomial fit to the $I(0)/N$ values to estimate the match-point (this preliminary fit is not guaranteed to have a single root like Eq. (5)). It then takes the square root of all $I(0)/N$ values and changes the sign of $I(0)/N$ values for $f_{D_2O} > f_{MP}^{estimated}$. Eq. (7) is then applied to estimate $\sigma\left(\sqrt{I(0)/N}\right) = \sigma(I(0))/\left(2N\sqrt{I(0)/N}\right)$. Weighted linear regression ($y = ax + b$) is then applied and the match-point is calculated as $f_{MP} = -b/a$. The uncertainty of f_{MP} is then estimated from the calculated uncertainties of a and b accounting for correlation between the two parameters.

7. This analysis could be done using a range of commercially available statistical packages. However, as *R* is actively developed and freely available. The following set of commands will perform the necessary fitting,

```
#Set the path to the working directory (this will need to be
   edited)
setwd("~/DataAnalysis")
#Load the comma separated file containing the I0 data into a data
   frame
I0.data = read.csv("I0_data.txt", header = TRUE)
#If spaces, not commas, separate the data, uncomment the line below
#I0.data = read.table("I0_data.txt", header = TRUE)
#Fit the constrained quadratic, where fMP is the match-point
I0.fit = nls(I0/N ~ a*(fD20 - fMP)^2, weights = (N/sigI0)^2,
start = list(a = 1, fMP = 0.5), data = I0.data)
#Print the model parameters
summary(I0.fit)
```

The commands above assume that the file I0_data.txt has the form

```
fD2O, I0, sigI0, N
0.0, 1.263, 0.055, 2.05
0.2, 0.637, 0.025, 2.03
0.4, 0.179, 0.006, 2.10
0.8, 0.110, 0.015, 2.01
1.0, 0.452, 0.016, 1.98
```

8. The contrast module of MULCh will calculate the contrast of a two-component molecule, accounting for composition of the buffer, the sequence of the macromolecule (protein, RNA or DNA, and ligands), stoichiometry, deuteration levels, volumes (will be estimated if not explicitly provided), and accessibility of labile hydrogens. It will provide the average contrast of the entire molecule, and of each component of the molecule for any contrast point. Parameters such as the subunit volumes and accessible exchangeable hydrogen positions should not be varied by more than 5% from typical values or estimates. If large corrections are required to get agreement between the theoretical and experimental match-points, you should revisit the data reduction (i.e., sample pathlengths may have been specified incorrectly), sample concentrations should be checked, H/D ratios of the solutions and deuteration levels of the biomolecules should be confirmed, and assumptions regarding the stoichiometry investigated (including whether the stoichiometry changes with f_{D_2O}).

3. Analysis of the radius of gyration

The radius of gyration (R_g) yields basic information on the shape of a molecule. The slope of the Guinier plot is proportional to R_g^2 (Guinier & Fournet, 1955), consequently, it is a parameter that is readily extracted from the data. For a homogeneous object, R_g is independent of contrast, but for a molecule composed of two regions with differing scattering density, R_g is dependent on contrast, and this dependence reveals information regarding the shape of the molecule and the arrangement of the components within the molecule. In recent times, the analysis has become a somewhat ignored part of the examination of contrast variation data, and this is no doubt due to its seemingly abstract nature, with more emphasis placed on obtaining low resolution structural models that are more appealing to the generalist structural biologist. However, analysis of the R_g dependence from a contrast series will identify problematic measurements at different contrast points and help guide the construction of sensible starting models.

Given a real space distribution of scattering volume elements, there are several ways to calculate and interpret R_g. The first interpretation is that R_g is the root-half-mean-square distance between scattering elements in a molecule weighted by the contrast of those scattering elements. Thus, R_g^2 can be expressed as a complicated double integral, or more simply as half the normalized second moment of the pair-distance distribution function, $p(r)$

$$R_g^2 = \frac{\iint \Delta\rho(\mathbf{r}_1)\Delta\rho(\mathbf{r}_2)(\mathbf{r}_1 - \mathbf{r}_2)^2 d\mathbf{r}_1 d\mathbf{r}_2}{2\iint \Delta\rho(\mathbf{r}_1)\Delta\rho(\mathbf{r}_2) d\mathbf{r}_1 d\mathbf{r}_2} = \frac{\int_0^{D_{\max}} p(r) r^2 dr}{2\int_0^{D_{\max}} p(r) dr}. \quad (9)$$

R_g can also be represented as the root-mean-square distance of scattering volume elements from the center of excess scattering length density ($\mathbf{r}_0$) of the molecule, weighted by the contrast of the scattering elements

$$R_g^2 = \frac{\int \Delta\rho(\mathbf{r})(\mathbf{r} - \mathbf{r}_0)^2 d\mathbf{r}}{\int \Delta\rho(\mathbf{r}) d\mathbf{r}}, \quad (10)$$

where $\mathbf{r}_0$ is defined such that

$$\int \Delta\rho(\mathbf{r})(\mathbf{r} - \mathbf{r}_0)d\mathbf{r} = 0. \tag{11}$$

For an object of uniform scattering length density, the contrast terms can be removed, and Eq. (11) is satisfied when r_0 is the geometric center of the object, and R_g is simply the root mean-square radius of the object. For an object composed of two distinct scattering length densities, this interpretation no longer holds, and further, the location of $\mathbf{r}_0$ is not constant across the contrast series.

One method to analyze the contrast dependence of R_g (Ibel & Stuhrmann, 1975), involves making the following substitution into Eq. (9),

$$\Delta\rho(\mathbf{r}) = \Delta\overline{\rho} + \rho_F(\mathbf{r}). \tag{12}$$

Eq. (12) splits the contrast at a given point in the molecule, $\Delta\rho(r)$, into the average contrast of the entire molecule, $\Delta\overline{\rho}$, plus some fluctuation of the scattering length density from the mean at that point, $\rho_F(r)$. Working through this substitution, it can be shown that

$$R^2_{g,obs} = R^2_C + \frac{\alpha}{\Delta\overline{\rho}} - \frac{\beta}{\Delta\overline{\rho}^2}, \tag{13}$$

where,

$$\alpha = \frac{1}{V}\int \rho_F(\mathbf{r})\mathbf{r}^2 d\mathbf{r}, \text{and } \beta = \frac{1}{V^2}\left(\int \rho_F(\mathbf{r})\mathbf{r}d\mathbf{r}\right)^2. \tag{14}$$

A plot of $R^2_{g,obs}$ as a function of $1/\Delta\overline{\rho}$ is known as a Sturhrmann plot and will generally be parabolic in form, where R_C is the radius of gyration of the molecule if it had a homogeneous contrast, α is the second moment of the scattering length density fluctuations, and β is the square of the first moment of the scattering length density fluctuations. Consider what the expressions in Eq. (14) reveal about the molecule of interest if it is composed of two regions, each with a distinct contrast. By definition, one region has a contrast higher than the average molecule contrast, or a positive ρ_F; and one region has a contrast lower than the average molecular contrast, or a negative ρ_F. It follows from Eq. (14) that:

- As $\mathbf{r}^2$ is always positive, $\rho_F(\mathbf{r})$ is the only term that can change the sign of α, and given that $\rho_F(\mathbf{r})$ is effectively weighted by $\mathbf{r}^2$, regions further from the center of the molecule more strongly influence the sign of α. Thus, if α is positive (Fig. 2A, C, and D), then the component with a positive ρ_F lies further from the center than the component with a negative ρ_F;

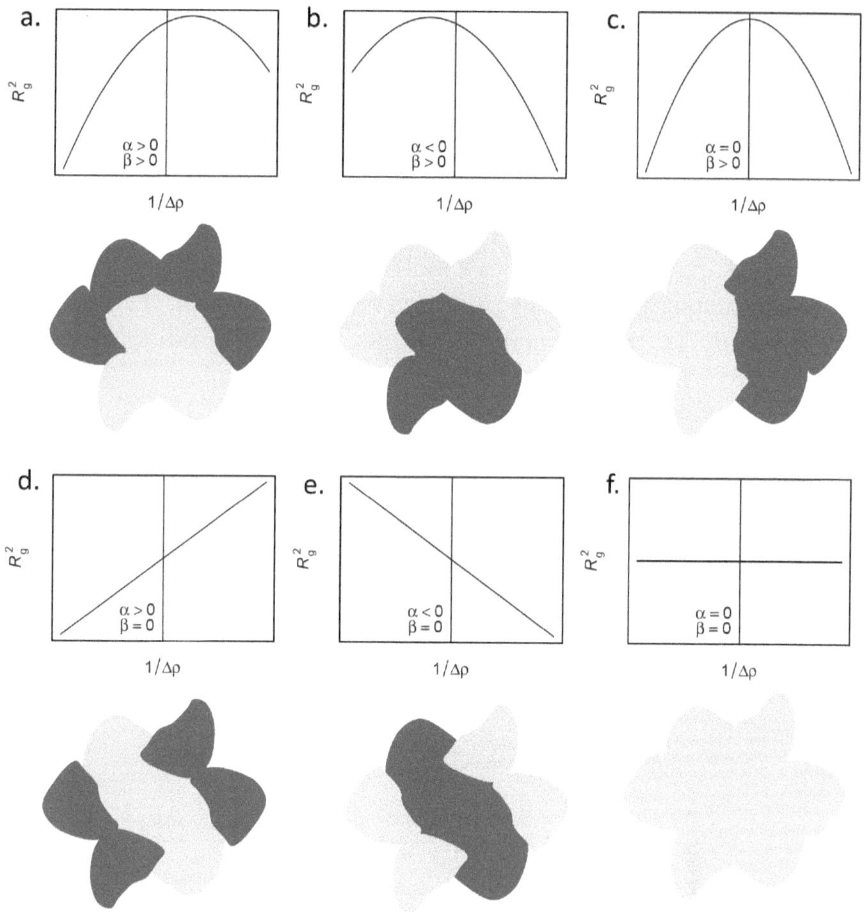

Fig. 2 Representative Stuhrmann plots for complexes and corresponding pictorial interpretations, where the component with $\rho_F > 0$ in shown in dark gray, and the component with $\rho_F < 0$ is shown in light gray. (A) Non-symmetric complex with $\rho_F > 0$ component toward the periphery, (B) non-symmetric complex with $\rho_F < 0$ component toward the periphery, (C) complex where each component has a similar size and shape, (D) symmetric complex with $\rho_F > 0$ component toward the periphery, (E) symmetric complex with $\rho_F < 0$ component toward the periphery, (F) two components with the same scattering length densities.

and if α is negative (Fig. 2B and E), the component with a negative ρ_F lies further from the center than the component with a negative ρ_F. If $\alpha \approx 0$ and $\beta > 0$ (Fig. 2C) then both components of the complex have a similar size and shape. If $\alpha \approx 0$ and $\beta \approx 0$ (Fig. 2F) the molecule has a homogeneous $\Delta\rho$.

- As β is the square of a given quantity, it is always greater than or equal to 0. Looking at term integrated, it can be seen that $\mathbf{r}$ and $\rho_F(\mathbf{r})$ can both be negative or positive. For asymmetric molecules (Fig. 2A–C), $\rho_F(\mathbf{r}) \neq \rho_F(-\mathbf{r})$, thus, $\rho_F(\mathbf{r}) \times \mathbf{r}$ and $\rho_F(-\mathbf{r}) \times -\mathbf{r}$ will not cancel in the integral. For symmetric molecules (Fig. 2D and E), $\rho_F(\mathbf{r}) = \rho_F(-\mathbf{r})$, thus, $\rho_F(\mathbf{r}) \times \mathbf{r}$ and $\rho_F(-\mathbf{r}) \times -\mathbf{r}$ will cancel in the integral. Thus, β is indicative of the extent of asymmetry of the molecule. For $\beta \approx 0$, the Stuhrmann plot will appear linear (Fig. 2D and E), while $\beta > 0$ will give a concave down parabola (note the negative sign in front of β in Eq. (13)), with the magnitude of the curvature increasing as the asymmetry increases (Fig. 2A–C).

Another approach to analyzing the radius of gyration is the use of the parallel axis theorem (Damaschun, Fichtner, Purschel, & Reich, 1968; Serdyuk, 1979; Serdyuk & Fedorov, 1973). The theorem breaks the integral from Eq. (10) into contributions from component 1 and component 2, and redefines the center of excess scattering density of each component relative to $\mathbf{r}_0$ (i.e., $\mathbf{r}_1 = \mathbf{r}_{0,1} + \mathbf{r}_{1,1}$ and $\mathbf{r}_2 = \mathbf{r}_{0,2} + \mathbf{r}_{2,2}$) (Fig. 3)

$$\begin{aligned} R_{g,obs}^2 &= \frac{\int \Delta\rho(\mathbf{r}_{1,1})|\mathbf{r}_{0,1} + \mathbf{r}_{1,1}|^2 d\mathbf{r}_{1,1} + \int \Delta\rho(\mathbf{r}_{2,2})|\mathbf{r}_{0,2} + \mathbf{r}_{2,2}|^2 d\mathbf{r}_{2,2}}{\int \Delta\rho(\mathbf{r}_1)d\mathbf{r}_1 + \int \Delta\rho(\mathbf{r}_2)d\mathbf{r}_2} \\ &= \frac{\Delta\bar{\rho}_1 V_1 R_{g,1}^2 + \Delta\bar{\rho}_2 V_2 R_{g,2}^2 + \Delta\bar{\rho}_1 V_1 |\mathbf{r}_{0,1}|^2 + \Delta\bar{\rho}_2 V_2 |\mathbf{r}_{0,2}|^2}{\Delta\bar{\rho}_1 V_1 + \Delta\bar{\rho}_2 V_2} \\ &= f_1 R_{g,1}^2 + f_2 R_{g,2}^2 + f_1|\mathbf{r}_{0,1}|^2 + f_2|\mathbf{r}_{0,2}|^2 \end{aligned} \tag{15}$$

where $f_1 = \Delta\bar{\rho}_1 V_1/(\Delta\bar{\rho}_1 V_1 + \Delta\bar{\rho}_2 V_2)$, $f_2 = 1 - f_1$ and $R_{g,1}$ and $R_{g,2}$ are the radii of gyration of each component of the complex. As Eq. (11) must be satisfied, it can be shown that $f_1\mathbf{r}_{0,1} = -f_2\mathbf{r}_{0,2}$. Consequently, if f_1 and f_2 have the same sign (i.e., both components have positive or both components have negative contrast), then vectors $\mathbf{r}_{0,1}$ and $\mathbf{r}_{0,2}$ point in opposite directions (Fig. 3A), thus, the origin of the whole molecule lies between the origin of each component. However, if the contrast of the two components differs in sign, then vectors $\mathbf{r}_{0,1}$ and $\mathbf{r}_{0,2}$ point in the same direction, meaning that the origin of the molecule does not lie between the origin of each component, and in fact, the origin may lie outside the envelope of the molecule (Fig. 3B). This point may appear obscure, but it does have relevance to the calculation of R_g from rigid bodies and will be briefly revisited

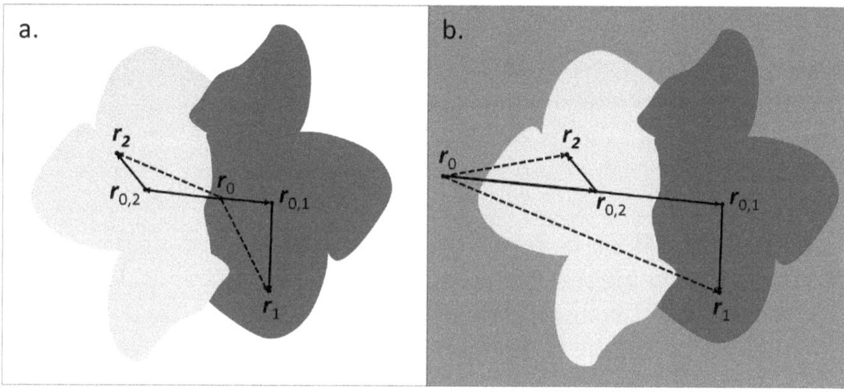

Fig. 3 Pictorial representation of the redefinition of the vector **r** relative to the origin of the molecule $\mathbf{r}_0$ in terms of the origin of each component in the complex, $\mathbf{r}_{0,1}$ and $\mathbf{r}_{0,2}$, (A) In solution conditions where the contrast of each component of the complex has the same sign, $\mathbf{r}_0$ lies between $\mathbf{r}_{0,1}$ and $\mathbf{r}_{0,2}$, (B) In solution conditions where the contrast of one component is positive and the other is negative, $\mathbf{r}_0$ does not lie between $\mathbf{r}_{0,1}$ and $\mathbf{r}_{0,2}$.

in Section 5. As the origin of the molecule moves depending on the relative contrast, the distance between the two components must be defined as a vector, $\mathbf{D}=\mathbf{r}_{0,1}-\mathbf{r}_{0,2}$. From this it follows that $\mathbf{r}_{0,1}=f_2\mathbf{D}$ and $\mathbf{r}_{0,2}=f_1\mathbf{D}$. Substituting into Eq. (15),

$$\begin{aligned} R^2_{g,obs} &= f_1 R^2_{g,1} + f_2 R^2_{g,2} + f_1 f_2^2 |\mathbf{D}|^2 + f_2 f_1^2 |\mathbf{D}|^2 \\ &= f_1 R^2_{g,1} + f_2 R^2_{g,2} + f_1 f_2 D^2 (f_1 + f_2) \\ &= f_1 R^2_{g,1} + f_2 R^2_{g,2} + f_1 f_2 D^2 \end{aligned} \quad (16)$$

Eq. (16) is intuitive, in that if the contrast of component 2 is zero, then $f_1=1$ and $f_2=0$, and $R_{obs}=R_{g,1}$ as would be expected, and vice versa. For a symmetric molecule, both $r_{0,1}$ and $r_{0,2}$ are coincident and $D=0$, thus, D is related to β from the Stuhrmann analysis. In fact, expressions to determine $R_{g,1}$, $R_{g,2}$ and D from the coefficients of the Stuhrmann analysis have been derived (Olah, Rokop, Wang, Blechner, & Trewhella, 1994). However, given accurate estimates of $R_{g,obs}$ multiple contrast points, multiple linear regression can be used to provide a direct route to the parameters of interest.

3.1 Programs

- The Contrast tool in the MULCh package (Whitten et al., 2008)
- R (R Development Core Team, 2013), running in RStudio (RStudio Team, 2020)

3.2 Procedure

1. Determine R_g together with an estimate of the uncertainty in R_g for each contrast point via Guinier analysis. Again, these parameters and their uncertainties can also be estimated using IFT to obtain $p(r)$ (see note 1).
2. Using these R_g values and estimates of the total molecule contrast, perform Stuhrmann analysis.
   a. If using MULCh, proceed to the Rg module from the contrast module (doing this will automatically fill the functional form of the contrast and volume). Enter f_{D_2O} values (between 0 and 1) into the table and click the "Get Contrast Values" button, which will automatically populate the contrast and volume fractions at each contrast point. Values for R_g, $\sigma(R_g)$ are then manually entered (see note 2). When executed, the tool will return estimates of R_c, α and β together with uncertainties. It will also return a table containing $1/\Delta\overline{\rho}$, ${R_{obs}}^2$, $\sigma({R_{obs}}^2)$ and ${R_{calc}}^2$, and these data should be plotted to show discrepancies or problems with the fit. If there are issues, simple restraints may be required (see note 3). or some data sets may need to be excluded. If the agreement between the model and the fit are acceptable, proceed to step 3.
   b. A statistical package such as R can be also used to do the fitting. Load a file containing $\Delta\overline{\rho}$, R_g, and $\sigma(R_g)$. Then fit a function of the form $R_{obs}^2 = R_C^2 + \alpha(1/\Delta\overline{\rho}) - \beta(1/\Delta\overline{\rho})^2$ (see note 4). As was the case for option (a), data should be plotted to highlight discrepancies or problems with the fit. Again, if there are issues, simple restraints may be required (see note 3). or some data sets may need to be excluded.
3. In the same calculation, MULCh will also apply the parallel axis theorem to obtain estimates of R_1, R_2, and D with uncertainties. These values can then be used to build or validate starting models. It is possible to use R to perform this multiple linear regression calculation, however, contrasts and volumes for each component would need to be loaded. For those comfortable with fitting in R, this would not be a challenge, but MULCh is purpose built for this calculation, and an easier option for those who are unfamiliar with R.

3.3 Notes

1. If R_g is reported with a negative sign, it indicates that it is in fact R_g^2 that is negative. Referring to Eq. (13), when the term $-\beta/\Delta\overline{\rho}^2$ (which is always ≤ 0) dominates ${R_C}^2$ and $\alpha/\Delta\overline{\rho}$, then R_g^2 can become negative. This will

happen for small values of $\Delta\overline{\rho}$ in asymmetric complexes with significant density fluctuations (see Fig. 2A–C, where the quadratic function will become negative at the extremities of the plot). For a symmetric complexes, it occurs when $\alpha/\Delta\overline{\rho}$ is negative and dominates the $R_C{}^2$ term. This will happen when $\alpha > 0$ and $\Delta\overline{\rho}$ is small and negative, or when $\alpha < 0$ and $\Delta\overline{\rho}$ is small and positive (see Fig. 2D and E). In practice, a negative value for R_g^2 occurs when data are collected in the vicinity of the match-point of the entire complex (i.e., a contrast point located between the match-points of the two components). Here, the contrast of one component is less than 0, and the other component has a contrast larger than 0, which for an asymmetric molecule, results in a large value $-\beta/\Delta\overline{\rho}^2$.

2. Applying Eq. (7), the uncertainty of R_g^2 can be calculated as

$$\sigma\left(R_g^2\right) \approx \left|2R_g\sigma\left(R_g\right)\right| \tag{17}$$

3. To accurately define the parabola in the Stuhrmann plot, contrast variation data must be measured on either side of the match-point. If data is not measured at either side of the match-point, then you may encounter numerical instabilities, and parabolas with positive curvatures may be obtained. To force the curvature of the parabola to be negative a restraint may be required. The simplest restraint is the incorporation of a dummy data point into the fitting with a large uncertainty. This data point should be chosen such that it forces the parabola to have a negative curvature, but a large uncertainty should be chosen as not to heavily influence the fit to the other points.
4. The following set of commands will perform the necessary fitting,

```
#Set the path to the working directory (this will need to be
   edited)
setwd("~/DataAnalysis")
#Load the comma separated file containing the Rg data into a data
   frame
Rg.data = read.csv("Rg_data.txt", header = TRUE)
#If spaces, not commas, separate the data, uncomment the line below
#Rg.data = read.table("Rg_data.txt", header = TRUE)
#Fit the Rg data
Rg.fit = lm(Rg^2 ~ I(1/contrast) + I(-1/contrast^2), weights = (1/
   (2*Rg*sigRg))^2,
   data = Rg.data)
```

```
#Print the model parameters
summary(Rg.fit)
```

The commands above assume that the file Rg_data.txt has the form

```
contrast, Rg, sigRg
   3.35, 36.3, 0.2
   2.20, 34.6, 0.3
   1.00, 24.3, 1.1
-1.15, 35.6, 0.5
-2.30, 37.0, 0.2
```

4. Composite scattering functions

For a 2-component complex, it is theoretically possible to fully characterize the low-resolution structure with 3 contrast points: At the match-point of component 1, which defines the structure of component 2; at the match-point of component 2, which defines the structure of component 1; and at a contrast point well away from the match-point of either component, which defines the structure of the whole molecule. However, in practice, it is preferable to have as many well-spaced contrast points as practical, as the data at each contrast point will contain experimental uncertainties. Scattering data collected at contrast match-points renders one of the components practically "invisible," and the resulting scattering data can be intuitively interpreted as being from the "visible" component of the complex. However, for those unfamiliar with contrast variation, interpreting scattering curves measured at contrast points away from the match-point of either component is not intuitive. An alternative to interpreting each contrast variation point separately is to decompose contrast variation data into a core set of scattering functions that summarize the information present in the experimental data.

One of the first approaches of this kind can be attributed to Stuhrmann and co-workers (Stuhrmann, 1973; Stuhrmann et al., 1975). The scattering amplitude, $A(q)$ (a complex function dependent on the structure of a molecule and defined such that $I(q) = \langle A(q) \rangle^2)$, can be partitioned into a shape component $A_C(q)$, and a fluctuation term $A_F(q)$. This partitioning is analogous to that shown for the contrast in Eq. (12) and allows the total scattering to be expressed in terms of three basic (Stuhrmann & Fuess, 1976) or fundamental (Hjelm et al., 1977) scattering functions,

$$I(q) = \langle \Delta\overline{\rho} A_C(q) + A_F(q) \rangle^2$$
$$= (\Delta\overline{\rho})^2 I_C(q) + (\Delta\overline{\rho}) 2 I_{CF}(q) + I_F(q). \quad (18)$$

Here, $I_C(q)$ is the scattering of a molecule with homogeneous contrast, $I_{CF}(q)$ is a convolution of the shape and the internal density fluctuations, and $I_F(q)$ is the scattering due to the internal scattering-length density fluctuations alone. Given a contrast series composed of many well-spaced datasets and given an estimate of $\Delta\overline{\rho}$ at each contrast point, the three scattering functions $I_C(q)$, $I_{CF}(q)$, and $I_F(q)$ can be obtained using multiple linear regression. This approach is entirely general, and involves no approximations, thus, in the absence of experimental uncertainty, $I_C(q)$ would be a faithful representation of the scattering curve of a homogeneous molecule and could in fact be modeled directly using *ab initio* approaches. For a protein or protein complex (with no isotopic labeling), the density fluctuations are small, and $I_{CF}(q)$, and $I_F(q)$ will also be small. For a protein complex where one component is labeled with deuterium, these terms will be significant and will encode important information regarding the internal structure of the complex. However, the nature of $I_{CF}(q)$ and $I_F(q)$ are not straightforward to model. A useful approach[a] for the analysis of 2-component systems is to assume the scattering amplitude of the density fluctuations, $A_F(q)$, are small, and to partition the total scattering amplitude into two homogeneous components, $A_1(q)$ and $A_2(q)$ representing the scattering amplitudes of each component of the complex (Olah et al., 1994; Zhao, Hoye, Boylan, Walsh, & Trewhella, 1998),

$$I(q) \approx \langle \Delta\overline{\rho}_1 A_2(q) + \Delta\overline{\rho}_2 A_2(q) \rangle^2$$
$$= (\Delta\overline{\rho}_1)^2 I_{11}(q) + (\Delta\overline{\rho}_1 \Delta\overline{\rho}_2) 2 I_{12}(q) + (\Delta\overline{\rho}_2)^2 I_{22}(q) \quad (19)$$

Here, $\Delta\overline{\rho}_1$ and $\Delta\overline{\rho}_2$ are the average contrasts of components 1 and 2 respectively, $I_{11}(q)$ and $I_{22}(q)$ represent the scattering intensity profiles of components 1 and 2, and $I_{12}(q)$ is the cross-term contribution to the scattering profile that encodes information about the arrangement of the two components relative to each other. Considering this expression at the match-point of either component, the scattering function for that term becomes zero,

[a] A survey of the literature prior to 2005 will reveal that the term "basic scattering functions" refers to both sets of scattering functions described in Eq. (18) (Stuhrmann & Fuess, 1976) and Eq. (19) (Olah et al., 1994). To avoid further potential confusion, the online MULCh package and associated publication (Whitten et al., 2008) refer to the functions from Eq. (19) as "composite scattering functions" (in that measured contrast variation data are composites of these three scattering functions).

as does the cross term, leaving only the scattering function from the visible component. In a similar fashion to that described for Eq. (18), with many contrast variation curves, and given estimates of $\Delta\bar{\rho}_1$ and $\Delta\bar{\rho}_2$ for each dataset point, the three scattering functions $I_{11}(q)$, $I_{12}(q)$, and $I_{22}(q)$ can be obtained by multiple linear regression using Eq. (19). The interpretation of $I_{11}(q)$ and $I_{22}(q)$ from this analysis is clear, and these curves can be analyzed in a conventional manner (Guinier plots, IFT) and modeled using established modeling programs. The interpretation of $I_{12}(q)$ is less obvious; the resulting $p(r)$ can be interpreted as the distribution of pair-distances *between* the two components (Whitten et al., 2007) and can be useful for model building in some circumstances. An approach more conducive to the use of available data analysis and modeling packages is to combine the $I_{12}(q)$ with $I_{11}(q)$ and $I_{22}(q)$ to approximate the scattering from a homogeneous molecule,

$$I_C(q) \approx I_{11}(q) + 2I_{12}(q) + I_{22}(q) \tag{20}$$

Composite scattering function analysis provides useful insight into the structure of the complex and its components and can also reveal problems with the SANS-CV data. Specifically, this approach tends to fail in situations where there are insufficient datasets, or the contrast points are not well spread. In such cases, one of the scattering-functions may be well defined, while the other two functions are uninterpretable. If the procedure cannot extract three well-defined scattering curves, modeling against the contrast variation data should be approached with caution, unless performed in combination with complementary data (such as SAXS data or crystal structures for rigid-body modeling).

4.1 Programs

- The Compost tool in the MULCh package (Whitten et al., 2008)

4.2 Procedure

1. Check the q-range and q-spacing for each data file. As the calculation of the composite scattering functions is done independently at each q-value, the q scale in each data file must be identical. If the q scale differs, the datasets will need to be re-binned (see note 1).
2. Check the incoherent background for each contrast point. This can be done in a number of ways (both procedures may become increasingly difficult for extended/anisotropic or flexible molecule systems):

a. At high-q, data should display a q^{-4} dependence (i.e., $I(q) \sim Aq^{-4} + B$), where A is a scale factor and B is the background. If both sides of the equation are multiplied by q^4 to yield $I(q)q^4 \sim A + Bq^4$, a plot of $I(q)q^4$ vs q^4 at large q-values will have a slope of B, and this value can be subtracted from the data to remove the contribution from incoherent background.

b. When performing IFT, incorrect background levels are compensated by deformations of the $p(r)$ function at small values of r (see note 2). By releasing the $p(0) = 0$ constraint when computing the $p(r)$, a background correction can be applied such that $p(0) = 0$ is satisfied without the use of a constraint.

3. Proceed to the Compost module from the Rg module. This will automatically fill the functional form of the contrast and volume. Enter f_{D_2O} values (between 0 and 1) into the table and click the "Get Contrast Values" button, which will automatically populate the table with contrast and volume fractions at each contrast point. Load each data file into the web interface. At this point you will need to make a choice whether to let the program refine the scale factor for each contrast point (a check box next to the f_{D_2O} value), or to enter the scale factor manually. As a general rule, scale factor optimization should be avoided (see note 3). If all data was measured at the same concentration, it should be possible to leave scale factors at 1.0 and not refine the scale. If the concentrations of each sample differ, either normalize each SANS contrast variation dataset by the respective concentration or enter a scale factor of 1/concentration.

4. Click "Submit" to run the calculation and save the three composite scattering functions. The output includes a table of differences between the model and the data, which can help identify any problematic data sets. If any datasets are problematic, they should be removed from the analysis and repeat this step as required.

5. Perform preliminary analysis on the composite scattering functions such as Guinier analysis, IFT, Porod analysis (see note 4), and check the parameters obtained agree with what is known about the molecule and its components. The molecular mass of the complex can also be estimated from $I_C(0)$ (Furlong et al., 2018), using $\Delta\rho = 1 \times 10^{10}\,\mathrm{cm}^{-2}$, but the corresponding value of the concentration will depend on whether data have been scaled or normalized by concentration before or during the calculation of the composite scattering functions.

4.3 Notes

1. In ideal circumstances, data are collected sequentially in a common configuration. However, things don't always go to plan, and a contrast series may be composed of data measured at different times, or even on different instruments. Thus, re-binning data (Manalastas-Cantos et al., 2021) to a common q-scale is often necessary.
2. Consider the expression that relates the scattering intensities to real-space scattering-pair distance distribution $I(q) = 4\pi \int_0^{D_{max}} p(r) \frac{\sin qr}{qr} dr$. The limit of the term being integrated for small values of r is $\lim_{r \to 0} \left[p(r) \frac{\sin qr}{qr} \right] = p(0)$. As this term is independent of q, it demonstrates that deformation of the $p(r)$ function at small values of r can provide a *de facto* background correction. By releasing the $p(0) = 0$ constraint and adjusting the background in a trial-and-error approach, a background level where $p(0) = 0$ can be estimated.
3. Compost estimates the scale factor by performing a Guinier fit to each data set (the upper qR_g limit can be set in the web-interface to whatever value required). It then divides each $I(0)$ value by the corresponding value of $\Delta\overline{\rho}^2$. Then the nth dataset is scaled relative to the first using a scale factor of $(I_1(0)/I_n(0)) \times \left(\Delta\overline{\rho}_n^2/\Delta\overline{\rho}_1^2\right)$.
4. Porod analysis to determine molecular volume requires the contrast of the object is approximately homogeneous. Except for data collected at component match-points, this is not the case for SANS-CV data. However, Porod analysis (de Oliveira Neto et al., 2022; Manalastas-Cantos et al., 2021; Porod, 1951) can be performed on the composite scattering functions $I_{11}(q)$, $I_{22}(q)$ and $I_C(q)$.

5. Dummy-atom (bead) modeling against contrast variation data

The information content of an isotropic scattering profile measured from a dilute, monodisperse and ideal (non-interacting) sample of macromolecules isotropically tumbling in solution can be described as *single* molecule scattering, spherically averaged over all orientations. Consequently, it is possible to generate single molecule 3D models, from which 1D scattering profiles can be calculated and then compared to the experimental data. One of the most common *ab initio* approaches for modeling small-angle scattering data is the optimization of a distribution of dummy atoms (or beads) such that the scattering profile of the distribution matches the measured scattering data. From a distribution of beads, the small-angle scattering profile can be calculated as (Chacón, Morán, Díaz, Pantos, & Andreu, 1998; Debye, 1915; Whitten, Jeffries, Harris, & Trewhella, 2008)

$$I(q) \propto \sum_{i}^{N} \sum_{j}^{N} f_i(q) f_j(q) \frac{\sin(qr_{ij})}{qr_{ij}}. \tag{21}$$

N is the number of beads and r_{ij} is the distance between beads i and j, and $f_i(q)$ and $f_j(q)$ are the scattering amplitudes of beads i and j. Thus, it is conceptually straightforward to calculate a scattering profile from a distribution of beads. In practice, a Monte-Carlo approach is typically used to optimize the distribution, where the number of evaluations made in an optimization is large and that computing time scales with N^2 for full evaluation of Eq. (21) (and scales with N for each step where a single grid point is altered). A more efficient approach, implemented in the program MONSA (Svergun, 1999 ; Svergun & Nierhaus, 2000), involves using a finite expansion of superimposed spherical harmonic functions to calculate the scattering profile of a multi-phase volume-element or dummy-atom (DAM) bead model. The DAM approach relies on taking a large search volume, typically with maximum dimensions equal to or greater than the maximum molecule dimension and filling it completely with bead elements from K different phases. The intensity profile of the multiphase DAM can be calculated as

$$I(q) = \left\langle \sum_{k=1}^{K} \Delta\rho_k A_k(q) \right\rangle^2, \tag{22}$$

Where $\Delta\rho_k$ and $A_\mathrm{k}(q)$ are the contrast and scattering amplitude of the k_th phase respectively. At the beginning of the slow-cool simulated annealing procedure, the beads are allowed to "switch phases" with relative ease. As the system cools, the target function is minimized, and as the fit to the data is improved, the distribution of each of the phases becomes contiguous. This occurs as there are restraint terms in the target function that enforce connectivity (ensuring that each bead is connected to at least one other) and compactness (ensuring each bead contacts as many other beads as possible).

An important consideration for neutron scattering is the resolution function. For a pinhole instrument, neutron scattering employs large divergent beam, and this beam is composed of a distribution of wavelengths.

Thus, at a given q-value (q_i), the scattering is a weighted averaged of the scattering from q-values around that point (q_j),

$$I(q_i) = \int W_i\left(q_i, q_j\right) I\left(q_j\right) dq_j, \tag{23}$$

Where $W_i(q_i, q_j)$ is the resolution function at q_j for data point q_i. The resolution function combines contributions from beam divergence, beam size, wavelength distribution and detector pixel size (Pedersen, Posselt, & Mortensen, 1990). For monotonically decreasing intensity profiles, the effect of pinhole smearing is small, however, for curves where the intensity changes rapidly, the instrumental resolution noticeably "smears" the scattering curve (Fig. 4A vs Fig. 4B). Importantly, the resolution function will account for any averaging effects that occur due to the choice of bin width when reducing the data. Thus, it is good practice to incorporate instrumental resolution when performing any structural modeling against SANS data.

5.1 Programs

- MONSA (and DAMESV) (Svergun, 1999; Svergun & Nierhaus, 2000)

5.2 Procedure

The most widely used bead modeling software for contrast variation data is MONSA. The program can model up to 4-phases (see note 1) against contrast variation data to produce a low-resolution bead model of the molecule. MONSA fits directly to the experiment scattering data, correcting for instrument resolution. The program is run from the command line and

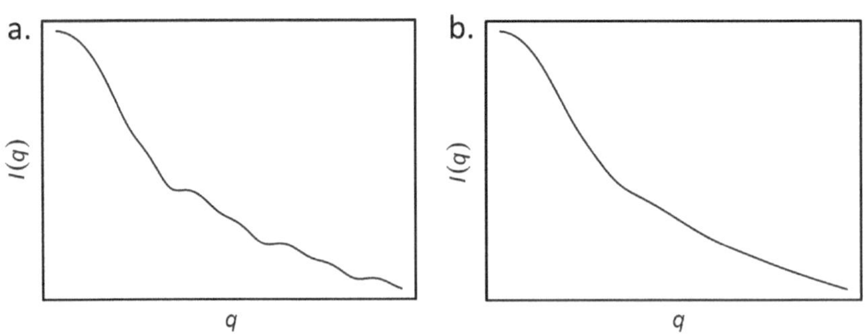

Fig. 4 The effect of instrument resolution on an arbitrary form-factor, (A) No resolution function applied to the scattering curve, (B) A resolution function corresponding to a wavelength distribution of $\Delta\lambda/\lambda = 5\%$ (1 standard deviation) applied to the scattering curve.

requires a number of input files to run correctly. The online manual for the software provides a reference to the file configurations required to run the program, while ATSAS online provides a means to perform MONSA modeling using an intuitive interface and to write all of the configuration files as templates for future use. The following procedure will go into a little more depth for setting up a MONSA run.

1. Create a file containing a search volume of spheres using DAMESV. The search volume will be defined by "half axis" length in dimensions A, B, and C (or x, y, and z). If nothing is known about the molecule, then the safe option is to set A, B, and C approximately 10% larger than $D_{max}/2$ (see note 2). The "Packing radius" must also be defined—in the first instance, a large packing radius (5–10% of D_{max}) may be chosen. The symmetry of the distribution is then required, followed by the number of phases.
2. Create the instrument resolution files (smear.res). The resolution function generated in MONSA uses the equations set out by Pedersen et al. (1990). The first line of the file is the diameter of the source aperture in cm (the first beam defining aperture immediately after the last guide segment will typically be between 3 and 5 cm). The second line is the diameter of the sample aperture in cm (the beam defining aperture immediately before the sample will typically be between 0.5 and 1.5 cm). The third line is the collimation distance in cm (the distance between the source and sample apertures will typically be between 150 and 1500 cm). The fourth line is the sample-to-detector distance in cm (typically be between 400 and 1500 cm). These four lines define the beam size and divergence at the detector. The fifth line is the wavelength of radiation used in Ångströms (typically ranges from 4 to 8 Å). The sixth line is the wavelength resolution (see note 3) (typically ranges from 0.05–0.15). The seventh line is the pixel size of the detector in cm (typically ranges from 0.5–1.0 cm). The last line is an averaging error that should be set to 0.0, as it is incorporated into the pixel size correction (see note 4). If datasets are collected using different instrument configurations, each configuration can be defined in an additional column, where column 1 refers to configuration 1, and column 2 refers to configuration 2, etc. A sample input file containing two different instrument settings is of the form:

```
5.00   4.00
1.25   1.00
1200.0   800.0
```

```
1250.0  850.0
5.00   6.00
0.10   0.05
0.75   0.50
0.00   0.00
```

3. Create the control files (*.con). The first line of the control file contains the name of the resolution file (in quotes) and the corresponding instrument setting (i.e., "smear.res" 1). The second line is stated to be an output file name for the fits but appears to be unused. A dummy name such as "neutron1.out" should be entered in quotes. The third line is a title for the control file, where any explanatory text is acceptable. The fourth line is the number of points in each data file (see note 5). The subsequent lines contain information for each dataset in the format: "filename," contrast of phase 1, contrast of phase 2, contrast of phase 3, contrast of phase 4 (see note 6), data multiplier (to take account of data measured at different concentrations), constant to be subtracted from data (to account for incoherent background from hydrogen) (see note 7). A sample control file is of the form:

```
'smear.res' 1
'neutron1.out'
Neutron data for the X-Y complex
230
'0.dat' 5.67 2.30 0.00 0.00 1.00 0.00 1.00
'42.dat' 3.28 -0.03 0.00 0.00 1.00 0.00 1.00
'60.dat' 2.26 -1.03 0.00 0.00 1.00 0.00 1.00
'80.dat' 1.12 -2.14 0.00 0.00 1.00 0.00 1.00
'100.dat' -0.01 -3.52 0.00 0.00 1.00 0.00 1.00
```

4. Create the master file (*.mst). The first line contains a description of the experiment. The second line contains the volumes of each of the four phases, and these can be entered as 0 if unknown (see note 5). The third line is the theoretical radii of gyration (R_g in Ångströms) of each phase, and again these can be entered as 0 if unknown. The fourth line is the connectivity information of each phase; if a single phase is connected, enter a value of 1 (or any value $\neq 0$), while if a given phase is not connected, enter a value of 0. This is a consideration for complexes with stoichiometries higher than 1:1, where a given phase may no longer be connected (Fig. 2D and E). The subsequent lines contain each control file name in quotes, together with a flag indicating whether a Guinier fit for each file listed in the control file will be performed. If specified,

the R_g values will be used as a restraint in the modeling, where a value of -1 indicates that no R_g restraints will be used, while a positive integer value flags the number of points to be used in the Guinier fit.

```
Dummy atom modelling of the X-Y complex
160000.0 30000.0 0.0 0.0
0.0 0.0 0.0 0.0
1 0 0 0
'neutron1.con' -1
'neutron2.con' -1
'xray.con' -1
```

5. Run the MONSA program. This command line program will ask for an output log file name, a title, the name of the master file, the order of the spherical harmonic expansion used (see note 8), the PDB file containing the initial search volume and phases, and the symmetry of the molecule. After this there are a number of parameters presented that can be modified, but in the first instance should be left at their default values—for example there are a range of penalty weights that define the importance of each restraint (continuity, looseness, volume, R_g). Changes to the weighting parameters can be trialed if issues with the resulting models and/or fits are obtained. There should be no reason to alter the annealing parameters except perhaps to increase the maximum number of iterations if the search volume is large, and/or the packing radius is small.
6. After the initial tests, a suitable search volume, packing radius will have been determined. *Re*-run step 1 and redefine an appropriate minimal search volume by reducing parameters *A*, *B*, and/or *C*. At the same time, the packing radius can be decreased, which will increase the number of beads. Optimization times are related to the search volumes, and packing radius, and large search volumes and small values for the packing radius will yield long optimization times. The initial optimizations with large packing radii help to define a minimal search volume so that smaller packing radii can be used for the final model.
7. Using the default values, or custom restraint weights, the calculation should be run using the new search volume between 10 and 20 times to establish the model variability. There are no tools available to average multiple structures, as is typically done for single phase bead models (Volkov & Svergun, 2003) although it is possible to spatially align each individual model using SUPALM (Konarev, Petoukhov, & Svergun, 2016). Instead, for publication, a single representative model should be chosen and the fits for this model presented. An average fit parameter

for all models (such as the χ^2 and its standard deviation can also be presented). However, while a single model may be described in the publication, all models can be submitted to the SASBDB (Kikhney, Borges, Molodenskiy, Jeffries, & Svergun, 2020). At this point in time, it is difficult to upload and display a full neutron contrast variation data as primary data. However, the contrast variation data, analysis and modeling can be linked to an associated primary SANS or SAXS entry.

5.3 Notes

1. A molecule of 3 or 4 phases, can be considered as multiple 2-phase molecules. The general experimental approach is to isotopically label different combinations of subunits and measure a contrast variation series on each.
2. There are some shapes with a given D_{max} value that cannot fit within a sphere with a diameter of D_{max} (i.e., a triangle). For these cases, or if nothing is known about the shape of the molecule, the initial dimensions of A, B, and C, should be chosen to be 10% larger than $D_{max}/2$. If preliminary modeling step still show that the model appears constrained by the search volume, then the at least one of the dimensions should be further increased. The choice of which dimension to increase is arbitrary if no symmetry is applied to the molecule, however, if symmetry is applied, be aware that the n-fold axis is parallel to the C axis (or z-axis in cartesian coordinates).
3. The wavelength distribution from a velocity selector is usually triangular and the resolution ($\Delta\lambda/\lambda$) is characterized in terms of the full width at half maximum height (FWHM). Most data modeling software use a Gaussian approximation and require the wavelength resolution in terms of the standard deviation of the distribution. The relationship between the two is $FWHM = 2\sqrt{2\ln 2}\sigma$, which means that $\sigma(\Delta\lambda/\lambda) \approx FWHM(\Delta\lambda/\lambda)/2.36$.
4. Often data from two different configurations, each with a different resolution function, are merged together. Thus, the data is described by two distinct resolution functions. In such cases, the instrument configuration for the low-q setting should be used for the merged data set. This is because at low-q, the resolution function is dominated by beam size and divergence, while at high-q, the resolution function is dominated by the wavelength distribution. Given the wavelength distribution will be common to both configurations, the low-q setting should adequately describe the entire merged data set.

5. Each set of data collected using a different configuration or with a different number of points or q-spacing requires a separate control file. If X-ray scattering data is included in the modeling, then instead of specifying an instrument resolution file in the first line of the .con file, "Point collimation 1" can be specified to prevent any instrumental smearing calculations for this dataset.
6. MONSA does not expect contrast values or volumes with specific units. Relative contrasts and volumes are sufficient.
7. Often modeling software gives an option for the optimization of a constant background (to account for incoherent scattering from hydrogen). Be aware that MONSA does not perform an automatic background correction, and any corrections need to be applied directly to the data before modeling, or a constant must be included explicitly in the control file.
8. The default order of the spherical harmonic expansion L_{max} is 14. Larger values will be required for large values of q_{max} (although values of $q > 0.3\,\text{Å}^{-1}$ shouldn't be used for bead modeling as features in the scattering curves above this q-value will be dominated by ρ_F, which is not well represented by a bead model). The order of the expansion required can be estimated as $L_{max} > \pi D_{max} q_{max}/12$, thus for a 200 Å molecule with a data measured to $q_{max} = 0.3$, $L_{max} > 16$.

6. Rigid body modeling against contrast variation data

Rigid-body modeling provides a means of combining a range of biophysical and biochemical data to produce appealing, interpretable models. The premise of rigid body is to use high resolution structures (obtained from crystallography, NMR, 3-dimension structure prediction software, even cryo-EM) as "rigid" building blocks that come together to give a structure that is consistent with the scattering data. The optimization procedure can be performed by manually shifting components and computing the scattering profile (Whitten et al., 2008) or using automatic optimization algorithms (Petoukhov & Svergun, 2006). The theory of calculating scattering profiles for rigid bodies is entirely analogous to that presented for bead modeling, but instead of having a fixed grid of dummy-atoms, there are subunit structures composed of different atoms. Thus, it is also possible to use the Debye formula to calculate the scattering profile (Hofmann & Whitten, 2014; Schneidman-Duhovny, Hammel, Tainer, & Sali, 2013), but in Eq. (21), $f_i(q)$ is now the scattering amplitude of each atom in the structure.

The hydration layer can also be included by the inclusion dummy-atoms at the surface of the molecule with dimensions and contrast representative of the hydration layer The program SASREF (Petoukhov & Svergun, 2006) (like MONSA) approximates the scattering amplitude of each phase by an expansion of superimposed spherical harmonic functions, and the coordinate system of these scattering amplitudes, $A_k(q)$, is readily transformed to match that corresponding rigid-body as it is translated and rotated during the optimization to give $A_k^*(q)$. Thus, for a model with K rigid-bodies, the scattering profile can be calculated as

$$I(q) = \left\langle \sum_{k=1}^{K} A_k^*(q) \right\rangle^2. \tag{24}$$

For clarity, it is assumed that $A_k^*(q)$ includes the hydration shell of the kth rigid-body (Svergun et al., 1998). This approach permits inclusion of the hydration shell in the modeling calculation in a straightforward manner, yet a shortcoming of this approach is that the hydration shell is not correctly described where two rigid bodies meet (at this point the hydration layer for one rigid body will overlap with the neighboring rigid body). The overall effect of this is expected to be small and neglecting the hydration shell entirely would be expected to significantly more bias in the model[b]. The relative movements of the rigid bodies relative to each other can be restrained by information from biophysical and/or biochemical data that defines proximity of given regions of the different rigid bodies (e.g., chemical cross-linking) (Christie et al., 2012).

A complicating factor for the application of rigid body modeling to neutron contrast variation data is that changing the D_2O content in the solvent affects the composition of each rigid body. The reason being that labile H in the macromolecule exchanges with the solvent, which changes the scattering amplitude of each phase. Further, labile hydrogen positions at the surface

[b] Interestingly, many SANS studies have found that explicit treatment of the hydration shell is not required to model neutron scattering data (Perkins, 2001; Spiteri et al., 2021; Whitten et al., 2007), but the reasons for this are not well described or understood. The generally accepted model of the hydration shell is a layer of water molecules with a thickness of ~3 Å and a density higher than the bulk solvent (Svergun et al., 1998). Because the density of the hydration shell differs from the bulk solvent, the accepted model suggests this layer must contribute to the scattering data at any contrast point (except where the scattering length density of the solvent is zero). Molecular dynamics simulations of proteins in aqueous solvents have shown the scattering length density distribution of the hydration layer is perhaps more complicated than the accepted model (Chen & Hub, 2014; Kim et al., 2016), and continued work in this area may shed further light on the nature of the hydration shell and the best way to implement this feature into structural models.

of the protein exchange rapidly with the solvent, and probability of finding a deuterium atom at these labile positions reflects the H/D ratio in the solvent. Slower exchanging hydrogen positions (typically backbone amides in the core of proteins) do not fully reflect the H/D ratio in the solvent. SASREF effectively deals with all these complications in the calculation of scattering profiles of a given rigid-body structure at multiple contrasts.

6.1 Programs

- CRYSON (Svergun et al., 1998)
- SASREF7 or SASREFCV (Petoukhov & Svergun, 2006)

6.2 Procedure

The most widely used rigid-body modeling software for contrast variation data is SASREF7/SASREFCV. The program fits directly to the experimental scattering data, correcting for instrument resolution. The program can be run from the command line and requires a number of input files to run correctly. The online manual for the software provides a reference to the file configurations required to run the program, while ATSAS online provides a means to perform SASREFCV modeling using an intuitive interface and to write all of the configuration files as templates for future use. The following procedure will go into a little more depth for setting up a SASREF run.

1. Define the rigid bodies in the structure and create a PDB file containing each rigid subunit (see notes 1 and 2).
2. Create the instrument resolution files (smear.res) in the same format as described for the bead modeling (Section 5.2).
3. Using the data from the R_g analysis and composite scattering functions (together with any biochemical data), attempt to build a starting model that is consistent with the radius of gyration of each component, the D_{max} of each component, and the distance between the two components. Use CRYSON to compute the scattering curves accounting for any deuterium labeling, the D_2O content of the solvent (see note 3), and the instrument resolution. This step may be repeated until a reasonable model of the structure is obtained.
4. Generate a contacts file (*.cnd), which contains distance restraints for the rigid body modeling. There are a number of distance restraints that can be defined. The first type of restraint is between the end of one rigid subunit and the start of another, and this will occur for modular proteins

composed of multiple rigid domains connected by short linkers, and these have the form

```
dist 3.8
1 0 0 2 1 1
```

This can be interpreted as a maximum distance of 3.8 Å between the last residue (described by a 0) in subunit 1 to the first residue in subunit 2. The residue number is repeated (i.e., 0 0 and 1 1) as it is possible to define a range of residues. The other type of distance restraint occurs between regions of different domains, where biochemical or biophysical data has shown these points are in close proximity, and these have the form

```
dist 12.0
1 38 45 2 10 21
```

This can be interpreted as a maximum distance of 12.0 Å between a residue in the range of 38–45 in subunit 1 to a residue in the range of 10–21 in subunit 2. In some cases, it may be known that a subunit makes contact with a given residue type. If there is more than one residue of this type on the surface of a subunit, a restraint can be defined so that only one instance need be satisfied

```
dist 5.0
1 1 0 2 10 10
1 1 0 2 32 32
```

This can be interpreted as a maximum distance of 12.0 Å between subunit 1 (1 0 define a range encompassing the entire subunit) and residue 10 in subunit 2 or residue 32 in subunit 2. If each were defined with a separate distance, the software would attempt to enforce both restraints.

5. Create the data control file (data.con). The first line of this file contains the number of contrast points. Each following line contains the name of the data file, the fraction of D_2O in the solvent (-1 is used to indicate X-ray scattering data), the symmetry of the molecule in those solution conditions, a flag indicating the angular units of the data file (1 for $Å^{-1}$ and 2 for nm^{-1}), the fraction of the curve to be fitted (i.e., the higher angle data can be excluded by setting a value less than 1.0), the resolution setting, the weight of this dataset in the optimization (see note 4), and a Y/N flag indicating whether a constant term will be optimized to adjust background levels. A sample data control file is of the form:

```
6
0.dat 0.00 P1 1 1.0 1 1.00 y
20.dat 0.20 P1 1 1.0 1 1.00 y
```

```
60.dat 0.60 P1 1 1.0 1 1.00 y
80.dat 0.80 P1 1 1.0 1 1.00 y
100.dat 0.99 P1 1 1.0 1 1.00 y
xray.dat -1.00 P1 1 1.0 0 1.00 y
```

6. Create the subunits control file (subs.con). The first line of this file contains the number of subunits. Each following line contains the name of the subunit, a Y/N flag indicating whether the subunit should be moved to the original and the start of the optimization, a Y/N flag indicating whether the subunit should be fixed in place, and the symmetry of the subunit in the complex. A sample subunit control file is of the form:

```
4
Nterm.pdb n n P1
link.pdb n n P1
Cterm.pdb n n P1
Ligand.pdb n n P1
```

7. Create the cross-table control file (table.con). It has a number of columns equal to the number of subunits, and a number of rows equal to the number of contrast points. Each number in the table refers to the deuteration level of the given subunit. If the subunit is not present in that condition, −1.0 should be entered, and for an X-ray dataset the value should be 0.0 if the subunit is present. A cross table control file is of the form:

```
0.00 0.00 0.00 0.80
0.00 0.00 0.00 0.80
0.00 0.00 0.00 0.80
0.00 0.00 0.00 0.80
0.00 0.00 0.00 0.80
0.00 0.00 0.00 0.00
```

8. With these files, run SASREF using the default parameters in the first instance. Depending on the results, you may need to tweak the contact list, and perhaps the restraint weights. Once you have settled upon a modeling strategy run the model 10–20 times and choose a representative structure for presentation. As was described for MONSA, all models can be deposited in SASBDB, linked to a primary SANS or SAXS entry.

6.3 Notes

1. At present, hybrid rigid body/bead modeling software (such as BUNCH or CORAL) for modeling flexible systems, or systems with unstructured regions are not compatible with SANS contrast variation data. In these

cases, SASREF can be used where the flexible linkers are defined as a chain of rigid links. As a last resort, BUNCH or CORAL could be used with the composite scattering functions I_{11}, I_{22} and $I_C = I_{11} + 2I_{12} + I_{22}$. However, be aware that these programs use X-ray scattering amplitudes, that systematically differ from the neutron scattering amplitudes, especially at high-q where the difference in the density fluctuations of the X-ray and neutron scattering length density distributions will be most obvious. The contribution of the hydration layer to the scattering data may also be incorrect as the contrast of the hydration layer relative to the macromolecule will differ for X-rays and neutrons.

2. Like all rigid body modeling, it is important that each subunit is not missing significant parts of the structure, i.e., all of the mass is accounted for and that the structure contains all side chain atoms. Hydrogen atoms are included automatically accounted for in the calculation of scattering amplitudes from molecules composed of amino acids and nucleic acids using CRYSON, but this is not necessarily true for post-translational modifications such as glycosylation. Given the accessibility of structure prediction severs, such as AlphaFold (Jumper et al., 2021; Varadi et al., 2022) or iTasser (Yang et al., 2015), it is relatively straightforward to reliably determine the structure of subunits, even when significant portions may be missing. Such programs also yield structures with the correct sequence (compared with homologous structures from different organisms), which is helpful when defining structural restraints.
3. When using CRYSON to generate neutron scattering curves from protein complexes, be aware that "`Rg ( Atoms - Excluded volume + Shell )`" is computed relative to the geometric center of the whole molecule, thus, it can often differ significantly from "`Rg from the slope of net intensity`". When using CRYSON to obtain theoretical radii of gyration, it is generally advisable to use the latter value as it is not dependent on any choice of origin.
4. When modeling SANS-CV data (potentially with SAXS data), you may find that the optimization provides an excellent fit to some contrast points at the expense of others. The target function that is minimized during a global refinement includes the sum of the χ^2 values for each contrast point. To work effectively, all data sets should have accurately estimated uncertainties, such that the final model yields $\chi^2 \approx 1$ for each contrast point. It is not uncommon to find that the uncertainties on the intensity measurements for SANS-CV are overestimated (yielding $\chi^2 < 1$), and the relative extent to which they are overestimated is larger for weakly

scattering samples. The reason for this is in-part due to the subtraction of blocked beam and empty cell measurements from both the sample and solvent during data reduction. When the solvent scattering is subsequently subtracted from the sample scattering, the blocked beam and empty cell contributions largely cancel, but the contribution they make to propagated uncertainty estimates may remain (depending on the data reduction algorithm). If the uncertainty estimates for any contrast point are artificially high, then these contrast points are effectively down-weighted in the optimization procedure. The contrast points with smaller relative uncertainties will then largely define the model. In such cases, individual data sets may need to be appropriately down-weighted to provide a more balanced fit.

7. Summary

The analyses presented in this chapter focus on binary complexes, but this should not limit the readers imagination, or suggest that contrast variation cannot be done on more complicated systems. Systems with more than two phases can be considered as the combinations of multiple of 2-phase systems. For example, consider generalization of the composite scattering functions for a 3-phase system; 6 composite scattering functions would be required to describe the system (I_{11}, I_{22}, I_{33}, I_{12}, I_{13}, I_{23}), and the practical way to approach this would be to collect a CV series with one labeling strategy, and the collect another CV series with a different labeling strategy. In this manner, a 3-phase system is described by 2 conventional CV series. Likewise, for a 4-phase system, 10 composite scattering functions would be required to describe the system (I_{11}, I_{22}, I_{33}, I_{44}, I_{12}, I_{13}, I_{14}, I_{23}, I_{24}, I_{34}), which could be approached as 3 different CV series with 3 different labeling strategies. Indeed, it is important that SANS be used to study complicated and important biologically relevant molecules, as this not only advances our understanding of biological systems, but it spurs technological advances in instrumentation and data analysis software, that will facilitate future research.

Having multiple related datasets, as is the case for neutron contrast variation, is a mixed blessing. It provides additional structural information, but unintended sample variability can compromise the data analysis. This variation may arise from differences in the way each sample was handled, or it may be due to subtle variations in the way molecules interact with each other in the presence of deuterium. Thus, the single largest barrier to successful analysis of SANS-CV data is measurement from a consistent set of monodisperse samples, where dissociation, aggregation and flexibility represent unique challenges for data analysis. A rigorous approach to the analysis of SANS-CV data assists in the identification of inconsistencies between samples and other potential issues before they compromise the outcomes of any modeling performed using the CV data. The publication standards outlined in 2017 (Trewhella et al., 2017), recommends inclusion of $I(0)$ plots to demonstrate sample homogeneity, Stuhrmann plots to show the R_g values display the expected behavior, and composite scattering functions to indicate the information content of the data. Thus, these procedures are not only a source of structural information, but they also demonstrate the reliability of the data and corresponding models. Having successfully completed the $I(0)$, R_g and composite function analyses, one can confidently move on to three-dimensional structural modeling with multiple phase dummy atoms or rigid body atomistic components.

Acknowledgments

The authors acknowledge the support of the Australian Centre for Neutron Scattering, Australian Nuclear Science and Technology Organisation, the Australian Government through the National Collaborative Research Infrastructure Strategy, and the European Union's Horizon 2020 research and innovation programme "iNEXT Discovery" grant agreement No 871037. The authors also wish to acknowledge the helpful comments of Jill Trewhella in the preparation of this manuscript.

References

Anthis, N. J., & Clore, G. M. (2013). Sequence-specific determination of protein and peptide concentrations by absorbance at 205 nm. *Protein Science*, *22*(6), 851–858. https://doi.org/10.1002/pro.2253.

Bacon, G. E. (1975). *Neutron diffraction* (3rd ed.). Oxford: Clarendon Press.

Breyton, C., Gabel, F., Lethier, M., Flayhan, A., Durand, G., Jault, J.-M., et al. (2013). Small angle neutron scattering for the study of solubilised membrane proteins. *The European Physical Journal E*, *36*(7), 71. https://doi.org/10.1140/epje/i2013-13071-6.

Brosey, C. A., & Tainer, J. A. (2019). Evolving SAXS versatility: Solution X-ray scattering for macromolecular architecture, functional landscapes, and integrative structural biology. *Current Opinion in Structural Biology*, *58*, 197–213. https://doi.org/10.1016/j.sbi.2019.04.004.

Chacón, P., Morán, F., Díaz, J. F., Pantos, E., & Andreu, J. M. (1998). Low-resolution structures of proteins in solution retrieved from X-ray scattering with a genetic algorithm. *Biophysical Journal*, *74*(6), 2760–2775. https://doi.org/10.1016/S0006-3495(98)77984-6.

Chen, P.-C., & Hub, J. S. (2014). Validating solution ensembles from molecular dynamics simulation by wide-angle X-ray scattering data. *Biophysical Journal*, *107*(2), 435–447. https://doi.org/10.1016/j.bpj.2014.06.006.

Christie, M. P., Whitten, A. E., King, G. J., Hu, S.-H., Jarrott, R. J., Chen, K.-E., et al. (2012). Low-resolution solution structures of Munc18:Syntaxin protein complexes indicate an open binding mode driven by the Syntaxin N-peptide. *Proceedings of the National*

Academy of Sciences of the United States of America, 109(25), 9816–9821. https://doi.org/10.1073/pnas.1116975109.

Clifton, L. A., & Myatt, D. P. n.d. Protein scattering length density calculator (version 1.0): Science and Technology Facilities Council. http://psldc.isis.rl.ac.uk/Psldc/.

Da Vela, S., & Svergun, D. I. (2020). Methods, development and applications of small-angle X-ray scattering to characterize biological macromolecules in solution. *Current Research in Structural Biology, 2*, 164–170. https://doi.org/10.1016/j.crstbi.2020.08.004.

Damaschun, G., Fichtner, P., Purschel, H. V., & Reich, J. G. (1968). Studies on quaternary structure of proteins by small angle scattering of X-rays .2. On structure of complex of trypsin with soyabean inhibitor. *Acta Biologica Et Medica Germanica, 21*(3), 309.

de Oliveira Neto, M., de Freitas Fernandes, A., Piiadov, V., Craievich, A. F., de Araújo, E. A., & Polikarpov, I. (2022). SAXSMoW 3.0: New advances in the determination of the molecular weight of proteins in dilute solutions from SAXS intensity data on a relative scale. *Protein Science, 31*(1), 251–258. https://doi.org/10.1002/pro.4227.

Debye, P. (1915). Zerstreuung von Röntgenstrahlen. *Annalen der Physik, 351*(6), 809–823. https://doi.org/10.1002/andp.19153510606.

Furlong, E. J., Choudhury, H. G., Kurth, F., Duff, A. P., Whitten, A. E., & Martin, J. L. (2018). Disulfide isomerase activity of the dynamic, trimeric Proteus mirabilis ScsC protein is primed by the tandem immunoglobulin-fold domain of ScsB. *Journal of Biological Chemistry, 293*(16), 5793–5805. https://doi.org/10.1074/jbc.RA118.001860.

Gabel, F., Engilberge, S., Perez, J., & Girard, E. (2019). Medical contrast media as possible tools for SAXS contrast variation. *IUCrJ, 6*(4), 521–525. https://doi.org/10.1107/S2052252519005943.

Gasteiger, E., Hoogland, C., Gattiker, A., Duvaud, S.e., Wilkins, M. R., Appel, R. D., et al. (2005). Protein identification and analysis tools on the ExPASy server. In J. M. Walker (Ed.), *The proteomics protocols handbook* (pp. 571–607). Totowa, NJ: Humana Press.

Glatter, O. (1977). A new method for the evaluation of small-angle scattering data. *Journal of Applied Crystallography, 10*(5), 415–421. https://doi.org/10.1107/S0021889877013879.

Grishaev, A., Anthis, N. J., & Clore, G. M. (2012). Contrast-matched small-angle X-ray scattering from a heavy-atom-labeled protein in structure determination: Application to a lead-substituted calmodulin–peptide complex. *Journal of the American Chemical Society, 134*(36), 14686–14689. https://doi.org/10.1021/ja306359z.

Guinier, A., & Fournet, G. (Eds.). (1955). *Small-angle scattering of X-rays*. New York: Wiley.

Hjelm, R. P., Kneale, G. G., Suau, P., Baldwin, J. P., Bradbury, E. M., & Ibel, K. (1977). Small-angle neutron-scattering studies of chromatin subunits in solution. *Cell, 10*(1), 139–151.

Hofmann, A., & Whitten, A. E. (2014). Two practical Java software tools for small-angle X-ray scattering analysis of biomolecules. *Journal of Applied Crystallography, 47*(2), 810–815. https://doi.org/10.1107/S1600576714004737.

Ibel, K., & Stuhrmann, H. B. (1975). Comparison of neutron and X-ray-scattering of dilute myoglobin solutions. *Journal of Molecular Biology, 93*(2), 255–265.

Jacques, D. A., & Trewhella, J. (2010). Small-angle scattering for structural biology—Expanding the frontier while avoiding the pitfalls. *Protein Science, 19*(4), 642–657. https://doi.org/10.1002/pro.351.

Jacrot, B. (1976). Study of biological structures by neutron-scattering from solution. *Reports on Progress in Physics, 39*(10), 911–953.

Jeffries, C. M., Graewert, M. A., Blanchet, C. E., Langley, D. B., Whitten, A. E., & Svergun, D. I. (2016). Preparing monodisperse macromolecular samples for successful biological small-angle X-ray and neutron-scattering experiments. *Nature Protocols, 11*(11), 2122–2153. https://doi.org/10.1038/nprot.2016.113.

Jumper, J., Evans, R., Pritzel, A., Green, T., Figurnov, M., Ronneberger, O., et al. (2021). Highly accurate protein structure prediction with AlphaFold. *Nature, 596*(7873), 583–589. https://doi.org/10.1038/s41586-021-03819-2.

Kikhney, A. G., Borges, C. R., Molodenskiy, D. S., Jeffries, C. M., & Svergun, D. I. (2020). SASBDB: Towards an automatically curated and validated repository for biological scattering data. *Protein Science*, *29*(1), 66–75. https://doi.org/10.1002/pro.3731.

Kim, H. S., Martel, A., Girard, E., Moulin, M., Härtlein, M., Madern, D., et al. (2016). SAXS/SANS on supercharged proteins reveals residue-specific modifications of the hydration Shell. *Biophysical Journal*, *110*(10), 2185–2194. https://doi.org/10.1016/j.bpj.2016.04.013.

Kirste, R. G., & Stuhrman, H. B. (1967). Elimination of intraparticular base scattering in X-ray small angle scattering on compact particles 3. Verification on solutions of myoglobin. *Zeitschrift Fur Physikalische Chemie-Frankfurt*, *56*(5–6), 338–341.

Konarev, P. V., Petoukhov, M. V., & Svergun, D. I. (2016). Rapid automated superposition of shapes and macromolecular models using spherical harmonics. *Journal of Applied Crystallography*, *49*(3), 953–960. https://doi.org/10.1107/S1600576716005793.

Krauss, E. M., & Cowburn, D. (1981). Hydrogen-deuterium-exchange kinetics of the amide protons of oxytocin studied by nuclear magnetic resonance. *Biochemistry*, *20*(4), 671–679. https://doi.org/10.1021/bi00507a001.

Krueger, S. (2017). Designing and performing biological solution small-angle neutron scattering contrast variation experiments on multi-component assemblies. In B. Chaudhuri, I. G. Muñoz, S. Qian, & V. S. Urban (Eds.), *Biological small angle scattering: Techniques, strategies and tips* (pp. 65–85). Singapore: Springer Singapore.

Lapinaite, A., Carlomagno, T., & Gabel, F. (2020). Small-angle neutron scattering of RNA–protein complexes. In V. Arluison, & F. Wien (Eds.), *RNA spectroscopy: Methods and protocols* (pp. 165–188). New York, NY: Springer US.

Mahieu, E., & Gabel, F. (2018). Biological small-angle neutron scattering: Recent results and development. *Acta Crystallographica Section D*, *74*(8), 715–726. https://doi.org/10.1107/S2059798318005016.

Manalastas-Cantos, K., Konarev, P. V., Hajizadeh, N. R., Kikhney, A. G., Petoukhov, M. V., Molodenskiy, D. S., et al. (2021). ATSAS 3.0: Expanded functionality and new tools for small-angle scattering data analysis. *Journal of Applied Crystallography*, *54*(1), 343–355. https://doi.org/10.1107/S1600576720013412.

Mason, S. A., Bentley, G. A., & McIntyre, G. J. (1984). Deuterium exchange in lysozyme at 1.4-Å resolution. In B. P. Schoenborn (Ed.), *Neutrons in biology* (pp. 323–334). Boston, MA: Springer US.

Murphy, L. R., Matubayasi, N., Payne, V. A., & Levy, R. M. (1998). Protein hydration and unfolding—Insights from experimental partial specific volumes and unfolded protein models. *Folding and Design*, *3*(2), 105–118. https://doi.org/10.1016/S1359-0278(98)00016-9.

Olah, G. A., Rokop, S. E., Wang, C. L. A., Blechner, S. L., & Trewhella, J. (1994). Troponin-I encompasses an extended troponin-C in the Ca2+-bound complex—A small-angle X-ray and Neutron-scattering study. *Biochemistry*, *33*(27), 8233–8239.

Pabit, S. A., Chen, Y.-L., Usher, E. T., Cook, E. C., Pollack, L., & Showalter, S. A. (2020). Elucidating the role of microprocessor protein DGCR8 in bending RNA structures. *Biophysical Journal*, *119*(12), 2524–2536. https://doi.org/10.1016/j.bpj.2020.10.038.

Pedersen, J. S., Posselt, D., & Mortensen, K. (1990). Analytical treatment of the resolution function for small-angle scattering. *Journal of Applied Crystallography*, *23*, 321–333.

Perkins, S. J. (2001). X-ray and neutron scattering analyses of hydration shells: A molecular interpretation based on sequence predictions and modelling fits. *Biophysical Chemistry*, *93*(2), 129–139. https://doi.org/10.1016/S0301-4622(01)00216-2.

Persson, F., & Halle, B. (2015). How amide hydrogens exchange in native proteins. *Proceedings of the National Academy of Sciences*, *112*(33), 10383–10388. https://doi.org/10.1073/pnas.1506079112.

Petoukhov, M. V., & Svergun, D. I. (2006). Joint use of small-angle X-ray and neutron scattering to study biological macromolecules in solution. *European Biophysics Journal, 35*(7), 567–576. https://doi.org/10.1007/s00249-006-0063-9.

Porod, G. (1951). Die Röntgenkleinwinkelstreuung von dichtgepackten kolloiden Systemen. *Kolloid Zeitschrift, 124*(2), 83–114. https://doi.org/10.1007/BF01512792.

R Development Core Team. (2013). *R: A language and environment for statistical computing.* Vienna, Austria: R Foundation for Statistical Computing. Retrieved from http://www.R-project.org/.

RStudio Team. (2020). *RStudio: Integrated development for R.* Boston, MA: RStudio, PBC. Retrieved from http://www.rstudio.com/.

Sarachan, K. L., Curtis, J. E., & Krueger, S. (2013). Small-angle scattering contrast calculator for protein and nucleic acid complexes in solution. *Journal of Applied Crystallography, 46*(6), 1889–1893. https://doi.org/10.1107/S0021889813025727.

Schneidman-Duhovny, D., Hammel, M., Tainer, J. A., & Sali, A. (2013). Accurate SAXS profile computation and its assessment by contrast variation experiments. *Biophysical Journal, 105*(4), 962–974. https://doi.org/10.1016/j.bpj.2013.07.020.

Schoenborn, B. P. (1972). A neutron diffraction analysis of myoglobin. III. Hydrogen-deuterium bonding in side chains. *Cold Spring Harbor Symposia on Quantitative Biology, 36*, 569–575. https://doi.org/10.1101/sqb.1972.036.01.071.

Schuck, P., Perugini, M. A., Gonzales, N. R., Howlett, G. J., & Schubert, D. (2002). Size-distribution analysis of proteins by analytical ultracentrifugation: Strategies and application to model systems. *Biophysical Journal, 82*(2), 1096–1111. https://doi.org/10.1016/S0006-3495(02)75469-6.

Serdyuk, I. N. (1979). A method of joint use of electromagnetic and neutron scattering: A study of internal ribosomal structure. In K. M. A. L. Grossman (Ed.), *Vol. 59. Methods in enzymology: Nucleic acids and protein synthesis part G* (pp. 750–775). Academic Press.

Serdyuk, I. N., & Fedorov, B. A. (1973). New method of studying structure of block copolymers in solution. *Journal of Polymer Science Part C-Polymer Letters, 11*(10), 645–649.

Spiteri, V. A., Goodall, M., Doutch, J., Rambo, R. P., Gor, J., & Perkins, S. J. (2021). Solution structures of human myeloma IgG3 antibody reveal extended Fab and Fc regions relative to the other IgG subclasses. *Journal of Biological Chemistry, 297*(3), 100995. https://doi.org/10.1016/j.jbc.2021.100995.

Stuhrmann, H. B. (1973). Comparison of the three basic scattering functions of myoglobin in solution with those from the known structure in crystalline state. *Journal of Molecular Biology*, 77(3), 363–369. https://doi.org/10.1016/0022-2836(73)90444-0.

Stuhrmann, H. B., & Fuess, H. (1976). A neutron small-angle scattering study of hen egg-white lysozyme. *Acta Crystallogr. Sect. A, 32*(1), 67–74. https://doi.org/10.1107/S0567739476000132.

Stuhrmann, H. B., Tardieu, A., Mateu, L., Sardet, C., Luzzati, V., Aggerbeck, L., et al. (1975). Neutron-scattering study of human-serum low-density lipoprotein. *Proceedings of the National Academy of Sciences of the United States of America*, 72(6), 2270–2273.

Svergun, D. I. (1999). Restoring low resolution structure of biological macromolecules from solution scattering using simulated annealing. *Biophysical Journal, 76*(6), 2879–2886. https://doi.org/10.1016/S0006-3495(99)77443-6.

Svergun, D. I., & Nierhaus, K. H. (2000). A map of protein-rRNA distribution in the 70 S Escherichia coli ribosome*. *Journal of Biological Chemistry, 275*(19), 14432–14439. https://doi.org/10.1074/jbc.275.19.14432.

Svergun, D. I., Richard, S., Koch, M. H. J., Sayers, Z., Kuprin, S., & Zaccai, G. (1998). Protein hydration in solution: Experimental observation by x-ray and neutron scattering. *Proceedings of the National Academy of Sciences, 95*(5), 2267–2272. https://doi.org/10.1073/pnas.95.5.2267.

Svergun, D. I., Semenyuk, A. V., & Feigin, L. A. (1988). Small-angle-scattering-data treatment by the regularization method. *Acta Crystallographica*, *A44*, 244–250.

Tardieu, A., Mateu, L., Sardet, C., Weiss, B., Luzzati, V., Aggerbeck, L., et al. (1976). Structure of human-serum lipoproteins in solution. 2. Small-angle X-Ray-scattering study of Hdl3 and Ldl. *Journal of Molecular Biology*, *101*(2), 129–153.

Tokuda, J. M., Pabit, S. A., & Pollack, L. (2016). Protein-DNA and ion-DNA interactions revealed through contrast variation SAXS. *Biophysical Reviews*, *8*(2), 139–149. https://doi.org/10.1007/s12551-016-0196-8.

Trewhella, J. (2022). Recent advances in small-angle scattering and its expanding impact in structural biology. *Structure*, *30*(1), 15–23. https://doi.org/10.1016/j.str.2021.09.008.

Trewhella, J., Duff, A. P., Durand, D., Gabel, F., Guss, J. M., Hendrickson, W. A., et al. (2017). 2017 publication guidelines for structural modelling of small-angle scattering data from biomolecules in solution: An update. *Acta Crystallographica Section D: Structural Biology*, *73*(9), 710–728. https://doi.org/10.1107/S2059798317011597.

Varadi, M., Anyango, S., Deshpande, M., Nair, S., Natassia, C., Yordanova, G., et al. (2022). AlphaFold protein structure database: Massively expanding the structural coverage of protein-sequence space with high-accuracy models. *Nucleic Acids Research*, *50*(D1), D439–D444. https://doi.org/10.1093/nar/gkab1061.

Volkov, V. V., & Svergun, D. I. (2003). Uniqueness of ab initio shape determination in small-angle scattering. *Journal of Applied Crystallography*, *36*, 860–864.

Whitten, A. E., Cai, S., & Trewhella, J. (2008). MULCh: ModULes for the analysis of small-angle neutron contrast variation data from bio-molecular assemblies. *Journal of Applied Crystallography*, *41*, 222–226.

Whitten, A. E., Jacques, D. A., Hammouda, B., Hanley, T., King, G. F., Guss, J. M., et al. (2007). The structure of the kina-Sda complex suggests an allosteric mechanism of histidine kinase inhibition. *Journal of Molecular Biology*, *368*(2), 407–420. https://doi.org/10.1016/j.jmb.2007.01.064.

Whitten, A. E., Jeffries, C. M., Harris, S. P., & Trewhella, J. (2008). Cardiac myosin-binding protein C decorates F-actin: Implications for cardiac function. *Proceedings of the National Academy of Sciences of the United States of America*, *105*(47), 18360–18365. https://doi.org/10.1073/pnas.0808903105.

Whitten, A. E., & Trewhella, J. (2009). Small-angle scattering and neutron contrast variation for studying bio-molecular complexes. In R. S. Foote, & J. W. Lee (Eds.), *Vol. 544. Micro and nano technologies in bioanalysis* (pp. 307–323). Totowa, NJ: Humana Press.

Yang, J., Yan, R., Roy, A., Xu, D., Poisson, J., & Zhang, Y. (2015). The I-TASSER suite: Protein structure and function prediction. *Nature Methods*, *12*(1), 7–8. https://doi.org/10.1038/nmeth.3213.

Zhao, J., Hoye, E., Boylan, S., Walsh, D. A., & Trewhella, J. (1998). Quaternary structures of a catalytic subunit-regulatory subunit dimeric complex and the holoenzyme of the cAMP-dependent protein kinase by neutron contrast variation *. *Journal of Biological Chemistry*, *273*(46), 30448–30459. https://doi.org/10.1074/jbc.273.46.30448.

CHAPTER FOUR

Observing protein degradation in solution by the PAN-20S proteasome complex: Astate-of-the-art example of bio-macromolecular TR-SANS

Frank Gabel*
Université Grenoble Alpes, Institut de Biologie Structurale, Grenoble, France
Commissariat à l'Energie Atomique et aux Energies Alternatives, Direction de la Recherche Fondamentale, Institut de Biologie Structurale, Grenoble, France
Centre National de la Recherche Scientifique, Institut de Biologie Structurale, Grenoble, France
*Corresponding author: e-mail address: frank.gabel@ibs.fr

Contents

Abstract

In the present book chapter we illustrate the state-of-the-art of time-resolved small-angle neutron scattering (TR-SANS) by a concrete example of a dynamic bio-macromolecular system, i.e., regulated protein degradation by the archaeal PAN-20S proteasome complex. We present the specific and unique structural information that can be obtained by this approach, in combination with bio-macromolecular deuteration and online spectrophotometric measurements of a fluorescent substrate (GFP). The complementarity with atomic-resolution structural biology techniques (SAXS, NMR, crystallography and cryo-EM) and with the advent of atomic structure prediction are discussed, as well as the respective limitations and future perspectives.

Methods in Enzymology, Volume 678
ISSN 0076-6879
https://doi.org/10.1016/bs.mie.2022.09.016

1. The interest of TR-SANS for dynamic bio-macromolecular systems

Bio-macromolecular structural biology has witnessed two major revolutions recently: on the one hand the Cryo-EM resolution revolution about a decade ago allowing this method to join NMR and crystallography as the third technique to provide atomic resolution (Bai, McMullan & Scheres, 2015; Cheng, 2015). On the other hand the very recent game-changing developments of in silico atomic structure predictions, based on bio-macromolecular (protein, RNA, DNA) sequences and the use of artificial intelligence (AI) (Baek et al., 2021; Jumper et al., 2021).

Both approaches have specific advantages and shortcomings. A strength of cryo-EM (apart from reaching atomic resolution in favorable cases) is the identification of transient, complex assemblies of several partners in a natural environment (Ognjenović, Grisshammer, & Subramaniam, 2019) and the possibility of classification into different conformational categories (Scheres, 2016). On the other hand there is a size limit of about 40 kDa for obtaining atomic resolution information (Wu & Lander, 2020) and cryo-EM has difficulty for regions displaying flexibility. In silico structure predictions have some shortcomings as well, and the majority of current applications are focused on isolated, compact structures (Perrakis & Sixma, 2021), even though macromolecular assemblies are expected to be addressed increasingly in the near future. At present, neither approach can provide time-resolved, global information from interacting dynamic macromolecular systems in vitro, involving changes of the conformation, oligomeric state, and the population of different species.

SANS, as a solution technique allowing labeling (i.e., deuteration) of specific components in a complex biological mixture, can follow the evolution of conformational states and interactions of individual biomolecules in vitro as a function of buffer conditions (Jeffries et al., 2021; Mahieu & Gabel, 2018). Encoding information on both rigid and flexible structural elements, short- and long-range distances, small and large molecules (see chapter "Time-resolved small-angle neutron scattering (TR-SANS) for structural biology of dynamic systems: Principles, recent developments, and practical guidelines" by Martel and Gabel in volume 677 of Methods in Enzymology), SANS is particularly complementary to solution-state NMR (see chapter "Combining NMR, SAXS and SANS to characterize the structure and dynamics of protein complexes" by Delhommel et al. in the same volume). The structural information provided by time-resolved SANS of general biological samples, including cellular ultra-structures, has been discussed in an associated book chapter (see chapter "Time-resolved small-angle neutron scattering (TR-SANS) for structural biology of dynamic systems: Principles, recent developments, and practical guidelines" by Martel and Gabel in volume 677 of Methods in Enzymology). In the present chapter, we focus on the kinetics of conformational changes of *individual* bio-macromolecules, their interaction between each other, and the evolution of different populations of molecular species. As an example, we review and discuss recent TR-SANS results of substrate degradation by the archaeal PAN-20S proteasome complex (Ibrahim et al., 2017; Mahieu et al., 2020). Even though a concrete example is chosen to illustrate the capacity of TR-SANS to obtain unique insight into individual partners in a complex biological system, the approach presented and discussed here is applicable to a diverse range of biological processes: remodeling of the proteome (Guo & Gross, 2014) and genome (Lans, Marteijn, & Vermeulen, 2012), chaperone/disaggregase activity (Rosenzweig, Nillegoda, Mayer, & Bukau, 2019), cytoskeleton remodeling (Marquardt, Chen, & Bi, 2019), and others.

2. Specific protein degradation in biological cells

The ensemble of proteins in living cells (i.e., the proteome) is under constant transformation, subject to both internal and external stimuli. During the different phases of the live cycle of a cell, proteins need to be produced at specific times and at specific locations in the appropriate amounts, and eliminated in a controlled way during cell death (apoptosis) (Thiede & Rudel, 2004). Likewise, each cell has to react and adapt the composition of its proteome as a function of external stimuli: availability of nutriments, changes of physico-chemical parameters (including temperature, humidity, pH, salinity, pressure and radiation) (Kültz, 2003), and the presence and interaction with other organisms, including symbionts, parasites, bacteria and viruses (Jean Beltran, Federspiel, Sheng, & Cristea, 2017). In multi-cellular organisms, cellular differentiation and organismal development (Rigbolt Kristoffer et al., 2011), as well as inter-cellular communication processes (Beck et al., 2011), equally require specific modifications of the proteomes.

The production of specific proteins in appropriate amounts and locations, on the one hand, is a highly coordinated and regulated process at the level of genome transcription and translation (Platt, 1986). On the other hand, a mechanism to eliminate proteins in a controlled way needs to exist for several reasons, including the following: specific proteins may be expressed in a faulty way, are being (partially) degraded or mis/unfolding over time, or are simply no longer needed at a specific moment for the cell's activities (Olivares, Baker, & Sauer, 2016). The presence and accumulation of such faulty and potentially dangerous proteins has to be avoided at any costs since it could lead to cellular dysfunction, and to neurodegenerative diseases in higher organisms (Rubinsztein, 2006). This process needs to be both efficient and highly specific, i.e., targeting only faulty proteins and sparing well-folded and functional proteins required for the cellular activities.

In *eukaryotic organisms*, the specificity of protein degradation is assured by a complex network of protagonists, with the ubiquitin pathway being the central element (Ciechanover, 2005). Indeed, the regulated attachment of poly-ubiquitin tags assures the specificity of the process of protein degradation by directing labeled proteins toward the proteasome complex, a 2.6 MDa complex constituted of $\approx$50 proteins (Tomko & Hochstrasser, 2013). In the proteasome complex, cleavage of proteins into smaller pieces takes place in the catalytic chamber of the 20S core particle (CP) (Fig. 1, left). Access to this chamber is controlled by a regulatory particle (RP), and ATP-driven molecular machine that unfolds protein substrates and directs them into the proteolytic chamber. *Bacteria* possess a simplified version of a protein degradation network, with the ClpXP particle replacing the eukaryotic proteasome complex (Gottesman, Roche, Zhou, & Sauer, 1998). In analogy to the eukaryotic system, ClpXP is constituted by a gate-keeping part (ClpP) and a proteolytic part (ClpX). However, a ubiquitin mechanism is not known in bacteria and proteins to be specifically degraded by ClpXP (and ClpAP) are in general labeled by the attachment of small molecular tags (Gottesman et al., 1998; Karzai, Roche, & Sauer, 2000). Organisms from the third domain of Life, *archaea*, have a protein-degradation system closer to the eukaryotic proteasome, but involving less co-factors and an architecture of simpler topology (Majumder et al., 2019) (Fig. 1, right). The hetero-oligomeric eukaryotic RPT (regulatory particle triphosphatase) particle is in particular replaced by a homo-hexameric PAN (proteasome-activating nucleotidase) particle. Recent research suggests that archaea have an analogous pathway to the eukaryotic ubiquitination system called sampylation (Humbard et al., 2010; Maupin-Furlow, 2012). Bacterial ssrA tags, on the other hand, have been shown to work for protein degradation by archaeal PAN-proteasome particles (Navon & Goldberg, 2001). For more details on the general mechanisms of protein degradation and

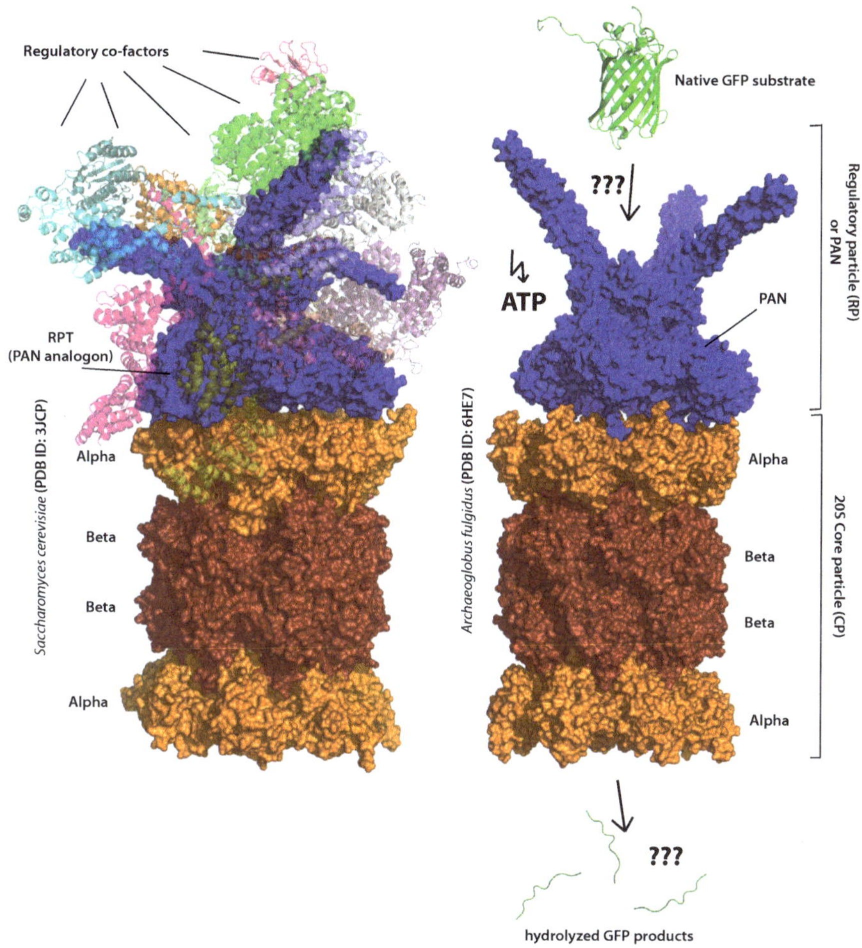

Fig. 1 Schematic comparison of a eukaryotic and an archaeal proteasome. *Left*: Yeast 26S complex, generated from the PDB ID 3JCP (Luan et al., 2016), using PyMOL. The 20S core particle (CP), consisting of alpha and beta-subunit rings, is depicted in surface mode. Regulatory co-factors are depicted in cartoon mode. *Right*: 20S proteasome/PAN complex from *Archaeoglobus fulgidus*, generated from PDB ID 6HE7 (Majumder et al., 2019) using PyMOL. A native GFP substrate (including an ssrA tag), as well as putative hydrolyses products are schematically represented. The 20S core particle (CP), consisting of alpha and beta-subunit rings, is depicted in surface mode.

the comparison between the three domains of Life, the readers are referred to comprehensive reviews (Maupin-Furlow, 2012; Sauer & Baker, 2011).

Exciting new insights into the atomic details of the protein degradation mechanism by eukaryotic and archaeal proteasome particles has been obtained recently by Cryo-EM from substrate-stalled complexes (de la Peña, Goodall, Gates, Lander, & Martin, 2018; Dong et al., 2019; Majumder et al., 2019). Biophysical studies, including single-molecule techniques have provided insight into the mechanics and dynamics of protein unfolding on bacterial ClpXP systems (Aubin-Tam, Olivares, Sauer, Baker, & Lang, 2011; Nager, Baker, & Sauer, 2011) while NMR studies on the AAA+ unfoldase VAT have allowed residue-specific information during the active unfolding process in solution (Augustyniak & Kay, 2018; Huang et al., 2016). A more comprehensive overview on the state-of-the art of structural and biophysical studies on protein degradation systems can be found in recent reviews (Olivares et al., 2016; Sauer & Baker, 2011).

While Cryo-EM and crystallography provide unique and unequaled insight into atomic details of the degradation process, both techniques are limited to stalled intermediate states and cannot quantify the substrate populations and the different time-scales of the active unfolding and degradation processes in solution. On the other hand, single-molecule techniques provide exciting information on the time-scales, probability and geometry of individual unfolding events. However, these biophysical approaches lack the average information of dynamic ensembles in solution as well as the overall structure of the unfolding substrates. Likewise, NMR reports residue-resolved information on structural conformations and interactions of flexible regions involved in protein degradation, but has difficulties to capture the large-scale conformational changes of the mega-Dalton assemblies during the process.

While SANS has a number of shortcomings, in particular a conformational and temporal averaging processes and the unattainability of atomic resolution (see chapter "Time-resolved small-angle neutron scattering (TR-SANS) for structural biology of dynamic systems: Principles, recent developments, and practical guidelines" by Martel and Gabel in volume 677 of Methods in Enzymology), it also has a number of strengths that make it complementary to Cryo-EM, crystallography, NMR and single-molecule techniques for studying the molecular details of protein degradation. In particular, SANS captures structural information on all relevant length-scales involved in protein-degradation processes, in particular short distances within tenths of kilo-Dalton substrates, but also longest distances in (multi) mega-Dalton proteasome complexes. Moreover, SANS encodes structural information from flexible, unfolded regions, and is extremely sensitive to the occurrence of protein aggregation. In addition, SANS averages over the ensemble of solubilized particles at a given instant and can thus access time-resolved information of different substrate populations and proteasome states at different moments during the active process in solution. Finally, deuterium labeling (Haertlein et al., 2016) (see chapter "Deuteration for biological SANS: Case studies, success and challenges in chemistry and biology" by Duff et al. in volume 677 of Methods in Enzymology) allows to distinguish between different partners involved in the reaction (in particular between substrate and the proteasome complex and even its individual constituents). We illustrate the unique strengths and the state-of-the-art of TR-SANS by the example of GFPssrA degradation by an archaeal PAN-proteasome complex in Sections 3–5.

3. A concrete example of bio-macromolecular TR-SANS: Insight into structural dynamics of substrate processing by the archaeal PAN-proteasome system

Several important structural and kinetic parameters of the substrate-processing proteasome system are difficult to access by the above-mentioned techniques. They include the following:

- The global (including long-range) *conformational motions of the regulatory particle* (RP) and the catalytic particle (CP) during the active unfolding process of a substrate
- The *kinetics* of these conformational motions as a function of substrate: proteasome:ATP stoichiometry and physico-chemical parameters in vitro (temperature, pH, buffer composition, etc.)
- The different *populations of substrate states* (native, in the process of unfolding and/or degradation, hydrolyzed products, aggregates...) at a given moment during the reaction, as a function of substrate: proteasome:ATP stoichiometry and physico-chemical parameters in vitro
- The time-evolution of these substrate states and the correlation with ATP consumption

As indicated in the accompanying book chapter, TR-SANS studies of bio-macromolecular systems require milligram amounts of material in a monodisperse and stable state, as well as the possibility of a triggering mechanism (see chapter "Time-resolved small-angle neutron scattering (TR-SANS) for structural biology of dynamic systems: Principles, recent developments, and practical guidelines" by Martel and Gabel in volume 677 of Methods in Enzymology). Due to challenges in their in vitro reconstitution process (Tomko et al., 2015) and transient assembly state (Bedford, Paine, Sheppard, Mayer, & Roelofs, 2010) eukaryotic proteasome systems are not very suited for such studies. *Archaeal* proteasomes, on the other hand, are in general closer to eukaryotic systems than the analogs existing in bacteria, regarding both the topology and architecture of the respective macromolecular assemblies (Kish-Trier & Hill, 2013), see also Fig. 1. Moreover, their ease of purification (Kocabıyık, Özdemir, Zwickl, & Özdoğan, 2010; Reuter Christopher & Maupin-Furlow Julie, 2004) and robustness (Franzetti, Schoehn, Garcia, Ruigrok, & Zaccai, 2002; Leuschner & Antranikian, 1995) has made them very popular "workhorses" for structural and biophysical studies (Benaroudj & Goldberg, 2000). Finally, since archaea are often thermophilic or hyper-thermophilic organisms (Stetter, 1996), *temperature* can readily be used to *trigger and regulate enzymatic activity* of proteins from such systems. Here, we present and discuss such a temperature-triggered TR-SANS study, which uses an ssrA-tagged green fluorescent protein (GFP) as a substrate, and the PAN-20S proteasome system from the deep sea thermophilic archaeon *Methanocaldococcus jannaschii*. The specific spectrophotometric properties of GFP allowed to record fluorescent data in parallel to the neutron scattering in situ, which report on the folding state of this protein substrate.

4. Sample conditions, instrumental setup and data reduction

An overview of sample conditions is provided in Table 1, full details on all sample preparations, deuteration and purification protocols are presented in the original publications (Ibrahim et al., 2017; Mahieu, Covès, et al., 2020). Three different major sample conditions were studied ("d" = deuterated, "h" = hydrogenated): dGFP + hPAN, hGFP + dPAN and dGFP + hPAN:h20S. All samples were recorded in 42% D_2O buffers, where the signals of the hydrogenated (h) proteins are invisible (i.e., "matched"). When ATP was present in the buffers, the concentrations were chosen relatively high (100 mM) in order to assure the perpetuation of the unfolding process during the neutron experiments which lasted up to 1 h (see below). The use of ATP-regeneration systems (Kim & Swartz, 1999) was not considered since it would have complicated the setup of the SANS experiment and the contrast-matching conditions. Control samples with the non-hydrolysable ATP analog ATP-γS and in the absence of nucleotides were also recorded, but are not presented here (the interested reader is referred to the original publications cited above). Proteins were fully deuterated (i.e., "per-deuterated") at the deuteration lab (DLAB) facility at the Institut Laue-Langevin (ILL), following published protocols (Haertlein et al., 2016) in order to obtain a maximum contrast and signal strength.

Table 1 Overview of SANS sample preparations.

Component (h = hydrogenated, d = deuterated)	Volume (stock concentration)	Final concentration	Buffer	Temperature	Figures
d-PAN	90 μL (10 mg/mL)	10 μM (3 mg/mL)	H_2O	60 °C	Fig. 2
h-GFPssrA	72 μL (20 mg/mL)	173.5 μM (4.8 mg/mL)	H_2O		
ATP	66 μL (455 mM)	100 mM	D_2O		
$MgCl_2$	12 μL (2500 mM)	100 mM	H_2O		
Buffer	60 μL	–	D_2O		
Total	300 μL	–	42% D_2O		
d-GFPssrA	12.4 μL (50 mg/mL)	72 μM (2 mg/mL)	42% D_2O	55 °C	Figs. 3–5
h-PAN	93 μL (20 mg/mL)	20 μM (6 mg/mL)	42% D_2O		
h-20S	111.8 μL (45 mg/mL)	20 μM (16 mg/mL)	42% D_2O		
ATP	68 μL (455 mM)	100 mM	42% D_2O		
$MgCl_2$	24.8 μL (2.5 M)	200 mM	42% D_2O		
Total	310 μL	–	42% D_2O		

The table presents the detailed composition of each SANS sample presented in this book chapter, including the temperature and the figures where the data are shown.

All samples (V ≈ 300 μL) were measured in Hellma® QS quartz cells with a path length of 1 mm in a thermostated sample holder at either 55 °C or 60 °C on the D22 SANS instrument at ILL (Grenoble, France), at a neutron wavelength $\lambda = 6$ Å (±10%). Detector/collimator setups were either chosen as 2 m/2 m, 2 m/2.8 m, 5.6 m/5.6 m or 8 m/8 m, adapted to the expected size and dimensions of the visible (i.e., non-matched) molecules and structures in the samples, covering a maximum exploitable q-range from 0.008 to 0.35 Å^{-1}.

All samples and buffers were kept on ice (≈4 °C) before the neutron experiments and transferred to the thermostated (55–60 °C) sample holder (Fig. 4C) by hand. After an initial ≈15 s dead time, required for the experimenters to leave the neutron hutch, each sample was exposed to neutrons through a rectangular hole (≈1.5 × 2.5 cm^2) in the upper part of the sample holder in a series of individual 30 s frames during 45–60 min. In parallel, samples were illuminated (excited) by UV light (Ocean Optics 65000pro), filtered at a wavelength of 400 nm, and the fluorescence (emission) signal measured at 509 nm, through a second, circular hole (diameter ≈ 0.5 cm), in the lower part of sample holder (Fig. 4C).

The two-dimensional SANS data of samples and buffers were corrected for detector efficiency and background noise, and normalized against the intensity of the direct neutron beam. Both sample and buffer data were averaged over concentric rings around the direct beam center and the resulting one-dimensional curves, corrected for transmission, subtracted from each other to yield the final 1D curves I(q). Radii of gyration (R_G) and forward scattered intensities I(0) were extracted by Guinier fits (Guinier, 1939). Low resolution shapes were obtained by the program DAMMIF (Franke & Svergun, 2009), pair distance distribution functions p(r) determined by GNOM (Konarev, Volkov, Sokolova, Koch, & Svergun, 2003), atomic models were scored against the SANS data using the program CRYSON (Svergun et al., 1998), and populations of different species were fitted with OLIGOMER (Konarev et al., 2003).

5. Experimental TR-SANS results and mechanistic model of protein degradation

5.1 The PAN-GFP system

The SANS samples containing a reaction mixture of PAN, GFPssrA and ATP in 42% D_2O allow to focus either on the conformational changes of PAN during active substrate unfolding (d-PAN and h-GFP) or on the evolution of GFP substrate states over time (h-PAN and d-GFP), in the absence of the proteolytic 20S core particle.

As illustrated in Fig. 2 deuterated (d-) *PAN undergoes a reversible conformational contraction* during the active unfolding process in solution, the signal from all h-GFPssrA species and states being matched in 42% D_2O (Ibrahim et al., 2017). Indeed, the radius of gyration decreases rapidly by ≈10% from an initial value of ≈66 to ≈60 Å after about 6–7 min of the reaction, before retrieving the initial R_G in a slower relaxation process toward

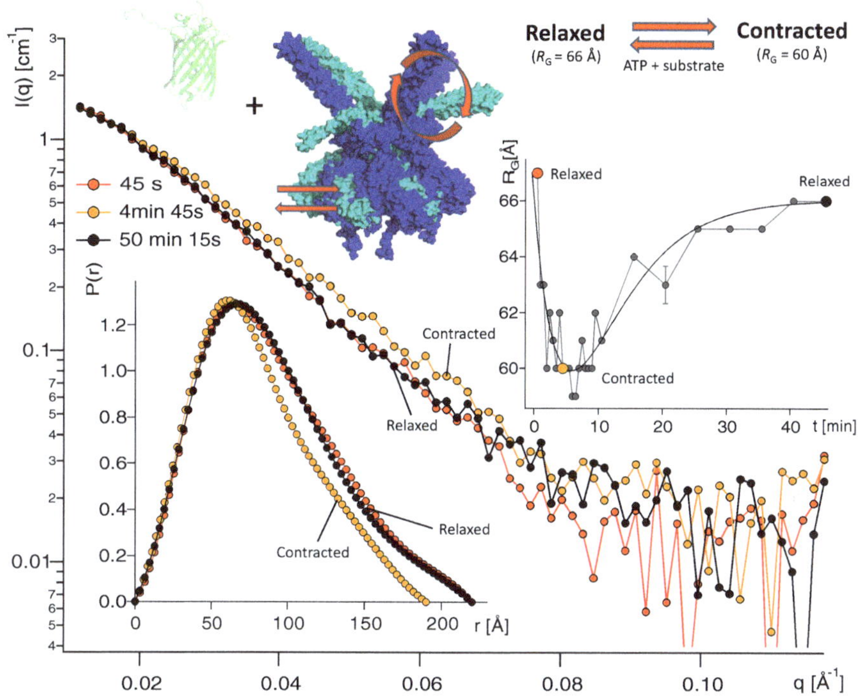

Fig. 2 Structural parameters of PAN during SANS experiment. SANS curves, R_G (*inset right*) and p(r) (*inset bottom*) at 45 s, 4 min 45 s and 50 min 15 s of the reaction in solution, in the presence of ATP and GFP at 60 °C in 42% D_2O in the presence of GFP substrates. *Top*: Schematic illustration of PAN conformational changes during the unfolding process.

the end of the reaction after ≈40 min. In parallel, the p(r) distribution function undergoes a reversible change from an initial maximum particle dimension (D_{max}) of ≈220 to ≈190 Å before retrieving the initial value at the end of the reaction. Together, these data suggest that PAN particles undergo a minor, reversible conformational change from an initial "relaxed" state to a "contracted" state during the active unfolding process, before retrieving the relaxed state at the end of the reaction. It is important to stress that the SANS curves reflect both an ensemble average of all PAN particles in solution, and a time average during each individual 30 s exposure frame. In addition, it is likely that the transfer from 4 to 60 °C takes several seconds to affect the entire sample volume (≈300 μL) homogeneously, thus potentially "delaying" the triggering mechanism by a couple of seconds. The averaging processes and the dephasing of the triggering mechanism imply that the conformational changes of *individual* PAN molecules may occur on different time-scales than the ensemble, in particular on shorter time-scales as suggested by single-molecule experiments on bacterial systems (Aubin-Tam et al., 2011; Nager et al., 2011).

When the PAN signal is matched (h-PAN in 42% D_2O), the SANS curves can be exclusively attributed to the d-GFPssrA substrate states. At the very beginning of the reaction ($t < 1$ min), the SANS curves are in very good agreement with those calculated from the atomic structure of a natively folded GFP protein (Fig. 3A, bottom data points and continuous fit). However, within the first ≈5 min of the reaction, the SANS curves evolve very strongly: the most prominent feature is an increase of the intensity at very small angles by two orders of magnitude (note the logarithmic scale), indicating a *strong and rapid aggregation process of GFP*. Concomitantly, the scattered intensity decreases in an intermediate q-range (≈0.05... 0.18 $Å^{-1}$). Both phenomena, as well as the simultaneous increase of R_G (Fig. 4A), can be explained by the rapid buildup of very extended, filamentous, and possibly branched, aggregates (Fig. 3A, top, low-resolution shapes determined by DAMMIF (Franke & Svergun, 2009)). Importantly, due to the presence of a beam stop, q_{min} was limited to ≈0.007 $Å^{-1}$ at a detector distance of 8 m, corresponding to a maximum dimension of ≈450 Å that can be reliably determined from the data (see chapter "Time-resolved small-angle neutron scattering (TR-SANS) for structural biology of dynamic systems: Principles, recent developments, and practical guidelines" by Martel and Gabel in volume 677 of Methods in Enzymology) (Koch, Vachette, & Svergun, 2003). Indeed, the back-fit with GNOM (Dmitri I. Svergun, 1992) against the experimental data at 50 min, suggests a maximum dimension >1000 Å, represented by the extended aggregate generated by DAMMIF (and using GNOM input), displayed for 50 min on the top of Fig. 3A. Moreover, the aggregates are visible by eye (Fig. 4D, middle). The shapes of the aggregates are not clump-like, bulky structures, but rather of filamentous nature with a diameter of the "fiber" corresponding to the dimension of a compact, native-like GFP molecule. This specific type of aggregation could therefore be due to unfolded GFP substrates being released by PAN in a form that, while partially refolding (i.e., a "molten globule"), exposes hydrophobic patches that lead to a first phase of oligomerization in the form of a "pearl necklace."

Overall the results from the apo-system (PAN + GFP) reveal that PAN particles undergo reversible conformational contractions during the active unfolding process, which are in phase with the characteristic time of the growth of GFP aggregates, as well as the ATP consumption and the loss

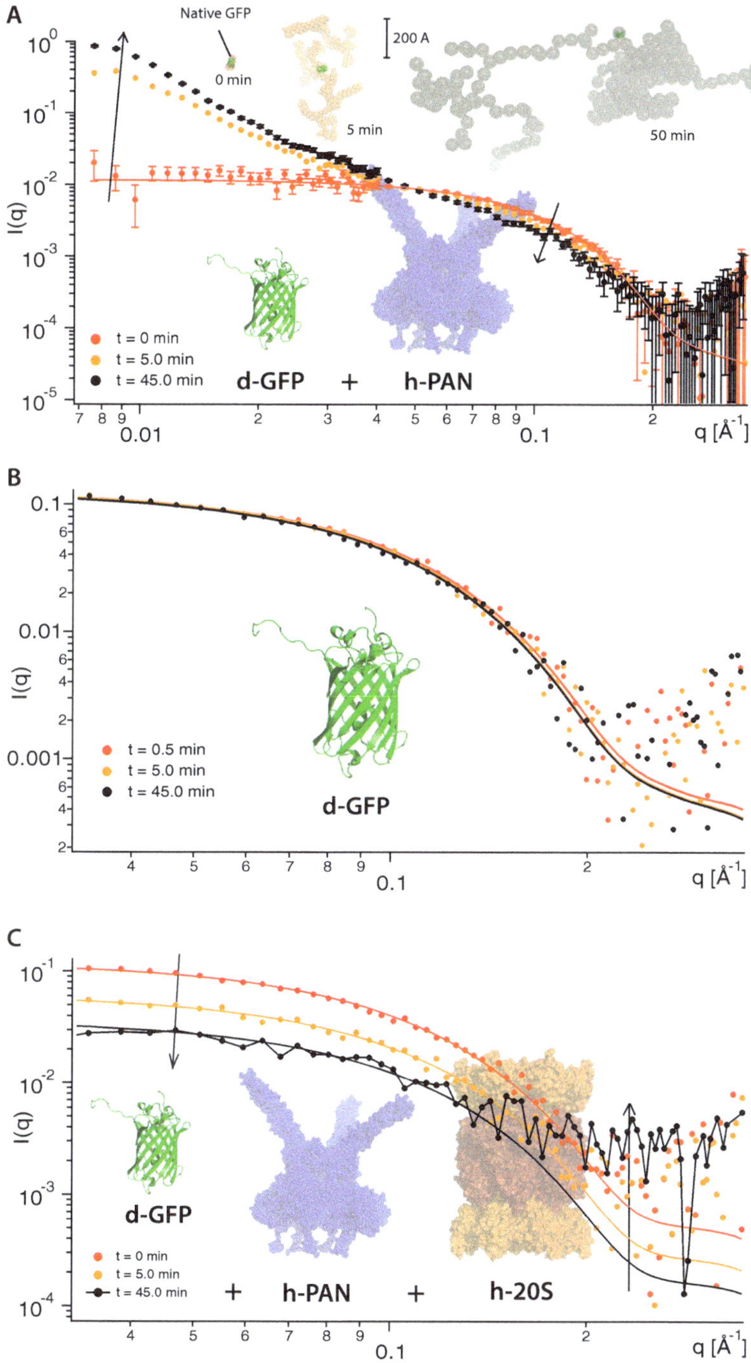

Fig. 3 SANS curves of several GFP systems in 42% D_2O and 100 mM ATP at 55°C. (A) d-GFP + h-PAN. In the presence of PAN (invisible in 42% D_2O) GFP oligomerizes very rapidly and strongly, building extended, filamentous aggregates. (B) Isolated d-GFP. (C) d-GFP + h-PAN + h-20S. Continuous lines in all figures represent fits with the native GFP atomic model. All samples were measured in the presence of 100 mM ATP.

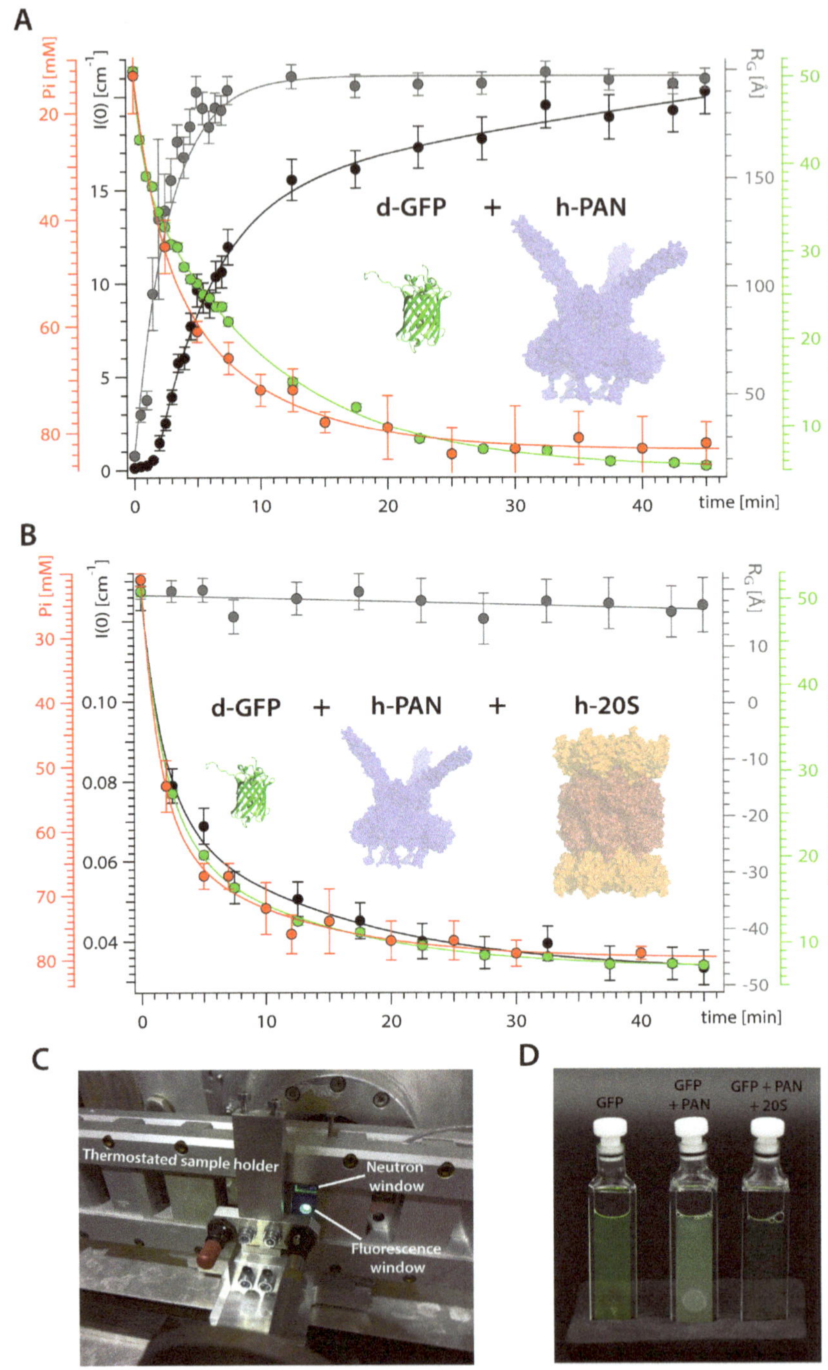

Fig. 4 SANS model-free parameters I(0) and R_G, online fluorescence at 509 nm, and ATP consumption in 42% D_2O and 100 mM ATP at 55 °C. (A) d-GFP + h-PAN. (B) d-GFP + h-PAN + h-20S. The ATP consumption in both cases was monitored offline by detecting inorganic phosphate (Pi) using a colorimetric assay (Mahieu, Covès, et al., 2020), under identical conditions as the SANS experiments. (C) Experimental setup of online fluorescence (lower, circular hole), simultaneously recorded to the SANS data (upper, rectangular hole). (D) Macroscopic sample states after ≈1 h of SANS experiments.

of GFP fluorescence (Fig. 4A). The quick and strong build-up of these potentially harmful aggregates by isolated PAN in vitro suggest the existence of a regulatory fail-safe mechanism in vivo that takes the substrates into charge immediately after the unfolding process and thus prevents the release of aggregation-prone forms. An essential element of this mechanism is indeed the 20S proteolytic core particle. Its presence in solution changes the GFP degradation process and the substrate states substantially, as presented in detail in the next paragraph. Finally, it is important to stress that the elevated sample temperature during the SANS experiment (55 °C) is *not* at the origin of GFP substrate aggregation, as demonstrated by a control experiment with isolated GFP (but in the presence of 100 mM ATP) (Figs. 3B and 4D, left).

5.2 The PAN-20S-GFP system

The presence of the 20S core particle in the reaction solution (h-PAN + h-20S + d-GFP in 42% D_2O, 100 mM ATP) induces several qualitative changes in the structural parameters and the time-evolution of the GFP substrate states. Most importantly, the *formation of GFP aggregates is completely suppressed*, with neither an increase at I(0) intensities nor an increase in the slope (i.e., R_G) (Figs. 3C and 4B). Indeed, one of the major strengths of small-angle scattering is its high sensitivity toward the presence of larger particles in solution, due to the fact that I(0) is proportional to the volume squared of solubilized particles (see chapter "Time-resolved small-angle neutron scattering (TR-SANS) for structural biology of dynamic systems: Principles, recent developments, and practical guidelines" by Martel and Gabel in volume 677 of Methods in Enzymology). Thus, even a minor presence (e.g., 0.1% of the total population) in the form of a putative "100mer" GFP aggregate would have resulted in a readily measurable modification of the SANS curves at low q-values, and in particular a 10% increase of I(0). The absence of any such modification, for all SANS frames during the entire length of the experiment (Fig. 3C), indicates that the presence of 20S, at the given stoichiometric ratio of the reaction solution (Table 1), is sufficient to prevent the nascence of aggregates at a very early stage. It is important to stress that this capacity and sensitivity to validate *and* quantify the absence of substrate aggregates, in the presence of a high-molecular weight complex (PAN:20S ≈ 1 MDa), is a unique strength of SANS, in combination with deuteration and contrast variation. Such an approach would indeed be impossible by SAXS (Mahieu, Ibrahim, et al., 2020). Finally, the absence of aggregation is confirmed by an inspection of the macroscopic sample after the reaction, which displays a clear, transparent solution (Fig. 4D).

The SANS data of d-GFP in the presence of h-PAN and h-20S (both invisible in 42% D_2O) provide also information on the *time-evolution of native GFP states, degradation products*, and allow to assess the *presence of a putative intermediate unfolded substrate state* (de la Peña et al., 2018; Dong et al., 2019). Indeed, the I(0) intensities of the SANS curves, rather than increasing as in the case of aggregation, diminish continuously while the radii of gyration remain relatively constant (Fig. 4B), and the intensities at higher angles increase (Fig. 3C, arrows). The loss of I(0) intensities is a measure for the progressive diminution of the initial pool of natively folded GFP substrates, while the relative stability of R_G indicates that the native state remains the largest (in terms of extension and molecular weight), and thus dominant, species in solution during the entire reaction. In parallel, the concomitant increase of scattering intensity at higher angles ($q > 0.14\,\text{Å}^{-1}$) with respect to the curve of native GFP suggests the progressive appearance of smaller GFP reaction products (Fig. 3C, right arrow). The loss of natively folded, fluorescent population of GFP is also evident at a macroscopic level with a less fluorescent sample after the SANS experiments, as compared to the reference sample of isolated GFP at the same (initial) concentration (Fig. 4D). Finally, the rate of diminution of the native GFP state (measured via the I(0) intensity) is perfectly in phase with the loss of fluorescence and the ATP consumption (Fig. 4B).

Based on these observations, the individual SANS frames were fitted with the program oligomer (Konarev et al., 2003), using *three different substrate populations* (Fig. 5): (i) natively folded GFP, including an ssrA tag, (ii) a putative, partially unfolded substrate state, constructed based on Cryo-EM and single molecule data (de la Peña et al., 2018; Nager et al., 2011), and (iii) an average decamer product, representative for the distribution of hydrolyzed products detected by mass-spectrometry measurements (Mahieu, Covès, et al., 2020). The evolution of the three populations reveals a continuous decrease of the native GFP substrate, and the concomitant increase of GFP hydrolysis products. The third species fitted, a putative, partially unfolded GFP intermediate, was generally very weakly present (population $< 5\%$) at any time during the reaction. Together, these experiments suggest that the putative intermediate unfolding states of the substrate, revealed by Cryo-EM installed eukaryotic proteasome complexes (de la Peña et al., 2018; Dong et al., 2019), are very transient and weakly populated at any time during the reaction process in solution.

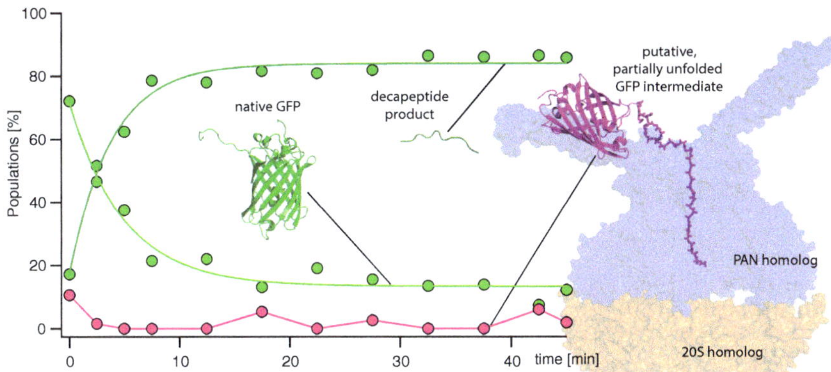

Fig. 5 Oligomer fits with native GFP, a representative decapeptide product and a putative, partially unfolded GFP intermediate. The decapeptide product was built from a GFP sequence, based on average values of reaction products obtained by mass spectrometry (Mahieu, Covès, et al., 2020). The partially unfolded intermediate was created based on Cryo-EM (de la Peña et al., 2018) and single molecule (Nager et al., 2011) data.

5.3 A mechanistic model of protein degradation based on the TR-SANS results

The ensemble of TR-SANS and online fluorescence (Fig. 4A and B) data from the apo-(PAN + GFP) and holo-(PAN + 20S + GFP) systems reviewed here, and presented in more detail elsewhere (Ibrahim et al., 2017; Mahieu, Covès, et al., 2020; Mahieu, Ibrahim, et al., 2020), can be integrated to propose a mechanistic model of protein substrate degradation by the archaeal PAN-20S proteasome system (Fig. 6):

(1) In the reaction mixture containing GFP, PAN, 20S and 100 mM ATP, PAN and 20S can initially be either present as a stable or as a transient complex, i.e., associate and dissociate at a certain rate (deciding on whether the complex is stable or transient, i.e., estimating its k_D, would be possible by using SANS data where the GFP signal is masked. However, due to the fact that the GFP used here was perdeuterated to maximize its signal, it cannot be masked, even in a 100% D_2O buffer (Mahieu & Gabel, 2018)).

(2) The native GFP substrates can interact, at a certain rate, with the assembled PAN-20S in solution, and the unfolding process may be initiated during these, potentially transient, interactions. Should unsuccessful unfolding attempts occur (Aubin-Tam et al., 2011; Cordova et al., 2014) they must be of such a nature that the released GFP substrate

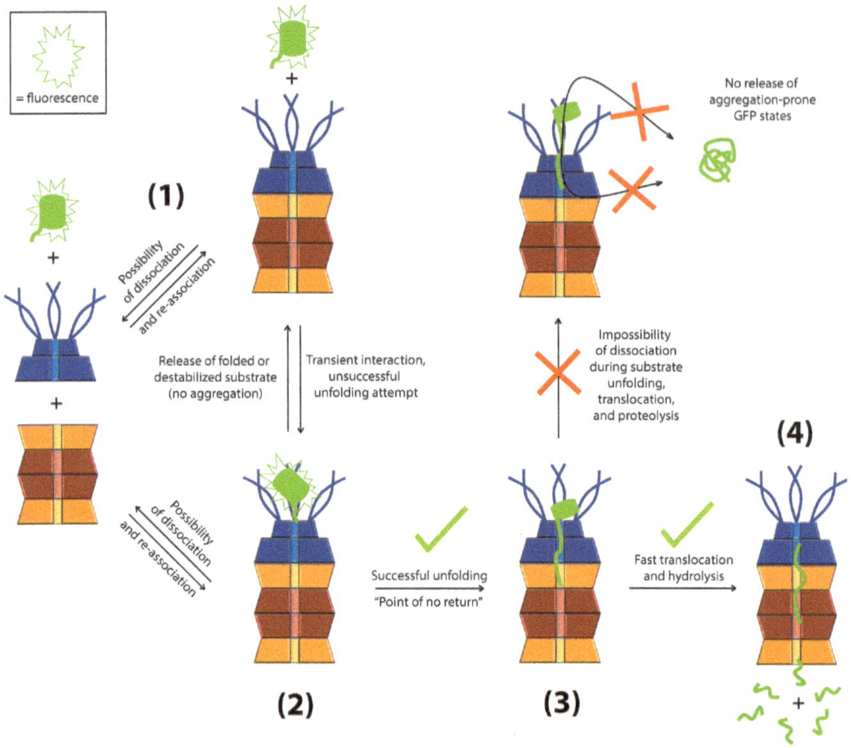

Fig. 6 Schematic representation of a mechanistic substrate processing model by the PAN-20S complex in solution, based on TR-SANS and fluorescence data. The different steps (1)–(4) are explained in the main text. GFP, PAN and 20S are represented schematically in the same color codes (or gray-scale) as in the previous figures.

is not conformationally altered (e.g., partially unfolded or incorrectly refolded) in such a way that it is in an aggregation-prone state. The very rapid buildup of GFP aggregates in the presence of PAN (but in the absence of 20S, Fig. 3A) suggests that GFP only interacts with the assembled PAN-20S complex, or, if it interacts with a dissociated PAN molecule, that 20S is recruited very quickly once the unfolding process is initiated. More accurate information on this step of the mechanism could be obtained by altering the stoichiometry GFP:PAN:20S (Table 1) systematically, e.g., by lowering the 20S concentration progressively in a series of TR-SANS experiments, while keeping the GFP and PAN concentrations constant.

(3) Once an unfolding process is successfully initiated and the GFP unfolding has progressed beyond a certain stage, a "point of no return" is reached and the substrate can no longer be released by PAN. Rather, it is being transferred through the central PAN channel toward the 20S catalytic particle, as suggested by Cryo-EM snapshots from substrate-stalled eukaryotic proteasome complexes (de la Peña et al., 2018; Dong et al., 2019). During this substrate transfer phase, the absence of GFP aggregates in the SANS data (Fig. 3C) indicate that 20S must be tightly bound to PAN in solution. It is conceivable that the extended substrate polypeptide chain, spanning from the PAN central channel to the 20S catalytic chamber, acts as a "tether" and assures (or reinforces) the stability of the PAN:20S complex during the duration of the unfolding and hydrolysis reactions.

(4) Once the substrate polypeptide chain reaches the catalytic chamber of 20S, it is hydrolyzed into polypeptide chains of variable length (Mahieu, Covès, et al., 2020), which are subsequently released and build up progressively in solution (Fig. 5), at a same rate as ATP consumption (Fig. 4B).

6. Conclusions and outlook

The PAN:20S:GFP example presented and discussed in this book chapter illustrates the state-of-the-art of extracting structural information from a dynamic and complex bio-macromolecular system in solution by TR-SANS, and its complementarity with atomic resolution techniques. It demonstrates that relevant structural information, beyond model-free parameters such as R_G and I(0), can be extracted for relatively small macromolecules (< 30 kDa) on the sub-minute time-scale, if measured at the high flux SANS instruments, and if the molecule has sufficient contrast (e.g., per-deuterated GFP in 42% D_2O). If the exposure times are further reduced (to a couple of seconds), the spread and the noise level of the SANS data will increase and the precision of the structural parameters that are extracted will diminish, as illustrated by the evolution of the d-GFP data (2 mg/mL) shown in Fig. 7.

While the R_G values remain consistent and stable, the associated errors increase. Likewise, while a fit against an atomic model yields a satisfactory agreement, even at 3 s exposure times, the capacity of the SANS data to discriminate between minor conformational changes in a macromolecule, or the possibility to apply fits including populations of multiple, different species (Fig. 5), will decrease. Therefore, a reasonable minimum exposure time at present-day high-flux SANS instruments for the extraction of reliable

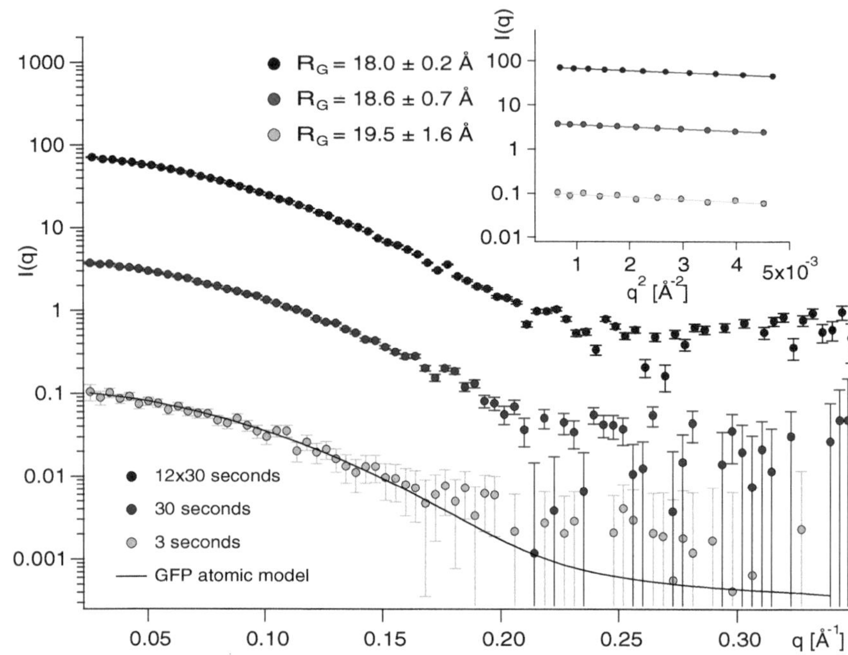

Fig. 7 Quality of d-GFP SANS data in a 42% D_2O buffer as a function of neutron exposure time. SANS curves for 3 s, 30 s and $12 \times 30 = 360$ s exposure times (frames), and respective Guinier-fits.

structural information from a protein the size of GFP seems to be of the order of 10–30 s. A better time-resolution can be obtained when repeating the experiment several times and averaging the information, if contrast is higher (e.g., perdeuterated protein in 100% H_2O or hydrogenated protein in 100% D_2O), and/or if the biological molecule of interest is significantly bigger (>100 kDa). Under these circumstances, a time resolution of 1 s seems realistic.

Even though the PAN:20S:GFP system was chosen here to illustrate the TR-SANS technique for biomacromolecular systems, the approach presented and discussed is applicable to a broad range of biological processes. Indeed, any biological process that aims at following the populations and conformations of a substrate and/or a macromolecular machine can be studied by this approach. However, certain conditions need to be fulfilled: (i) A triggering mechanism must exist, (ii) the possibility to label (deuterated) parts of the system and reconstitute the mixture in solution, (iii) the possibility to mix and reconstitute the essential partners in vitro necessary for the correct functioning of the complex biological process, and (iv) the macromolecules to be studied must be big enough (>5–10 kDa) and the expected structural changes (conformational or association) must be substantial. Biological systems/processes that have been recently studied and fulfill these requirements include the following: several remodeling processes of the proteome (Guo & Gross, 2014) and genome (Lans et al., 2012), chaperone and disaggregase activity (Rosenzweig et al., 2019), and cytoskeleton remodeling (Marquardt et al., 2019).

Finally, there are a number of recent instrumental and sample environment developments that have the potential to improve the quality of TR-SANS data in the near future, which are presented and discussed in the accompanying chapter of this book edition (see chapter "Time-resolved small-angle neutron scattering (TR-SANS) for structural biology of dynamic systems: Principles, recent developments, and practical guidelines" by Martel and Gabel in volume 677 of Methods in Enzymology).

Acknowledgments

Financial support from the "Agence Nationale de la Recherche" (Grant "PROTstretch" ANR-15-CE11-0026) is acknowledged. The author would like to thank Dr. Anne Martel (ILL) for reprocessing the 3 s SANS dataset shown in Fig. 7.

References

Aubin-Tam, M.-E., Olivares, A. O., Sauer, R. T., Baker, T. A., & Lang, M. J. (2011). Single-molecule protein unfolding and translocation by an ATP-fueled proteolytic machine. *Cell*, *145*(2), 257–267. https://doi.org/10.1016/j.cell.2011.03.036.

Augustyniak, R., & Kay, L. E. (2018). Cotranslocational processing of the protein substrate calmodulin by an AAA+ unfoldase occurs via unfolding and refolding intermediates. *Proceedings of the National Academy of Sciences of the United States of America*, *115*(21), E4786–E4795. https://doi.org/10.1073/pnas.1721811115.

Baek, M., DiMaio, F., Anishchenko, I., Dauparas, J., Ovchinnikov, S., Lee Gyu, R., et al. (2021). Accurate prediction of protein structures and interactions using a three-track neural network. *Science*, *373*(6557), 871–876. https://doi.org/10.1126/science.abj8754.

Beck, M., Schmidt, A., Malmstroem, J., Claassen, M., Ori, A., Szymborska, A., et al. (2011). The quantitative proteome of a human cell line. *Molecular Systems Biology*, *7*(1), 549. https://doi.org/10.1038/msb.2011.82.

Bedford, L., Paine, S., Sheppard, P. W., Mayer, R. J., & Roelofs, J. (2010). Assembly, structure, and function of the 26S proteasome. *Trends in Cell Biology*, *20*(7), 391–401. https://doi.org/10.1016/j.tcb.2010.03.007.

Benaroudj, N., & Goldberg, A. L. (2000). PAN, the proteasome-activating nucleotidase from archaebacteria, is a protein-unfolding molecular chaperone. *Nature Cell Biology*, *2*(11), 833–839. Retrieved from https://doi.org/10.1038/35041081.

Cheng, Y. (2015). Single-particle Cryo-EM at crystallographic resolution. *Cell*, *161*(3), 450–457. https://doi.org/10.1016/j.cell.2015.03.049.

Ciechanover, A. (2005). Proteolysis: From the lysosome to ubiquitin and the proteasome. *Nature Reviews Molecular Cell Biology*, *6*(1), 79–87. https://doi.org/10.1038/nrm1552.

Cordova, J. C., Olivares, A. O., Shin, Y., Stinson, B. M., Calmat, S., Schmitz, K. R., et al. (2014). Stochastic but highly coordinated protein unfolding and translocation by the ClpXP proteolytic machine. *Cell*, *158*(3), 647–658. https://doi.org/10.1016/j.cell.2014.05.043.

de la Peña, A. H., Goodall, E. A., Gates, S. N., Lander, G. C., & Martin, A. (2018). Substrate-engaged 26S proteasome structures reveal mechanisms for ATP-hydrolysis–driven translocation. *Science*, *362*(6418), eaav0725. https://doi.org/10.1126/science.aav0725.

Dong, Y., Zhang, S., Wu, Z., Li, X., Wang, W. L., Zhu, Y., et al. (2019). Cryo-EM structures and dynamics of substrate-engaged human 26S proteasome. *Nature*, *565*(7737), 49–55. https://doi.org/10.1038/s41586-018-0736-4.

Franke, D., & Svergun, D. I. (2009). DAMMIF, a program for rapid ab-initio shape determination in small-angle scattering. *Journal of Applied Crystallography*, *42*, 342–346.

Franzetti, B., Schoehn, G., Garcia, D., Ruigrok, R. W. H., & Zaccai, G. (2002). Characterization of the proteasome from the extremely halophilic archaeon *Haloarcula marismortui*. *Archaea (Vancouver, B.C.)*, *1*, 601719. https://doi.org/10.1155/2002/601719.

Gottesman, S., Roche, E., Zhou, Y., & Sauer, R. T. (1998). The ClpXP and ClpAP proteases degrade proteins with carboxy-terminal peptide tails added by the SsrA-tagging-system. *Genes & Development*, *12*(9), 1338–1347. Retrieved from http://genesdev.cshlp.org/content/12/9/1338.abstract.

Guinier, A. (1939). La diffraction des Rayons X aux Tres Faibles Angles: Applications a l'Etude des Phenomenes Ultra-microscopiques. *Annales de Physique*, *12*, 161–236.

Guo, M. S., & Gross, C. A. (2014). Stress-induced remodeling of the bacterial proteome. *Current Biology*, *24*(10), R424–R434. https://doi.org/10.1016/j.cub.2014.03.023.

Haertlein, M., Moulin, M., Devos, J. M., Laux, V., Dunne, O., & Trevor Forsyth, V. (2016). Chapter five—Biomolecular deuteration for neutron structural biology and dynamics. In Z. Kelman (Ed.), *Vol. 566. Methods in enzymology* (pp. 113–157). Academic Press.

Huang, R., Ripstein, Z. A., Augustyniak, R., Lazniewski, M., Ginalski, K., Kay, L. E., et al. (2016). Unfolding the mechanism of the AAA+ unfoldase VAT by a combined cryo-EM, solution NMR study. *Proceedings of the National Academy of Sciences*, *113*(29), E4190. https://doi.org/10.1073/pnas.1603980113.

Humbard, M. A., Miranda, H. V., Lim, J.-M., Krause, D. J., Pritz, J. R., Zhou, G., et al. (2010). Ubiquitin-like small archaeal modifier proteins (SAMPs) in Haloferax volcanii. *Nature*, *463*(7277), 54–60. https://doi.org/10.1038/nature08659.

Ibrahim, Z., Martel, A., Moulin, M., Kim, H. S., Härtlein, M., Franzetti, B., et al. (2017). Time-resolved neutron scattering provides new insight into protein substrate processing by a AAA+ unfoldase. *Scientific Reports*, 7, 40948. https://doi.org/10.1038/srep40948.

Jean Beltran, P. M., Federspiel, J. D., Sheng, X., & Cristea, I. M. (2017). Proteomics and integrative omic approaches for understanding host–pathogen interactions and infectious diseases. *Molecular Systems Biology*, *13*(3), 922. https://doi.org/10.15252/msb.20167062.

Jeffries, C. M., Ilavsky, J., Martel, A., Hinrichs, S., Meyer, A., Pedersen, J. S., et al. (2021). Small-angle X-ray and neutron scattering. *Nature Reviews Methods Primers*, *1*(1), 70. https://doi.org/10.1038/s43586-021-00064-9.

Jumper, J., Evans, R., Pritzel, A., Green, T., Figurnov, M., Ronneberger, O., et al. (2021). Highly accurate protein structure prediction with AlphaFold. *Nature*, *596*(7873), 583–589. https://doi.org/10.1038/s41586-021-03819-2.

Karzai, A. W., Roche, E. D., & Sauer, R. T. (2000). The SsrA–SmpB system for protein tagging, directed degradation and ribosome rescue. *Nature Structural Biology*, 7(6), 449–455. https://doi.org/10.1038/75843.

Kim, D.-M., & Swartz, J. R. (1999). Prolonging cell-free protein synthesis with a novel ATP regeneration system. *Biotechnology and Bioengineering*, *66*(3), 180–188. https://doi.org/10.1002/(SICI)1097-0290(1999)66:3<180::AID-BIT6>3.0.CO;2-S.

Kish-Trier, E., & Hill, C. P. (2013). Structural biology of the proteasome. *Annual Review of Biophysics*, *42*(1), 29–49. https://doi.org/10.1146/annurev-biophys-083012-130417.

Kocabıyık, S., Özdemir, İ., Zwickl, P., & Özdoğan, S. (2010). Molecular cloning and co-expression of Thermoplasma volcanium proteasome subunit genes. *Protein Expression and Purification*, *73*(2), 223–230. https://doi.org/10.1016/j.pep.2010.05.004.

Koch, M. H., Vachette, P., & Svergun, D. I. (2003). Small-angle scattering: A view on the properties, structures and structural changes of biological macromolecules in solution. *Quarterly Reviews of Biophysics*, *36*(2), 147–227.

Konarev, P. V., Volkov, V. V., Sokolova, A. V., Koch, M. H. J., & Svergun, D. I. (2003). PRIMUS: A windows PC-based system for small-angle scattering data analysis. *Journal of Applied Crystallography*, *36*, 1277–1282.

Kültz, D. (2003). Evolution of the cellular stress proteome: From monophyletic origin to ubiquitous function. *Journal of Experimental Biology*, *206*(18), 3119–3124. https://doi.org/10.1242/jeb.00549.

Lans, H., Marteijn, J. A., & Vermeulen, W. (2012). ATP-dependent chromatin remodeling in the DNA-damage response. *Epigenetics & Chromatin*, *5*(1), 4. https://doi.org/10.1186/1756-8935-5-4.

Leuschner, C., & Antranikian, G. (1995). Heat-stable enzymes from extremely thermophilic and hyperthermophilic microorganisms. *World Journal of Microbiology and Biotechnology*, *11*(1), 95–114. https://doi.org/10.1007/BF00339139.

Luan, B., Huang, X., Wu, J., Mei, Z., Wang, Y., Xue, X., et al. (2016). Structure of an endogenous yeast 26S proteasome reveals two major conformational states. *Proceedings of the National Academy of Sciences*, *113*(10), 2642–2647. https://doi.org/10.1073/pnas.1601561113.

Mahieu, E., Covès, J., Krüger, G., Martel, A., Moulin, M., Carl, N., et al. (2020). Observing protein degradation by the PAN-20S proteasome by time-resolved neutron scattering. *Biophysical Journal*, *119*(2), 375–388. https://doi.org/10.1016/j.bpj.2020.06.015.

Mahieu, E., & Gabel, F. (2018). Biological small-angle neutron scattering: Recent results and development. *Acta Crystallographica Section D*, *74*(8), 715–726. https://doi.org/10.1107/S2059798318005016.

Mahieu, E., Ibrahim, Z., Moulin, M., Härtlein, M., Franzetti, B., Martel, A., et al. (2020). The power of SANS, combined with deuteration and contrast variation, for structural studies of functional and dynamic biomacromolecular systems in solution. *EPJ Web of Conferences*, *236*. https://doi.org/10.1051/epjconf/202023603002. Retrieved from.

Majumder, P., Rudack, T., Beck, F., Danev, R., Pfeifer, G., Nagy, I., et al. (2019). Cryo-EM structures of the archaeal PAN-proteasome reveal an around-the-ring ATPase cycle. *Proceedings of the National Academy of Sciences*, *116*(2), 534. https://doi.org/10.1073/pnas.1817752116.

Marquardt, J., Chen, X., & Bi, E. (2019). Architecture, remodeling, and functions of the septin cytoskeleton. *Cytoskeleton*, *76*(1), 7–14. https://doi.org/10.1002/cm.21475.

Maupin-Furlow, J. (2012). Proteasomes and protein conjugation across domains of life. *Nature Reviews Microbiology*, *10*(2), 100–111. https://doi.org/10.1038/nrmicro2696.

Bai, X.-C., McMullan, G., & Scheres, S. H. W. (2015). How cryo-EM is revolutionizing structural biology. *Trends in Biochemical Sciences*, *40*(1), 49–57. https://doi.org/10.1016/j.tibs.2014.10.005.

Nager, A. R., Baker, T. A., & Sauer, R. T. (2011). Stepwise unfolding of a β-barrel protein by the AAA+ ClpXP protease. *Journal of Molecular Biology*, *413*(1), 4–16. https://doi.org/10.1016/j.jmb.2011.07.041.

Navon, A., & Goldberg, A. L. (2001). Proteins are unfolded on the surface of the ATPase ring before transport into the proteasome. *Molecular Cell*, *8*(6), 1339–1349. https://doi.org/10.1016/S1097-2765(01)00407-5.

Ognjenović, J., Grisshammer, R., & Subramaniam, S. (2019). Frontiers in cryo electron microscopy of complex macromolecular assemblies. *Annual Review of Biomedical Engineering*, *21*(1), 395–415. https://doi.org/10.1146/annurev-bioeng-060418-052453.

Olivares, A. O., Baker, T. A., & Sauer, R. T. (2016). Mechanistic insights into bacterial AAA + proteases and protein-remodelling machines. *Nature Reviews. Microbiology*, *14*(1), 33–44. https://doi.org/10.1038/nrmicro.2015.4.

Perrakis, A., & Sixma, T. K. (2021). AI revolutions in biology. *EMBO Reports*, *22*(11), e54046. https://doi.org/10.15252/embr.202154046.

Platt, T. (1986). Transcription termination and the regulation of gene expression. *Annual Review of Biochemistry*, *55*(1), 339–372. https://doi.org/10.1146/annurev.bi.55.070186.002011.

Reuter Christopher, J., & Maupin-Furlow Julie, A. (2004). Analysis of proteasome-dependent proteolysis in Haloferax volcanii cells, using short-lived green fluorescent proteins. *Applied and Environmental Microbiology*, *70*(12), 7530–7538. https://doi.org/10.1128/AEM.70.12.7530-7538.2004.

Rigbolt Kristoffer, T. G., Prokhorova Tatyana, A., Akimov, V., Henningsen, J., Johansen Pia, T., Kratchmarova, I., et al. (2011). System-wide temporal characterization of the proteome and phosphoproteome of human embryonic stem cell differentiation. *Science Signaling*, *4*(164), rs3-rs3. https://doi.org/10.1126/scisignal.2001570.

Rosenzweig, R., Nillegoda, N. B., Mayer, M. P., & Bukau, B. (2019). The Hsp70 chaperone network. *Nature Reviews Molecular Cell Biology*, *20*(11), 665–680. https://doi.org/10.1038/s41580-019-0133-3.

Rubinsztein, D. C. (2006). The roles of intracellular protein-degradation pathways in neurodegeneration. *Nature*, *443*(7113), 780–786. https://doi.org/10.1038/nature05291.

Sauer, R., & Baker, T. (2011). AAA+ proteases: ATP-fueled machines of protein destruction. *Annual Review of Biochemistry*, *80*, 587–612.

Scheres, S. H. W. (2016). Chapter six—Processing of structurally heterogeneous Cryo-EM data in relion. In R. A. Crowther (Ed.), *Vol. 579. Methods in enzymology* (pp. 125–157). Academic Press.

Stetter, K. O. (1996). Hyperthermophilic procaryotes. *FEMS Microbiology Reviews*, *18*(2–3), 149–158. https://doi.org/10.1111/j.1574-6976.1996.tb00233.x.

Svergun, D. I. (1992). Determination of the regularization parameter in indirect-transform methods using perceptual criteria. *Journal of Applied Crystallography*, *25*, 495–503.

Svergun, D. I., Richard, S., Koch, M. H. J., Sayers, Z., Kuprin, S., & Zaccai, G. (1998). Protein hydration in solution: Experimental observation by x-ray and neutron scattering. *Proceedings of the National Academy of Sciences of the United States of America*, *95*(5), 2267–2272.

Thiede, B., & Rudel, T. (2004). Proteome analysis of apoptotic cells. *Mass Spectrometry Reviews*, *23*(5), 333–349. https://doi.org/10.1002/mas.10079.

Tomko, R. J., & Hochstrasser, M. (2013). Molecular architecture and assembly of the eukaryotic proteasome. *Annual Review of Biochemistry*, *82*(1), 415–445. https://doi.org/10.1146/annurev-biochem-060410-150257.

Tomko, R. J., Taylor, D. W., Chen, Z. A., Wang, H.-W., Rappsilber, J., & Hochstrasser, M. (2015). A single α Helix drives extensive remodeling of the proteasome lid and completion of regulatory particle assembly. *Cell*, *163*(2), 432–444. https://doi.org/10.1016/j.cell.2015.09.022.

Wu, M., & Lander, G. C. (2020). How low can we go? Structure determination of small biological complexes using single-particle cryo-EM. *Current Opinion in Structural Biology*, *64*, 9–16. https://doi.org/10.1016/j.sbi.2020.05.007.

CHAPTER FIVE

Extracting structural insights from soft X-ray scattering of biological assemblies

Sintu Rongpipi[a], Joshua T. Del Mundo[b], Enrique D. Gomez[b,c,*], and Esther W. Gomez[b,d,*]
[a]Advanced Light Source and Materials Science Division, Lawrence Berkeley National Laboratory, Berkeley, CA, United States
[b]Department of Chemical Engineering, The Pennsylvania State University, University Park, PA, United States
[c]Department of Materials Science and Engineering and Materials Research Institute, The Pennsylvania State University, University Park, PA, United States
[d]Department of Biomedical Engineering, The Pennsylvania State University, University Park, PA, United States
*Corresponding authors: e-mail address: edg12@psu.edu; ewg10@psu.edu

Contents

Methods in Enzymology, Volume 678
ISSN 0076-6879
https://doi.org/10.1016/bs.mie.2022.09.017

Abstract

Resonant soft X-ray scattering (RSoXS), a technique that combines X-ray absorption spectroscopy and X-ray scattering, can probe the nano- and meso-scale structure of biological assemblies with chemical specificity. RSoXS experiments yield scattering data collected at several photon energies, for example across an elemental absorption edge of interest. Collecting a near-edge X-ray absorption fine structure (NEXAFS) spectrum complements RSoXS experiments and determines X-ray energies that are best suited for RSoXS measurements. The analysis of RSoXS data is similar in many ways to analysis of small angle X-ray scattering using hard X-rays, with an added dimension that includes an X-ray energy dependence. This chapter discusses procedures for predicting scattering contrast and thereby identifying energies suitable for RSoXS measurements using NEXAFS spectra, analyses of 2D RSoXS images through integration into 1D profiles, and strategies for elucidating the origin of RSoXS scattering features. It also discusses existing and potential methods for interpretation of RSoXS data to gain detailed structural insights into biological systems.

1. Introduction

Resonant soft X-ray scattering (RSoXS) probes the nano- to meso-scale structure of soft matter systems including biological assemblies (Collins & Gann, 2021; Ingham et al., 2015, 2016; Ingham, Kirby, Wang, Brady, & Carr, 2018; Rongpipi, Del Mundo, Gomez, & Gomez, 2021; Ye et al., 2018). It combines spectroscopy and scattering to enable sensitivity to chemical composition and molecular bond orientation. The absorption edges of light elements, such as C, N, and O, are accessible in the soft X-ray regime (X-ray with energies between 100 and 1500 eV). This makes RSoXS a promising tool for chemically sensitive characterization of biological systems, such as proteins and biopolymers. RSoXS measurements of biological assemblies that incorporate metal ions, such as calcium, have also been used to examine the structure of plant primary cell walls and casein protein in milk (Ingham et al., 2015, 2016, 2018; Ye, Kiemle, et al., 2018).

RSoXS data consist of scattering and spectroscopy information and is called hyperspectral data (Collins & Gann, 2021). Data obtained from near edge X-ray absorption fine structure (NEXAFS) spectroscopy measurements can be used for predicting scattering contrast in RSoXS experiments between different components of a sample. Scattering data often consist of 2D images of scattering collected at several energies, for example spanning across an elemental edge of interest.

RSoXS data can be affected by instrumentation and beamline related noise and artifacts. As such, 2D images must be corrected before conversion into 1D data. Some of the corrections commonly applied to 2D data follow similar procedures as small angle X-ray scattering with hard X-rays (Pauw, 2013; Pauw, Smith, Snow, Terrill, & Thünemann, 2017), and include dark image subtraction, geometric correction for solid angle of scattering, and masking of CCD edges and the beam stop. In addition, with RSoXS, data must be normalized by the direct beam flux to account for intensity fluctuations with energy. Once the instrument-related corrections are applied to 2D images, they are reduced into 1D data. Converting 2D data into 1D enables quantitative extraction of structural information, such as the length scales of structural features, from the location of scattering peaks with respect to wave vector.

The interpretation of scattering data often relies on fitting to models of structure (structure factor) or shape (form factors), and extracting information about the morphology, size, and separation distance of domains. Nevertheless, scattering techniques fail to provide a unique structural solution because multiple structures can provide the same scattering profile. RSoXS can alleviate this problem through the modulation of contrast with X-ray energy. For example, the scattering contrast of two or more domains can be predicted using NEXAFS data. Matching the total scattering intensity from RSoXS data collected at various energies to the predicted scattering contrast supports that measured scattering originates from the domains chosen to predict the contrast.

Obtaining the correct energy dependence of scattering profiles is crucial for interpretation of RSoXS data. This is also the most challenging step in RSoXS data analysis because the energy dependence of scattering intensity is often affected by several factors other than resonance effects. These factors include variation of the direct beam flux, absorption, fluorescence, and radiation damage. Furthermore, local order within molecules or domains (e.g., alignment of dipoles) can also affect RSoXS intensities when the beam is linearly polarized (Collins et al., 2012). The effect of variation in the direct beam flux is accounted for by normalizing each scattering profile collected at a particular energy by the direct beam flux at the corresponding energy measured at a time close to the time of measuring the sample. The contribution of absorption and fluorescence effects to scattering intensity can be separated through a spectral model of RSoXS (Ferron, Pope, & Collins, 2017). Radiation damage can also influence the energy dependence of scattering profiles, and it is often monitored by comparing the scattering patterns at a few energies and at a few different locations on the sample before and after long exposure times. The damage is mitigated by collecting data for short exposure times or by spreading out the scans across similar locations on the sample or similar samples (Collins & Gann, 2021). The critical dose for radiation damage can also be determined by fitting the total scattering intensity versus radiation dose using an equation previously described for electron microscopy of polymers (Martin & Thomas, 1995). Using this method, the critical dose for bovine serum albumin in solution was determined to be 1.7×10^7 Gy at 288.3 eV (corresponding to 150 s of exposure time with an X-ray flux of 10^{15} photons/cm^2/s) (Ye, Le, et al., 2018). The role of local order on polarized RSoXS remains challenging to interpret despite many recent efforts (Collins et al., 2012; Litofsky et al., 2019; Mukherjee et al., 2021), although the effect of a polarized soft X-ray beam has not been demonstrated to be important for studies of biological molecules.

In this chapter, we describe how to use NEXAFS data to plan RSoXS experiments, reduce 2D images into 1D data, visualize and interpret RSoXS data, and predict the structure of biological molecules and assemblies from RSoXS data. We also highlight some studies where the discussed RSoXS analyses have been implemented to reveal the structure of biological systems. Finally, we present analytical approaches for analysis of RSoXS data which have not previously been implemented for biological systems but hold potential for future applications toward analysis of the structure of biomolecules.

2. Predicting RSoXS scattering contrast from NEXAFS spectra

NEXAFS is an X-ray absorption spectroscopy technique that is sensitive to electronic environment, chemical composition, and molecular bond orientation. NEXAFS measurements can be collected in three modes: transmission, electron yield (EY), and fluorescence yield (FY) (Stöhr, 2013). In transmission NEXAFS, the X-ray beam is normal to the sample, and the X-ray absorption of the sample is measured by tracking the amount of forward-scattered X-rays. Electron yield (EY), including total electron yield (TEY) and partial electron yield (PEY), and FY modes of NEXAFS are often collected in a reflection geometry. Thus, transmission NEXAFS spectroscopy probes the entire thickness of the sample while EY and FY NEXAFS modes are surface sensitive. Consequently, transmission NEXAFS is typically used when bulk information is desired, and EY and FY NEXAFS are used when surface information is warranted. For example, at the carbon K-edge EY measurements probe ~1 to 10 nm into the sample (DeLongchamp, Lin, & Fischer, 2005; Gainar, Stevens, Jaye, Fischer, & Schroeder, 2015; Samant et al., 1996; Su, Patel, Pemmaraju, Prendergast, & Chabinyc, 2017) while FY measurements probe ~100 nm into a hydrocarbon (organic) sample (Arvanitis, Dobler, Wenzel, Baberschke, & Stöhr, 1986; DeLongchamp, Lin, & Fischer, 2005).

NEXAFS measurements complement RSoXS experiments by enabling prediction of the contrast between domains. As such, NEXAFS data can help determine important energies at which contrast is high, and these energies can then be used for performing scattering measurements at optimal contrast conditions. It is crucial to obtain NEXAFS measurements of pure materials for predicting contrast between domains, and to reveal contrast between domains versus vacuum for different materials that can lead to significant scattering from the surface roughness or voids, and for obtaining a quantitative prediction of how contrast varies with energy.

Here, we discuss using NEXAFS spectra for calculating the predicted scattering contrast between different components in a biological assembly. We will highlight this procedure through the example of RSoXS experiments of plant primary cell wall in onion outer epidermal peel. The predicted scattering contrast between cellulose and calcium cross-linked pectin has been previously calculated in an RSoXS study of onion outer epidermal peel (Ye, Kiemle, et al., 2018).

2.1 Materials

- NEXAFS data (photodiode scan, EY scan, or FY intensity from sample) of pure components. For example, NEXAFS data of cellulose and calcium-crosslinked pectin are required to calculate the predicted scattering contrast between these components (Fig. 1)
- NEXAFS of blank sample (photodiode scan as a function of X-ray energy)
- KKcalc software (Watts, 2014) for calculating optical constants
- Data analysis and visualization tool (e.g., Microsoft excel) or technical computing software (e.g., Matlab, Mathematica) for calculating the scattering contrast

2.2 Procedure

1. The method for normalization of NEXAFS data differs based on the mode of NEXAFS data collection (see note 1 and note 2).

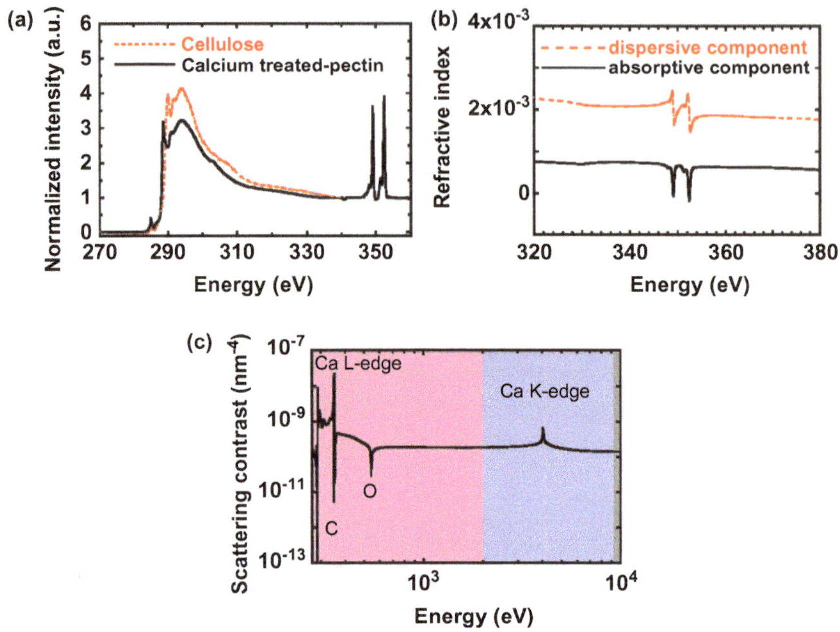

Fig. 1 NEXAFS data can be used to predict scattering contrast between components within a biological sample. (A) Experimentally measured NEXAFS spectra of cellulose (red, dashed) and calcium-treated pectin (black, solid). Cellulose NEXAFS data are from (Gainar et al., 2015) and calcium-treated pectin data are from (Ye, Kiemle, et al., 2018). (B) Dispersive (red, dashed) and absorptive (black, solid) components of the refractive index for calcium treated-pectin. The dispersive component is calculated using Kramers-Kronig transformation of the absorptive component that is obtained from experimental NEXAFS data. (C) Predicted scattering contrast between cellulose and calcium treated-pectin. The soft, tender, and hard X-ray regimes are indicated by pink, purple, and gray shading, respectively. *Panels (A) and (C) are adapted from Ye, D., Kiemle, S. N., Rongpipi, S., Wang, X., Wang, C., Cosgrove, D. J., Gomez, E. W., & Gomez, E. D. (2018). Resonant soft X-ray scattering reveals cellulose microfibril spacing in plant primary cell walls.* Scientific Reports, *8(1), 12449. doi: 10.1038/s41598-018-31024-1.*

For NEXAFS data collected in EY and FY modes, normalize the EY signal and FY intensity by the incident X-ray intensity I0.

$$Normalized\ EY(or\ FY) = \frac{EY(or\ FY)}{I0} \tag{1}$$

2. The pre- and post-edge regions of the spectra should be flat for modest energy ranges (about 10 eV). If there is an apparent linear background, correct NEXAFS data for background from the sample substrate by

fitting the pre-edge data with a linear function and subtracting it from the entire NEXAFS spectra (Gainar et al., 2015; Ye, Kiemle, et al., 2018). As an example, Fig. 1A shows the normalized NEXAFS spectra of cellulose and calcium-treated pectin (Ye, Kiemle, et al., 2018). For details on how the data were collected, see note 3.

3. The absorptive component β and dispersive component δ of the refractive index of two materials are required to calculate the predicted scattering contrast between them. Experimentally collected NEXAFS data represents the absorptive component β of the refractive index. Input the normalized experimental NEXAFS data into KKCalc software to merge the experimentally obtained data with the β obtained from KKcalc, which can provide a more accurate estimate near the absorption edges of interest (see note 4). This step requires providing the chemical composition and density of the material in KKcalc.
4. Compute the dispersive component δ from the absorptive component β using KKcalc for both of the components for which the scattering contrast is to be calculated. The software does the transformation using Kramers-Kronig relations (Watts, 2014). As an example, Fig. 1B shows the dispersive and absorptive components of calcium treated-pectin obtained from KKcalc using experimental NEXAFS data. In a similar way, dispersive and absorptive components of cellulose can be obtained.
5. The scattering contrast between two components, i and j, in a system can be predicted as (Virgili, Tao, Kortright, Balsara, & Segalman, 2007):

$$\textit{Scattering contrast} = \frac{\left(\delta_i(\lambda) - \delta_j(\lambda)\right)^2 + \left(\beta_i(\lambda) - \beta_j(\lambda)\right)^2}{\lambda^4} \tag{2}$$

where λ is the wavelength of incident X-ray, i and j denote two species, and β and δ are the absorptive and dispersive component of the refractive index $n(\lambda)$, respectively, where $n(\lambda) = 1 - \delta(\lambda) + i\beta(\lambda)$. Using Eq. (2) along with β and δ, calculate the predicted scattering contrast between the two materials. The scattering contrast between cellulose and calcium-treated pectin is shown in Fig. 1C. Here, the predicted scattering contrast near the calcium L-edge (~350 eV) is about two orders of magnitude higher than the scattering contrast in the hard X-ray region (~10,000 eV). RSoXS measurements were collected at energies where the scattering contrast between the two components varies with energy, such that for primary cell wall samples from onion epidermal peel X-ray energies near 350 eV were used.

2.3 Notes

1. For NEXAFS data collected in transmission mode, calculate the transmittance (T) by normalizing the transmitted X-ray intensity (I) by the incident X-ray intensity (I0)

$$T = \frac{I}{I0} \tag{3}$$

Transmission NEXAFS data is generally presented in the form of absorbance (A) from the sample, expressed as:

$$A = -Log(T) \tag{4}$$

Normalize the transmittance (or absorbance) from the sample by the transmittance (or absorbance) through a blank substrate to account for the fluctuation of the X-ray beam with beamline energy.

2. The NEXAFS signal (I) and the intensity of the incident X-ray beam (I0) are generally saved in the same file. Software programs such as QANT (Gann, McNeill, Tadich, Cowie, & Thomsen, 2016), Athena (Ravel & Newville, 2005), and Larch (Newville, 2013) are available for NEXAFS data loading and analysis.
3. Ideally, NEXAFS data should be collected in a single mode (transmission, PEY or FY) for simplicity. In this example, the NEXAFS spectra for cellulose is from published literature and was collected in PEY mode (Gainar et al., 2015). Commercially available citrus pectin was used to obtain the NEXAFS spectra of calcium cross-linked pectin (Ye, Kiemle, et al., 2018). Because of the low calcium content in citrus pectin, the calcium signal was suppressed in the NEXAFS spectra, so the carbon K-edge NEXAFS spectra from citrus pectin collected in transmission mode was stitched with the calcium L-edge NEXAFS spectra from calcium-treated onion epidermal peel collected in TEY mode. For stitching, it was assumed that the non-pectin components of the cell wall do not complex with calcium and therefore represent the NEXAFS spectra near the calcium L-edge (340–355 eV) as a line that connects the pre-edge (340 eV) and post-edge (355 eV). This straight line was then subtracted from the NEXAFS spectra of calcium-treated cell wall and divided by the mass fraction of pectin in cell wall (assumed to be 0.4). The subtracted calcium L-edge spectra from calcium-treated cell wall and carbon K-edge spectra from citrus pectin were then merged at 320 eV.
4. KKcalc can be used to predict the X-ray absorption of any material based only on the atomic composition and density, but it fails to capture the fine absorption features near an absorption edge. Inputting an

experimentally measured NEXAFS spectra of a material allows KKcalc to merge the experimentally measured imaginary part of the refractive index near the edge with the theoretical values away from the edge. This provides the imaginary part of the refractive index for an extended energy range and the real part is then calculated using the Kramers–Kronig relationship (Watts, 2014).

3. Reduction of RSoXS 2D data into 1D

RSoXS data can consist of 2D detector images collected at several photon energies across an elemental edge of interest. To extract quantitative structural information from 2D images, one approach is to convert to 1D data through integration of pixel intensities and to represent the data as profiles as a function of scattering vector or X-ray energy. 2D RSoXS images can be reduced to 1D data using software packages available for conventional hard X-ray scattering, such as the Nika package with Igor Pro (Ilavsky, 2012), FIT2D (Hammersley, 2016), SASview (*SasView for Small Angle Scattering Analysis*), and ATSAS (Manalastas-Cantos et al., 2021). Customized RSoXS analysis panels are also available for different RSoXS beamlines that can be integrated with the Nika package in Igor software. In this section, we will discuss the procedure for reducing RSoXS 2D images into 1D data through the example of RSoXS of onion epidermal peel. Fig. 2A shows 2D RSoXS data of onion epidermal peel collected near the calcium L-edge (352.6 eV, on-edge).

3.1 Materials

- RSoXS sample image files at energies on and off the absorption edge of interest. For analysis of onion epidermal peels, two on-resonance energies (349.3 eV and 352.6 eV) and two off-resonance energies (345 eV and 355 eV) near the calcium L-edge are used as an example
- RSoXS dark image files
- RSoXS calibrant image files
- Direct beam photodiode scan
- Software for analysis of conventional hard X-ray scattering data such as Nika package (Ilavsky, 2012) with Igor, FIT2D (Hammersley, 2016), SASview (*SasView for Small Angle Scattering Analysis*), and ATSAS (Manalastas-Cantos et al., 2021)
- Customized RSoXS analysis panel to be integrated with Nika package in Igor (Ferron, 2018; Gann, 2022)

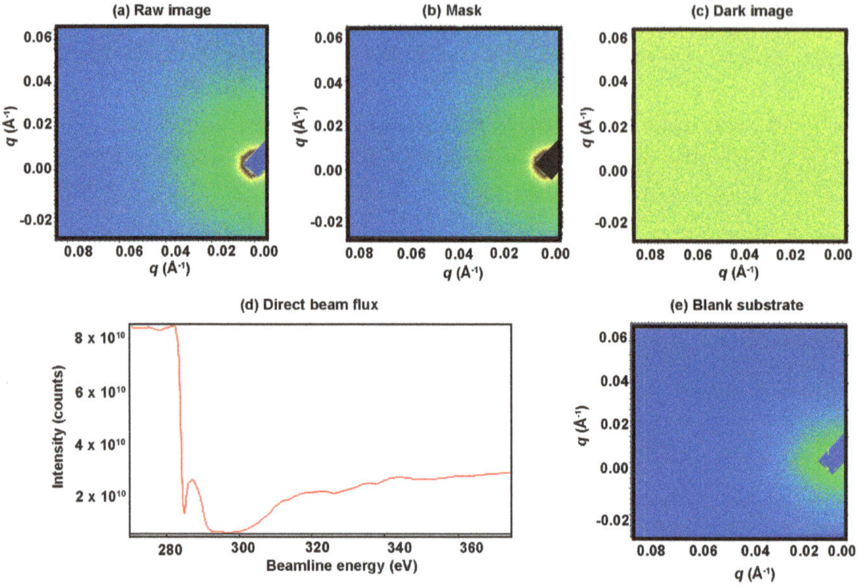

Fig. 2 Analysis steps for RSoXS data collected for plant primary cell wall sample. (A) RSoXS 2D data from plant primary cell wall of onion epidermal peel collected near the calcium L-edge (352.6 eV, on-edge). (B) Image showing masking of the beam-stop and edges of the CCD image for RSoXS data from panel (A). (C) Dark image collected at the same exposure time as RSoXS data shown in panel (A). (D) Variation of direct beam (I0) flux with beamline energy. (E) RSoXS 2D data from a blank substrate (silicon nitride window) collected at 352.6 eV.

3.2 Procedure

1. Using RSoXS calibrant image files, refine the beam center and the sample-to-detector distance to obtain correct q (wave-vector) values of the scattering data (see note 1). Software programs (mentioned in Section 3.1) for scattering data analysis can be used for refinement of the beam center and sample-to-dector distance.
2. Draw a mask on regions of the 2D images with dead or damaged pixels to remove these from subsequent analyses (see note 2). Dead or damaged pixels are found in regions of the detector that are pinned to the maximum or minimum value of intensity (Pauw, 2013). Typically, the pixels on the edges of the CCD detector and the pixels covered by the beam stop and the beam stop holder are also masked. For example, the shaded regions in Fig. 2B are the masked regions before data reduction. Software programs (mentioned in Section 3.1) for scattering data analysis have built-in capabilities for masking 2D images.

3. Correct the 2D RSoXS data for dark counts by subtracting a dark image, which is an image file with the same exposure time but collected with the X-ray shutter closed (see note 3). Fig. 2C shows a dark image with the same exposure time as the RSoXS data shown in Fig. 2A.
4. Correct the 2D RSoXS data for beam fluctuation with energy by normalizing by the direct beam intensity (I0) at each energy. Fig. 2D shows the I0 intensity as a function of energy used for normalization of the RSoXS data shown in Fig. 2A.
5. Correct the 2D RSoXS data for exposure time and solid angle. Most scattering analysis software programs have built-in capabilities for these corrections.
6. Reduce 2D detector images into 1D data by integrating pixel intensities over all polar angles. Azimuthal integration of the 2D images involves averaging circularly over the entire image (see note 4).
7. Background correct the azimuthally integrated 1D data by subtracting integrated data from a blank substrate (see note 5). Example 2D scattering data from a blank substrate is shown in Fig. 2E.
8. Visualize the energy dependence of the scattering profiles by plotting all 1D profiles from different energies together for comparison. For example, background corrected, azimuthally averaged scattering profiles from a calcium-treated onion epidermal peel at 345, 349.3, 352.6, and 355 eV are shown in Fig. 3.

3.3 Notes

1. Several materials such as silver behenate (AgB), highly oriented pyrolytic graphite (HOPG), and polystyrene 100 nm (PS100) or 300 nm (PS300) spheres are used as calibration samples. Because of the known resonance peaks in carbon K-edge NEXAFS spectra, HOPG and PS spheres are often used for calibration of beamline energy as well. The NEXAFS analysis software QANT (Gann et al., 2016) can be used to calibrate beamline energy for NEXAFS spectra.
2. Regions on 2D images with unwanted streaks or spots, due to artifacts from sample preparation, are also masked.
3. Procedures for dark count corrections can depend on the software available. For example, at the time of writing the manuscript, the dark count correction of RSoXS data from beamline 11.0.1.2 at the Advanced Light Source (ALS) requires manual selection of a corresponding dark image file for a RSoXS sample image file (Ferron, 2018) while the dark count correction for RSoXS data from beamline 7-ID-1 (SST1) is done automatically by the customized RSoXS analysis panel (Gann, 2022).

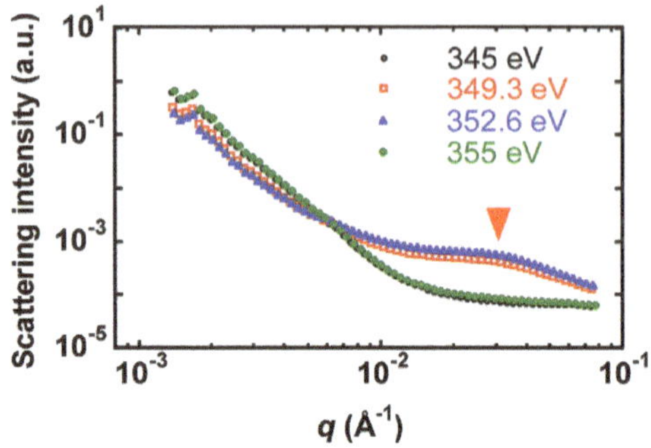

Fig. 3 1D RSoXS reduced data from a calcium-treated onion epidermal peel at two on-resonance energies (349.3 eV and 352.6 eV) and two off-resonance energies (345 eV and 355 eV). RSoXS profiles at on-resonance energies reveal a feature at about 0.03 Å^{-1} (shown with arrowhead) that is not seen in data collected at off-resonance energies. *Adapted from Ye, D., Kiemle, S. N., Rongpipi, S., Wang, X., Wang, C., Cosgrove, D. J., Gomez, E. W., & Gomez, E. D. (2018). Resonant soft X-ray scattering reveals cellulose microfibril spacing in plant primary cell walls.* Scientific Reports, 8*(1), 12449. doi: 10.1038/s41598-018-31024-1.*

4. Typically, images with isotropic patterns are circularly averaged while images with anisotropic patterns are sector averaged over the sector of interest.
5. The subtraction of background scattering (using data acquired from a blank substrate) can be done through (i) subtraction of the blank substrate data from the 2D image file of the RSoXS sample and (ii) subtraction of blank substrate 1D data from reduced 1D data obtained from the RSoXS sample.

4. Interpretation of scattering data

Interpretation of scattering data often relies on fitting scattering data to models of structure or form factors of systems with known morphology and geometries. The 1D scattering data I(q), intensity (I) as a function of scattering vector (q), consists of two parts—a form factor and structure factor. The form factor provides information about the size and shape of the individual particles, while the structure factor provides information on interactions between neighboring particles. Many X-ray scattering studies involve determination of form factor or structure factor.

More advanced analyses in X-ray scattering of biomacromolecules such as proteins in solution may include determining pair distribution functions and three-dimensional (3D) envelopes of protein structures (Hura et al., 2009; Liu & Zwart, 2012; Murugova et al., 2015; Skou, Gillilan, & Ando, 2014). The pair distribution function, P(r), is calculated through indirect Fourier inversion of the 1D scattering intensity I(q) using the program GNOM (Svergun, 1992) and the distance at which the P(r) gradually falls off to zero is the maximum dimension of the scattering particles. 3D envelope functions can be generated using ab initio modeling approaches in software programs such as GASBOR (Svergun, Petoukhov, & Koch, 2001). Several 3D structures generated by GASBOR are aligned using a software program called SUPCOMB (Kozin & Svergun, 2001) and averaged using another software program called DAMAVER (Volkov & Svergun, 2003). The generated 3D envelopes can then be visualized using the molmap command in Chimera software (Pettersen et al., 2004). Similar to analysis for hard X-ray scattering, these approaches can also be applied to soft X-ray scattering data. For example, experimentally measured NEXAFS spectra for bovine serum albumin (BSA) have been used to predict the scattering contrast between BSA and a solution of phosphate buffered saline (PBS) near the carbon K-edge (Ye, Le, et al., 2018). Based on the peaks in scattering contrast, RSoXS measurements were collected at 285.3 eV and 288.3 eV. RSoXS data collected at 285.3 eV, 288.3 eV, and 10 keV reveal differences in scattering across different energies. The maximum dimension of scattering particles and 3D reconstructions show differences in protein shape resulting from scattering from different moieties within the protein structure. Details for computing pair distribution functions, maximum dimensions of scattering particles, and envelope functions can be found in previously published methods and reviews (Da Vela & Svergun, 2020; Gräwert & Svergun, 2020; Mertens & Svergun, 2010; Skou et al., 2014).

Attempting to model one-dimensional scattering data from three-dimensional models can lead to overfitting because the information content in a one-dimensional scattering profile is much less than the equivalent information content in a three-dimensional structure of identical resolution (Chen et al., 2019; Lattman, Grant, & Snell, 2018). Very often, overfitting is checked by goodness-of-fit parameters, such as chi-squared. Reduced chi-squared values are required for a fitting to be considered good, but a chi-squared less than 1 indicates overfitting (Rambo & Tainer, 2013). In general, two approaches have been adopted to reduce the risk of overfitting. The first approach is called rigid-body modeling or normal mode fitting that involves constraining nearly all degrees of freedom of the biomolecule during refinement (Gorba, Miyashita, & Tama, 2008; Petoukhov & Svergun, 2005).

The second approach includes complementing physical information to the low information content scattering data, for example, by coupling a force field-based molecular dynamics (MD) simulation to the data (Bjorling, Niebling, Marcellini, van der Spoel, & Westenhoff, 2015; Kimanius, Pettersson, Schluckebier, Lindahl, & Andersson, 2015). Use of multiple scattering data sets has also been suggested for controlling against overfitting (Chen et al., 2019; Hub, 2018). RSoXS data that varies with energies may also help alleviate overfitting problems, by providing multiple sets to be reconciled to a single structural model. More generally, large data sets allow dividing the data into independent training and test sets. Thus, allowing the structural model to be fitted using only the training set, while overfitting can be excluded by cross-validation against the test set. This strategy, however, will work only when the information in the test set is not used during model fitting (Hub, 2018).

In this section, a general approach for analyzing RSoXS data will be described. We briefly describe how to fit small-angle scattering data to form or structure factors, although more detailed descriptions can be found elsewhere (Boldon, Laliberte, & Liu, 2015; Pedersen, 1997, 2002; Svergun, Koch, Timmins, & May, 2013). As an example, soft X-ray scattering data for plant primary cell wall using energies near the calcium L-edge will be discussed.

4.1 Materials

- Azimuthally integrated 1D RSoXS data
- Data fitting software for conventional X-ray scattering, such as the Irena package with Igor (Ilavsky & Jemian, 2009), FIT2D (Hammersley, 2016), and SASfit (Bressler, Kohlbrecher, & Thunemann, 2015)

4.2 Procedure

1. For samples where scattering objects are dilute or isolated, scattering is likely dominated by the shape and size of the sample, and therefore data can often be fit to a form factor. Examples include fitting form factors to scattering profiles from proteins or lipid vesicles in solution and has been described previously (Cheng et al., 2021; Finet, Skouri-Panet, Casselyn, Bonneté, & Tardieu, 2004; Fogarty, Arjunwadkar, Pandit, & Pan, 2015; Komorowski et al., 2018; Zhang et al., 2007), including for RSoXS (Ye, Le, et al., 2018). Fit the I(q) scattering data to the characteristic form factors of known shapes to identify the form factor of the

scattering particles. The form factor represents the scattering that results from a system of dilute particles with no interaction between them. A form factor for a well-defined shape and size has a characteristic slope and fringe like pattern at higher q in the I(q) profile. The 1D scattering data can be fit to form factors of known shapes using the fitting software for conventional hard X-ray scattering data.

2. In the example discussed here of plant primary cell wall, the large amount of cellulose microfibrils suggests a large concentration of scattering objects, such that the correlations in spacing between objects and therefore the structure factor will dominate scattering. In such cases, scattering intensity profiles as a function of scattering vector will exhibit a peak corresponding to the periodicity in the location of scattering objects. This peak in scattering intensity may be in the form of a broad shoulder in not highly ordered systems, such as some biological assemblies. The spacing, d, between the scattering domains is extracted from the position of the peak in the I(q) profile as described below:

$$d = \frac{2\pi}{q} \tag{5}$$

where q is the position of the scattering peak. For example, for calcium-treated onion epidermal peel as shown in Fig. 3, a scattering feature near $q \sim 0.03\,\text{Å}^{-1}$ is seen for on-resonance energies and no feature is seen for off-resonance energies (see note 1). This feature corresponds to a spacing of ~20 nm (see note 2).

3. Alternatively, scattering profiles can be fit to models of structure factors, as has been discussed for small angle scattering data previously (Lombardo, Calandra, & Kiselev, 2020; Pedersen, 1997, 2002; Zhang et al., 2007).

4.3 Notes

1. The same study found that the scattering feature is weak in an unextracted onion epidermal peel, and the feature disappears in a pectate lyase-treated onion epidermal peel. In both of these cases, calcium concentration is expected to be less than the calcium-treated sample (Ye, Kiemle, et al., 2018).
2. This distance is supported by results from complimentary techniques such as atomic force microscopy, nuclear magnetic resonance spectroscopy, and small angle neutron scattering (Pérez García et al., 2011; Thimm, Burritt, Ducker, & Melton, 2000; Ye, Kiemle, et al., 2018; Zhang, Zheng, & Cosgrove, 2016).

5. Identification of the real-space structure that leads to an observed scattering profile

One of the challenges of interpreting scattering data is the lack of unique structural solutions (Chacón, Morán, Díaz, Pantos, & Andreu, 1998; Collins & Gann, 2021; Glatter, 2018; Pauw, 2013; Rongpipi et al., 2021; Ye, Kiemle, et al., 2018; Ye, Le, et al., 2018). This arises because multiple structures can lead to the same scattering pattern. For example, in multicomponent systems, multiple components can be arranged with similar spacings, making it challenging to identify which set of domains is the origin of the scattering that is observed. In this section, we will discuss how this problem can be alleviated by matching total scattering intensity (TSI) to the predicted scattering contrast between different components.

5.1 Materials

- Azimuthally integrated 1D scattering data
- Experimental NEXAFS data of single components of a sample (e.g., cellulose and calcium-treated pectin)
- Data analysis and visualization tool (e.g., Microsoft excel) or technical computing software (e.g., Matlab, Mathematica) for calculating the scattering contrast and total scattering intensity

5.2 Procedure

1. Calculate the total scattering intensity (TSI) using the azimuthally integrated 1D scattering data as given below after normalizing scattering data by direct beam flux, exposure time, and sample transmittance (see note 1):

$$TSI = \int I(q)q^2 dq \tag{6}$$

   The TSI obtained from 1D RSoXS data of calcium-treated primary cell wall in onion epidermal peel near the calcium L-edge is shown in Fig. 4 (red squares).

2. Calculate the predicted scattering contrast between different pairs of components of a sample (see Section 2.2) using experimental NEXAFS data of single components for determining the optical constants of the components in the software KKcalc. Predicted scattering contrast for different

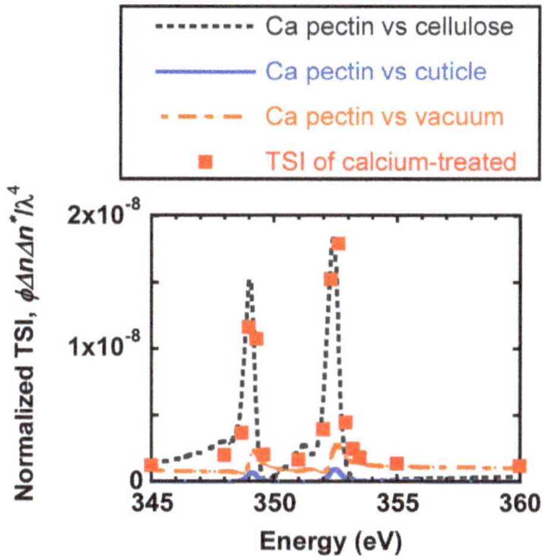

Fig. 4 Total scattering intensities (TSI) from RSoXS data for plant primary cell wall at energies near the calcium L-edge match with the predicted scattering contrast ($\varphi \Delta n \Delta n^*$) between different components found within the primary cell wall. The TSI matches well with the predicted contrast between Ca-treated pectin and cellulose suggesting that the RSoXS feature originates from the contrast between these two components. *Adapted from Ye, D., Kiemle, S. N., Rongpipi, S., Wang, X., Wang, C., Cosgrove, D. J., Gomez, E. W., & Gomez, E. D. (2018). Resonant soft X-ray scattering reveals cellulose microfibril spacing in plant primary cell walls.* Scientific Reports, *8(1), 12449. doi: 10.1038/s41598-018-31024-1.*

pairs of primary cell wall components (cellulose, Ca-pectin, and cuticle) estimated near the calcium L-edge using experimentally measured NEXAFS spectra of pure components are shown in Fig. 4.

3. Compare the TSI obtained from 1D RSoXS data with the predicted scattering contrast between different pairs of components (see note 2). If the TSI shape resembles the predicted scattering contrast between any two components, it suggests that the scattering profile originates from the contrast between those two components (see note 3). The TSI from 1D RSoXS data from primary cell wall matches well with the predicted scattering contrast between Ca-pectin and cellulose (Fig. 4). This suggests that the RSoXS scattering feature originates from contrast between cellulose and calcium cross-linked pectin (see note 4).

5.3 Notes

1. For RSoXS measurements collected above the absorption edge of a constituent element that is present at high concentration (such as the carbon edge for polymer and biological samples), RSoXS 1D data should be

corrected for fluorescence before calculating the TSI. The fluorescence correction is done by subtracting the predicted (Ferron et al., 2017) or measured (Ferron, Grabner, McAfee, & Collins, 2020) fluorescence intensity from the sample.

2. Calculating absolute scattering intensities in RSoXS experiments can be challenging. Therefore, the TSI can be scaled to match the predicted scattering contrast. A least squares regression is best for this approach.
3. In a heterogeneous system, it is possible that a combination of scattering contrasts between different sets of components may contribute to a certain scattering pattern. In such cases, a linear combination of scattering contrasts from different sets of components can be fit to the TSI using linear regression.
4. A similar approach has been used for analyzing RSoXS data of BSA in 1× PBS solution. The computed TSI from RSoXS data at four energies near the carbon K-edge was found to match strongly with the predicted scattering contrast, suggesting that the observed scattering profiles originate from contrast between BSA and 1× PBS (Ye, Le, et al., 2018).

6. Opportunities for application of new RSoXS analysis approaches to biological assemblies

RSoXS is sensitive to the molecular orientation in the sample. Molecular orientation within domains can lead to anisotropic scattering patterns in RSoXS data, particularly when working near a resonance to an anisotropic orbital (e.g., π-bond). Anisotropic RSoXS 2D data have been reported for polymers including a conjugated *block* copolymer system of poly(3-hexylthiophene)-*block*-poly((9,9-dioctylfluorene)-2,7-diyl-*alt*-[4,7-bis (thiophen-5-yl)-2,1,3-benzothiadiazole]-2′,2″-diyl) (P3HT-b-PFTBT) and PFTBT derivatives (Litofsky et al., 2019), semiconducting polymer poly(2,5-bis(3-alkylthiophen-2-yl)thieno[3,2-*b*]thiophene) (PBTTT) (Collins et al., 2012), and polystyrene grafted gold nanoparticles (Mukherjee et al., 2021).

Azimuthal sector averaging of scattering intensity along specific polar angles can highlight the anisotropy in scattering and reveal structural information regarding orientation correlations. One method to quantify the anisotropy is to calculate a parameter called the anisotropy ratio as a function of q and energy (E) using the following formula:

$$A(q, E) = \frac{I_p(q, E) - I_s(q, E)}{I_p(q, E) + I_s(q, E)} \tag{7}$$

where I_p is the scattering in the direction parallel to the photon polarization and I_s is the scattering perpendicular to the photon polarization.

Computational approaches to simulate RSoXS patterns can be used to complement experimental RSoXS measurements. For polystyrene grafted gold nanoparticles, a computational scattering pattern simulation was used to measure the thickness of the anisotropic region of the polymeric corona and the extent of chain orientation within it (Mukherjee et al., 2021). Furthermore, NEXAFS spectroscopy is sensitive to polarization of incident X-rays, and it can characterize orientational states of peptides, proteins, and nucleic acids (Lee, Nguyen, Grainger, Gamble, & Castner, 2007; Liu et al., 2006; Polzonetti et al., 2006). These findings together suggest that RSoXS could potentially be used to probe orientational correlation between different domains in biological systems.

7. Conclusion and outlook

The use of RSoXS for the study of plant primary cell wall and proteins in solution suggests many opportunities for the application of soft X-ray scattering in studies of the structure of biological assemblies. In addition to the two biological systems discussed here, RSoXS appears promising for probing complex biological assemblies consisting of multiple components, such as biological membranes (consisting of lipids, proteins, and carbohydrates) and viruses (consisting of nucleic acids and proteins). The components of these biological assemblies mostly consist of low-Z elements and have similar electron densities, so conventional hard X-ray scattering is limited in extracting structural information about the structure of individual components within the assemblies. The chemical heterogeneity, however, may enable RSoXS to extract chemically sensitive structural information by tuning the X-ray energy near the absorption edges of chemical moieties unique to a particular component.

Because the technique combines spectroscopy and scattering, analysis of RSoXS data includes analysis of both spectroscopic and scattering data. Here, we have outlined the procedures for analysis of NEXAFS data, identification of energies for RSoXS measurements based on NEXAFS, reduction and interpretation of RSoXS data, and identification of the source of scattering based on variations in scattering with energy. We have also highlighted emerging analytical approaches, such as analysis of polarization dependent scattering, that may provide unique insights for biological systems.

Acknowledgments

This work was supported as part of the Center for Lignocellulose Structure and Formation, an Energy Frontier Research Center funded by the US Department of Energy, Office of Science, Basic Energy Sciences under award no. DE-SC0001090. This work is based upon research conducted at the (1) Advanced Light Source, supported by the Director, Office of Science, Office of Basic Energy Sciences, of the U.S. Department of Energy under Contract No. DE-AC02-05CH11231 and (2) National Synchrotron Light Source II, supported by the Office of Science, Office of Basic Energy Sciences, of the U.S. Department of Energy under Contract No. DE-SC0012704.

References

Arvanitis, D., Dobler, U., Wenzel, L., Baberschke, K., & Stöhr, J. (1986). A new technique for submonolayer NEXAFS: Fluorescence yield at the carbon K edge. *Le Journal de Physique Colloques*, *47*(C8), 173–178. https://doi.org/10.1051/jphyscol:1986832.

Bjorling, A., Niebling, S., Marcellini, M., van der Spoel, D., & Westenhoff, S. (2015). Deciphering solution scattering data with experimentally guided molecular dynamics simulations. *Journal of Chemical Theory and Computation*, *11*(2), 780–787. https://doi.org/10.1021/ct5009735.

Boldon, L., Laliberte, F., & Liu, L. (2015). Review of the fundamental theories behind small angle X-ray scattering, molecular dynamics simulations, and relevant integrated application. *Nano Reviews*, *6*, 25661. https://doi.org/10.3402/nano.v6.25661.

Bressler, I., Kohlbrecher, J., & Thunemann, A. F. (2015). SASfit: A tool for small-angle scattering data analysis using a library of analytical expressions. *Journal of Applied Crystallography*, *48*(5), 1587–1598. https://doi.org/10.1107/S1600576715016544.

Chacón, P., Morán, F., Díaz, J. F., Pantos, E., & Andreu, J. M. (1998). Low-resolution structures of proteins in solution retrieved from X-ray scattering with a genetic algorithm. *Biophysical Journal*, *74*(6), 2760–2775. https://doi.org/10.1016/s0006-3495(98)77984-6.

Chen, P. C., Shevchuk, R., Strnad, F. M., Lorenz, C., Karge, L., Gilles, R., et al. (2019). Combined small-angle X-ray and neutron scattering restraints in molecular dynamics simulations. *Journal of Chemical Theory and Computation*, *15*(8), 4687–4698. https://doi.org/10.1021/acs.jctc.9b00292.

Cheng, R., Li, J., Ríos De Anda, I., Taylor, T. W. C., Faers, M. A., Anderson, J. L. R., et al. (2021). Protein–polymer mixtures in the colloid limit: Aggregation, sedimentation, and crystallization. *The Journal of Chemical Physics*, *155*(11), 114901. https://doi.org/10.1063/5.0052122.

Collins, B. A., & Gann, E. (2021). Resonant soft X-ray scattering in polymer science. *Journal of Polymer Science*, *60*(7), 1199–1243. https://doi.org/10.1002/pol.20210414.

Collins, B., Cochran, J., Yan, H., Gann, E., Hub, C., Fink, R., et al. (2012). Polarized X-ray scattering reveals non-crystalline orientational ordering in organic films. *Nature Materials*, *11*(6), 536–543. https://doi.org/10.1038/nmat3310.

Da Vela, S., & Svergun, D. I. (2020). Methods, development and applications of small-angle X-ray scattering to characterize biological macromolecules in solution. *Current Research in Structural Biology*, *2*, 164–170. https://doi.org/10.1016/j.crstbi.2020.08.004.

DeLongchamp, D. M., Lin, E. K., & Fischer, D. A. (2005). Organic semiconductor structure and chemistry from near-edge X-ray absorption fine structure (NEXAFS) spectroscopy. In *Organic Field-Effect Transistors IV*. San Diego, CA: Proceedings of SPIE - the International Society for Optical Engineering.

Ferron, T., & Gann, E. (2018). *RSoXS data reduction for ALS 11.0.12*. https://labs.wsu.edu/carbon/xray-analysis-tools/. (Accessed 16 October 2022).

Ferron, T., Grabner, D., McAfee, T., & Collins, B. (2020). Absolute intensity calibration for carbon-edge soft X-ray scattering. *Journal of Synchrotron Radiation*, *27*(6), 1601–1608. https://doi.org/10.1107/s1600577520011066.

Ferron, T., Pope, M., & Collins, B. A. (2017). Spectral analysis for resonant soft X-ray scattering enables measurement of interfacial width in 3D organic nanostructures. *Physical Review Letters*, *119*(16), 167801–167806. https://doi.org/10.1103/physrevlett.119.167801.

Finet, S., Skouri-Panet, F., Casselyn, M., Bonneté, F., & Tardieu, A. (2004). The Hofmeister effect as seen by SAXS in protein solutions. *Current Opinion in Colloid & Interface Science*, *9*(1), 112–116. https://doi.org/10.1016/j.cocis.2004.05.014.

Fogarty, J. C., Arjunwadkar, M., Pandit, S. A., & Pan, J. (2015). Atomically detailed lipid bilayer models for the interpretation of small angle neutron and X-ray scattering data. *Biochimica et Biophysica Acta (BBA) - Biomembranes*, *1848*(2), 662–672. https://doi.org/10.1016/j.bbamem.2014.10.041.

Gainar, A., Stevens, J. S., Jaye, C., Fischer, D. A., & Schroeder, S. L. M. (2015). NEXAFS sensitivity to bond lengths in complex molecular materials: A study of crystalline saccharides. *The Journal of Physical Chemistry B*, *119*(45), 14373–14381. https://doi.org/10.1021/acs.jpcb.5b07159.

Gann, E. (2022). Eliot's-NIKA. GitHub. Retrieved October 16, 2022, from https://github.com/EliotGann/Eliots-NIKA.

Gann, E., McNeill, C. R., Tadich, A., Cowie, B. C. C., & Thomsen, L. (2016). Quick AS NEXAFS tool (QANT): A program for NEXAFS loading and analysis developed at the Australian Synchrotron. *Journal of Synchrotron Radiation*, *23*(1), 374–380. https://doi.org/10.1107/S1600577515018688.

Glatter, O. (2018). Chapter 3—The inverse scattering problem. In O. Glatter (Ed.), *Scattering methods and their application in colloid and interface science* (pp. 33–74). Elsevier. https://doi.org/10.1016/B978-0-12-813580-8.00003-1.

Gorba, C., Miyashita, O., & Tama, F. (2008). Normal-mode flexible fitting of high-resolution structure of biological molecules toward one-dimensional low-resolution data. *Biophysical Journal*, *94*(5), 1589–1599. https://doi.org/10.1529/biophysj.107.122218.

Gräwert, T. W., & Svergun, D. I. (2020). Structural modeling using solution small-angle X-ray scattering (SAXS). *Journal of Molecular Biology*, *432*(9), 3078–3092. https://doi.org/10.1016/j.jmb.2020.01.030.

Hammersley, A. (2016). FIT2D: A multi-purpose data reduction, analysis and visualization program. *Journal of Applied Crystallography*, *49*(2), 646–652. https://doi.org/10.1107/S1600576716000455.

Hub, J. S. (2018). Interpreting solution X-ray scattering data using molecular simulations. *Current Opinion in Structural Biology*, *49*, 18–26. https://doi.org/10.1016/j.sbi.2017.11.002.

Hura, G. L., Menon, A. L., Hammel, M., Rambo, R. P., Poole Ii, F. L., Tsutakawa, S. E., et al. (2009). Robust, high-throughput solution structural analyses by small angle X-ray scattering (SAXS). *Nature Methods*, *6*(8), 606–612. https://doi.org/10.1038/nmeth.1353.

Ilavsky, J. (2012). Nika: Software for two-dimensional data reduction. *Journal of Applied Crystallography*, *45*(2), 324–328. https://doi.org/10.1107/S0021889812004037.

Ilavsky, J., & Jemian, P. R. (2009). Irena: Tool suite for modeling and analysis of small-angle scattering. *Journal of Applied Crystallography*, *42*(2), 347–353. https://doi.org/10.1107/S0021889809002222.

Ingham, B., Erlangga, G. D., Smialowska, A., Kirby, N. M., Wang, C., Matia-Merino, L., et al. (2015). Solving the mystery of the internal structure of casein micelles. *Soft Matter*, *11*(14), 2723–2725. https://doi.org/10.1039/C5SM00153F.

Ingham, B., Smialowska, A., Erlangga, G. D., Matia-Merino, L., Kirby, N. M., Wang, C., et al. (2016). Revisiting the interpretation of casein micelle SAXS data. *Soft Matter*, *12*(33), 6937–6953. https://doi.org/10.1039/C6SM01091A.

Ingham, B., Kirby, N., Wang, C., Brady, M., & Carr, A. (2018). Elemental fingerprinting of mineral species in iron-fortified milk: Anomalous small-angle X-ray scattering and resonant soft X-ray scattering studies. *Journal of Synchrotron Radiation*, *25*(4), 1106–1112. https://doi.org/10.1107/S1600577518007774.

Kimanius, D., Pettersson, I., Schluckebier, G., Lindahl, E., & Andersson, M. (2015). SAXS-guided metadynamics. *Journal of Chemical Theory and Computation*, *11*(7), 3491–3498. https://doi.org/10.1021/acs.jctc.5b00299.

Komorowski, K., Salditt, A., Xu, Y., Yavuz, H., Brennich, M., Jahn, R., et al. (2018). Vesicle adhesion and fusion studied by small-angle X-ray scattering. *Biophysical Journal*, *114*(8), 1908–1920. https://doi.org/10.1016/j.bpj.2018.02.040.

Kozin, M. B., & Svergun, D. I. (2001). Automated matching of high- and low-resolution structural models. *Journal of Applied Crystallography*, *34*(1), 33–41. https://doi.org/10.1107/S0021889800014126.

Lattman, E. E., Grant, T. D., & Snell, E. H. (2018). Theoretical background. In E. E. Lattman, T. D. Grant, & E. H. Snell (Eds.), *Biological small angle scattering: Theory and practice* (pp. 13–34). Oxford University Press.

Lee, C.-Y., Nguyen, P.-C. T., Grainger, D. W., Gamble, L. J., & Castner, D. G. (2007). Structure and DNA hybridization properties of mixed nucleic acid/maleimide-ethylene glycol monolayers. *Analytical Chemistry*, *79*(12), 4390–4400. https://doi.org/10.1021/ac0703395.

Litofsky, J. H., Lee, Y., Aplan, M. P., Kuei, B., Hexemer, A., Wang, C., et al. (2019). Polarized soft X-ray scattering reveals chain orientation within nanoscale polymer domains. *Macromolecules*, *52*(7), 2803–2813. https://doi.org/10.1021/acs.macromol.8b02198.

Liu, H., & Zwart, P. H. (2012). Determining pair distance distribution function from SAXS data using parametric functionals. *Journal of Structural Biology*, *180*(1), 226–234. https://doi.org/10.1016/j.jsb.2012.05.011.

Liu, X., Jang, C.-H., Zheng, F., Jürgensen, A., Denlinger, J. D., Dickson, K. A., et al. (2006). Characterization of protein immobilization at silver surfaces by near edge X-ray absorption fine structure spectroscopy. *Langmuir*, *22*(18), 7719–7725. https://doi.org/10.1021/la060988w.

Lombardo, D., Calandra, P., & Kiselev, M. A. (2020). Structural characterization of biomaterials by means of small angle X-rays and neutron scattering (SAXS and SANS), and light scattering experiments. *Molecules*, *25*(23), 5624. https://doi.org/10.3390/molecules25235624.

Manalastas-Cantos, K., Konarev, P. V., Hajizadeh, N. R., Kikhney, A. G., Petoukhov, M. V., Molodenskiy, D. S., et al. (2021). ATSAS 3.0: Expanded functionality and new tools for small-angle scattering data analysis. *Journal of Applied Crystallography*, *54*(1), 343–355. https://doi.org/10.1107/S1600576720013412.

Martin, D. C., & Thomas, E. L. (1995). Experimental high-resolution electron microscopy of polymers. *Polymer*, *36*(9), 1743–1759. https://doi.org/10.1016/0032-3861(95)90922-O.

Mertens, H. D. T., & Svergun, D. I. (2010). Structural characterization of proteins and complexes using small-angle X-ray solution scattering. *Journal of Structural Biology*, *172*(1), 128–141. https://doi.org/10.1016/j.jsb.2010.06.012.

Mukherjee, S., Streit, J. K., Gann, E., Saurabh, K., Sunday, D. F., Krishnamurthy, A., et al. (2021). Polarized X-ray scattering measures molecular orientation in polymer-grafted nanoparticles. *Nature Communications*, *12*(1), 4896. https://doi.org/10.1038/s41467-021-25176-4.

Murugova, T., Vlasov, A., Ivankov, O., Rogachev, A., Ryzhykau, Y., Soloviov, D., et al. (2015). Low resolution structural studies of apoferritin via SANS and SAXS: The effect of concentration. *Journal of Optoelectronics and Advanced Materials*, *17*, 1397–1402.

Newville, M. (2013). Larch: An analysis package for XAFS and related spectroscopies. *Journal of Physics: Conference Series*, *430*, 012007. https://doi.org/10.1088/1742-6596/430/1/012007.

Pauw, B. R. (2013). Everything SAXS: Small-angle scattering pattern collection and correction. *Journal of Physics: Condensed Matter, 25*(38), 383201. https://doi.org/10.1088/0953-8984/25/38/383201.

Pauw, B. R., Smith, A. J., Snow, T., Terrill, N. J., & Thünemann, A. F. (2017). The modular small-angle X-ray scattering data correction sequence. *Journal of Applied Crystallography, 50*(6), 1800–1811. https://doi.org/10.1107/s1600576717015096.

Pedersen, J. S. (1997). Analysis of small-angle scattering data from colloids and polymer solutions: Modeling and least-squares fitting. *Advances in Colloid and Interface Science, 70*, 171–210. https://doi.org/10.1016/S0001-8686(97)00312-6.

Pedersen, J. S. (2002). *Modelling of small-angle scattering data from colloids and polymer systems in neutrons, X-rays and light*. Elsevier Science BV.

Pérez García, M., Zhang, Y., Hayes, J., Salazar, A., Zabotina, O. A., & Hong, M. (2011). Structure and interactions of plant cell-wall polysaccharides by two- and three-dimensional magic-angle-spinning solid-state NMR. *Biochemistry, 50*(6), 989–1000. https://doi.org/10.1021/bi101795q.

Petoukhov, M. V., & Svergun, D. I. (2005). Global rigid body modeling of macromolecular complexes against small-angle scattering data. *Biophysical Journal, 89*(2), 1237–1250. https://doi.org/10.1529/biophysj.105.064154.

Pettersen, E. F., Goddard, T. D., Huang, C. C., Couch, G. S., Greenblatt, D. M., Meng, E. C., et al. (2004). UCSF chimera - A visualization system for exploratory research and analysis. *Journal of Computational Chemistry, 25*(13), 1605–1612. https://doi.org/10.1002/jcc.20084.

Polzonetti, G., Battocchio, C., Iucci, G., Dettin, M., Gambaretto, R., Di Bello, C., et al. (2006). Thin films of a self-assembling peptide on TiO2 and Au studied by NEXAFS, XPS and IR spectroscopies. *Materials Science and Engineering: C, 26*(5), 929–934. https://doi.org/10.1016/j.msec.2005.09.062.

Rambo, R. P., & Tainer, J. A. (2013). Accurate assessment of mass, models and resolution by small-angle scattering. *Nature, 496*(7446), 477–481. https://doi.org/10.1038/nature12070.

Ravel, B., & Newville, M. (2005). Athena, artemis, hephaestus: Data analysis for X-ray absorption spectroscopy using IFEFFIT. *Journal of Synchrotron Radiation, 12*(4), 537–541. https://doi.org/10.1107/s0909049505012719.

Rongpipi, S., Del Mundo, J. T., Gomez, E. D., & Gomez, E. W. (2021). Resonant X-ray scattering of biological assemblies. *MRS Communications, 11*(1), 1–17. https://doi.org/10.1557/s43579-021-00020-4.

Samant, M. G., Stöhr, J., Brown, H. R., Russell, T. P., Sands, J. M., & Kumar, S. K. (1996). NEXAFS studies on the surface orientation of buffed polyimides. *Macromolecules, 29*(26), 8334–8342. https://doi.org/10.1021/ma951820c.

Skou, S., Gillilan, R. E., & Ando, N. (2014). Synchrotron-based small-angle X-ray scattering of proteins in solution. *Nature Protocols, 9*(7), 1727–1739. https://doi.org/10.1038/nprot.2014.116.

Stöhr, J. (2013). *NEXAFS spectroscopy. Vol. 25*. Springer Science & Business Media.

Su, G. M., Patel, S. N., Pemmaraju, C. D., Prendergast, D., & Chabinyc, M. L. (2017). First-principles predictions of near-edge X-ray absorption fine structure spectra of semiconducting polymers. *The Journal of Physical Chemistry C, 121*(17), 9142–9152. https://doi.org/10.1021/acs.jpcc.7b01353.

Svergun, D. I. (1992). Determination of the regularization parameter in indirect-transform methods using perceptual criteria. *Journal of Applied Crystallography, 25*(4), 495–503. https://doi.org/10.1107/S0021889892001663.

Svergun, D. I., Koch, M. H. J., Timmins, P. A., & May, R. P. (2013). Small angle X-ray and neutron scattering from solutions of biological macromolecules. *International Union of Crystallography Texts on Crystallography*. Oxford University Press.

Svergun, D. I., Petoukhov, M. V., & Koch, M. H. J. (2001). Determination of domain structure of proteins from X-ray solution scattering. *Biophysical Journal*, *80*(6), 2946–2953. https://doi.org/10.1016/s0006-3495(01)76260-1.

Thimm, J. C., Burritt, D. J., Ducker, W. A., & Melton, L. D. (2000). Celery (*Apium graveolens* L.) parenchyma cell walls examined by atomic force microscopy: Effect of dehydration on cellulose microfibrils. *Planta*, *212*(1), 25–32. https://doi.org/10.1007/s004250000359.

Virgili, J. M., Tao, Y., Kortright, J. B., Balsara, N. P., & Segalman, R. A. (2007). Analysis of order formation in block copolymer thin films using resonant soft X-ray scattering. *Macromolecules*, *40*(6), 2092–2099. https://doi.org/10.1021/ma061734k.

Volkov, V. V., & Svergun, D. I. (2003). Uniqueness of ab initio shape determination in small-angle scattering. *Journal of Applied Crystallography*, *36*(3), 860–864. https://doi.org/10.1107/s0021889803000268.

Watts, B. (2014). Calculation of the Kramers-Kronig transform of X-ray spectra by a piecewise Laurent polynomial method. *Optics Express*, *22*(19), 23628–23639. https://doi.org/10.1364/OE.22.023628.

Ye, D., Kiemle, S. N., Rongpipi, S., Wang, X., Wang, C., Cosgrove, D. J., et al. (2018). Resonant soft X-ray scattering reveals cellulose microfibril spacing in plant primary cell walls. *Scientific Reports*, *8*(1), 12449. https://doi.org/10.1038/s41598-018-31024-1.

Ye, D., Le, T. P., Kuei, B., Zhu, C., Zwart, P. H., Wang, C., et al. (2018). Resonant soft X-ray scattering provides protein structure with chemical specificity. *Structure*, *26*(11), 1513–1521.e1513. https://doi.org/10.1016/j.str.2018.07.018.

Zhang, F., Skoda, M. W. A., Jacobs, R. M. J., Martin, R. A., Martin, C. M., & Schreiber, F. (2007). Protein interactions studied by SAXS: Effect of ionic strength and protein concentration for BSA in aqueous solutions. *The Journal of Physical Chemistry B*, *111*(1), 251–259. https://doi.org/10.1021/jp0649955.

Zhang, T., Zheng, Y., & Cosgrove, D. J. (2016). Spatial organization of cellulose microfibrils and matrix polysaccharides in primary plant cell walls as imaged by multichannel atomic force microscopy. *The Plant Journal*, *85*(2), 179–192. https://doi.org/10.1111/tpj.13102.

CHAPTER SIX

Reconstruction of 3D density from solution scattering

Thomas D. Grant*
Department of Structural Biology, Jacobs School of Medicine and Biomedical Sciences, SUNY University at Buffalo, Buffalo, NY, United States
*Corresponding author: e-mail address: tdgrant@buffalo.edu

Contents

Methods in Enzymology, Volume 678
ISSN 0076-6879
https://doi.org/10.1016/bs.mie.2022.09.018

Abstract

Ab initio modeling methods have proven to be powerful means of interpreting solution scattering data. In the absence of atomic models, or complementary to them, ab initio modeling approaches can be used for generating low-resolution particle envelopes using only solution scattering profiles. Recently, a new ab initio reconstruction algorithm has been introduced to the scientific community, called DENSS. DENSS is unique among ab initio modeling algorithms in that it solves the inverse scattering problem, i.e., the 1D scattering intensities are directly used to determine the 3D particle density. The reconstruction of particle density has several advantages over conventional uniform density modeling approaches, including the ability to reconstruct a much wider range of particle types and the ability to visualize low-resolution density fluctuations inside the particle envelope. In this chapter we will discuss the theory behind this new approach, how to use DENSS, and how to interpret the results. Several examples with experimental and simulated data will be provided.

1. Introduction

1.1 History of modeling solution scattering data

Decades of effort have resulted in the development of dozens of algorithms for the extraction of useful structural information from solution scattering experiments (Bernadó, Mylonas, Petoukhov, Blackledge, & Svergun, 2007; Chacon, Moran, Diaz, Pantos, & Andreu, 1998; Kratky & Pilz, 1972; Petoukhov & Svergun, 2005; Stuhrmann, 1970; Svergun, Barberato, & Koch, 1995; Zheng & Tekpinar, 2011). A variety of different approaches have been used, largely clustered into ab initio methods (Chacon et al., 1998; Franke & Svergun, 2009; Kratky & Pilz, 1972; Stuhrmann, 1970; Svergun, 1999; Svergun, Petoukhov, & Koch, 2001; Svergun & Stuhrmann, 1991; Virtanen, Makowski, Sosnick, & Freed, 2011) and those requiring atomic models (Bernadó et al., 2007; Grudinin, Garkavenko, & Kazennov, 2017; Jiménez-García, Pons, Svergun, Bernadó, & Fernández-Recio, 2015; Knight & Hub, 2015; Panjkovich & Svergun, 2016; Perkins et al., 2016; Petoukhov & Svergun, 2005; Schneidman-Duhovny, Hammel, Tainer, & Sali, 2013, 2016; Svergun et al., 1995; Zheng & Tekpinar, 2011). Atomic modeling approaches have proven extremely powerful for discriminating between multiple plausible atomic models often generated from X-ray crystallography, cryoelectron microscopy, or computational modeling. In the absence of plausible atomic models (or complementary to them), ab initio approaches have shown to be useful for the visualization of low-resolution particle shapes directly from the solution scattering data, often leading to the generation of new hypotheses. In all modeling approaches described to date the forward scattering problem is solved, where models are generated, scattering profiles are calculated from these models, the quality of the fits to the experimental data are evaluated, and the model is subsequently adjusted to improve the fit. Until recently, no algorithm has solved the inverse scattering problem, where the intensities from the scattering profile are directly used to determine the 3D object.

1.2 Ab initio modeling

Historically, ab initio modeling has been applied with stringent assumptions about the particles in question. Initially, 3D models were described by simple shapes, such as spheres, cylinders, core shells, ellipsoids, etc., where the object is assumed to be homogenous (Henderson, 1996; Kratky & Pilz, 1972). To model the experimental data, the dimensions of the shapes are adjusted until the calculated scattering profile best matches the solution scattering profile. This approach applies primarily at very low resolutions describing only the overall particle size and gross shape features (e.g., elongated, globular, etc.). Variations on this approach have proven powerful and are commonly employed to describe particles whose shapes are approximately known (Doucet et al., 2017). An alternative development used spherical harmonics to describe the homogenous particle shapes (Stuhrmann, 1970; Svergun & Stuhrmann, 1991). This had the advantage of describing objects that were not simple geometric shapes, increasing the complexity of the models that could be obtained. The late 1990s brought the advent of coarse-grained bead modeling algorithms (Chacon et al., 1998; Svergun, 1999; Svergun et al., 2001; Walther, Cohen, & Doniach, 2000), which have since come to dominate the field of ab initio modeling. These approaches have the advantage of describing almost arbitrarily complex shapes and have proven to be surprisingly capable of reconstructing such shapes at low resolution from quantifiably low information content inherent in the 1D scattering profiles.

1.3 Ab initio density reconstruction

An important limitation of the ab initio approaches developed so far has been the implicit assumption of representing an object as a uniform, homogenous density within the particle envelope. The methods described above all attempt to model the *shape* of the particle, i.e., the envelope, and do not attempt to model the internal density fluctuations within the particle. A new approach that has been developed recently directly aims to address this point by calculating the 3D density function of the object (Grant, 2018). This approach, developed in our lab, is the first to solve the inverse scattering problem, i.e., where the intensities of the scattering profile directly determine the 3D density function. This chapter aims to describe the theory behind this approach, known as DENSS (DENsity from Solution Scattering), how to use DENSS, and how to interpret the results from DENSS.

2. Theory

2.1 Iterative phase retrieval

Detectors in scattering experiments record the intensity of radiation (most often X-rays or neutrons) scattered from an object (Chaudhuri, Muñoz, Qian, & Urban, 2017; Feigin & Svergun, 1987; Glatter, Kratky, & Kratky, 1982; Guinier, Fournet, & Yudowitch, 1955; Lattman, Grant, & Snell, 2018; Svergun, Koch, Timmins, & May, 2013). If one could collect the full complex valued structure factors in three dimensions, consisting of both the amplitudes and phases of the scattered waves, then one could reconstruct the 3D density of the object. However, in practice detectors detect the real valued intensities of the scattered wave, not the complex valued structure factors, yielding information about the amplitudes of the structure factors, but no information regarding the phases. This results in the so-called phase problem. One common algorithm to solving the phase problem is known as iterative phase retrieval which reconstructs an image knowing only the magnitudes of the object transform (Elser, 2003; Fienup, 1982, 2013). Iterative phase retrieval is used in fields as diverse as coherent diffractive imaging, single particle diffraction, and astronomy, even including the recently launched James Webb Space Telescope (Dean, Aronstein, Smith, Shiri, & Acton, 2006; Elser, 2003; Fienup, 1982, 2013; Fienup & Dainty, 1987; Miao, Sayre, & Chapman, 1998). In this approach, random starting phases are combined with observed amplitudes and an inverse Fourier transform is calculated to generate an image of the object. This initial image is often noisy and not representative of the actual object since the phases are random. However, knowledge about the real space properties of the object can be imposed in the real space image domain. One of the primary restraints imposed in real space is the fact that the object is typically finite in size. To impose this restraint the envelope of the object is identified, known as the object support, often from previous experiments (Marchesini et al., 2003), and the region outside the support is flattened, i.e., it is set to a constant value. Additional real space restraints can also be imposed such as positivity of the object density. After real space density modification is performed, the forward Fourier transform is calculated to determine the new structure factors. The structure factor amplitudes will no longer match the experimentally derived amplitudes, due to the imposed real space restraints, thus the structure factors are rescaled to match the observed amplitudes. This process is repeated iteratively for many cycles until convergence, at which point in successful cases the correct phases will have been retrieved, along with the object image.

2.2 Overview of DENSS iterative structure factor retrieval

In solution scattering experiments, particles tumbling in random orientations scatter radiation in all directions. This summation of many individual anisotropic single particle scattering patterns results in an isotropic scattering pattern collected on a 2D area detector. Whereas in a single molecule X-ray diffraction image the 2D pattern will fluctuate in intensity throughout the image, in a solution scattering experiment the 2D detector image no longer exhibits anisotropic fluctuations, but instead only exhibits a gradual decay in intensity as a function of the distance from the beam center corresponding to the scattering angle. The 2D detector image is radially integrated and averaged to reduce the data to a 1D scattering profile of intensity, I, as a function of scattering angle, most often transformed into momentum transfer $q = 4\pi \sin(\theta)/\lambda$, where 2θ is the scattering angle and λ is the wavelength of the scattered radiation. The spherical averaging of the intensity results in a significant loss of information in solution scattering experiments. In particular, small angle scattering (SAS) experiments are not only low resolution (by nature of the limiting small angles), but are also low information, and often contain between 10 and 30 unique pieces of information (Damaschun, Mueller, & Pürschel, 1968; Shannon, 1948; Svergun & Koch, 2003; Taupin & Luzzati, 1982). Despite the limited information content in SAS profiles, modeling algorithms have proven capable of reconstructing low-resolution 3D shapes from the 1D experimental profiles. In solution scattering not only are the 3D phases of the structure factors not known, but the 3D amplitudes are also unknown. Only the 1D spherically averaged intensities are measured. Due to this loss of information, iterative phase retrieval cannot be directly applied for reconstructing a particle shape using the 1D intensity profile. DENSS operates based on a similar but modified version of iterative phase retrieval, called iterative structure factor retrieval (Grant, 2018). This approach, like iterative phase retrieval, solves the inverse scattering problem by directly calculating the 3D density function using the 1D intensities.

To describe how this is possible, the problem is first divided into the two relevant domains, the real space domain containing the object density, and the reciprocal space domain, containing the structure factors. These two domains are related by the forward and inverse Fourier transforms. Each domain is represented as a 3D image, i.e., a 3D array of points (Fig. 1). For computational efficiency, we use a Fast Fourier Transform (FFT (Cooley & Tukey, 1965)) to relate the two images and typically set the number of grid points on a side to be a power of two (though this is not required). Additionally, we assume the objects in the real space domain have real valued density, and thus enforce this property on the real space image. The reciprocal space domain however is inherently complex valued, where each grid point represents a single structure factor consisting of the amplitude and phase at that point. Since the object image is real valued, the corresponding reciprocal space image is conjugate symmetric. Since the two domains are related by the FFT, certain properties of an image in one domain can be derived from properties of the corresponding image in the other domain. For example, the real space dimensions of the image (i.e., the side length, L_x) containing the particle density directly determine the voxel size in the Fourier domain (i.e., the q-sampling, Δq), where $\Delta q = 2\pi/L_x$. Similarly, the voxel size in real space (i.e., the real space sampling, Δx) determines the reciprocal space image size, L_q, where $\Delta x = 2\pi/L_q$. This can often be thought of more intuitively as the real space resolution (or more accurately the sampling) being inversely proportional to the maximum q value. The initial set up of the real space grid, in particular the side length L_x, must be given to the algorithm, since the algorithm itself has no knowledge of the overall size of the object. A balance must be struck between having a real space image size large enough to contain the object and having a voxel size small enough to accurately represent the contours of the object features. There exists a fundamental lower limit to the size of the real space image that can be derived from Shannon information theory (Shannon, 1948). From Shannon information theory it can be surmised that the image size must be at least twice the maximum dimension of the particle, D_{max}, to capture the relevant frequencies in the Fourier

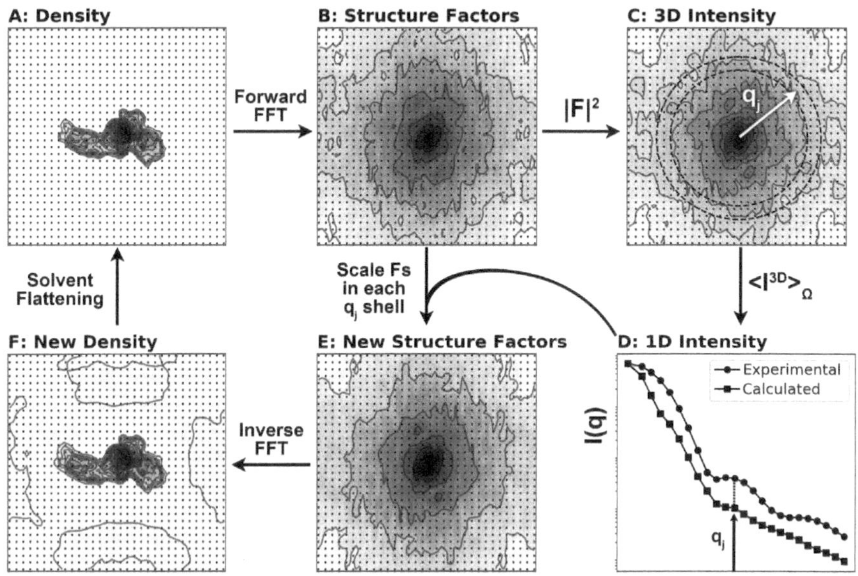

Fig. 1 Iterative structure factor retrieval scheme. (A) The starting 3D density for the iterative retrieval cycle is shown as a flattened image and contour map on a 2D grid for illustration purposes. Each step begins with the solvent flattened density output by the last step, except for the first step which begins with random density (not shown). (B) Using a forward FFT the structure factors, F, are calculated from the starting density (for illustration purposes only the amplitudes are shown). (C) The 3D intensities are then calculated as the square of the amplitudes. The dashed circles outline the concentric q shells used to bin grid points for subsequent spherical averaging. (D) The 1D spherically averaged intensities (denoted by $\langle I^{3D} \rangle_\Omega$) are calculated and compared with the experimental 1D scattering profile, and used to calculate scale factors for each q shell. (E) Each spherical shell of structure factors is scaled using the scale factor, k, calculated in the previous step, resulting in a new set of structure factors. (F) The inverse FFT is performed to obtain a new density map whose corresponding scattering profile now matches the experimental 1D scattering profile. Real space restraints such as solvent flattening are then applied to the new density and the cycle is repeated.

domain (Damaschun et al., 1968; Taupin & Luzzati, 1982). Thus, in addition to the experimental scattering profile, DENSS also requires the particle D_{max} to be given to set up the grid, though DENSS can also estimate this directly from the data if desired (described below). DENSS calculates the grid size using two parameters, D_{max} and the number of times larger than D_{max} the grid size should be, called the oversampling. The oversampling should thus be a minimum of two. By default, oversampling is set to three to guard against underestimates of D_{max}. Thus, the real space side length is calculated to be $L_x = \text{oversampling} * D_{max}$, and each real space grid point has a corresponding real space coordinate, (x, y, z). The reciprocal space image size is then calculated along with the corresponding coordinates for each grid point in reciprocal space, (q_x, q_y, q_z). The magnitude of the q vector for each grid point is $q = |\vec{q}| = \sqrt{q_x^2 + q_y^2 + q_z^2}$.

Once the system is set up, the next step is to perform the iterative structure factor retrieval. Fig. 1 shows the outline of the iterative structure factor retrieval process. Each step of the cycle begins with the real space density (Fig. 1A). The real space density is the output of the previous iteration, however for the first step we fill the real space image with uniformly distributed random density by assigning a random value to each grid point. Next, the reciprocal space image, i.e., the structure factors, F, are calculated using the forward FFT (Fig. 1B). The amplitudes of the structure factors, |F|, can be then calculated as the magnitudes of the complex valued structure factors and squared to obtain the 3D intensities (Fig. 1C). If we had a 3D diffraction volume, as in single particle X-ray diffraction experiments (Hajdu, 2000), we could directly scale the 3D structure factors such that the calculated amplitudes match the observed amplitudes (i.e., the square root of the observed 3D intensities). However, in solution scattering experiments we do not measure 3D intensities, we only measure the spherically averaged 1D intensity profile. Thus, we lack the necessary information to scale the individual structure factors in 3D. However, we can still use the low information content 1D scattering profile to modify our structure factors. To do this, we bin each 3D grid point according to the magnitude of the q vector, i.e., its distance from the origin of reciprocal space. This binning results in a series of concentric q shells (region outlined by dashed circles in Fig. 1C). We then calculate the corresponding 1D spherically averaged intensity profile by taking the average of all 3D intensities that fall within each bin. This generates the 1D calculated scattering profile which can be compared with the 1D experimental scattering profile (Fig. 1D). The key step is the following. We are ultimately interested in a set of structure factors whose corresponding 1D spherically averaged intensities matches the experimental solution scattering profile. To obtain this, using the 1D experimental scattering profile we scale the structure factors such that the corresponding spherically averaged intensities match the observed 1D solution scattering profile (Fig. 1E). Since we do not have 3D intensity information, we cannot obtain individual scale factors for each grid point within a bin. Instead, we determine a single scale factor for an entire bin and multiply the structure factor of each grid point within that bin by that same scale factor. The scale factor for a particular bin is calculated as the square root of the ratio of the experimental to the calculated 1D intensity for the q value corresponding to that bin. Mathematically this can be explained as follows.

Let F_i be the calculated structure factor for a particular grid point, i. The 3D calculated intensity of the grid point is then $|F_i|^2$. Let $I_c(q_j)$ be the spherically averaged 1D intensity for bin j containing N_j grid points, where each grid point has a corresponding q magnitude falling within the jth bin. The calculated 1D intensity can be expressed as the average intensity of all grid points in the bin,

$$I_c\left(q_j\right) = \frac{1}{N_j} \sum_{i \in j}^{N_j} |F_i|^2$$

The goal is to scale each structure factor F_i by a scale factor, k_i, such that the spherical average in the bin matches the experimental 1D intensity, $I_e(q_j)$ (in practice, the experimental scattering profile is interpolated to the q values corresponding to the center of each bin). This can be expressed as follows:

$$I_e\left(q_j\right) = \frac{1}{N_j} \sum_{i \in j}^{N_j} |k_i F_i|^2$$

$$I_e\left(q_j\right) = \frac{1}{N_j} \sum_{i \in j}^{N_j} k_i^2 |F_i|^2$$

However, as mentioned above, we do not have enough information to determine each k_i. Instead, we assume the same scale factor, k, applies to each grid point, thus $k = k_i$. We can now solve for k as follows:

$$I_e\left(q_j\right) = \frac{1}{N_j} \sum_{i \in j}^{N_j} k^2 |F_i|^2$$

$$I_e\left(q_j\right) = k^2 \frac{1}{N_j} \sum_{i \in j}^{N_j} |F_i|^2$$

$$I_e\left(q_j\right) = k^2 I_c\left(q_j\right)$$

$$k^2 = \frac{I_e\left(q_j\right)}{I_c\left(q_j\right)}$$

$$k = \sqrt{\frac{I_e\left(q_j\right)}{I_c\left(q_j\right)}}$$

We now multiply each structure factor, F_i, by the scale factor k. We perform this calculation independently for each q bin. This results in a new set of structure factors whose corresponding 1D intensities match the experimental data.

Next the new real space object image can be calculated using the inverse FFT (Fig. 1F). Near the beginning of the reconstruction, this will generate a very noisy density that does not resemble the object, even though the corresponding scattering profile matches the experimental data. This is because there is a nearly infinite number of density functions that yield the same scattering profile. Hence, while the spherically averaged intensity matches the data, the 3D intensities within each q bin are still randomly distributed, as are the phases, due to the initial random density. The additional information required to reconstruct the object density comes through real space restraints. As in iterative phase retrieval, the finite size of the object can be utilized as a restraint to improve the density of the object. By identifying which grid points are likely to belong to the particle, called the object support, the voxels outside the support can then be flattened, i.e., set to a constant value. The method for identifying the object support is described in the next section. Additionally, if applicable, positivity of the object density can also be enforced, where any negative density points are set to zero. After the real space restraints are applied, the object density no longer has a corresponding calculated 1D scattering profile that matches the data. Thus, the forward FFT is calculated to determine the new structure factors, and the cycle is repeated iteratively until convergence. After completion, the final set of structure factors has a corresponding 1D intensity function that agrees with the experimental data and a corresponding real space density function that agrees with the real space restraints. The real space density function can then be visualized, interpreted, and used for further analysis. This approach solves the inverse scattering problem for the first time by using the experimental 1D intensities to directly determine the 3D structure factors to generate the object density.

2.3 Determination of object support using Shrinkwrap

In coherent diffractive imaging or single particle X-ray diffraction an estimate of the object support is often known ahead of time (Marchesini et al., 2003). However, in solution scattering this is often not the case, as solution scattering experiments are often performed with the express purpose of determining the object shape. In DENSS, we must therefore determine the object support ab initio, as well as the object density. To do this, DENSS utilizes an approach known as Shrinkwrap developed previously and used commonly in coherent diffractive imaging (Marchesini et al., 2003). Shrinkwrap itself is an iterative method, as it requires regularly updating the object support as the object density improves. To begin, DENSS makes a copy of the real space image (Fig. 2A) and uses a Gaussian low-pass filter to blur the object density (Fig. 2B). This low-pass filter has the effect of making neighboring points of noisy density more similar to each other. Next, a threshold is set, typically about 20% of the maximum value of the filtered density. Grid points where the filtered density is greater than this threshold are selected as part of the object support (Fig. 2C). As the iterative structure factor retrieval progresses, the object support is updated using Shrinkwrap regularly, typically every 20 steps or so. The density gradually increases inside the support as the noise outside the support decreases as the solvent region is flattened (Fig. 2D), and subsequently the size of the support gradually decreases. In the early steps of the iterative structure factor retrieval process the density is very noisy, thus the low-pass filter uses a very broad Gaussian for blurring, and the resulting support is very large. As the object density improves, the width of the Gaussian is narrowed to ensure finer features of the object density are included in the object support, until a set minimum is obtained, at which point the width of the Gaussian no longer changes (though the support is still regularly updated with Shrinkwrap).

2.4 Solvent considerations

In the atomic modeling and ab initio modeling approaches described above, an important contribution to the scattering profile that must be considered and calculated is the contribution from the solvent (Grudinin et al., 2017; Knight & Hub, 2015; Petoukhov & Svergun, 2005; Schneidman-Duhovny et al., 2013, 2016; Svergun, 1999; Svergun et al., 1995, 2001; Virtanen et al., 2011). Virtually all solution scattering experiments are not performed with the particle in a vacuum, but instead the particle is

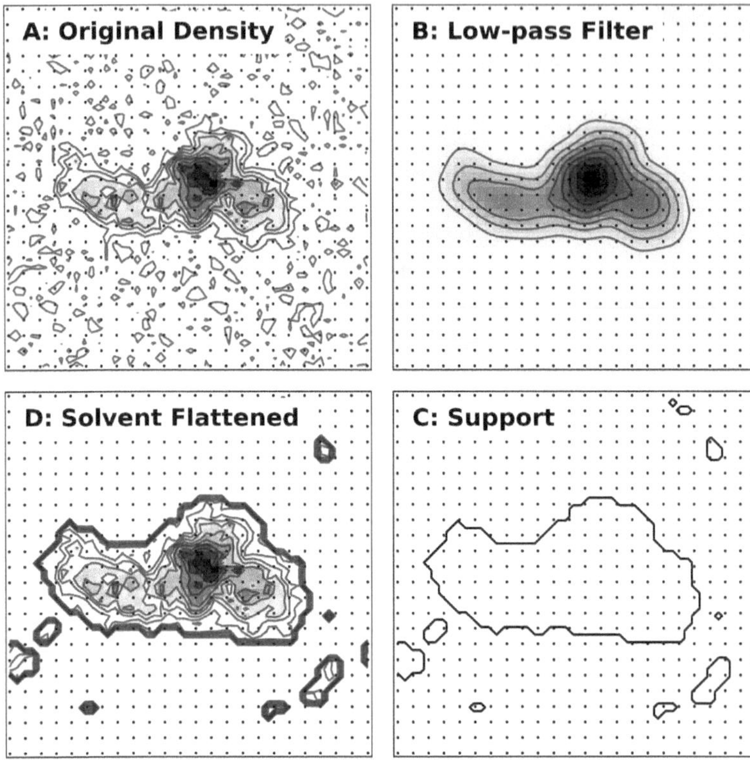

Fig. 2 Shrinkwrap procedure. (A) The noisy 3D density is shown as a flattened 2D image and a contour map on a 2D grid of voxels, represented by black dots. For illustration purposes the noise is artificially increased. (B) A low-pass Gaussian filter is applied to a copy of the original density. (C) The object support, calculated as the set of filtered voxels above a given threshold, is shown outlined by a solid black line. (D) Solvent flattening is performed by setting the density of voxels outside of the support to zero.

embedded in a solvent, often aqueous solvent. To deal with this average bulk solvent scattering, solution scattering experiments require collecting a second scattering pattern from the bulk solvent only. This bulk solvent scattering profile is subtracted from the scattering profile of the particle in the solvent. However, this bulk solvent scattering subtraction does not result in a profile that emulates the particle in vacuum, but instead represents the *contrast* of the object density relative to the average bulk solvent density. This affects not only the scale of the scattering intensity, but also the shape of the profile. Additionally, solvent often associates with the particle surface resulting an approximately 10–15% higher average density near the surface for proteins, often called the hydration shell

(Svergun et al., 1995). Modeling algorithms must account for these additional solvent scattering terms by modeling them typically by fitting free parameters (Schneidman-Duhovny et al., 2016; Schneidman-Duhovny, Kim, & Sali, 2012; Svergun et al., 1995). This is not trivial, and often the discrepancies in modeling the scattering profiles, particularly at larger scattering angles, are caused by inaccurate solvent models (Schneidman-Duhovny et al., 2012; Tuukkanen, Kleywegt, & Svergun, 2016). One advantage of DENSS is that it implicitly deals with these solvent contributions by representing the object as a 3D density, rather than an atomic or coarse-grained model. The output of DENSS will inherently be the object contrast relative to the bulk solvent with the hydration shell accounted for implicitly. Since the bulk solvent scattering is subtracted from the particle scattering, the solvent region of the image (i.e., the region outside of the support) is set to zero during the solvent flattening step (Fig. 2D).

2.5 Uniqueness

The number of unique pieces of information required to describe an object scales with the cube of the resolution of the image. The number of unique pieces of information in a scattering profile scales linearly with resolution. Thus, the relative amount of information that can theoretically be reconstructed from a scattering profile gets significantly worse with increasing resolution. The practical effect of this is that increasing the experimentally determined q range to wider scattering angles does not directly increase the resolution of a reconstruction. The resolution of DENSS reconstructions, which will be discussed in a later section in detail, is limited by the low information content inherent in a solution scattering profile. This is the so-called uniqueness problem and has previously been quantified for solution scattering (Grant, 2018). DENSS begins with assigning random values to the object image to generate a random starting density. Since the solution scattering profile does not contain sufficient information to reconstruct the object density uniquely, each time DENSS is run the resulting object density will be different, due to a different initial random density. However, running DENSS multiple times results in object densities that are similar at low resolution, but whose density fluctuations at high resolution are often inconsistent. To obtain a single consistent low-resolution object density, each of the individual reconstructions are aligned to one another and averaged. This step helps to reduce the density of features that are not common between different reconstructions and helps to amplify features that are common. The details of the alignment and averaging procedure will be given in a later section. As will be seen later in Section 6.4, the resolution achievable by DENSS is fundamentally limited by the nature of a solution scattering experiment due to this uniqueness problem and low information content, and is ultimately largely dependent on the particle size, which dictates the amount of information present in the data.

3. DENSS software

3.1 DENSS software suite

DENSS is written to be as user friendly as possible and consists of a suite of scripts including both the core functions of DENSS as well as several helper scripts for assistance with setup and analysis. DENSS is written in Python using NumPy and SciPy modules for speed and efficiency (van der Walt, Colbert, & Varoquaux, 2011; Virtanen et al., 2020). DENSS uses a series of command line scripts for users to interface with the core functionality of DENSS. The script titles are most often of the form "denss.*.py," with the specific script function placed in the script name. DENSS is open source under the GPLv3 license. The open-source nature of DENSS has allowed it to be integrated into other software, including the popular BioXTAS RAW software for SAS data reduction and processing (Hopkins, Gillilan, & Skou, 2017). RAW offers a convenient GUI interface along with a binary installer for most operating systems and does not require a separate installation of DENSS to run. The most up to date details for downloading, installing, running and interpreting DENSS results are available online at both the GitHub repository (https://www.github.com/tdgrant1/denss) and the DENSS website (https://www.tdgrant.com/denss). A single reconstruction with DENSS takes approximately 3 min on a modern laptop computer (on a 2.9 GHz Intel Core i9 processor), whereas the full pipeline including running 20 reconstructions, alignment and averaging takes anywhere from approximately 20 to 90 min to complete, depending on the computational hardware available. Multicore processing and GPU acceleration are implemented to further speed up the calculations as well.

The input to DENSS is a 1D scattering profile and the output is a 3D MRC-formatted density map which can be visualized in several common graphical software programs such as UCSF Chimera, PyMOL, and COOT (Emsley, Lohkamp, Scott, & Cowtan, 2010; Pettersen et al., 2004;

Schrodinger, LLC, 2015). The basic outline of steps to run the complete DENSS pipeline to achieve a useful result is given below. After outlining the steps, we will then describe each step in more detail.

3.2 DENSS workflow

DENSS is quite simple to run. It will be useful to first provide a general workflow for running DENSS, before going into each step in more detail, including various options available. The basic workflow for running DENSS can be summarized as follows:

1. Run denss.fit_data.py to fit raw data with a smooth function and estimate D_{max}.
2. Run denss.py to reconstruct a single density map and optimize parameters.
   a. Assess fit to data and convergence criteria.
   b. Modify denss.py parameters if necessary and repeat step 2 until reconstruction is acceptable.
3. Run denss.all.py to perform full averaging of 20 reconstructions.
   a. Assess resolution, average R_g, etc.
   b. Optionally run refinement of average density with denss.refine.py.
4. Visualize final density in Chimera/PyMOL/etc.
5. Optionally align PDBs, visualize in Chimera/PyMOL/etc.

4. Preparing the data with denss.fit_data.py

As mentioned above, DENSS must interpolate a given scattering profile to the calculated q values in the centers of the concentric q bins, which are often separated in q space much more widely than the experimentally measured q values. To reduce the effect of noise during this interpolation step, DENSS does not work directly with raw experimental data but instead uses a smooth fit to the noisy experimental profile. Using the smooth fit to the experimental data provides two important properties for DENSS. First, the smooth fit provides an estimate of the forward scattering at q equal to zero, $I(q=0)$. This forward scattering intensity is required for DENSS to reconstruct an accurate object density, yet is not measured in an experimental scattering profile due to its coincidence with the incident beam. Secondly, using this smooth fit takes advantage of the fact that scattering profiles are typically highly oversampled, making the interpolated values far more precise. For example, in a standard biological small angle

X-ray scattering (SAXS) experiment, the number of data points is typically on the order of 500 or more. By applying Shannon theory to solution scattering (Damaschun et al., 1968; Moore, 1980; Rambo & Tainer, 2013; Shannon, 1948; Taupin & Luzzati, 1982), it can be determined that the number of unique pieces of information is equal to the q range of the data divided by the width of so-called Shannon channels. The width of a Shannon channel is π/D_{max}, where D_{max} is the maximum dimension of the particle. For a protein with a D_{max} of, say, 100 Å, the instrument setup at a standard synchrotron SAXS beamline may result in q range from 0.01 to 0.50 $Å^{-1}$. Therefore, the approximate number of unique pieces of information in the SAXS profile (i.e., the number of Shannon channels) is $(q_{max} - q_{min})/(\pi/D_{max}) \approx 15$. For a typical SAXS profile containing some 500 data points or more, each unique piece of information therefore has been oversampled some 33 times. By fitting a smooth function to the noisy experimental profile, DENSS takes advantage of the oversampling when it interpolates, resulting in more widely spaced q values that have all the precision offered by the original raw experimental profile.

The most common way to fit a scattering profile with a smooth function is using an indirect Fourier transform (IFT), which produces a pair distribution function (P(r)) in addition to the smooth fit to the 1D intensities. Originally proposed by Glatter (Glatter, 1977), several algorithms for performing an IFT have been proposed (Hansen & Pedersen, 1991; Moore, 1980; Rambo & Tainer, 2013; Semenyuk & Svergun, 1991; Vestergaard & Hansen, 2006). DENSS also offers its own script to calculate the IFT, called denss.fit_data.py. This script uses a modified IFT based on the method of Moore (Grant, 2022; Moore, 1980) to fit the raw data with a trigonometric series. The smooth fit to the data is stored in a *.fit* file, which can be used in DENSS, which reads the D_{max} directly from the file header. Running denss.fit_data.py opens a graphical user interface (GUI) showing the raw data, the fit to the data, and the corresponding P(r) curve (Fig. 3). The script will attempt to automatically estimate the optimal parameters, including D_{max} and the smoothing parameter, α, which can then be manually optimized if desired.

Arguably one of the most popular software programs for performing the IFT is the GNOM program available from the ATSAS software package (Manalastas-Cantos et al., 2021; Semenyuk & Svergun, 1991). Due to its common use, DENSS allows the input of GNOM *.out* files directly. In addition to providing the smooth fit, DENSS will also read D_{max} directly from the file, thus not requiring the separate input of D_{max} at the command

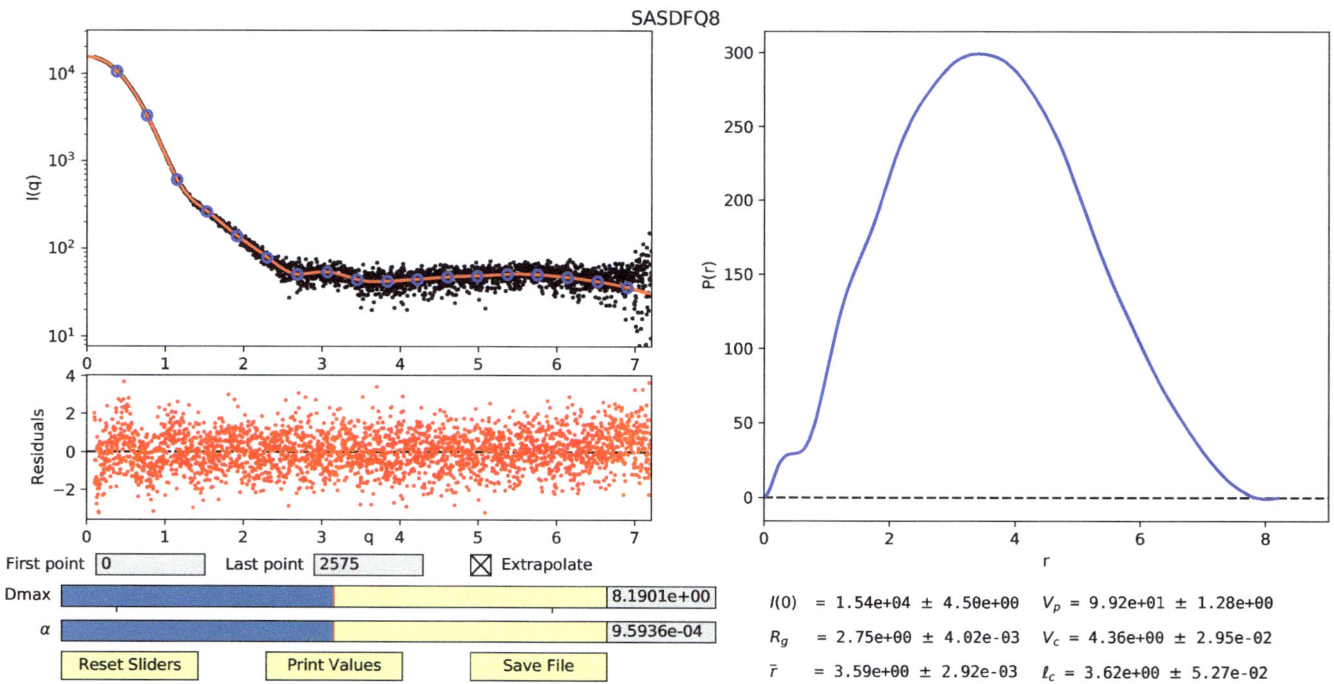

Fig. 3 Fitting experimental data with denss.fit_data.py GUI. A screenshot of the interactive GUI of denss.fit_data.py is shown. The upper left panel shows the experimental scattering profile in black dots, the fit to the data as a red line, and the intensities at the Shannon points as blue circles. The Residuals between the experimental data and the fit are shown below the intensities. The upper right panel shows the P(r) curve corresponding to the fit. Sliders for adjusting Dmax and α are shown on the lower left. Various parameters describing the size of the particle calculated from the fit are displayed on the bottom right.

line. However, it is recommended not to use the default settings of GNOM when preparing fits for use in DENSS. By default, GNOM limits the maximum q range of the fit to $q^*R_g < 8$, due to the inability of subsequent bead modeling algorithms to accurately model the solvent at higher q ranges (Franke & Svergun, 2009; Svergun, 1999; Svergun et al., 2001; Tuukkanen et al., 2016). However, since DENSS deals explicitly with density, it is fully capable of utilizing the full experimental q range. Thus, it is recommended to manually run GNOM, or alternatively run denss.fit_data.py, to fit the full experimental q range when preparing data for density reconstruction. While GNOM and denss.fit_data.py are the two most common IFTs used prior to running DENSS, any algorithm that prepares an adequate fit to the raw data can be used, such as the popular Scatter software (Rambo, 2022), formatting the fit as a simple three column ASCII text file with columns q, intensity, and error bars.

5. Running a single reconstruction with denss.py

5.1 Overview of basic parameters and modes of denss.py

The method for running a single reconstruction with DENSS is the denss.py script. The minimal input to the denss.py script is the 1D scattering profile. denss.py requires the D_{max} to set up the real space image, which can be set by the user, read from the input file, or estimated automatically. As mentioned above, DENSS does not use the raw experimental profile, but the smooth fit to the profile. Thus, the input to denss.py can be either the GNOM *.out* file, the denss.fit_data.py *.fit* file, or a *.dat* file. If the *.out* or *.fit* files are given, then the D_{max} is read directly from the file. If a *.dat* file is given, then denss.py checks whether the scattering profile is raw data or a smooth fit to the data by checking whether or not intensity at $q=0$ exists (since the fit will have intensity at $q=0$, but experimental data will not). If the data are raw data, then denss.py will use the algorithm in denss.fit_data.py to estimate D_{max} and generate a fit to the curve automatically. While the automated IFT in DENSS is quite robust, it is always preferable to manually run denss.fit_data.py first to ensure the fit is correct.

After initiating denss.py, DENSS will proceed to set up the real space and reciprocal space images as described above. To begin, DENSS creates a 3D cubic box as an array of points to represent the real space image. The real space dimensions of the box are then calculated based on the D_{max} given and the oversampling ratio, described above. The length of the side of the box is set to D_{max} times the oversampling ratio. The default oversampling ratio is three, such that a particle D_{max} of 100 Å will result in a box size of 300 Å on each side. It is important to note that D_{max} is not used internally by DENSS for anything other than setting the size of the box. Thus, even if D_{max} is incorrect, the oversampling ratio, set to three by default, will typically compensate for the error and allow DENSS to run adequately.

DENSS provides numerous parameters that can be edited by the user. For convenience, DENSS provides three modes of use, each of which sets the various parameters to suitable defaults for the given mode. The three modes are SLOW, FAST, and MEMBRANE. By default, DENSS runs in SLOW mode. SLOW mode is distinguished from FAST mode primarily by the number of grid points (i.e., the number of "samples"), N, in each dimension of the real space image. The size of each voxel is then dictated by N and the box size, such that the voxel size equals the box size divided by the number of samples. The FFT is most efficient when N is an integer power of two, such as 32, 64, or 128 (though this is not required). In SLOW mode, the number of points for each dimension is 64, this typically provides a reasonably small voxel size adequate for ensuring appropriate contouring of features in the particle density. Despite the name, SLOW mode runs fairly quickly, with a single reconstruction taking about 3 min on a modern laptop computer. The speed of the algorithm is largely dictated by N. FAST mode uses $N=32$ to generate particle density maps quickly, only taking approximately 15 s. While FAST mode allows for more rapid assessment, the voxel sizes are twice as large in length (and thus eight times as large in volume) as in SLOW mode, and may result in lower resolution reconstructions that are more "pixelated" when viewing the density maps in graphical viewers (Fig. 4). Hence, SLOW mode is recommended, and FAST mode can be used for a quick initial assessment before running the reconstruction in SLOW mode. In addition to running single reconstructions faster, FAST mode also results in smaller file sizes, and requires less time during the alignment and averaging steps later.

SLOW mode is the default as it is typically appropriate for most particle shapes, even complex particle shapes. Despite the name, SLOW mode runs reasonably fast on a modern laptop, completing a typical reconstruction in approximately 1–2 min. FAST mode can occasionally be utilized for simple particle shapes, but even in such cases it is recommended to use SLOW mode. Fig. 4 shows some examples of single reconstructions of standard proteins. These examples use experimental data taken from the Benchmark

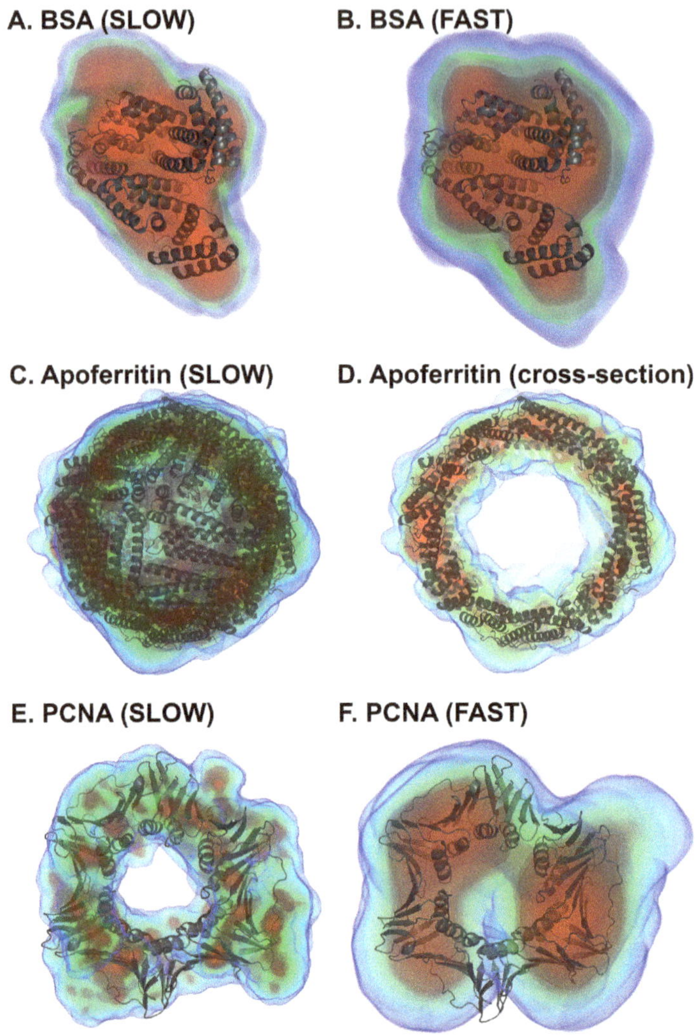

Fig. 4 Single reconstructions of density using DENSS. (A) A single reconstruction of BSA from experimental data using the default SLOW mode of DENSS. (B) A single reconstruction of BSA in FAST mode. (C) A single reconstruction of apoferritin from experimental data using SLOW mode. (D) A cross-section of the density displayed in (C) showing the interior cavity in apoferritin. (E) A single reconstruction of PCNA in SLOW mode showing the distinctive central hole. (F) A single reconstruction of PCNA in FAST mode. All density maps are displayed using the Volume representation in PyMOL colored according to density value (blue: 1σ, green: 5σ, red: 10σ). Atomic models taken from the SASBDB entry are displayed in gray cartoon representation.

section of the SASBDB (Valentini, Kikhney, Previtali, Jeffries, & Svergun, 2015). Fig. 4 shows the density map of monomeric bovine serum albumin (66 kDa, SASDCQ8 (Valentini et al., 2015)) reconstructed using both SLOW mode and FAST mode. Fig. 4C shows the density map of apoferritin (479 kDa, SASDFN8 (Graewert et al., 2020)) reconstructed using SLOW mode. Apoferritin contains a large cavity in its core, and Fig. 4D shows a cross-section through the middle to illustrate that DENSS is able to reconstruct the hollow cavity clearly. Objects that have more complex shapes, such as PCNA containing a distinctive central hole (90 kDa, SASBDB7 (Prakash, Moharana, Wallace, & Doublié, 2017)), may require SLOW mode to accurately reconstruct key features. As shown in Fig. 4E and F, the reconstruction of PCNA in SLOW mode shows the central hole more clearly than in FAST mode.

In addition to the FAST and SLOW modes, there also exists MEMBRANE mode. By default, SLOW and FAST modes enable an additional option called positivity. Positivity ensures that no density values are permitted to be negative by setting all negative density points to be zero each step. In most cases this is an appropriate restraint. However, in some cases the object may have regions that are lower density than the bulk solvent. Such cases may include, for example, lipid nanoparticles, detergent micelles, or membrane proteins embedded in such particles, that have low density interiors where the hydrophobic tails associate. The resulting average density in this region is often lower than the average bulk solvent density. In such cases the resulting contrast is negative. Thus, if enabled, the positivity restraint may result in artifacts in the density reconstructed by DENSS. MEMBRANE mode is nearly identical to SLOW mode except that it disables the positivity restraint, allowing for negative contrast to appear in the density map. This mode thus relies entirely on solvent flattening as a real space restraint. While largely intended for particles with negative contrast, this mode is also sufficient for reconstructing standard particles without negative contrast, though this may result in slightly noisier density maps (due to the removal of the additional positivity restraint). Thus, if it is unknown whether the particle has negative contrast, it is generally appropriate to use MEMBRANE mode. Fig. 5 shows a single DENSS reconstruction of a DDM micelle from experimental data using both SLOW and MEMBRANE modes (Gabel, Engilberge, Perez, & Girard, 2019). The micelle reconstruction using MEMBRANE mode shows the expected spherical shape with the negative contrast in the interior (Fig. 5A), while the SLOW mode reconstruction fails to show the expected spherical shape

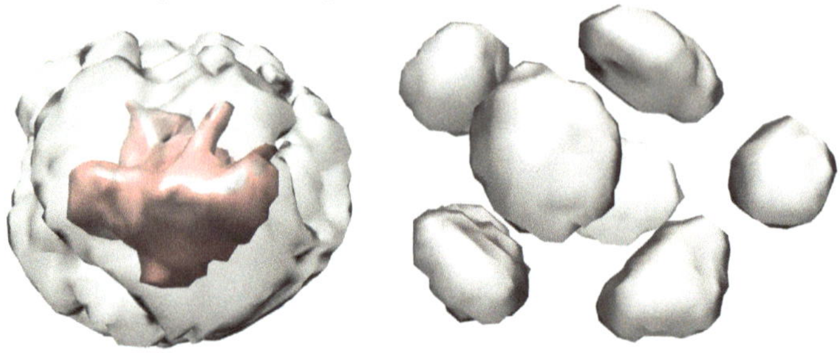

Fig. 5 DENSS reconstruction of a micelle with negative contrast. The density map reconstructed by DENSS using experimental data of a DDM micelle using either MEMBRANE mode (A) or SLOW mode (B). Density is displayed as an isocontour surface at $+1\sigma$ (gray transparent surface) and -1σ (pink surface). Note that the negative contrast is only seen in MEMBRANE mode.

and instead shows a collection of small disconnected regions of density (Fig. 5B). It is clear from this example that MEMBRANE mode is critical to reconstructing accurate particle densities when particles contain significant negative contrast. The ability of DENSS to reconstruct negative contrast is a unique feature compared to traditional uniform density modeling approaches. Notably, DENSS is capable of reconstructing particles with negative contrast using the entire experimental q range, in contrast to bead modeling algorithms which must significantly limit the maximum q range to approximate the object shape while ignoring negative contrast (Flayhan et al., 2018). Additionally, DENSS is capable of reconstructing particles with negative contrast from a single scattering profile, compared with even multi-phase bead modeling algorithms which require multiple scattering curves with different contrasts to generate accurate models (Gabel et al., 2019; Svergun, 1999).

5.2 Assessing the quality of reconstructions

DENSS runs typically for several thousands of steps. Each step, DENSS will calculate three values including χ^2 as a metric quantifying the quality of the fit of the calculated scattering profile to the experimental scattering profile, the radius of gyration (R_g) of the object density, and the volume of the support region. As the reconstruction progresses, the χ^2 value will generally decrease rapidly as the density map improves. After several thousand steps, the χ^2 will drop less and less rapidly to the point that the fit will no longer improve substantially even with many additional steps. At this point,

DENSS assumes that convergence has been achieved, i.e., that further iterations of the iterative structure factor retrieval cycle will not reduce the χ^2 substantially. Convergence is specifically calculated as the step where the standard deviation of χ^2 values over the previous 100 steps is less than 0.1% of the average χ^2 for those same 100 steps. This ensures that the variation in χ^2, and thus in the variation in the object density, is small. In some cases, this convergence criterion is not reached within a reasonable number of steps, and thus the maximum number of steps is 10,000. Once DENSS reaches 10,000 steps, it will stop the iterative structure factor retrieval cycle and write out the density map as an MRC file. The maximum number of steps can be changed. The progression of χ^2 as a function of step number, along with the progression of R_g and support volume, are stored and saved as a text file as well as plotted in a figure (an example of which is shown in Fig. 6A). To confirm that DENSS has indeed converged, despite reaching the maximum number of steps, the plots of χ^2, R_g, and support volume should be investigated to ensure that little change is occurring by the end of the reconstruction. Fig. 6A shows an example of a successful reconstruction that has converged, shown by the asymptotic decay of all three metrics. In the rare case of the maximum number of steps being reached before convergence was truly obtained (e.g., if χ^2 fluctuates significantly until the last step), the threshold for convergence can be adjusted when running denss.py or disabled entirely to run any desired number of steps.

As described above, DENSS uses the 1D scattering profile to directly determine the 3D structure factor amplitudes using the data to scale the structure factors in reciprocal space. Thus, the calculated scattering profile after this reciprocal space scaling step will exactly equal the interpolated experimental scattering profile. Therefore, the χ^2 value after this step will be zero. To calculate a useful nonzero χ^2, DENSS instead calculates χ^2 using the calculated scattering profile after real space restraints are applied (such as solvent flattening), rather than using the calculated scattering profile after the reciprocal space scaling is applied. This makes it possible to use χ^2 as a diagnostic tool for assessing convergence. Note however that the final step of DENSS is always reciprocal space scaling, such that the final density map will always reflect the interpolated experimental scattering profile exactly, having a corresponding χ^2 value of zero.

When evaluating the success of the reconstruction, one of the most important metrics to assess is the quality of the fit of the calculated scattering profile (after real space restraints are applied) to the experimental

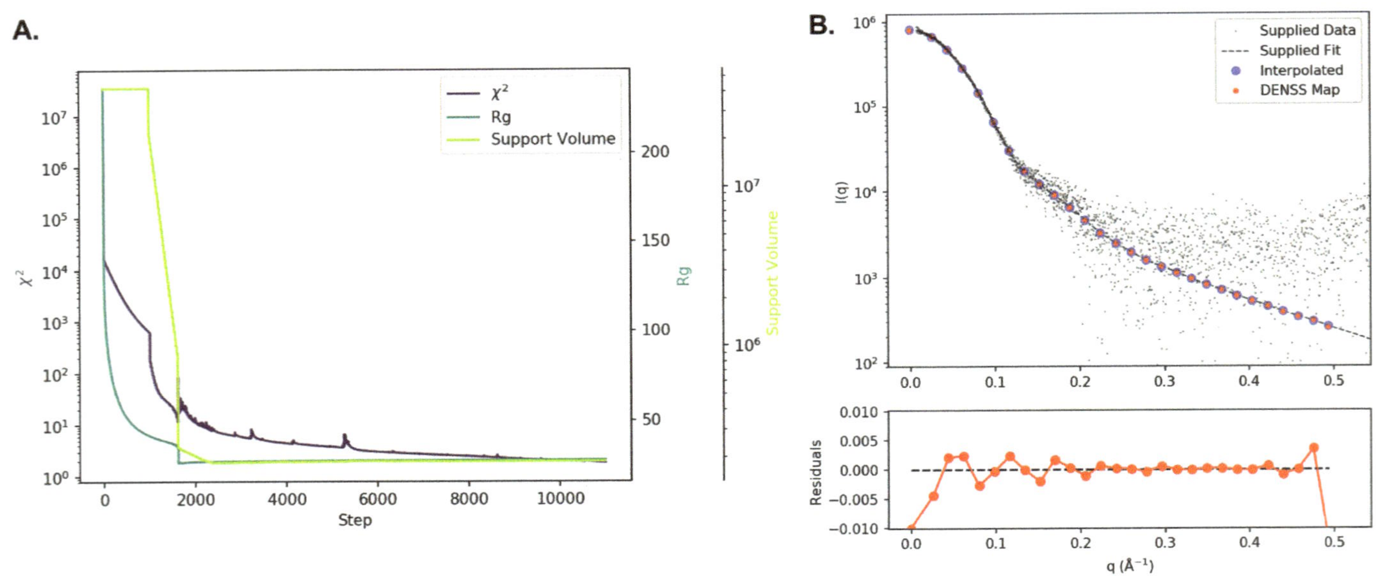

Fig. 6 Evaluating the success of DENSS reconstructions. (A) A plot generated by denss.py shows the three metrics used to assess convergence including χ^2, R_g, and support volume. Each value is plotted on a separate y-axis as a function of the number of steps completed of the iterative reconstruction procedure. (B) A plot generated by denss.py showing the quality of the fit of the calculated scattering profile from the final density map compared to the experimental data. As described above, the experimental data (black dots) are first fitted using a smooth curve (black dashed line, e.g., using denss.fit_data.py). The fit is then interpolated to the q values corresponding to the concentric q shells in reciprocal space (blue circles). The final calculated scattering profile from the DENSS reconstruction is shown as red circles, and the residuals between the interpolated experimental intensities and the calculated intensities is shown below.

scattering profile. While χ^2 is a useful metric for assessing convergence as a function of the number of steps, visual inspection of the fit is more appropriate for determining if the reconstruction converged to a solution that fits the data adequately (Fig. 6B). For small angle scattering data, *the quality of fit in the low q bins is often the most indicative of success*, as the size and shape of the particle is largely reflected at lower q values. Deviation from the experimental profile, particularly in the low q shells, indicates that the reconstruction failed to produce an object density that accurately reflects the true particle density. In such cases, editing one or more of the options described below can help to ensure DENSS reconstructs object density that matches the experimental data.

5.3 Advanced options

The three modes of denss.py are designed to reflect an optimal balance between speed and accuracy and are sufficient for the vast majority of use cases. However, in some cases it is appropriate to modify one or more of the options in an attempt to improve the ability of DENSS to reconstruct the object density. Here we will outline several of the more advanced options that may be useful in many cases.

5.3.1 Symmetry averaging

Proteins often exist as homomeric complexes in solution. This information can often be utilized by DENSS to enforce symmetry resulting in significantly improved reconstructions. DENSS imposes this restraint, called non-crystallographic symmetry, or NCS, using a method known as symmetry averaging by rotating copies of the object density and averaging the resulting densities. The first step is to select the axis of rotation. To do this, DENSS first calculates the principal axes of inertia of the object density since an axis of symmetry must coincide with one of the principal axes. However, it is generally not possible to select automatically which axis is the correct axis of symmetry directly from the reconstruction due to the limited information in the scattering profile. Thus, it is left to the user to select which axis is the axis of symmetry. The user also must define the amount of symmetry, i.e., the N-fold rotational symmetry (e.g., twofold for dimer, threefold for trimer, etc.). Once the N-fold axis of symmetry is defined, DENSS creates N copies of the density and rotates each copy about the axis by 360/N degrees (Fig. 7). As the rotations occur on a cubic grid, they may not fall exactly on grid points and the image copy will need to be interpolated onto the original grid. Finally, DENSS averages the N rotated copies of

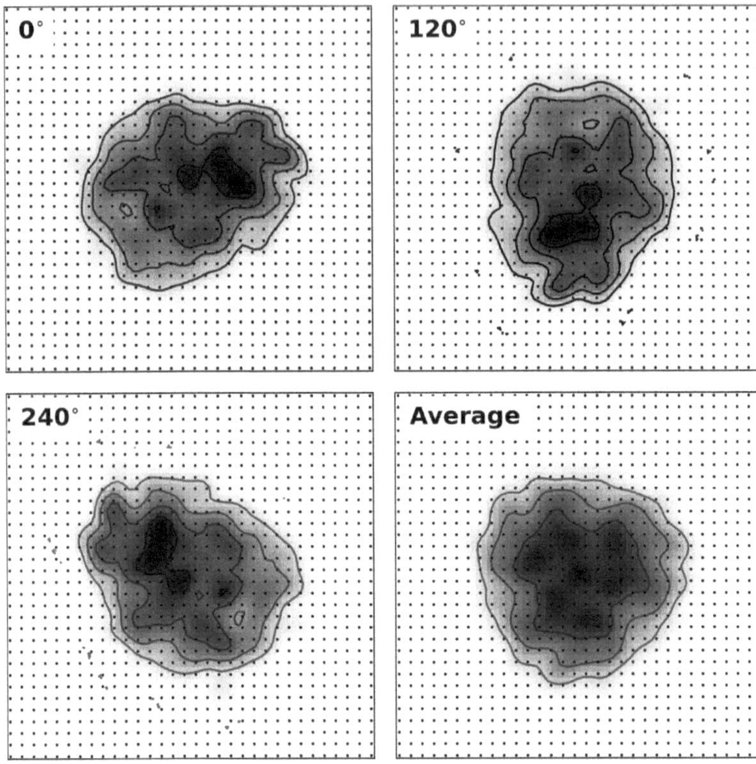

Fig. 7 Symmetry averaging procedure for threefold symmetry. During the symmetry averaging step, the initial density (0°) is rotated by 120° and 240°. The three symmetry related copies are then averaged to produce a final density with threefold symmetry.

the density to arrive at a final, symmetry-averaged density (Fig. 7). By default, DENSS will not strictly *enforce* symmetry by applying it at every step of the iteration, but will instead *bias* the result to be symmetric by applying the symmetry at only a few steps throughout the reconstruction process. Setting more or fewer steps allows one to control how much bias is imposed by symmetry averaging.

Fig. 8 illustrates an example of the effect that symmetry averaging can have on improving the accuracy of reconstructions. Fig. 8 shows the pentameric immunoglobulin M (IgM, SAXS profile simulated from PDB 2RCJ using FoXS, (Perkins, Nealis, Sutton, & Feinstein, 1991; Schneidman-Duhovny et al., 2016)) with and without symmetry averaging using C5 symmetry along the shortest principal axis. IgM has a flat, disc-like shape

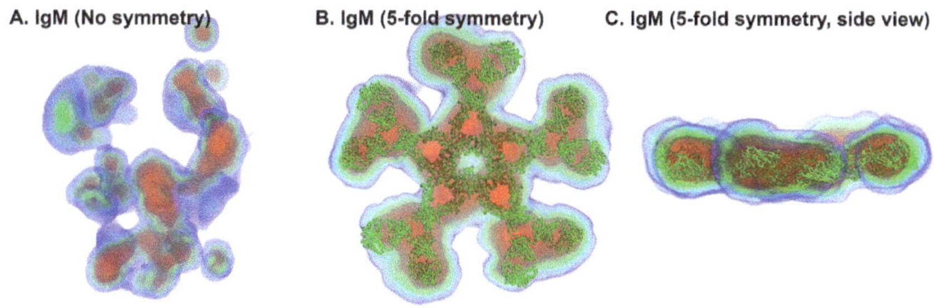

Fig. 8 Symmetry averaging significantly improves DENSS reconstructions of IgM. (A) DENSS reconstruction of IgM using default SLOW mode. (B) DENSS reconstruction adding fivefold symmetry. (C) Side view of density shown in (B). Density maps are displayed using the Volume representation in PyMOL colored according to density value (blue: 1σ, green: 5σ, red: 10σ). The atomic model was taken from the PDB entry 2RCJ and is displayed in green ribbon representation.

which is notoriously difficult for ab initio modeling algorithms to reconstruct due to its high ambiguity (Grant, 2018; Petoukhov & Svergun, 2015). However, applying fivefold symmetry has a dramatic impact on the ability of DENSS to reconstruct the difficult shape.

Additionally, there are two types of symmetry possible in DENSS. In addition to the standard cyclical symmetry described above, it is also possible to invoke dihedral symmetry. In addition to the axis of N-fold rotational symmetry, dihedral symmetry adds N twofold axes perpendicular to the N-fold axis of symmetry. Several well-known proteins adopt dihedral symmetry. For example, GroEL (Ludtke et al., 2008) has D7 symmetry, meaning that there are 14 subunits related by one 7-fold axis of symmetry and seven 2-fold axes of symmetry perpendicular to the 7-fold axis. Other common examples are tetramers possessing D2 symmetry such as in glucose isomerase (Henrick, Blow, Carrell, & Glusker, 1987), where the four subunits are related by two perpendicular twofold axes of symmetry, in contrast to other tetramers possessing C4 symmetry where the four subunits are instead related by a single fourfold axis of symmetry, such as neuraminidase (Russell et al., 2006).

5.3.2 Oversampling

In all modes the oversampling ratio defaults to three in denss.py. However, in some cases, particularly where the particle has a large aspect ratio, the default oversampling ratio may not be adequate for fitting the scattering profile. In such cases, one can increase the oversampling ratio from three to five or six, for example, to see if the resulting fit improves. However, as the oversampling ratio increases the box size increases, and therefore so does the voxel size. To keep a similar voxel size while increasing the oversampling ratio, one must also increase the number of samples, i.e., grid points. This will allow for greater oversampling and similar voxel size, but will also result in longer calculation times.

5.3.3 Shrinkwrap options

The Shrinkwrap algorithm is an important component of DENSS for calculating the object support. There are several parameters for running Shrinkwrap, each of which can be modified by the user. One possible cause of a poor fit to the scattering profile is an overly aggressive set of Shrinkwrap options. For example, Shrinkwrap uses a threshold for deciding which voxels of the low-pass filtered real space image should be considered the object support (Fig. 2). By default, this threshold is set to be 0.20 times the maximum filtered density. However, in cases where the calculated scattering profile does not fit the experimental scattering profile well, it may be appropriate to adjust this to a lower value, such as 0.15 or less. This will allow more voxels to be included in the object support, allowing for greater variation in the density of the object, and therefore a greater likelihood ofbeing able to reconstruct a density that matches the experimental profile. Consequently, editing options such as these may result in more noise in the reconstruction, which may be ameliorated by subsequently averaging over more reconstructions. Additional Shrinkwrap options can also be modified, such as the amount of filtering to start and end Shrinkwrap, how quickly to reduce the amount of filtering, and how frequently to run Shrinkwrap to update the support.

5.3.4 Enforce connectivity

In most cases, Shrinkwrap reduces the support adequately to a single finite collection of connected voxels. Occasionally for particles with large aspect ratios or with complex shapes, Shrinkwrap may fail to reduce the support to a single connected region, instead resulting in two or more disconnected regions, often with the additional regions containing small amounts of density far away from the primary object density (see Fig. 2 for example). To accelerate the removal of these regions and ensure only a single connected region exists in the final support, an additional option to enforce the connectivity of the support is invoked by default at one or a few steps during the reconstruction. When invoked, enforce connectivity will identify each separate region of the support and remove any regions disconnected from the primary support region, retaining only the region with the greatest amount of density. This feature helps to ensure DENSS produces a single connected density. This feature can be edited to change or add steps where it is invoked, or even disabled entirely.

6. Alignment and averaging

DENSS reconstructs a single object density by creating a 3D grid and filling it with random numbers to generate a uniformly distributed random density map. Given identical starting parameters, running denss.py multiple times will result in multiple density maps each with slightly different results. Importantly, the reconstructions are similar at low resolution (Grant, 2018), while higher resolution features of the maps are inconsistent between different reconstructions, due to the lack of information in the scattering profile. To generate a single low-resolution density map, the denss.all.py script runs multiple reconstructions (20 by default), aligns each density map to the others, and averages the results. This process tends to lower the density of regions that are not consistent between maps and increase the density of regions that are common between maps. In this section we will describe the basic steps in the alignment and averaging procedure.

6.1 Alignment

Several visualization tools have the ability to align two density maps. These tools rely on local minimization for alignment, i.e., they find the closest rotation and translation relative to the starting position that optimizes the alignment. Since DENSS starts with random density, the resulting reconstructions are randomly oriented relative to the coordinate axes. Thus, the alignment procedure must find the optimal global alignment, not just the optimal local alignment. To align two density maps globally, DENSS separates the process into two steps, coarse alignment to determine a starting, roughly optimal orientation, followed by a fine local minimization to optimize the rotation and translation. DENSS uses the real space correlation coefficient as the metric to quantify the quality of the alignment of two density maps. The correlation coefficient between two density maps, $CC(\rho_1,\rho_2)$, is calculated as

$$CC(\rho_1, \rho_2) = \frac{\sum \rho_1\rho_2}{\sqrt{\sum \rho_1^2}\sqrt{\sum \rho_2^2}}$$

where the summation is over all voxels on the grid. For the alignment of two maps, one map is chosen at random to be the reference map, ρ_1, while the other map, ρ_2, is reoriented to maximize CC. During the stage of coarse alignment, DENSS translates both maps to the center of the grid based on their centers of mass. DENSS determines a series of orientations uniformly separated by $\sim 20°$ around the unit sphere. DENSS then rotates the second map, ρ_2, by each of those rotations, interpolating the resultant rotation onto the grid of ρ_1. Next, DENSS calculates the CC for each of those orientations and selects the top five orientations with the highest CC values. For the fine alignment stage, DENSS performs a local minimization of all six degrees of freedom (three rotation and three translation) for each of the top five orientations found in the coarse alignment stage. DENSS then selects the highest CC of the five optimized alignments as the best alignment.

6.2 Averaging

In order to align all density maps, there must first be a common reference. The averaging procedure begins by generating a reference map using a binary tree to first select pairs of density maps. Next, each member of the pair is aligned to its corresponding partner and the two are averaged. Then, the resulting averages are paired up, and the process is repeated until only one average density remains. This approach, commonly used in other fields such as single particle tomography (Galaz-Montoya, Flanagan, Schmid, & Ludtke, 2015), is significantly faster than the alternative, which is to align every map to every other map, while still generating an unbiased reference. Of course, it is still possible to align each map to every other map if desired using the dedicated denss.align.py script.

After the reference map is generated, each of the 20 reconstructions is subsequently aligned to the reference map. Finally, the average density is calculated from all 20 maps. By default, outliers are removed from the average if their CC to the reference is more than three standard deviations lower than the mean. Fig. 9 shows 10 individual reconstructions of monomeric bovine serum albumin (BSA, SASDCQ8) compared to the final, averaged density superimposed with the atomic model. It is important to note that due to the nature of DENSS reconstructing actual object densities, the averaging procedure produces a true average. This is in contrast to the averaging procedure in popular bead modeling algorithms which instead produce a consensus bead model (Volkov & Svergun, 2003), since it is not possible to actually average bead models due to their binary nature.

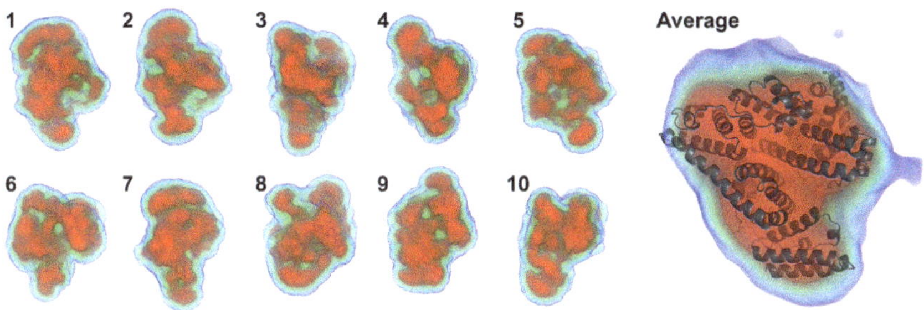

Fig. 9 Averaging of multiple individual DENSS reconstructions of BSA. Ten individual density reconstructions generated by denss.all.py are numbered and shown on the left. The final reconstruction from denss.all.py after alignment and averaging of all 20 individual reconstructions is shown on the right. All density maps are displayed using the Volume representation in PyMOL colored according to density value (blue: 1σ, green: 5σ, red: 10σ). The atomic model is taken from the SASBDB entry and displayed in gray cartoon representation.

In such procedures the beads that are most common among multiple individual bead reconstructions are saved in this consensus bead model, and information is thus typically lost regarding the relative frequency of the presence of beads. The DENSS averaging procedure on the other hand fully retains the information from each reconstruction, since the density maps are floating point values (rather than binary values) and thus are quantitatively averaged to produce an average density map.

6.3 Enantiomer selection

Two density maps that are mirror images of one another will yield identical scattering profiles. Thus, when DENSS reconstructs individual object densities, it is equally as likely to generate the left-handed image as it is the right-handed image. While it is not possible to distinguish which enantiomer has the correct hand solely from a low-resolution density map (since SAS is ambiguous to handedness), it is important to ensure that during the averaging procedure all reconstructions at least have the *same* handedness. To do this, for each of the 20 reconstructions, DENSS generates the opposite enantiomer. DENSS randomly selects a single reconstruction to be the temporary reference and aligns each enantiomer of the remaining maps to the temporary reference. For each map, the enantiomer with the highest CC value is selected as the optimal enantiomer and is used for the subsequent alignment and averaging steps.

6.4 Resolution estimation

After producing a fully averaged density reconstruction, it is useful to quantify the quality of the averaging procedure. A useful metric to do so is the resolution based on Fourier Shell Correlation (FSC) commonly used in other fields such as cryoEM for assessing resolution (Henderson et al., 2012; Scheres & Chen, 2012; Van Heel & Schatz, 2005), and has previously been used for assessing the resolution of ab initio bead models (Tuukkanen et al., 2016) from solution scattering. To calculate the FSC from the set of 20 density maps, the FFT of each map after alignment to the reference is calculated. Then, using the reference generated in the above averaging procedure, DENSS calculates the FSC for each map according to the following equation:

$$FSC_j(r) = \frac{\sum_{r_i \in r} F_{ref}(r_i) F_j^*(r_i)}{\sqrt{\sum_{r_i \in r} |F_{ref}(r_i)|^2 \sum_{r_i \in r} |F_j(r_i)|^2}}$$

where FSC_j is the FSC for the *j*th map, r is the resolution of the bin containing the voxels r_i, F_{ref} is the structure factor of the reference density map, and F_j^* is the complex conjugate of the structure factor of the *j*th density map. This results in an FSC curve as a function of resolution for each map relative to the reference. To generate a single FSC curve, DENSS calculates an average FSC curve from all 20 FSC curves from each map (an example output from denss.all.py is shown in Fig. 10A). This averaging step helps to reduce noise and shape features from each individual FSC that are not consistent across all maps. Finally, an estimate of the resolution is calculated based on where the averaged FSC curve falls below 0.5 (Fig. 10A). The resolution estimated in this procedure does not directly assess the accuracy of the averaged density relative to the true object density, but instead measures the consistency, or precision, between the 20 individual reconstructions. However, as shown previously (Grant, 2018), and further in Fig. 10B, this approach provides a reasonable proxy for resolution compared to the true object density. Fig. 10B shows a plot of the estimated resolution compared to the true resolution of ~1900 averaged densities using denss.all.py generated from simulated scattering profiles calculated from known structures of proteins from the PDB using FoXS (Berman et al., 2000; Schneidman-Duhovny et al., 2016). This plot shows that the resolution estimate from DENSS closely approximates the true resolution

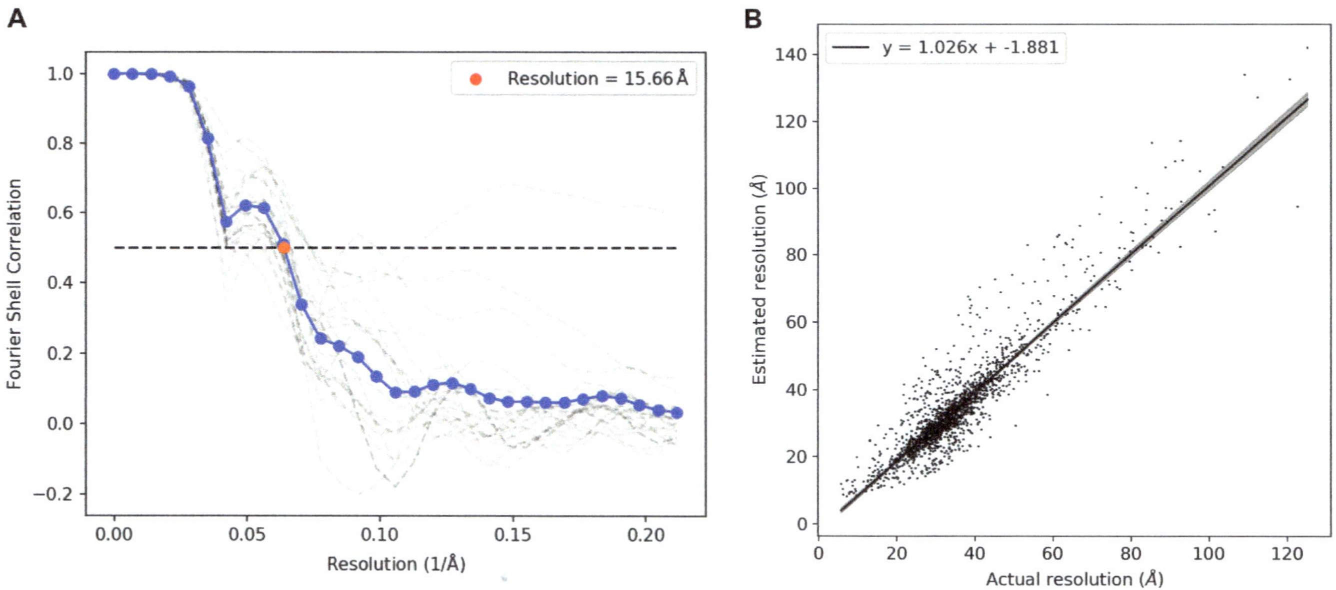

Fig. 10 Resolution estimation from DENSS. (A) An example of the Fourier Shell Correlation vs resolution plot, output by denss.all.py, is shown. FSC curves for each of the 20 reconstructions compared to the reference are shown as gray dashed lines, and the averaged FSC is shown as a blue line. The estimated resolution, where the averaged FSC curve crosses 0.5, is shown by the red circle. (B) ~1900 Simulated scattering curves from PDB entries were used by DENSS to generate averaged density maps using default settings. The resolution estimated by denss.all.py is shown on the y-axis. For each reconstruction, the averaged density map was aligned to the known structure using denss.align.py and the actual resolution was calculated using the denss.pdb2mrc.py and denss.calcfsc.py scripts, plotted on the x-axis. A linear trend was fit to the data and is shown by the solid black line (95% confidence intervals shown by the shaded gray region). The equation of the line is given in the legend, showing a slope of 1.026 and intercept of −1.881, demonstrating that the estimated resolution by DENSS closely approximates the true resolution.

when the structure is known and can thus be used to estimate the resolution when the structure is not known.

6.5 Refinement

While each individual reconstruction has a corresponding scattering profile that matches the experimental data, the average reconstruction often has a corresponding scattering profile that does not match the experimental data. While the averaged density typically has a better resolution than any individual reconstruction (Grant, 2021), it is possible to obtain a single reconstruction that matches the experimental scattering profile while still containing the information from the averaging procedure. This procedure is akin to refining the averaged density map using the experimental scattering profile and is provided by a script called denss.refine.py. This script runs similarly to denss.py, taking as input the 1D scattering profile. Additionally, denss.refine.py takes as input the averaged density map output from denss.all.py. While denss.py begins with a random density map, denss.refine.py instead begins with the given averaged density map, thus acting as a type of refinement of the averaged density map against the data. While refinement produces a density map from the averaged density that matches the data more closely, the resulting refined map does not improve the resolution relative to the averaged map without refinement when compared with a known structure (Grant, 2021). Thus, refinement should be considered an optional step.

7. Analysis and interpretation of results

DENSS produces MRC files, which are common file types used in other fields such as cryoEM to store electron density maps. MRC files are very similar in format to CCP4 map files as well. Many common software programs directly deal with MRC map files, including graphical viewers such as UCSF Chimera (Pettersen et al., 2004), PyMOL (Schrodinger, LLC, 2015), COOT (Emsley et al., 2010) and others (Fig. 11). Several of these programs are useful for not only visualizing the density maps and making figures, but also for performing analysis and modeling. In this section we will describe some of the tools and methods recommended for visualization, analysis, and interpretation of DENSS results.

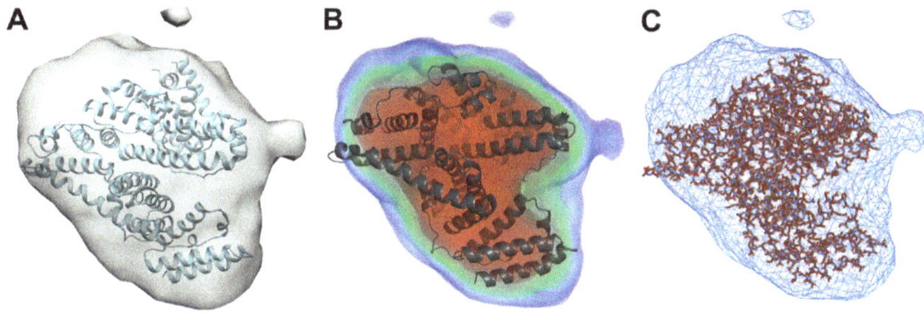

Fig. 11 Graphical viewers showing density maps of BSA from DENSS. (A) Density in UCSF Chimera shown as a gray transparent isosurface at 1σ. The aligned atomic model is shown in cartoon representation. (B) Density displayed using the Volume representation in PyMOL colored according to density value (blue: 1σ, green: 5σ, red: 10σ). The atomic model in gray cartoon representation. (C) Density displayed as isomesh in COOT contoured at 1σ. The atomic model is shown using the lines representation.

7.1 Visualization

There are several options for viewing density. To highlight the shape of an object, a common approach is to use isocontours, often represented as an isomesh (Fig. 11C) or isosurface (Fig. 11A). An isomesh or isosurface is a 3D surface representation connecting density of the same value, effectively the 3D equivalent of a contour line. Showing only a single isocontour makes it easy to visualize the density as a shape, however it reduces the amount of information the user actually sees. While one option is to show multiple isocontours (see, for example, Fig. 5A), some graphical viewers allow setting the contour value interactively (Emsley et al., 2010). This often makes it easy for users to get an intuitive sense of the full 3D density map, rather than viewing a single shape.

Contour levels can be calculated using exact density values in number of electrons per unit volume, $e^-/Å^3$ (as in Chimera), or using relative density values (as in PyMOL and COOT). Note that the density calculated by DENSS is in units of $e^-/Å^3$ and is only roughly accurate in scale if the total number of electrons is given as an additional option to DENSS (which, as mentioned above, actually calculates contrast, not absolute density). Relative density values are often calculated using the root mean square (RMS) of all voxel density values throughout the map. The RMS for a density map is calculated by DENSS and provided in the resulting log files but can also be calculated using the denss.get_info.py script and is typically calculated internally by visualization programs. Thus, a contour at 1σ in

Fig. 11A corresponds to the density at 1*RMS, whereas a contour at 10σ corresponds to 10*RMS. An important question arises as to the appropriate contour level, sometimes called the threshold or sigma value, to display the density. Lower contour levels will result in larger, more voluminous density maps and often contain more noise. Higher contour levels will result in focusing on the regions of the map with greater density and less noise, reflecting regions of the map with greater confidence. Selecting the appropriate contour level is a highly subjective process, and standards for representation will likely be useful for interpreting density maps for the community. We have chosen to show contours at 1σ, 5σ and 10σ throughout this chapter to represent most of the information in the density map (and even at −1σ when negative contrast is present).

In addition to representing density as a single contour or using multiple contours, it is also possible to use modern graphical visualizations to instead provide a gradient of colors to represent all density values in a single view. For example, in PyMOL the "Volume" coloring tools can be used to represent the density map as a continuous semitransparent cloud colored according to density (Fig. 11B). Particularly with the low-resolution density maps provided by solution scattering, this is a useful representation for viewing density maps that allows one to visualize the most information content in a single view. This representation simultaneously allows for visualization of the largest extent of the envelope of the particle, while also allowing identification of high-density "hot spots" in the interior, corresponding to regions of density with the greatest confidence or which may correspond to protein subunits or domains, enabling greater insight into the low-resolution object density. These interior density fluctuations are a unique feature of DENSS reconstructions, and have been used to guide rigid body modeling of subunits (Beyett et al., 2019; Florio, Lokareddy, Gillilan, & Cingolani, 2019; Karim et al., 2022; Niazi et al., 2020).

Different contour levels result in the isosurfaces with various internal volume estimates. For example, Chimera has a tool to calculate the volume of an isosurface while interactively adjusting the contour threshold value. Increasing the contour threshold generates smaller isosurfaces containing only high-density regions, resulting in decreasing volume estimates. Lowering the contour threshold results in larger isosurfaces and correspondingly larger volume estimates. Volume can also be directly calculated from the scattering profile using the Porod approximation (Glatter et al., 1982). This calculation can be done in many software programs and is also calculated in DENSS using denss.fit_data.py. One possible rationale for selecting an appropriate contour level may be the density value that results in an isosurface whose volume most closely matches the Porod volume calculated directly from the solution scattering profile. While a potential solution, volume estimates directly from scattering profiles are typically very noisy and prone to systematic error and may not be accurate enough for selecting the contour value. Note that the ambiguity in volume estimation is implicit to the notion of representing objects at low resolution. The volume of a particle is not simply the "dry volume" of the atoms making up the particle, but depends on several factors including resolution, hydration, dynamics, etc.

7.2 Comparison with atomic model

A common approach to interpreting density maps is to fit an atomic model to the density and visualize the resulting overlap (Fig. 11). DENSS provides a tool for performing the global alignment of a density map to an atomic model, a script called denss.align.py. DENSS aligns an atomic model, represented as a set of atomic coordinates in a PDB file, to a reconstructed density map by first calculating a density map directly from the atomic coordinates. To calculate a density map from an atomic model, DENSS can represent an atom as a simple Gaussian sphere (commonly used in electron microscopy software (Pettersen et al., 2004; Tang et al., 2007)), or using a more accurate real space atomic form factor defined as a sum of Gaussians described by nine parameters (Cromer-Mann coefficients, commonly used in crystallography software (Brown, Fox, Maslen, O'Keefe, & Willis, 2006; Kleywegt & Jones, 1994; Waasmaier & Kirfel, 1995; Winn et al., 2011)). The denss.align.py script, as well as the stand-alone denss.pdb2mrc.py, uses the more accurate atomic form factors by default. Additionally, the effect of resolution is calculated using a real space B-factor modification (Debye, 1913; Tickle, 2012). To align a density map to the PDB, DENSS uses the alignment algorithm described above to optimize the real space correlation of the density to the density calculated from the atomic model, and additionally checks both enantiomers of the DENSS reconstruction for best agreement with the atomic model.

In addition to the global alignment tool provided by DENSS, several software programs also provide their own local alignment tools, such as Chimera and COOT. These tools can be particularly useful for guiding the alignment with an initial manual guess. For example, in Chimera the atomic model can be manually rotated and translated independently of the DENSS density to generate a rough alignment. The Fit in Map tool can then be used for performing the local alignment based either on simple overlap (which is sensitive to the contour threshold level), or by calculating the real space correlation (which may or may not be sensitive to contour level, depending on the options selected). Similar alignment can also be performed using the rigid body modeling tools in COOT (Emsley et al., 2010).

7.3 Advanced modeling using DENSS reconstructions

One of the major advantages of DENSS producing standard MRC-formatted density maps is that many software tools that have been developed for cryoEM can be directly used with the low-resolution density maps from DENSS. Researchers can use common tools such as Chimera, COOT, phenix.real_space_refine, phenix.dock_in_map, SITUS and others (Afonine et al., 2018; Emsley et al., 2010; Liebschner et al., 2019; Pettersen et al., 2004; Wriggers, 2012), and several instances of this approach have been published to date (Beyett et al., 2019; Florio et al., 2019; Karim et al., 2022; Niazi et al., 2020). As described above, the use of density more accurately represents the true object than conventional bead modeling and allows for more accurate modeling directly into the 3D ab initio density. Virtually any approach developed for modeling low-resolution cryoEM maps can also be used directly with DENSS maps, such as rigid body modeling, flexible fitting, etc. Despite the lower resolution and limited information from SAS experiments, these modeling algorithms previously developed as part of other software suites have shown to be valuable tools for interpreting the density maps generated by DENSS. Perhaps some of the most novel uses of DENSS demonstrated so far are the direct 3D modeling examples listed above, where researchers have repeatedly used these density maps as a target for rigid body modeling. These examples have shown that the resulting calculated scattering profiles from the atomic models fit the data as well as conventional rigid body search algorithms that operate in reciprocal space to directly fit the data. This new paradigm of modeling enabled by using nonuniform density maps promises a new era in SAS modeling.

8. SANS

While most of the reconstructions described so far have utilized SAXS data, it is also possible to use small angle neutron scattering (SANS) data directly within DENSS as well (Guo, Sumner, & Qian, 2021; Murray,

Weiss, Stanley, Nagy, & Stroupe, 2021). Atomic or coarse-grained modeling approaches require accurate estimates of the atomic form factors for each atom or coarse-grained bead, which differ for X-ray and neutron scattering experiments. Thus, modeling algorithms must be developed specifically for modeling SANS data. In contrast, DENSS does not require or assume any form factors and relies solely on the Fourier transform of the real space and reciprocal space images for reconstruction. DENSS can thus be used directly with SANS data without modification. Fig. 12 shows an example of a DENSS reconstruction of a ligand-gated ion channel (GLIC) measured with SANS (SASDL33 (Lycksell et al., 2021)). When compared to the atomic model, the DENSS reconstruction closely matches, demonstrating the ability of DENSS to use SANS data without modification. Additionally, since DENSS implicitly reconstructs object contrast, it is possible to use SANS data collected in any solvent conditions, such as using varying percentages of D_2O (Murray et al., 2021), commonly done in SANS experiments to vary the bulk solvent density. Contrast matching is common practice in SANS experiments (and can be done with some effort with SAXS as well) as a means of matching the density of the solvent to a particular component, such as a lipid or detergent belt for example, thus making that component invisible. DENSS will directly reconstruct the object contrast regardless of the bulk solvent density value and can therefore be very useful for interpreting SANS profiles from such experiments. While DENSS

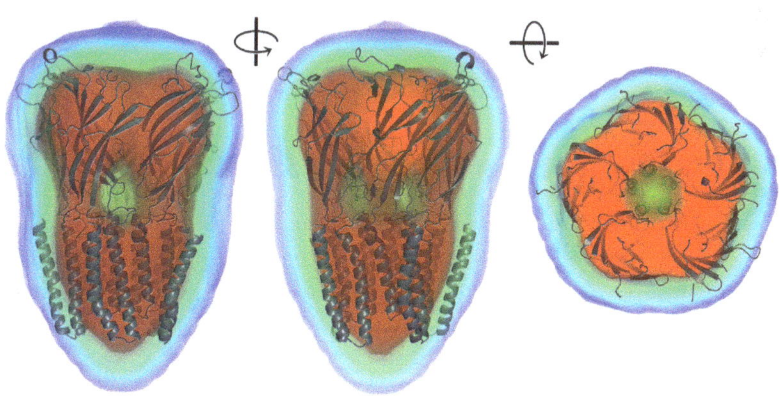

Fig. 12 DENSS reconstruction of GLIC protein from SANS data. GLIC was reconstructed using the denss.all.py script with C5 symmetry applied using default parameters. The density map is displayed using the Volume representation in PyMOL colored according to density value (blue: 1σ, green: 5σ, red: 10σ). The atomic model is shown in gray cartoon representation. The density maps on the right have each been rotated by 90°.

does not currently have the ability to simultaneously fit multiple scattering curves from multiple contrasts, this feature is in active development and will likely lead to an increase in the resolution of the reconstruction due to the increased information content from such experiments. Additionally, since DENSS does not require any atomic form factors, scattering profiles from samples that have been modified in some way, such as through deuteration of a component to aid in contrast matching, can also be directly used for reconstructions.

9. Materials science applications

All examples described so far have been biological samples in aqueous environments, mostly taken from the SASBDB. However, since DENSS does not implicitly assume any qualities of the sample or the environment, scattering profiles from nonbiological samples can be probed as well. Examples including organic or inorganic particles in aqueous or organic solvents have been described (Fujii & Sakurai, 2021; He et al., 2022). The only implicit requirements for the iterative structure factor retrieval process in DENSS are that the sample is monodisperse, the object is of finite size and the solvent can be approximated as a flat value. There are no implicit restrictions on particle size, type of incident radiation, or the makeup of the particle components, be it organic or inorganic. The lack of restrictions makes DENSS extremely versatile and likely to be used for a variety of new samples and previously unexploited experiments.

10. Publication guidelines and SASBDB deposition

Guidelines for publishing SAXS and SANS data and modeling have been developed by the leaders in the community (Trewhella et al., 2017). These developments have been valuable for researchers interested in not only publishing their data, but also for guidance on proper protocols and validation metrics for interpreting data and drawing conclusions. DENSS is a recent addition to the suite of available tools in the community and does not yet have guidelines agreed upon by the scientific community for publication standards. Based on the previous guidelines for ab initio modeling approaches and the methods laid out above, we propose Table 1 of metrics be presented in any publication reporting DENSS results,

Table 1 Example of metrics to report in publications using DENSS.

IFT algorithm	denss.fit_data.py
q_{max} (data, IFT fit, density) (1/Å)	0.531, 0.531, 0.782
D_{max}, oversampling, box size	167.3 Å, 3.0, 501.9 Å
Grid sampling, voxel size	$64 \times 64 \times 64$, 7.842 Å
DENSS mode	SLOW
Symmetry, axis	D7, long
N maps calculated, averaged, rejected	20, 20, 0
FSC resolution, mean CC	36.3 ± 4.9 Å, 0.89 ± 0.08
χ^2, R_g, support volume	7.4 ± 1.2, 67.5 ± 3.3 Å, 2.5e6 ± 2.2e5 Å^3
Final density mean, SD, RMSD	0.00296, 0.02988, 0.03002
Volume at 1, 5, 10 * rmsd (Å^3)	2.389e6, 1.656e6, 9.117e5

in addition to tables describing data collection and modeling previously described (Trewhella et al., 2017). As more users become familiar with DENSS results, Table 1 will likely be updated as guidelines become agreed upon by the community.

Since DENSS depends directly on the IFT algorithm used to fit the raw experimental data, which algorithm has been used (e.g., denss.fit_data.py, GNOM, Scatter, etc.) should be reported. Additionally, the maximum q value in the experimental data, the IFT fit, and the maximum q value reported by DENSS (corresponding to the largest q shell value) should also be reported. Information about the real space box characteristics, including D_{max}, oversampling ratio, box size, number of samples, and the resulting voxel size, should be reported. The mode of DENSS used (either SLOW, FAST, or MEMBRANE) should be reported, along with information about applied symmetry (if any) including the type (cyclical or dihedral) and the N-fold rotation as well as which principal axis the rotational symmetry was applied along. For averaging, the number of maps calculated as well as if any were rejected and the subsequent number averaged should be reported. DENSS provides useful aggregated statistics in the denss.all.py log file, including resolution (with standard deviation) and average correlation coefficient (with standard deviation), which should be reported. Additionally, DENSS reports the averages and standard deviations for χ^2, R_g, and Support Volume aggregated over all 20 individual reconstructions, which should be reported. Values describing the final averaged or refined density should be reported, including the mean density, standard deviation, and RMS (which are useful for selecting contour levels, as described above). Finally, we propose reporting the total envelope volume for isosurfaces calculated at the three recommended contour levels for display, including 1σ, 5σ and 10σ, which can be calculated using the Measure Volume and Area tool in UCSF Chimera. These volume estimates are useful as they can be compared with the reported Porod volume estimate.

As shown throughout this chapter, the SASBDB has become a useful public repository of SAS data and models, and is now commonly used for archiving published results (Valentini et al., 2015). Recently, SASBDB has enabled the deposition of MRC-formatted density maps from DENSS. As a general publication guideline, we strongly encourage all users of DENSS to deposit any published density maps in the SASBDB.

11. Summary

Solution scattering is continually growing in popularity in the scientific community and ab initio modeling has proven to be a powerful tool for generating hypotheses and visualizing 3D low-resolution structures of particles. Among these tools, DENSS is unique, capable of reconstructing 3D density from the 1D scattering profile and giving more information than a standard envelope or shape reconstruction. The ease and versatility of DENSS for using solution scattering to reconstruct almost any object of any size, biological or nonbiological, in almost any environment and using multiple forms of radiation, promises to open up new avenues for the application of solution scattering to broader fields and areas of research. Scientists are regularly coming up with new methods for interpreting DENSS maps, such as rigid body modeling directly into density maps. More than a new modeling algorithm, DENSS introduces the first solution to the inverse scattering problem, the iterative structure factor retrieval method. This provides the foundation for a general approach for solving the problem of image reconstruction when not knowing the 3D amplitudes of the structure factors in addition to not knowing the 3D phases. As such, this approach may in the future be applied to other areas of research, such as fiber diffraction and fluctuation scattering (Donatelli, Zwart, & Sethian, 2015; Millane, 1990), which similarly suffer from unknown 3D amplitudes. As the software is under active development and new features are being released regularly, DENSS will continue to grow as a powerful tool for the community to interpret solution scattering data.

Acknowledgments

The author would like to thank Frank Gabel for providing data for the DDM micelle shown in Fig. 5 (Gabel et al., 2019). The author also would like to thank Stephen Moore for performing the density reconstructions used in Fig. 10 for resolution estimation.

Funding

This work is supported by the National Institutes of Health National Institute for General Medical Sciences under award number R01GM133998 and by the National Science Foundation BioXFEL Science and Technology Center 1231306.

References

Afonine, P. V., Poon, B. K., Read, R. J., Sobolev, O. V., Terwilliger, T. C., Urzhumtsev, A., et al. (2018). Real-space refinement in PHENIX for cryo-EM and crystallography. *Acta Crystallographica. Section D: Structural Biology, 74*(Pt. 6), 531–544. https://doi.org/10.1107/S2059798318006551.

Berman, H. M., Westbrook, J., Feng, Z., Gilliland, G., Bhat, T. N., Weissig, H., et al. (2000). The protein data bank. *Nucleic Acids Research, 28*(1), 235–242. https://doi.org/10.1093/nar/28.1.235.

Bernadó, P., Mylonas, E., Petoukhov, M. V., Blackledge, M., & Svergun, D. I. (2007). Structural characterization of flexible proteins using small-angle X-ray scattering. *Journal of the American Chemical Society, 129*(17), 5656–5664.

Beyett, T. S., Fraley, A. E., Labudde, E., Patra, D., Coleman, R. C., Eguchi, A., et al. (2019). Perturbation of the interactions of calmodulin with GRK5 using a natural product chemical probe. *Proceedings of the National Academy of Sciences of the United States of America, 116*(32), 15895–15900. https://doi.org/10.1073/pnas.1818547116.

Brown, P. J., Fox, A. G., Maslen, E. N., O'Keefe, M. A., & Willis, B. T. M. (2006). Intensity of diffracted intensities. *International Tables for Crystallography*, 554–595.

Chacon, P., Moran, F., Diaz, J., Pantos, E., & Andreu, J. (1998). Low-resolution structures of proteins in solution retrieved from X-ray scattering with a genetic algorithm. *Biophysical Journal, 74*(6), 2760–2775.

Chaudhuri, B., Muñoz, I. G., Qian, S., & Urban, V. S. (2017). *Biological small angle scattering: Techniques, strategies and tips. Vol. 1009*. Springer.

Cooley, J. W., & Tukey, J. W. (1965). An algorithm for the machine calculation of complex Fourier series. *Mathematics of Computation, 19*(90), 297–301.

Damaschun, G., Mueller, J., & Pürschel, H.-V. (1968). Über die Meß-Strategie bei der Untersuchung der Röntgenkleinwinkelstreuung von verdünnten monodispersen Lösungen von Makromolekülen, 1. Mitt. *Monatshefte für Chemie/Chemical Monthly, 99*(6), 2343–2348.

Dean, B., Aronstein, D., Smith, J. S., Shiri, R., & Acton, D. S. (2006). *Phase retrieval algorithm for JWST flight and testbed telescope. Vol. 6265*. SPIE.

Debye, P. (1913). Interference of x rays and heat movement. *Annalen der Physik, 43*, 49–95.

Donatelli, J. J., Zwart, P. H., & Sethian, J. A. (2015). Iterative phasing for fluctuation X-ray scattering. *Proceedings of the National Academy of Sciences of the United States of America, 112*(33), 10286–10291.

Doucet, M., Cho, J. H., Alina, G., Bakker, J., Bouwman, W., Butler, P., et al. (2017). *SasView version 4.1. Zenodo*. http://www.sasview.org.

Elser, V. (2003). Phase retrieval by iterated projections. *Journal of the Optical Society of America. A, Optics, Image Science, and Vision, 20*(1), 40–55.

Emsley, P., Lohkamp, B., Scott, W. G., & Cowtan, K. (2010). Features and development of Coot. *Acta Crystallographica. Section D: Biological Crystallography, 66*(4), 486–501.

Feigin, L., & Svergun, D. I. (1987). *Structure analysis by small-angle X-ray and neutron scattering. Vol. 1*. Springer.

Fienup, J. R. (1982). Phase retrieval algorithms: A comparison. *Applied Optics, 21*(15), 2758–2769.

Fienup, J. R. (2013). Phase retrieval algorithms: A personal tour. *Applied Optics, 52*(1), 45–56.

Fienup, C., & Dainty, J. (1987). Phase retrieval and image reconstruction for astronomy. *Image Recovery: Theory and Application, 231*, 275.

Flayhan, A., Mertens, H. D. T., Ural-Blimke, Y., Martinez Molledo, M., Svergun, D. I., & Löw, C. (2018). Saposin lipid nanoparticles: A highly versatile and modular tool for membrane protein research. *Structure, 26*(2). https://doi.org/10.1016/j.str.2018.01.007. 345–355.e5.

Florio, T. J., Lokareddy, R. K., Gillilan, R. E., & Cingolani, G. (2019). Molecular architecture of the inositol phosphatase Siw14. *Biochemistry, 58*(6), 534–545. https://doi.org/10.1021/acs.biochem.8b01044.

Franke, D., & Svergun, D. I. (2009). DAMMIF, a program for rapid ab-initio shape determination in small-angle scattering. *Journal of Applied Crystallography, 42*(2), 342–346.

Fujii, S., & Sakurai, K. (2021). Structural analysis of an octameric resorcinarene self-assembly in toluene and its morphological transition by temperature. *The Journal of Physical Chemistry Letters, 12*(28), 6464–6468. https://doi.org/10.1021/acs.jpclett.1c01978.

Gabel, F., Engilberge, S., Perez, J., & Girard, E. (2019). Medical contrast media as possible tools for SAXS contrast variation. *IUCrJ, 6*(4), 521–525. https://doi.org/10.1107/S2052252519005943.

Galaz-Montoya, J. G., Flanagan, J., Schmid, M. F., & Ludtke, S. J. (2015). Single particle tomography in EMAN2. *Journal of Structural Biology, 190*(3), 279–290. https://doi.org/10.1016/j.jsb.2015.04.016.

Glatter, O. (1977). A new method for the evaluation of small-angle scattering data. *Journal of Applied Crystallography, 10*(5), 415–421.

Glatter, O., Kratky, O., & Kratky, H. (1982). *Small angle X-ray scattering*. Academic Press.

Graewert, M. A., Da Vela, S., Gräwert, T. W., Molodenskiy, D. S., Blanchet, C. E., Svergun, D. I., et al. (2020). Adding size exclusion chromatography (SEC) and light scattering (LS) devices to obtain high-quality small angle X-ray scattering (SAXS) data. *Crystals, 10*(11), 975. Retrieved from https://www.mdpi.com/2073-4352/10/11/975.

Grant, T. D. (2018). Ab initio electron density determination directly from solution scattering data. *Nature Methods, 15*(3), 191–193.

Grant, T. D. (2021). Reply to: Limitations of the iterative electron density reconstruction algorithm from solution scattering data. *Nature Methods, 18*(3), 246–248. https://doi.org/10.1038/s41592-021-01083-w.

Grant, T. D. (2022). Describing small-angle scattering profiles by a limited set of intensities. *Journal of Applied Crystallography, 55*(5). https://doi.org/10.1107/S1600576722006598.

Grudinin, S., Garkavenko, M., & Kazennov, A. (2017). Pepsi-SAXS: An adaptive method for rapid and accurate computation of small-angle X-ray scattering profiles. *Acta Crystallographica. Section D: Structural Biology, 73*(5), 449–464.

Guinier, A., Fournet, G., & Yudowitch, K. L. (1955). *Small-angle scattering of X-rays*. New York: Wiley.

Guo, R., Sumner, J., & Qian, S. (2021). Structure of diisobutylene maleic acid copolymer (DIBMA) and its lipid particle as a "stealth" membrane-mimetic for membrane protein research. *ACS Applied Bio Materials*, *4*(6), 4760–4768. https://doi.org/10.1021/acsabm.0c01626.

Hajdu, J. (2000). Single-molecule X-ray diffraction. *Current Opinion in Structural Biology*, *10*(5), 569–573. https://doi.org/10.1016/S0959-440X(00)00133-0.

Hansen, S., & Pedersen, J. S. (1991). A comparison of three different methods for analysing small-angle scattering data. *Journal of Applied Crystallography*, *24*(5), 541–548. https://doi.org/10.1107/S0021889890013322.

He, B., Chen, S., Cui, Y., Chen, X., Lei, Y., & Sun, J. (2022). Hollow polymeric ionic liquid spheres with hierarchical electron distribution: A novel composite of g-C3N4 for visible light photocatalytic water splitting enhancement. *Chemical Engineering Journal*, *440*, 135625. https://doi.org/10.1016/j.cej.2022.135625.

Henderson, S. J. (1996). Monte Carlo modeling of small-angle scattering data from non-interacting homogeneous and heterogeneous particles in solution. *Biophysical Journal*, *70*(4), 1618–1627.

Henderson, R., Sali, A., Baker, M. L., Carragher, B., Devkota, B., Downing, K. H., et al. (2012). Outcome of the first electron microscopy validation task force meeting. *Structure*, *20*(2), 205–214.

Henrick, K., Blow, D. M., Carrell, H. L., & Glusker, J. P. (1987). Comparison of backbone structures of glucose isomerase from Streptomyces and Arthrobacter. *Protein Engineering, Design and Selection*, *1*(6), 467–469. https://doi.org/10.1093/protein/1.6.467.

Hopkins, J. B., Gillilan, R. E., & Skou, S. (2017). BioXTAS RAW: Improvements to a free open-source program for small-angle X-ray scattering data reduction and analysis. *Journal of Applied Crystallography*, *50*(5), 1545–1553. https://doi.org/10.1107/S1600576717011438.

Jiménez-García, B., Pons, C., Svergun, D. I., Bernadó, P., & Fernández-Recio, J. (2015). pyDockSAXS: Protein–protein complex structure by SAXS and computational docking. *Nucleic Acids Research*, *43*(W1), W356–W361.

Karim, R. M., Yang, L., Chen, L., Bikowitz, M. J., Lu, J., Grassie, D., et al. (2022). Discovery of dual TAF1-ATR inhibitors and ligand-induced structural changes of the TAF1 tandem bromodomain. *Journal of Medicinal Chemistry*, *65*(5), 4182–4200.

Kleywegt, G. J., & Jones, T. A. (1994). Detection, delineation, measurement and display of cavities in macromolecular structures. *Acta Crystallographica. Section D: Biological Crystallography*, *50*(2), 178–185.

Knight, C. J., & Hub, J. S. (2015). WAXSiS: A web server for the calculation of SAXS/WAXS curves based on explicit-solvent molecular dynamics. *Nucleic Acids Research*, *43*(W1), W225–W230.

Kratky, O., & Pilz, I. (1972). Recent advances and applications of diffuse X-ray small-angle scattering on biopolymers in dilute solutions. *Quarterly Reviews of Biophysics*, *5*(4), 481–537.

Lattman, E. E., Grant, T. D., & Snell, E. H. (2018). *Biological small angle scattering: Theory and practice*. *Vol. 29*. Oxford University Press.

Liebschner, D., Afonine, P. V., Baker, M. L., Bunkoczi, G., Chen, V. B., Croll, T. I., et al. (2019). Macromolecular structure determination using X-rays, neutrons and electrons: Recent developments in Phenix. *Acta Crystallographica. Section D: Structural Biology*, *75*(10), 861–877. https://doi.org/10.1107/S2059798319011471.

Ludtke, S. J., Baker, M. L., Chen, D.-H., Song, J.-L., Chuang, D. T., & Chiu, W. (2008). De novo backbone trace of GroEL from single particle electron cryomicroscopy. *Structure*, *16*(3), 441–448. https://doi.org/10.1016/j.str.2008.02.007.

Lycksell, M., Rovšnik, U., Bergh, C., Johansen, N. T., Martel, A., Porcar, L., et al. (2021). Probing solution structure of the pentameric ligand-gated ion channel GLIC by small-angle neutron scattering. *Proceedings of the National Academy of Sciences of the United States of America, 118*(37), e2108006118. https://doi.org/10.1073/pnas.2108006118.

Manalastas-Cantos, K., Konarev, P. V., Hajizadeh, N. R., Kikhney, A. G., Petoukhov, M. V., Molodenskiy, D. S., et al. (2021). ATSAS 3.0: Expanded functionality and new tools for small-angle scattering data analysis. *Journal of Applied Crystallography, 54*(1), 343–355.

Marchesini, S., He, H., Chapman, H. N., Hau-Riege, S. P., Noy, A., Howells, M. R., et al. (2003). X-ray image reconstruction from a diffraction pattern alone. *Physical Review B, 68*(14), 140101.

Miao, J., Sayre, D., & Chapman, H. N. (1998). Phase retrieval from the magnitude of the Fourier transforms of nonperiodic objects. *Journal of the Optical Society of America. A, Optics, Image Science, and Vision, 15*(6), 1662–1669. https://doi.org/10.1364/JOSAA.15.001662.

Millane, R. P. (1990). Phase retrieval in crystallography and optics. *Journal of the Optical Society of America. A, Optics, Image Science, and Vision*, 7(3), 394–411.

Moore, P. B. (1980). Small-angle scattering. Information content and error analysis. *Journal of Applied Crystallography, 13*(2), 168–175.

Murray, D. T., Weiss, K. L., Stanley, C. B., Nagy, G., & Stroupe, M. E. (2021). Small-angle neutron scattering solution structures of NADPH-dependent sulfite reductase. *Journal of Structural Biology, 213*(2), 107724. https://doi.org/10.1016/j.jsb.2021.107724.

Niazi, M., Florio, T. J., Yang, R., Lokareddy, R. K., Swanson, N. A., Gillilan, R. E., et al. (2020). Biophysical analysis of Pseudomonas-phage PaP3 small terminase suggests a mechanism for sequence-specific DNA-binding by lateral interdigitation. *Nucleic Acids Research, 48*(20), 11721–11736. https://doi.org/10.1093/nar/gkaa866.

Panjkovich, A., & Svergun, D. I. (2016). Deciphering conformational transitions of proteins by small angle X-ray scattering and normal mode analysis. *Physical Chemistry Chemical Physics, 18*(8), 5707–5719.

Perkins, S. J., Nealis, A. S., Sutton, B. J., & Feinstein, A. (1991). Solution structure of human and mouse immunoglobulin M by synchrotron X-ray scattering and molecular graphics modelling: A possible mechanism for complement activation. *Journal of Molecular Biology, 221*(4), 1345–1366. https://doi.org/10.1016/0022-2836(91)90937-2.

Perkins, S. J., Wright, D. W., Zhang, H., Brookes, E. H., Chen, J., Irving, T. C., et al. (2016). Atomistic modelling of scattering data in the Collaborative Computational Project for Small Angle Scattering (CCP-SAS). *Journal of Applied Crystallography, 49*(6), 1861–1875.

Petoukhov, M. V., & Svergun, D. I. (2005). Global rigid body modeling of macromolecular complexes against small-angle scattering data. *Biophysical Journal, 89*(2), 1237–1250.

Petoukhov, M. V., & Svergun, D. I. (2015). Ambiguity assessment of small-angle scattering curves from monodisperse systems. *Acta Crystallographica. Section D: Biological Crystallography, 71*(5), 1051–1058. https://doi.org/10.1107/S1399004715002576.

Pettersen, E. F., Goddard, T. D., Huang, C. C., Couch, G. S., Greenblatt, D. M., Meng, E. C., et al. (2004). UCSF chimera—A visualization system for exploratory research and analysis. *Journal of Computational Chemistry, 25*(13), 1605–1612.

Prakash, A., Moharana, K., Wallace, S. S., & Doublié, S. (2017). Destabilization of the PCNA trimer mediated by its interaction with the NEIL1 DNA glycosylase. *Nucleic Acids Research, 45*(5), 2897–2909. https://doi.org/10.1093/nar/gkw1282.

Rambo, R. P. (2022). *ScatterIV*. Retrieved from https://github.com/rambor/scatterIV.

Rambo, R. P., & Tainer, J. A. (2013). Accurate assessment of mass, models and resolution by small-angle scattering. *Nature, 496*(7446), 477–481.

Russell, R. J., Haire, L. F., Stevens, D. J., Collins, P. J., Lin, Y. P., Blackburn, G. M., et al. (2006). The structure of H5N1 avian influenza neuraminidase suggests new opportunities for drug design. *Nature, 443*(7107), 45–49. https://doi.org/10.1038/nature05114.

Scheres, S. H., & Chen, S. (2012). Prevention of overfitting in cryo-EM structure determination. *Nature Methods, 9*(9), 853–854.
Schneidman-Duhovny, D., Hammel, M., Tainer, J. A., & Sali, A. (2013). Accurate SAXS profile computation and its assessment by contrast variation experiments. *Biophysical Journal, 105*(4), 962–974.
Schneidman-Duhovny, D., Hammel, M., Tainer, J. A., & Sali, A. (2016). FoXS, FoXSDock and MultiFoXS: Single-state and multi-state structural modeling of proteins and their complexes based on SAXS profiles. *Nucleic Acids Research, 44*(W1), W424–W429.
Schneidman-Duhovny, D., Kim, S. J., & Sali, A. (2012). Integrative structural modeling with small angle X-ray scattering profiles. *BMC Structural Biology, 12*(1), 17. https://doi.org/10.1186/1472-6807-12-17.
Schrodinger, LLC. (2015). *The PyMOL Molecular Graphics System, Version 1.8.*
Semenyuk, A., & Svergun, D. I. (1991). GNOM—A program package for small-angle scattering data processing. *Journal of Applied Crystallography, 24*(5), 537–540.
Shannon, C. E. (1948). A mathematical theory of communication. *The Bell System Technical Journal, 27*(3), 379–423.
Stuhrmann, H. B. (1970). Interpretation of small-angle scattering functions of dilute solutions and gases. A representation of the structures related to a one-particle scattering function. *Acta Crystallographica. Section A: Crystal Physics, Diffraction, Theoretical and General Crystallography, 26*(3), 297–306.
Svergun, D. I. (1999). Restoring low resolution structure of biological macromolecules from solution scattering using simulated annealing. *Biophysical Journal, 76*(6), 2879–2886.
Svergun, D. I., Barberato, C., & Koch, M. H. (1995). CRYSOL—A program to evaluate X-ray solution scattering of biological macromolecules from atomic coordinates. *Journal of Applied Crystallography, 28*(6), 768–773.
Svergun, D. I., & Koch, M. H. J. (2003). Small-angle scattering studies of biological macromolecules in solution. *Reports on Progress in Physics, 66*(10), 1735–1782. https://doi.org/10.1088/0034-4885/66/10/r05.
Svergun, D. I., Koch, M. H., Timmins, P. A., & May, R. P. (2013). *Small angle X-ray and neutron scattering from solutions of biological macromolecules. Vol. 19.* Oxford University Press.
Svergun, D. I., Petoukhov, M. V., & Koch, M. H. (2001). Determination of domain structure of proteins from X-ray solution scattering. *Biophysical Journal, 80*(6), 2946–2953.
Svergun, D. I., & Stuhrmann, H. (1991). New developments in direct shape determination from small-angle scattering. 1. Theory and model calculations. *Acta Crystallographica. Section A: Foundations of Crystallography, 47*(6), 736–744.
Tang, G., Peng, L., Baldwin, P. R., Mann, D. S., Jiang, W., Rees, I., et al. (2007). EMAN2: An extensible image processing suite for electron microscopy. *Journal of Structural Biology, 157*(1), 38–46. https://doi.org/10.1016/j.jsb.2006.05.009.
Taupin, D., & Luzzati, V. (1982). Information content and retrieval in solution scattering studies. I. Degrees of freedom and data reduction. *Journal of Applied Crystallography, 15*(3), 289–300.
Tickle, I. J. (2012). Statistical quality indicators for electron-density maps. *Acta Crystallographica. Section D: Biological Crystallography, 68*(Pt. 4), 454–467. https://doi.org/10.1107/S0907444911035918.
Trewhella, J., Duff, A. P., Durand, D., Gabel, F., Guss, J. M., Hendrickson, W. A., et al. (2017). 2017 Publication guidelines for structural modelling of small-angle scattering data from biomolecules in solution: An update. *Acta Crystallographica. Section D: Structural Biology, 73*(9), 710–728.
Tuukkanen, A. T., Kleywegt, G. J., & Svergun, D. I. (2016). Resolution of ab initio shapes determined from small-angle scattering. *IUCrJ, 3*(6), 440–447.
Valentini, E., Kikhney, A. G., Previtali, G., Jeffries, C. M., & Svergun, D. I. (2015). SASBDB, a repository for biological small-angle scattering data. *Nucleic Acids Research, 43*(D1), D357–D363.

van der Walt, S., Colbert, S. C., & Varoquaux, G. (2011). The NumPy array: A structure for efficient numerical computation. *Computing in Science & Engineering*, *13*(2), 22–30. https://doi.org/10.1109/MCSE.2011.37.

Van Heel, M., & Schatz, M. (2005). Fourier shell correlation threshold criteria. *Journal of Structural Biology*, *151*(3), 250–262.

Vestergaard, B., & Hansen, S. (2006). Application of Bayesian analysis to indirect Fourier transformation in small-angle scattering. *Journal of Applied Crystallography*, *39*(6), 797–804. https://doi.org/10.1107/S0021889806035291.

Virtanen, P., Gommers, R., Oliphant, T. E., Haberland, M., Reddy, T., Cournapeau, D., et al. (2020). SciPy 1.0: Fundamental algorithms for scientific computing in Python. *Nature Methods*, *17*(3), 261–272. https://doi.org/10.1038/s41592-019-0686-2.

Virtanen, J. J., Makowski, L., Sosnick, T. R., & Freed, K. F. (2011). Modeling the hydration layer around proteins: Applications to small-and wide-angle x-ray scattering. *Biophysical Journal*, *101*(8), 2061–2069.

Volkov, V. V., & Svergun, D. I. (2003). Uniqueness of ab initio shape determination in small-angle scattering. *Journal of Applied Crystallography*, *36*(3), 860–864.

Waasmaier, D., & Kirfel, A. (1995). New analytical scattering-factor functions for free atoms and ions. *Acta Crystallographica. Section A: Foundations of Crystallography*, *51*(3), 416–431. https://doi.org/10.1107/S0108767394013292.

Walther, D., Cohen, F. E., & Doniach, S. (2000). Reconstruction of low-resolution three-dimensional density maps from one-dimensional small-angle X-ray solution scattering data for biomolecules. *Journal of Applied Crystallography*, *33*(2), 350–363.

Winn, M. D., Ballard, C. C., Cowtan, K. D., Dodson, E. J., Emsley, P., Evans, P. R., et al. (2011). Overview of the CCP4 suite and current developments. *Acta Crystallographica. Section D: Biological Crystallography*, *67*(Pt. 4), 235–242. https://doi.org/10.1107/S0907444910045749.

Wriggers, W. (2012). Conventions and workflows for using Situs. *Acta Crystallographica. Section D: Biological Crystallography*, *68*(4), 344–351.

Zheng, W., & Tekpinar, M. (2011). Accurate flexible fitting of high-resolution protein structures to small-angle x-ray scattering data using a coarse-grained model with implicit hydration shell. *Biophysical Journal*, *101*(12), 2981–2991.

CHAPTER SEVEN

Computational methods for the analysis of solution small-angle X-ray scattering of biomolecules: ATSAS

Haydyn D.T. Mertens*
European Molecular Biology Laboratory, Hamburg Unit, Hamburg, Germany
*Corresponding author: e-mail address: hmertens@embl-hamburg.de

Contents

Abstract

The ATSAS software suite provides a comprehensive set of programs for the processing, analysis and modeling of small-angle scattering data, tailored for but not limited to data acquired on biological macromolecules. In this review the major components and developments in the ATSAS package are described, with a focus on user driven application. Data reduction, analysis and modeling approaches and strategies will be introduced and discussed. At the time of writing the latest package, ATSAS 3.1, is freely available for academic users at: https://www.embl-hamburg.de/biosaxs/software.html.

Methods in Enzymology, Volume 678
ISSN 0076-6879
https://doi.org/10.1016/bs.mie.2022.09.019

1. Introduction

Small-angle X-ray scattering (SAXS) is a widely used technique in structural biology, with a multitude of specific applications developed for liquid state samples under near-native conditions (Brosey & Tainer, 2019; Da Vela & Svergun, 2020; Svergun, Koch, Timmins, & May, 2013; Trewhella, 2022). Although anisotropic methods are in use, primarily in the soft-matter community, the majority of biological solution SAXS experiments are conducted on isotropic solutions of biomacromolecules. As the requirements for robust sample preparation are arguably less demanding than for higher resolution methods, SAXS data can be readily acquired for many systems and provide a stand-alone approach to low-resolution characterization and an unmatched level of method complementarity. Indeed, a main strength of SAXS is in the validation and refinement of higher-resolution structural data, e.g., from Macromolecular X-ray crystallography (MX), Nuclear Magnetic Resonance (NMR) and cryogenic electron microscopy (cryoEM) (Kachala, Valentini, & Svergun, 2015; Kikhney & Svergun, 2015; Lipfert & Doniach, 2007; Mertens & Svergun, 2017). Additionally, SAXS is a powerful method for the characterization of conformationally flexible and disordered systems which are inaccessible to higher resolution structural techniques (Bernadó, Mylonas, Petoukhov, Blackledge, & Svergun, 2007; Hammel et al., 2021; Panjkovich & Svergun, 2016; Tria, Mertens, Kachala, & Svergun, 2015).

In a solution SAXS experiment a liquid sample typically under flow is exposed to a collimated X-ray beam. The scattered X-rays are normally isotropic and registered on a photon sensitive detector producing a 2D image. For biological SAXS experiments it is standard to azimuthally average the detector image and generate a 1D scattering intensity curve I(s), ($s = 4\pi \sin\theta/\lambda$, where 2θ is the angle between the scattered and the incident radiation, and λ is the X-ray wavelength). To improve data quality, replicate exposures are collected where possible and averaged. The average scattering of the background is then subtracted from that of the sample using a separate measurement of the solvent/buffer collected under identical instrumental conditions. This yields a difference scattering curve that describes the scattering of the particles of interest, with the contribution of the solvent, sample holder and parasitic scattering effects removed (Svergun et al., 2013).

The background-subtracted 1D scattering curve can be used to derive important structural characteristics of the particles in solution, such as the radius of gyration (Rg) (Guinier, 1939), maximum particle dimension (D_{max}), pair distance distribution function P(r) (Glatter, 1977; Hansen, 2012; Svergun, 1992), excluded particle/Porod volume (V_p) (Porod, 1951) and molecular weight (MW) (Hajizadeh, Franke, Jeffries, & Svergun, 2018; Mylonas & Svergun, 2007; Orthaber, Bergmann, & Glatter, 2000; Rambo & Tainer, 2013). The average electronic structure of the particles or form-factor can also be recovered from a high-quality curve. In addition, interparticle interactions described by the so-called Structure-factor can be characterized. Advanced modeling procedures have been developed to extend beyond the simple parametric characterization of SAXS data including low-resolution ab initio methods for three-dimensional shape reconstruction (Franke & Svergun, 2009; Grant, 2018; Svergun, 1999; Svergun, Petoukhov, & Koch, 2001). Hybrid modeling approaches incorporating high-resolution models from other techniques such as X-ray crystallography are also available and are routinely applied to obtain atomistic representations (Panjkovich & Svergun, 2016; Petoukhov & Svergun, 2005). Polydisperse samples exhibiting either particle conformational sampling/disorder (e.g., protein flexibility) or distributions in particle size (e.g., assembly/oligomerization processes) can also be characterized through advanced modeling and fitting procedures that treat the sample as a mixture of defined components with specific volume fractions (Konarev & Svergun, 2018; Tria et al., 2015).

A number of dedicated software tools and packages have been developed for the processing, analysis and modeling of SAXS data, both in the soft matter and biological fields (Alves, Pedersen, & Oliveira, 2017; Breßler, Kohlbrecher, & Thünemann, 2015; Bressler, Pauw, & Thünemann, 2015; Brookes, Vachette, Rocco, & Pérez, 2016; Hopkins, Gillilan, & Skou, 2017; Liu, Morris, Hexemer, Grandison, & Zwart, 2012; Perkins et al., 2016; Schneidman-Duhovny, Hammel, & Sali, 2010). The most comprehensive and popular software suite used for biological SAXS is the cross-platform ATSAS package (Franke et al., 2017; Konarev, Petoukhov, Volkov, & Svergun, 2006; Manalastas-Cantos et al., 2021; Petoukhov et al., 2012; Petoukhov, Konarev, Kikhney, & Svergun, 2007), developed with a focus on ease of use and application. ATSAS is under continuous development by the team at EMBL (Hamburg), benefiting from the developers also supporting experiments at the P12 synchrotron SAXS beamline (PETRA III, DESY, Hamburg). The individual ATSAS programs provide an extensive set of utilities that can be applied to monodisperse and also polydisperse biological systems. The suite has been freely available for academic use since 2003 and has been downloaded more than 100,000 times by over

18,000 unique users. An online interface to the most popular programs is also available and submission to date is approaching 50,000 jobs annually by almost 1000 unique users. Below we will introduce the key underlying components of the engine that drives the modeling procedures of the mature ATSAS suite and also describe a number of the important data analysis tools required by the SAXS user community.

2. Calculation and simulation of scattering data

2.1 Computation of SAXS intensities and model fitting

A key feature of ATSAS is the use of spherical harmonics for the rapid computation of scattering amplitudes and isotropic SAXS intensities from high-resolution atomic structures, and in fitting predicted scattering curves to experimental SAXS data (Svergun, Barberato, & Koch, 1995). The workhorse program for these operations is CRYSOL. CRYSOL provides scattering intensities in several convenient forms, the first in units proportional to electrons-squared, and the second on an absolute scale per unit concentration [$I_{abs}(s)$ (cm^{-1})/c (mg.mL^{-1})]. In addition, CRYSOL can incorporate wavelength dependent anomalous effects for the analysis of anomalous X-ray scattering experiments. The general execution of CRYSOL via the GUI or command line has been described in detail in our previous work and is also available via the online manual repository (https://www.embl-hamburg.de/biosaxs/manuals/), here we will outline how to execute CRYSOL for anomalous scattering calculations using the common command line interface.

The options for running an ATSAS tool such as CRYSOL can be obtained by adding the -h "help" flag to the executable:

```
$> crysol -h
Usage in the batch mode:
crysol [<Inp_File1>]...[<Inp_FileK>] [<Dat_File>]
[-param1 <param1>]...[-paramN <paramN>]
where the arguments Inp_File can be either the
names of a PDB format file (extension .pdb or .ent)
or CRYSOL save file (extension .sav). Wildcards
(e.g. m?.sav or 1a*.ent or *.pdb) are permitted, in
which case the program makes a loop over these files
Dat_File is experimental scattering data file name.
When no experimental file name is given the intensity
is predicted with default parameters (no fitting)
```

```
The following program options can be given as
keys with parameters
-lm Maximum order of harmonics (default 15)
-fb Order of Fibonacci grid (default 17)
-sm Maximum scattering vector (default 0.5)
-ns # points in computed curve (default 51)
-un Angular units
-dns Solvent density (default 0.334 e/A**3)
-dro Contrast of hydration shell (0.03 e/A**3)
-kp Whether to keep output files other than
  (i) in fitting mode: *.log and *.fit
  (ii) in prediction mode: *.log, *.alm and *.int
  (default No, i.e. other files will be deleted)
-cst Subtract constant (default no)
-eh Account for explicit hydrogens
-err=N Write old-style fit file without
  information on experimental errors
-old Read a PDB file with an old atom naming
  (pre 2008?)
-nmr # of NMR conformer
-cid Chain identifier(s) (0: all chains)
-el Absorbing element (for anomalous SAXS)
-en Energy, absorption edge (for ASAXS)
-ff File with anomalous form factors (ASAXS)
-p Prefix for the output file names
-v Print version information and exit
-h Print this help and exit
Examples:
crysol 6lyz.pdb -lm 20
Calculate scattering intensity from the PDB file
6lyz.pdb with Lmax = 20 and without fitting
crysol mod*.pdb exp.dat -un 2
Process PDB files with the names beginning with
"mod" and fit experimental data exp.dat with
scattering vector given in inverse nanometres.
crysol *.sav lyzexp.dat
Restore the scattering intensity from all
sav files in the current directory and fit
the experimental data in the file lyzexp.dat
```

From the detailed list of options one can see the relevant flags for anomalous scattering (i.e., -el, -en and -ff), alongside the standard options for model (*.pdb and *.cif format) and scattering data file (*.dat) input. Anomalous effects occur when the wavelength of the incident radiation causes electronic transitions of a particular element to occur, i.e., the radiation wavelength is close the absorption edge. When this condition is met the incident radiation is partially absorbed and electrons are excited to higher-energy states. This leads to a reduction in scattering intensity (James, 1950). From this anomalous effect distance information can be extracted and has been successfully used in both X-ray crystallography (Hendrickson, 2014) and SAXS (Miake-Lye, Doniach, & Hodgson, 1983; Stuhrmann & Notbohm, 1981), however, the change observed in the SAXS signal is typically very low and requires an optimized instrumental set-up. Below the command line arguments required for the prediction of anomalous scattering intensities from the metalloprotein parvalbumin (PDB: 4cpv), with calcium atoms substituted by terbium (absorption edge 7514 eV) is shown. The calculation is conducted at the absorption edge, and a two additional points far from it (3000 eV and 11,000 ev):

```
$> crysol -lm 50 -ns 1024 -el tb -en 7514 parvalbumin_Tb.pdb
$> crysol -lm 50 -ns 1024 -el tb -en 3000 parvalbumin_Tb.pdb
$> crysol -lm 50 -ns 1024 -el tb -en 11000 parvalbumin_Tb.pdb
```

The generation of a difference curve (Fig. 1), obtained by subtracting the high energy scattering profile (i.e., 11,000 eV) from that at the absorption edge (7514 eV), provides structural information on the distances between the two Tb metal centers in the protein.

A detailed description of using ASAXS for biomolecular studies is presented by Gruzinov et al. (2021), with a detailed discussion on the effect of anomalous scattering on SAXS parameters such the radius of gyration and the conversion via Fourier Transformation to real-space.

The previous example demonstrates the basic usage of CRYSOL for the computation of theoretical scattering curves, with the expansion of the program for ASAXS computation also shown. In the following example we demonstrate the fitting procedure of CRYSOL for the validation of a structural model of a protein from X-ray crystallography with the SAXS curve obtained in solution.

Here we compare the predicted scattering profile from the high-resolution crystallographic structure of Bovine Serum Albumin (BSA) 4F5S to the experimental data (bsa_sec.dat) collected at the P12 BIOSAXS beamline (PETRA III, Hamburg). The data for this sample was obtained

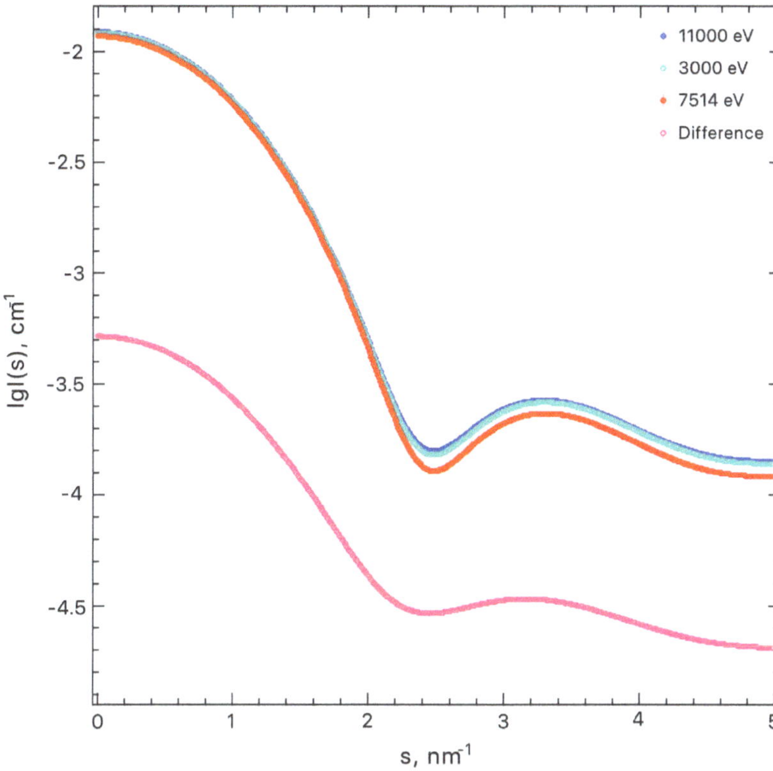

Fig. 1 ASAXS analysis of Parvalbumin. Predicted scattering curves from PDB ID. 4CPV at several energies, with both Calcium centers replaced by Terbium. The difference curve (pink circles) results from subtraction of the curve computed at the absorption edge of Terbium (7514 eV) from that computed far from the edge at 11000 eV.

by in-situ Size-exclusion chromatography (SEC) coupled SAXS, producing purified monomeric BSA during the measurement:

```
$> crysol -lm 50 -ns 1024 -cst 4F5S.pdb bsa_sec.dat
```

This command instructs CRYSOL to compute the predicted intensities from the 4F5S pdb structure file, using 50 spherical harmonics and 1024 points. The fit of the predicted profile to the experimental data is optimized through adjustment of the average atomic radius (adjusting the excluded volume used in the calculation, i.e., the volume occupied by the protein in a continuum of solvent) and the contrast of the thin water layer at the surface of the protein (the hydration layer of ordered water molecules, distance from the bulk solvent). An additional constant is applied to account for mismatch between the measured buffer and sample solutions (i.e., the buffer composition of both is not exactly the same) (Fig. 2).

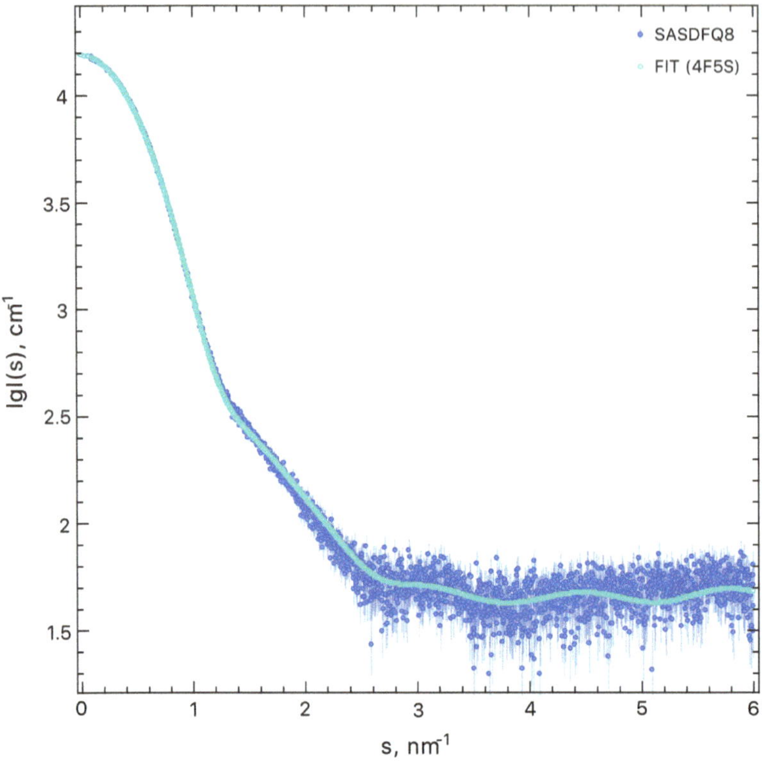

Fig. 2 CRYSOL fitting procedure for the comparison of the crystallographic structure of BSA with solution SAXS data. The experimental data was collected at the P12 SAXS beamline (EMBL, PETRA III, Hamburg) and is available at https://www.sasbdb.org/ (accession number: SASDFQ8).

The key component of the computation is the speed of the spherical harmonics approximation, operating in 3D polar (i.e., spherical) coordinates rather than the historical Debye point-to-point approach. The sum of these orthogonal angular functions $Y_{lm}(\Omega)$, the multipole expansion, provides an expression (Eq. 1) for the scattering density, $\rho(r)$ of a particle in solution that describes both gross structural features (lower order terms) and finer detailed structure (higher order terms) (Stuhrmann, 1970):

$$\rho(r) = \sum_{l=0}^{L} \sum_{m=-l}^{l} \rho_{lm}(r) Y_{lm}(\omega) \tag{1}$$

Where $(r,\omega) = (r, \theta, \varphi)$ are spherical coordinates and

$$\rho_{lm}(r) = \int_{\omega} \rho(r) Y_{lm}^{*}(\omega) d\omega \tag{2}$$

are the radial functions. The scattering amplitude is represented in reciprocal space as:

$$A(\mathbf{s}) = \sum_{l=0}^{L} \sum_{m=-l}^{l} A_{lm}(s) Y_{lm}(\Omega) \tag{3}$$

Where the momentum transfer $s = 4\pi \sin(\theta/2)/\lambda$ with θ the scattering angle and λ the wavelength, Ω is the solid angle in reciprocal space, and the truncation value L defines the resolution. The partial amplitudes $A_{lm}(s)$ are calculated as:

$$A_{lm}(s) = 4\pi i^{l} \sum_{j=1}^{N} f_j(s) j_l(sr_j) Y_{lm}^{*}(\omega_j) \tag{4}$$

where the sum runs over all elements with coordinates $rj = (rj, \omega j) = (rj, \theta j, \varphi j)$ and $fj(s)$ are the corresponding form factors, and $j_l(sr)$ are spherical Bessel (Stuhrmann, 1970) functions.

The orthogonal properties of the spherical harmonics allows for the following expression of the spherically averaged scattering intensity $I(s)$:

$$I(s) = 2\pi^2 \sum_{l=0}^{L} \sum_{m=-l}^{l} |A_{lm}(s)|^2 \tag{5}$$

Eq. (5) provides a convenient mechanism for the rapid computation of particle scattering intensity as a sum of independent contributions (e.g., from radially separated scattering centers with known atomic form factors). In addition, the multipole expansion eliminates cross-terms from the calculation. Eqs. (1)–(5) also facilitate the reverse procedure, going from a scattering pattern to low resolution structure.

2.2 Simulation of 2D SAXS intensities

ATSAS also provides tools for the simulation of SAXS intensities as recorded on a 2D detector. The image simulation program, IMSIM, provides SAXS patterns in 2D (frames/images), as would be registered from the scattering of a sample on a modern detector (Franke, Hajizadeh, & Svergun, 2020). These images are invaluable for the testing of data reduction and analysis/modeling approaches and also as an aid to experimental design and beamline optimization.

To generate a simulated 2D SAXS image of a protein molecule with IMSIM, the first step is to compute the 1D intensity profile on an absolute scale (cm^{-1}) using CRYSOL for example (see previous section). The simulation is based upon this curve and uses a statistical simulation approach such that the intensities and error estimates reflect that obtained from experimental data. A number of options are available as listed below:

```
$> imsim -h
Usage: imsim [OPTIONS] <ABSFILE>
Known Arguments:
  ABSFILE                  Calculated scattering intensity in I
[cm^-1] / conc[mg/ml]
Known Options:
      --seed=<INT>         Set the seed for the random number generator
      --background=<FILE>  Calculated background intensity in I
[cm^-1] / conc[mg/ml] (default: built-in flat background)
      --detector=<NAME>    Detector name, defines pixel size and
mask dimensions (default: Pilatus1M)
      --detector-distance=<m>Sample-detector-distance in meter
(default: 1.0)
      --flux=<N>           Number of incoming virtual photons per
second (default: 10^12)
  -p, --scattering-probabilitProbability that a virtual photons
scatters (default: 1e-6)
      --exptime=<T>        Exposure time in seconds (default: 1)
  -c, --concentration=<C>  Factor proportional to concentration
of sample (default: 0)
      --wavelength=<m>     X-ray wavelength in meter (default: 0.1e-9)
      --axis               Do not simulate but write axis for
radial averaging
      --mask               Do not simulate but write a centered
detector mask for radial averaging
      --mask-offset-x=<X>  Mask offset in horizontal direction
(default: 0)
      --mask-offset-y=<Y>  Mask offset in vertical direction
(default: 0)
      --header             Do not simulate but write header infor-
mation suitable for IM2DAT
```

```
  -o, --output=<FILE>       Output file name (default: basename of
the input file, extension depends on other options)
  -f, --format=<FORMAT>     Output format, one of: TIFF, EDF, CBF,
MSK, ...
      --atomic              Use atomic scattering without hydration
shell only
  -h, --help                Print usage information and exit
  -v, --version             Print version information and exit
```

Here we show an example for the simulation of the 2D SAXS image of BSA (PDB ID. 4F5S):

1. Computation of theoretical 1D SAXS intensities from PDB file, with a maximum s value of 1.0 Å^{-1}, 5001 points and 50 spherical harmonics.
   ```
   $> crysol -sm 1.0 -ns 5001 -lm 50 4F5S
   ```
2. Simulation of the 2D image from the 1D SAXS curve on absolute scale (4F5S00.abs), using the sim.sh helper script. The key options used such as X-ray wavelength, flux and exposure time, as well as desired output such as an axis calibration file and mask file are shown.
   ```
   $> bash sim.sh 4F5S00.abs
   --- * Helper script sim.sh * --
   #! /bin/bash
   # Shorthand for common options that are the same for all calls to
   IMSIM.
   PROFILE="--detector=pilatus6m      --detector-distance=3.1      --
   wavelength=0.124e-9 --flux=1e12 --scattering-probability=1e-3"
   SAMPLE_EXPTIME=01.00
   BUFFER_EXPTIME=10.00

   abs=${1}
   prefix=$(basename ${abs} .abs)

   # MASK and AXIS do not depend on random are the same everywhere.
   mask=${prefix}_mask.msk
   imsim ${PROFILE} ${abs} --mask --mask-offset-y=-1100 -o ${mask}
   axis=${prefix}_axis.dat
   imsim ${PROFILE} ${abs} --axis -o ${axis}

   for CONCENTRATION in 0.5 2.0 8.0 ; do
     # Prepare output file names.
     buffer=${prefix}_buffer_c${CONCENTRATION}_e${BUFFER_EXPTIME}
     sample=${prefix}_sample_c${CONCENTRATION}_e${SAMPLE_EXPTIME}
   ```

```
  sub=${prefix}_sub_c${CONCENTRATION}
  norm=${prefix}_norm_c${CONCENTRATION}

  # Do all simulations in parallel.
  imsim ${PROFILE} --exptime=${BUFFER_EXPTIME} -c 0.0 ${abs} -o
${buffer}.tiff &
  imsim ${PROFILE} --exptime=${BUFFER_EXPTIME} -c 0.0 ${abs} -o
${buffer}.txt --header &
  imsim ${PROFILE} --exptime=${SAMPLE_EXPTIME} -c ${CONCENTRATION}
${abs} -o ${sample}.tiff &
  imsim ${PROFILE} --exptime=${SAMPLE_EXPTIME} -c ${CONCENTRATION}
${abs} -o ${sample}.txt --header
  wait

  # Obtain the actual position of the incident beam.
  x=$(grep beam-center-x ${sample}.txt | cut -d: -f2 | tr -d [:
space:])
  y=$(grep beam-center-y ${sample}.txt | cut -d: -f2 | tr -d [:
space:])

  # Radially average two-dimensional images to one-dimensional
scattering data.
  im2dat -x ${x} -y ${y} -beamstop-mask ${mask} --axis-data ${axis}
${buffer}.tiff ${buffer}.txt -o ${buffer}.dat
  im2dat -x ${x} -y ${y} -beamstop-mask ${mask} --axis-data ${axis}
${sample}.tiff ${sample}.txt -o ${sample}.dat

  # Subtract background from sample and normalize by concentration.
  datop sub ${sample}.dat ${buffer}.dat -o ${sub}.dat
  datop div ${sub}.dat ${CONCENTRATION} -o ${norm}.dat

  # (optionally) clean up the resulting images to save some space.
  # rm -f *.tiff
done
```

Support for performing operations at the level of 2D images has been recently introduced into ATSAS, including the image operations program IMOP, which follows a similar strategy to that supplied by the data operations program DATOP for 1D scattering data (Franke et al., 2017). IMOP supports operations such as addition and subtraction of images, as well as AND, OR and XOR operations for binary masks. IMOP can be employed for cross-validation of data reduction operations.

Once a simulated 2D image has been generated, data reduction programs such as IM2DAT can be used to azimuthally average the image and provide 1D SAXS curves for standard analysis, with appropriate errors and statistics based on the desired input. The data reduction protocol is outlined in the following section.

3. Primary data processing

As mentioned in Section 1 it is typical in biological SAXS data analysis for the acquired intensities to be represented as 1D profiles/curves. A number of data reduction and processing steps are performed to obtain the desired SAXS profile, and the ATSAS package provides both an automated and semi-automated/manual workflow for these procedures. The standard data reduction steps are outlined in Fig. 3.

The initial step post 2D image acquisition is the azimuthal/radial integration of the scattering intensities. This is performed using the IM2DAT program (Konarev et al., 2006), where a calibration file based on an appropriate standard such as Silver Behenate ($AgC_{22}H_{43}O_2$, $d_{00}=5.84\,nm$) and a mask file (masking out of detector regions not contributing to the intensities, e.g., detector edges, beamstop shadow and space between modules) are used to reduce the dimensionality of the data. Error estimates are based on Poisson photon counting statistics, and outliers in each integration ring section detected through a transformation step approximating a normal distribution (Anscombe, 1948; Iglewicz & Hoaglin, 1993). Sub-pixel analysis (i.e., pixel-splitting) has not been implemented as we aim to avoid the introduction of correlations between neighboring intensity estimates. The resulting 1D SAXS profile/curve produced is then passed to programs specifically operating on 1D data sets with care taken to ensure reliable error propagation.

3.1 Primary data processing of 1D SAXS profiles/curves

Correct data handling is essential for the delivery of reliable 1D SAXS data sets (Kikhney, Borges, Molodenskiy, Jeffries, & Svergun, 2020). Cross-validation at various stages during primary processing helps to ensure that error propagation is correctly conducted.

3.1.1 Basic operations on 2D and 1D scattering data

Tools have been introduced to allow data reduction and processing to be conducted on both 2D and 1D data. For 2D images the program IMOP

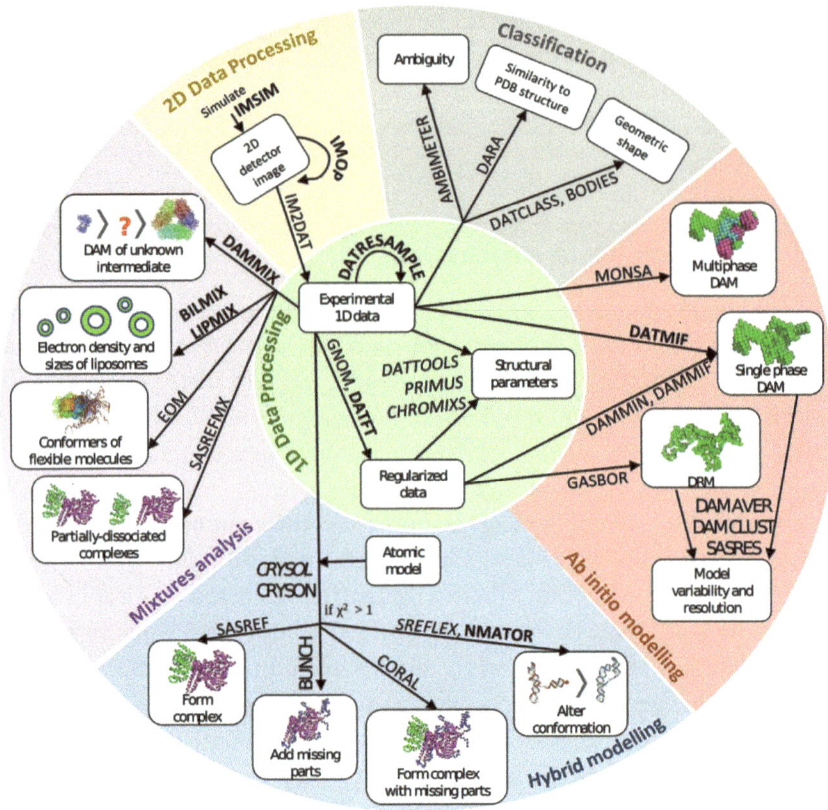

Fig. 3 The ATSAS suite of programs and procedures, showing both reduction/analysis/modeling procedures/steps and specific program modules. Also shown are the potential use cases involving monodisperse/polydisperse systems (DAM: dummy-atom model; DRM: dummy-residue model). *Reused with permission from IUCRJ Manalastas-Cantos, K., Konarev, P.V., Hajizadeh, N.R., Kikhney, A.G., Petoukhov, M.V., Molodenskiy, D.S., Panjkovich, A., Mertens, H.D.T., Gruzinov, A., Borges, C., Jeffries, C.M., Svergun, D.I., Franke, D. (2021). ATSAS 3.0: Expanded functionality and new tools for small-angle scattering data analysis.* Journal of Applied Crystallography 54:*343–355. https://doi.org/10.1107/S1600576720013412.*

(image operations) can be used for basic mathematical operations such as image addition and subtraction, and also provides functionality for the masking of pixels. As described above, the reduction of 2D images to 1D intensity profiles is facilitated by the program IM2DAT. Processing of 1D scattering profiles is conducted using the command line based DATOOLS suite, which has been designed for easy implementation into automated high-throughput small-angle scattering data processing procedures, such as the SASFLOW pipeline (Franke, Kikhney, & Svergun, 2012) employed at the P12 (EMBL, Hamburg) beamline. The DATOOLS (Franke et al., 2017) operations include functions for addition/subtraction/multiplication/division (DATOP, data operations) and for the extraction of SAXS invariants and other associated parameters as will be outlined in Sections 3.1.1–3.1.3. Additional helper tools are also available for the merging of data sets (DATMERGE, for data collected across different angular ranges, or the combining of low and high concentration data sets collected on the same samples), averaging (DATAVER, 1D data set averaging), and also for the statistical comparison of datasets (DATCMP, data comparison) to identify outliers. The components of the DATTOOLS suite have been integrated into the PRIMUS (Primary Data Analysis) GUI, such that those not familiar with command line usage can work visually with 1D SAXS data to perform all of the necessary data manipulations (Fig. 4).

The two primary SAXS invariant parameters extracted from SAXS data are the radius of gyration, R_g, and the forward scattering intensity (or intensity at zero angle), I_0. These parameters describe the overall average particle size and scattering mass, respectively. An example of using the command line DATRG protocol is shown below:

(i) Accessing the usage information/help:

```
$> datrg -h
Usage: datrg [OPTIONS] <DATAFILE(S)>
Radius of Gyration (Rg) from Guinier approximation
Known Arguments:
 DATAFILE             Data file
Known Options:
 -t, --type=<TYPE>    particle shape: GLOBULAR, ROD or FLAT (default
GLOBULAR)
   --first=<N>        index of the first point of the Guinier region
   --last=<N>         index of the last point of the Guinier region
 -h, --help           Print usage information and exit
 -v, --version        Print version information and exit
Mandatory arguments to long options are mandatory for short options
too.
Report bugs to <atsas@embl-hamburg.de>.
```

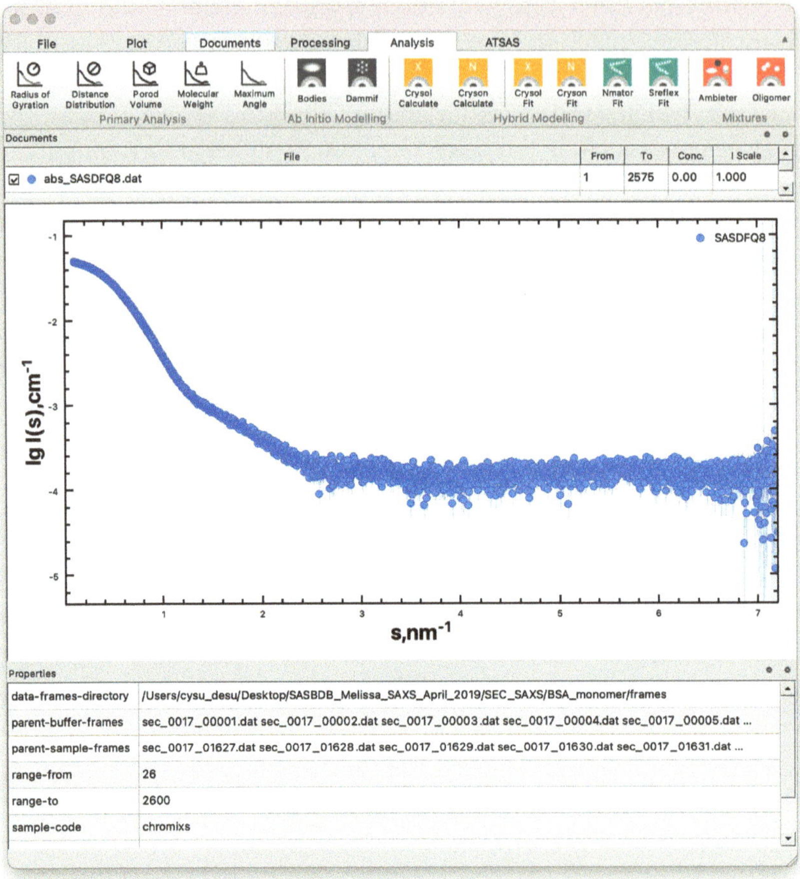

Fig. 4 PRIMUS GUI interface to ATSAS DATTOOLS suite of operations for data manipulation and analysis. Top panel shows the general GUI providing data manipulation and analysis tools alongside a data visualization canvas. Lower panels show the plot tools corresponding to the data visualization canvas (A), Guinier analysis/Radius of gyration analysis and corresponding residuals (B and E), real-space distance distribution (Indirect Fourier transformation) analysis and corresponding residuals (C and F), and the dimensionless Kratky plot representation (D).

(ii) Estimation of Rg and I0 parameters from an experimental data set (SASDFQ8.dat):

```
$> datrg --type globular --first 14 --last 135 abs_SASDFQ8.dat
 2.76215 0.188875E-02 0.499999E-01 0.202396E-04 0.123471E-01
```

The results are provided in columnar format as R_g, error R_g, I_0, error I_0 and fidelity.

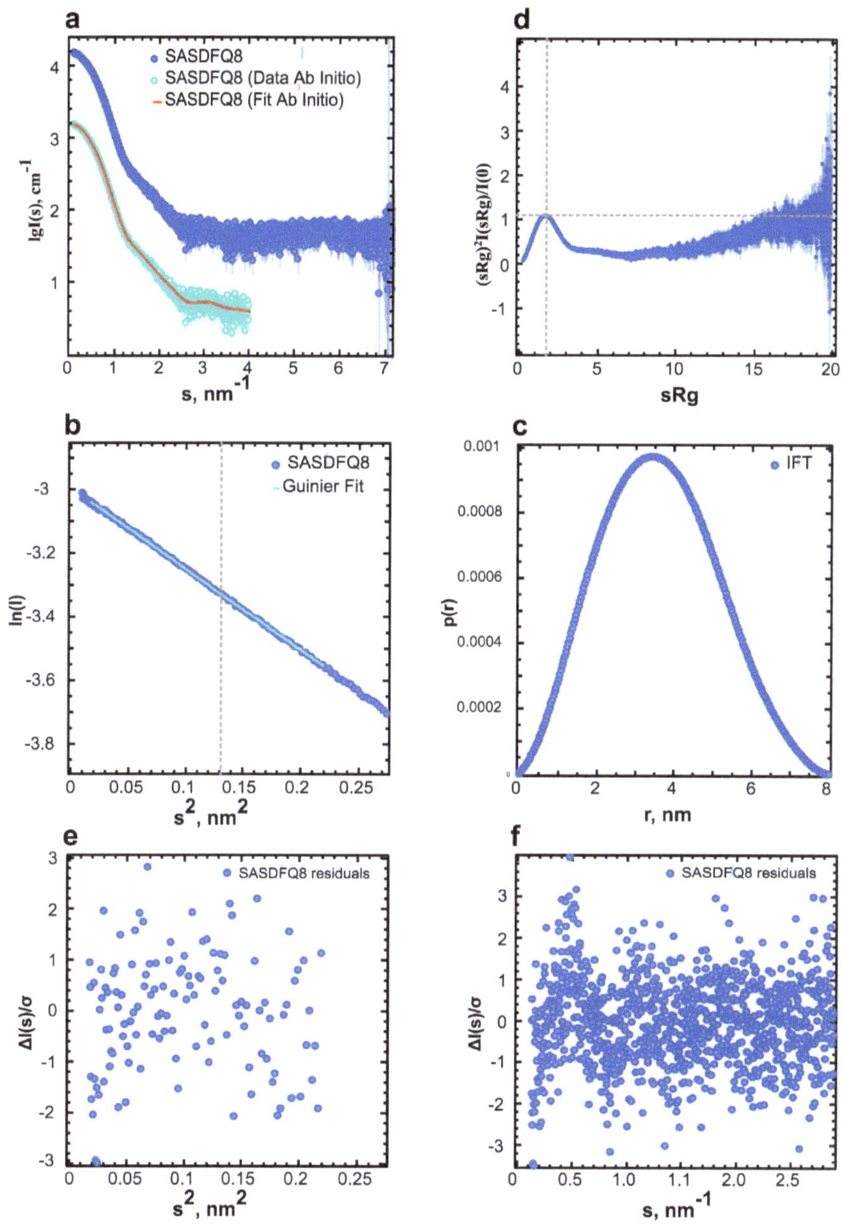
a
SASDFQ8
SASDFQ8 (Data Ab Initio)
SASDFQ8 (Fit Ab Initio)
lgI(s), cm^{-1}
s, nm^{-1}
d
(sRg)^{2}I(sRg)/I(0)
sRg
b
SASDFQ8
Guinier Fit
ln(I)
s^2, nm^2
c
IFT
p(r)
r, nm
e
SASDFQ8 residuals
ΔI(s)/σ
s^2, nm^2
f
SASDFQ8 residuals
ΔI(s)/σ
s, nm^{-1}

Fig. 4—cont'd

3.1.2 Indirect Fourier transformation and the real-space distance distribution

The SAXS intensities recorded in reciprocal space, I(s), are related to the scattering particles in real-space, *p(r)*, by Fourier transformation:

$$I(s) = 4\pi \int_0^{D_{\max}} p(r)\frac{\sin(sr)}{sr}\mathrm{d}r \tag{6}$$

The correlation of pointwise distances within the particles, weighted by the product of the excess scattering-length densities of each pair of points, can be described by the pair distance distribution function, *p(r)* (also referred to as the PDDF). The *p(r)* function is derived from *I(s)* by the inverse transform:

$$p(r) = \frac{r}{2\pi^2}\int_0^{\infty} sI(s)\sin(sr)\mathrm{d}s \tag{7}$$

As the measured intensities, *I(s)* are known only as a series of discrete points across a finite range ($0 < s < \infty$), *p(r)* is not typically determined directly from Eq. (2). The Indirect Fourier Transformation (IFT) approach, introduced by Glatter (1977), facilitates *p(r)* determination via the fitting of the experimental data with approximated functions prior to transformation. The ATSAS package expands upon this functionality in the program GNOM (Semenyuk & Svergun, 1991), where *p(r)* is parameterized by a set of analytical functions, and a regularization procedure is used to balance the fit to the experimental data and the smoothness of the real-space distribution. An example of the procedure applied to BSA is shown in Fig. 4. From the IFT approach, R_g and I_0 are also determined, providing an independent method to validate these parameters determined from Guinier analysis. In addition, IFT provides an estimate of the maximum particle dimension, D_{max}, and the hydrated particle volume (also referred to as the Porod volume, V_p).

Although, as previously stated, *p(r)* is not typically determined directly from Eq. (2), functionality in ATSAS is provided to perform a direct Fourier transform of I(s) with the help of additional constraints in the program DATFT. If I_0 and R_g can be reliably estimated via Guinier analysis, the truncated low angle data (lim → 0) can be extrapolated to 0. Extrapolation of the data at higher angles (lim → ∞) is approximated as $I(s) = s - n$ (where n is dependent on the particle form, with $n = 4$ for globular particles and $n = 2$ for fully flexible chain systems). These extrapolations are required to reduce artificial oscillations in the *p(r)* function caused by the finite data range (Harris, 1978). DATFT can operate on scattering data for which no preprocessing has been conducted, however, optimal results are found for data having low-noise and uniformly distributed points.

Several helper tools have been introduced to validate experimentally determined p(r) functions (PDDFFIT) and to facilitate easy extraction of the distribution and the regularized/fit curve for visualization/plotting (OUT2POFR, OUT2FIT).

3.1.3 Estimation of molecular weight of proteins from SAXS data

I_0 is one of the most reliable invariant parameters extracted from experimental SAXS data. This parameter is directly related to the volume of scatterers (i.e., electrons for X-rays) and thus the molecular weight of the particles in solution. If the SAXS data has been placed on an absolute scale, for example relative to pure water or another reference sample (Mylonas & Svergun, 2007; Orthaber et al., 2000), and the concentration, c of the macromolecule of interest is accurately known (ideally within 1–2% of the true value), then the molecular weight of the macromolecule can in principle be determined from I_0/c:

$$MW = \left(N_{av}{}^{*}I_0/c\right)/\left(\mathrm{PSV}^{*}\Delta P\right)^2, \tag{8}$$

Where, $N_{av} = 6.022\times10^{23}$, PSV is the partial specific volume (cm^3/g) and ΔP is the scattering contrast (cm^{-2}) between the macromolecule and the solvent.

The determined molecular weight is a valuable parameter for sample characterization. Aggregation, oligomeric state, complex association and dissociation can all be explored/verified by the molecular weight values obtained from the experiment. Several ATSAS tools are available for absolute calibration (DATABSOLUTE) relative to pure water and molecular weight estimation (DATABSMW). The following shows an example of molecular weight estimation of BSA (MW = 66,000 Da) from DATABSMW:

```
$> databsmw -h
Usage: databsmw [OPTIONS] [FILE(S)]
Molecular weight from absolute scaling.
Known Arguments:
 FILE         Data file on absolute scale, divided by the concentra-
tion [mg/ml]
```

```
Known Options:
  --psv=<X>          Partial Specific Volume (default: 0.7425)
  --contrast=<X>     Contrast in units of 10^10 (default: 2.8086)
  --i0=<X>           I(0) of the input file
 -h, --help          Print usage information and exit
 -v, --version       Print version information and exit
Mandatory arguments to long options are mandatory for short options
too.
Report bugs to <atsas@embl-hamburg.de>.
$> databsmw --psv 0.74 --contrast 2.90 --i0 0.051 abs_SASDFQ8.dat
66690
```

An accurate concentration estimation is not always available for every protein sample, and in size-exclusion chromatography coupled (SEC-SAXS) data collected without UV/RI-detection (UV = ultraviolet wavelength radiation, RI = refractive index) concentration is unavailable. In such cases concentration independent approaches can be used to obtain molecular weight estimates. These approaches can additionally be used to check molecular weight estimates derived from I_0/c. ATSAS provides several tools for concentration independent MW estimation, using scattering invariants independent of data scale. DATPOROD and SAXSMoW calculate the Porod invariant (Q_p) to provide an estimate of the volume (V_p) of the sample, where *MW* is estimated as the quotient of V_p and the protein partial specific volume (Porod, 1951). The final MW estimate is obtained following several empirical corrections (Petoukhov et al., 2012; Piiadov, Ares de Araújo, Oliveira Neto, Craievich, & Polikarpov, 2019) implemented within these programs. Although technically not an invariant, the parameter V_c, volume of correlation, correlates with *MW* with both protein and RNA, and *MW* estimation based on V_c (Rambo & Tainer, 2013) has been implemented in the ATSAS tool DATVC. DATCLASS, a machine-learning method based on protein structures in the PDB, combines classification of shape with D_{max} and *MW* estimation from scattering data to provide another concentration independent approach to *MW* estimation (Franke, Jeffries, & Svergun, 2018). All of the concentration independent approaches to *MW* estimation outlined can be combined to as input for the DATMW tool (Hajizadeh et al., 2018), a Bayesian approach to generate a consensus *MW* value with a probability estimate, enabling users to examine the interval of possible values.

4. Structural modeling from SAXS data

Modeling procedures based on SAXS data have evolved tremendously in the last decade, moving beyond simple parameter extraction and analytical from-factor fitting to three-dimensional modeling and the deconvolution of complex mixtures. The first generally applicable shape modeling procedures developed were those based on ab initio (i.e., from first principles) modeling. In these approaches the average structure of particles in solution is determined directly from the measured SAXS intensities through comparison to a shape consisting of spheres or dummy atoms. Hybrid methods incorporate high-resolution structural data (e.g., models deposited at the PDB) and other biophysical and biochemical constraint information, using SAXS data to drive iterative refinement to optimize a fit between the computed scattering of the models and the experimental data. Approaches have also been developed for the analysis of flexible or unstructured systems (e.g., intrinsically disordered proteins/regions, IDPs/IDRs) and for the characterization and deconvolution of polydisperse systems including oligomeric mixtures and association/dissociation processes. Membrane proteins and lipidic/detergent systems are also becoming a significant target of solution scattering studies, and as such ATSAS provides several tools applicable to the modeling of these multi-contrast macromolecular complexes. Below a number of modeling procedures used for the characterization of biological macromolecules incorporating SAXS data are outlined.

4.1 Ab initio (shape) methods

In the absence of structural information, models of macromolecules can be reconstructed directly from SAXS intensities and constrained by penalties designed to limit the search space to biophysically plausible solutions. ATSAS provides a number of tools for the generation of fits to geometrical shapes (BODIES) (Konarev, Volkov, Sokolova, Koch, & Svergun, 2003), reconstruction of bead/dummy-atom shape models (DAMMIN (Svergun, 1999)/DAMMIF (Franke & Svergun, 2009)/MONSA (Svergun & Nierhaus, 2000)/DATMIF (Manalastas-Cantos et al., 2021)), and reconstruction of dummy-residue models of proteins by means of a globbic approximation of the average scattering of amino-acids (GASBOR) (Svergun et al., 2001). These methods have been in use for more than a decade and have proven invaluable to many SAXS based data analyses.

4.1.1 Direct modeling from experimental data

Several of the ATSAS ab initio bead modeling applications (DAMMIN, DAMMIF, GASBOR) require a regularized scattering profile as input, such as that produced during the fitting routine of the indirect Fourier transformation procedure (e.g., from GNOM). This indirect approach is historical and has been replaced by a more direct modeling procedure implemented in the program DATMIF (Dummy Atom Minimization Fast). DATMIF generates bead models through a direct fitting of the scattering data and thus incorporates the experimental error estimates. In the older ATSAS ab initio programs the penalties used to guide the solution toward "biological-like" models include those penalizing discontinuity of beads, compatibility with R_g, implicit looseness/compactness, and explicit anisometry (when symmetry has been requested). In DATMIF, the only modeling penalty applied is the Akaike information criterion (AIC) for minimizing the number of beads. This results in DATMIF minimizing the volume of the final model and results in compact protein-like model generation (Fig. 5). An example of the command line execution of DATMIF for the protein BSA is shown below:

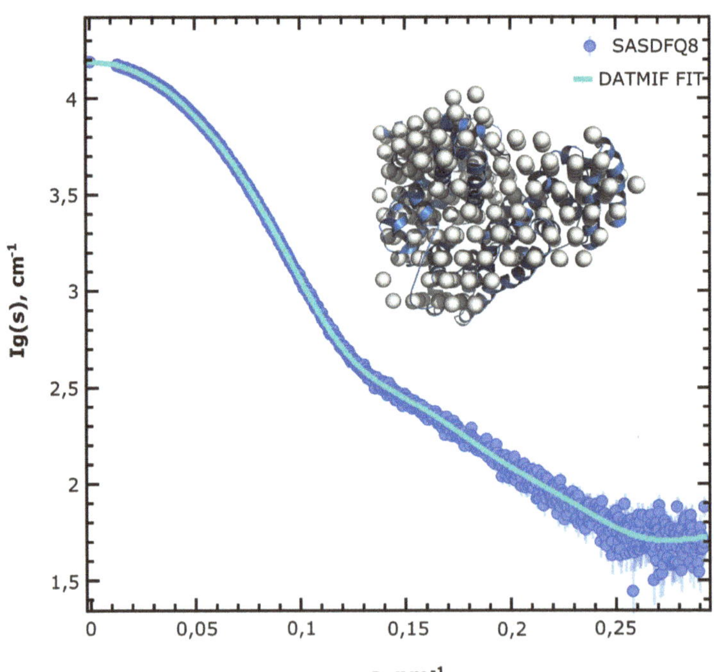

Fig. 5 DATMIF ab initio model of BSA. 1D scattering data (blue), overlaid with the corresponding DATMIF fit (cyan). Inset shows the DATMIF bead model superimposed on the high resolution crystallographic model of BSA (PDB ID 4F5S). Model generated from data deposited at www.sasbdb.org for entry SASDFQ8.

```
$> datmif -h
Usage: datmif [OPTIONS] <DATFILE>
Ab-initio shape determination of experimental small-angle scatter-
ing data
Known Arguments:
 DATFILE              Experimental data to model
Known Options:
 -h, --help            Print usage information and exit
 -v, --version         Print version information and exit
 -q, --quiet           Reduce verbosity level
  --seed=<INT>         Set the seed for the random number generator
 -u, --unit=<UNIT>     angular unit, inv. ANGSTROM, inv. NANOMETRE
or UNKNOWN (default: unknown)
  --constant=<VALUE>   constant to subtract (default: automatic)
  --first=<VALUE>      first data point (default: determined by
autorg)
  --last=<VALUE>       last data point (default: s * Rg <= 8)
 -p, --prefix=<PREFIX> output PREFIX prepended to any output
filename (default: datmif)
  --dmax=<D>           Estimate of the maximum overall distance? [A]
(required)
  --particle-bead-radius=dummy atom bead radius [A] (default: 4A)
 -s, --symmetry=<PXY>  particle symmetry
  --max-step=<ARG>     maximum number of temperature steps
(default: 500)
  --max-iter=<N>       maximum number of phase changes per temp.
step (default: #beads/10)
  --max-succ=<N>       maximum number of successful phase changes
before temp. is reduced (default: max-iter/10)
  --min-succ=<N>       minimum number of successful phase changes
before termination (default: max-iter/1000)
  --temp-schedule=<T>  temperature schedule factor (0.0-1.0)
(default: 0.95)
Mandatory arguments to long options are mandatory for short options
too.
$>datmif --first 14 --last 1024 --unit nanometer --dmax 80 --prefix
datmif_bsa SASDFQ8.dat
```

4.1.2 Multiphase modeling of membrane proteins in solution

Membrane proteins are commonly solubilized in solution using lipids, lipidic scaffolds (e.g., nanodiscs, SMALp, saliproNPs) and detergents (Denisov, Grinkova, Lazarides, & Sligar, 2004; Glover et al., 2001; Seddon, Curnow, & Booth, 2004; Tanford & Reynolds, 1976). This is required for stability in aqueous solvents as a significant amount of surface area is hydrophobic and must be shielded from the polar environment to retain structure and function. Membrane mimics provide the mixed polar/apolar environment required to stabilize integral membrane proteins, however, differences in electron density of the polar and apolar regions of the mimics and also that of the protein yield a multiple scattering contrast system. This multi-contrast situation can be difficult to model unambiguously using the aforementioned ab initio methods, however, the program MONSA (Svergun, 1999; Svergun & Nierhaus, 2000), facilitates ab initio modeling of systems consisting of multiple phases with distinct contrasts, and can be used to model, e.g., detergent-solubilized transmembrane proteins.

MONSA can reconstruct models of membrane proteins solubilized in membrane mimics, and solves the ill-posed problem through the addition of information about the chemistry of the system. Knowledge based constraints on the search volume used by MONSA are provided through extensions made to the tool BODIES, with the three phase initial MONSA search volume (1. Protein, 2. Hydrophobic lipid/detergent tails, 3. Hydrophilic lipid/detergent heads) constrained by user specified hydrophobic and hydrophilic thickness of the non-protein phases on the basis of the chemistry of the lipid/detergent used (Fig. 6). The phase assignment at the boundaries of each pair of phases, including that between the protein core and the solvent, is variable, facilitating better optimization during modeling in MONSA.

An example showing how bodies can be used to set up the input search volume for MONSA is shown below (Protein in DDM detergent):

```
$> bodies -h
Usage: bodies [OPTIONS] <FILE>
Fit or predict the small angle scattering data or generate a Dummy
Atom Model (DAM) of various body types. Predict and DAM mode are
not available through command-line arguments but through dialog
mode only (start without arguments).
Known Arguments:
 FILE      Data file to fit
```

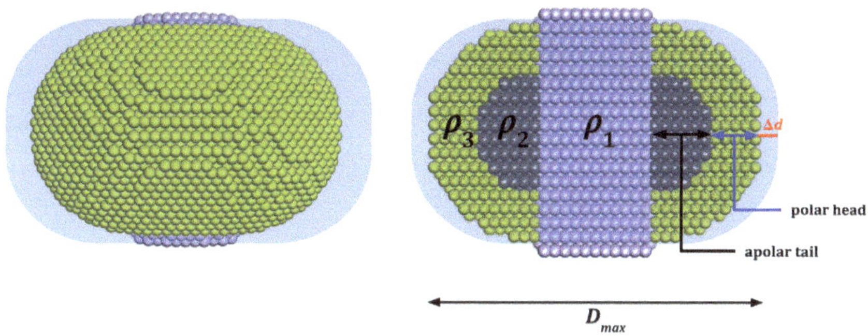

Fig. 6 Search volume created by BODIES for multiphase ab initio modeling with MONSA. The protein phase ρ_1 (blue), hydrophobic/apolar tail phase ρ_2 (gray) and hydrophilic/polar head phase ρ_3 (yellow) are shown within an ellipsoidal total search volume of maximum diameter D_{max}. Phase thicknesses and a variable boundary region Δd are specified by the user on the command line. A view of the surface and the internal structure of the phases are shown on the left and right, respectively.

```
Known Options:
 -p, --prefix=<PREFIX> output prefix for .log and .fit
  --body=<BODY>   Name of the BODY type
  --rg=<X>        Expected Rg (default: estimated from data)
  --i0=<X>        Expected I0 (default: estimated from data)
  --volume=<X>    Expected volume (default: estimated from data)
  --first=<N>     First point of the Guinier range (default: 1)
  --last=<N>      Last point of the fit range (default: ~4 Shannon
channels)
  --seed=<INT>    Set the seed for the random number generator
 -q, --quiet      Reduce verbosity level
 -h, --help       Print usage information and exit
 -v, --version    Print version information and exit
$> bodies
Operation mode? Select one of: (f) fit, (p) predict, (d) dam, (v)
searchvolume(default:fit)......................................:d
Body type? Select one of: (1) ellipsoid, (2) rotation-ellipsoid, (3)
cylinder, (4) elliptic-cylinder, (5) hollow-cylinder, (6)
parallelepiped, (7) hollow-sphere, (8) dumbbell, (9) liposome, (10)
memebrane protein (default: ellipsoid) ............................ :
10
```

```
Value of parameter 'rmemb' [Angstrom]? ............................ :
25
Value of parameter 'rtail' [Angstrom]? ............................ :
26
Value of parameter 'rhead' [Angstrom]? ............................ :
44
Value of parameter 'delta' [Angstrom]? ............................ : 5
Value of parameter 'zcorona' [Angstrom]? .......................... : 0
Dummy atom model symmetry? Select one of: (1) P1 (default: P1)
..... : p1
Number of dummy atoms? (default: 2000) ............................ :
5000
DAM output file name? ............................................. :
vol.pdb
```

The membrane protein specific options are:

R_{memb}: approximate radius of the protein phase from the origin.

R_{tail}: radial distance of the hydrophobic tail phase (distance from the end of protein phase)

R_{head}: radial distance of the hydrophilic head phase (distance from the end of the tail phase)

Delta: distance of variable boundaries between phases

Z_{corona}: offset of membrane from center of protein phase/origin (i.e. height of membrane relative to membrane protein)

4.2 Hybrid modeling procedures

Ab initio modeling methods are useful in the absence of prior structural information, but if such information is available more advanced modeling procedures can be applied. Hybrid modeling methods are provided in ATSAS for the modeling of rigid-body and flexible systems, where partial models of structured domains and/or additional distance/orientation/biophysical information is included in the model refinement procedure.

4.2.1 Rigid-body modeling

Rigid-body methods incorporate structural information in the form of high-resolution structures from the PDB or high-scoring homology models (e.g., models obtained from sophisticated algorithms such as those used by AlphaFold) (Jumper et al., 2021) into the computational refinement procedures. In the ATSAS programs BUNCH, SASREF and CORAL (Petoukhov et al., 2012; Petoukhov & Svergun, 2005), the rigid bodies introduced by the user are considered independent building blocks of known structure, and a solution is sought such that an optimal spatial arrangement of the building blocks is obtained that best fits the scattering data. The target function constructed for optimization additionally includes terms describing geometric criteria such as connectivity, symmetry and anisometry and penalties for steric clashes between backbone/main-chain atoms. SASREF provides functionality for simple rigid-body modeling, allowing refinement against many X-ray and neutron scattering data sets collected under a number of contrast conditions. BUNCH facilitates multidomain structure refinement with the addition of missing terminal and linker regions, modeled as dummy residues. CORAL (COmplexes with RAndom Loops) combines the methods of SASREF and BUNCH and facilitates a generally applicable hybrid-modeling approach, where rigid bodies for domains and missing region modeling can be combined with additional distance constraint information. CORAL allows users to specify fixed components and those free-to-move, and can impose symmetry as specified. An example of the input required for the program CORAL for the rigid body refinement of the hexameric transcriptional activator protein Rov C (Regulator Of Virulence interconnected with the Csr system, SASBDB ID SASDHP5) (Fig. 7.) is shown below.

(i) Configuration file (coral.con) listing the rigid bodies (PDB ID 6XZ5, split into four domains) and regions of missing sequence (NTER/LINK with number of residues) to be a modeled:

```
NTER 7
6XZ5_8to41.pdb
LINK 6
6XZ5_49to93.pdb
LINK 10
6XZ5_104to140.pdb
LINK 5
6XZ5_146to170.pdb
LINK 15
6XZ5_186to252.pdb
```

(ii) Options for CORAL (regular user mode), P32 symmetry and rigid body positions fixed:

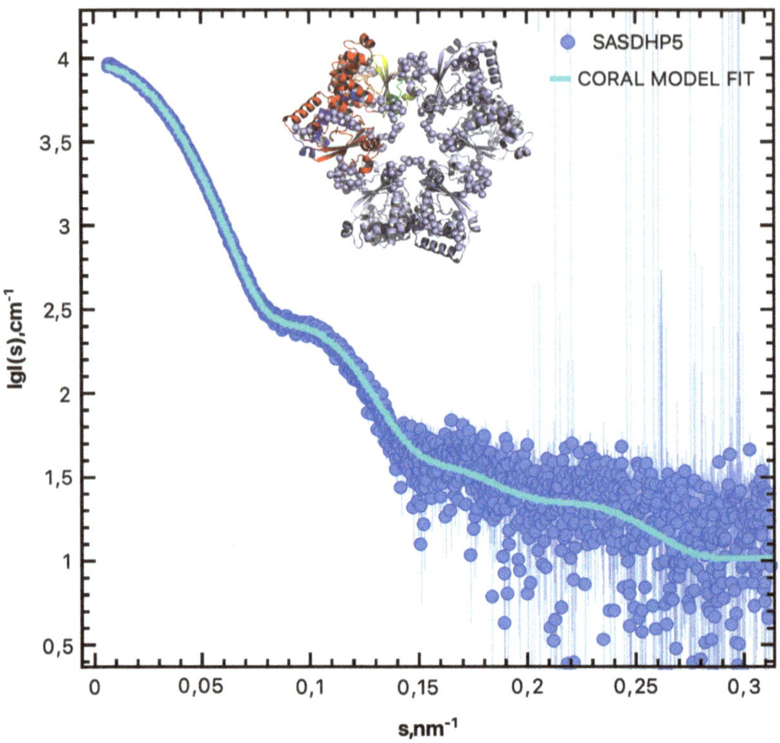

Fig. 7 CORAL rigid body refinement of RovC. 1D scattering data (blue), overlaid with the corresponding CORAL fit (cyan). Inset shows the hexameric model generated with P32 symmetry (Red, orange, yellow and blue cartoon shows the input rigid bodies for the refinement based on PDB ID 6XZ5 and modeled linker residues shown as beads). Model generated from data deposited at www.sasbdb.org for entry SASDHP5.

```
$> coral
*** Rigid body modelling of complexes with multidomain ***
*** proteins against multiple scattering curves.  ***
*** Symmetry/contacts can be taken into account.  ***
*** Linkers are represented as interconnected dummy ***
*** residues chains attached to the appropriate   ***
*** Ca atoms in rigid domains. Models with overlaps ***
*** and disconnected ones are penalized.     ***
*** Please reference: M.V. Petoukhov et al. (2012) ***
*** J. Appl. Cryst. 45, 342-350.       ***
*** Version 1.1 (r13837)        ***
*** Copyright (c) ATSAS Team        ***
```

```
*** EMBL, Hamburg Outstation, 2010-2015    ***
====== CORAL started on   29-Mar-2022 10:41:26
Computation mode (User or Expert) ...... <  User >: u
Log file name ......................... <  .log >: RovC
Project identifier ................................ : RovC
Enter project description ............. : RovC hexamer refinement
P32 symmetry
File name with objects info ........... <  .con >: coral.con
Coordinates  of  the  1-st  subunit  evaluated  from  .........  :
6XZ5_8to41.pdb
 296 atoms read, center at -30.90 26.67 -19.06
Fix the subunit at original position? [
Y / N ] .............................. <  No >: y
Coordinates  of  the  2-nd  subunit  evaluated  from  .........  :
6XZ5_49to93.pdb
 373 atoms read, center at -22.26 30.15 -13.19
Fix the subunit at original position? [
Y / N ] .............................. <  No >: y
Coordinates  of  the  3-rd  subunit  evaluated  from  .........  :
6XZ5_104to140.pdb
 305 atoms read, center at -10.53 26.03 -15.61
Fix the subunit at original position? [
Y / N ] .............................. <  No >: y
Coordinates  of  the  4-th  subunit  evaluated  from  .........  :
6XZ5_146to170.pdb
 125 atoms read, center at -6.84 21.32 -1.02
Fix the subunit at original position? [
Y / N ] .............................. <  No >: y
Coordinates  of  the  5-th  subunit  evaluated  from  .........  :
6XZ5_186to252.pdb
 563 atoms read, center at -40.84 8.40 -8.72
Fix the subunit at original position? [
Y / N ] .............................. <  No >: y
Number of backbone atoms generated ..................... : 252
Symmetry: P1...19 or Pn2 (n=1,..,12) ... <  P1 >: P32
Input total number of scattering curves <  1 >: 1
Enter file name, 1-st experimental data <  .dat >: SASDHP5.dat
Fitting range in fractions of Smax ..... <  1.000 >: 1.0
...
```

4.2.2 Flexible body modeling and ensembles

In rigid-body modeling the high-resolution components are considered immutable, with no internal structural changes. However, it can be that significant structural plasticity is present for such bodies in solution, and so flexible modeling procedures can be applied. The ATSAS program SREFLEX (Panjkovich & Svergun, 2016) provides functionality for the conformational rearrangement of input high-resolution protein structures against experimental scattering data. During refinement the protein is morphed along the available Cartesian normal modes in an optimization procedure, with the goal of finding a set of conformations that best fits the data. Normal Modes Analysis (NMA) has been used to characterize interdomain motions in a number of protein systems (Alexandrov et al., 2005; Dobbins, Lesk, & Sternberg, 2008; Krebs et al., 2002; Tama & Sanejouand, 2001; Tobi & Bahar, 2005; Wako & Endo, 2017), using connected harmonic motions around a starting equilibrium to sample the available conformational space (Goldstein, 1980). An extended normal modes approach is that provided by NMATOR (Normal Modes Analysis in Torsion angle space) (Manalastas-Cantos et al., 2021), where the coordinated motions are described in torsion angle space. NMATOR facilitates NMA based refinement and conformational sampling of both protein and non-protein biomolecular systems, morphing structure through rotations of dihedral angles. NMATOR enables normal mode computation in torsion angle space, structure refinement along normal modes to optimize a fit to experimental SAXS data, and the generation of ensembles of conformations which can be used as input for the modeling and characterization of flexible systems. The latter two methods are also available in Cartesian space using SREFLEX.

Ensemble based approaches allow for the characterization of flexible and unstructured systems, where a single model usually provides an inadequate description of the solution behavior. ATSAS provides the Ensemble Optimization Method (EOM) (Bernadó et al., 2007; Tria et al., 2015) tool for the analysis of such systems, which is particularly suited to Intrinsically Disordered Proteins (IDPs) and macromolecules with Intrinsically Disordered Regions (IDRs).

The scattering intensity of an ensemble of conformationally polydisperse particles can be approximated as:

$$I(s) = \sum_{k} v_k I_k(s) \tag{9}$$

where $I_k(s)$ is the scattering intensity from the kth component with volume fraction v_k (Konarev et al., 2006). For flexible protein systems an indirect approach is used to approximate the contribution of the many components to the overall scattering. In EOM a large pool of possible conformations is generated (based on the sequence and any available structures corresponding to the sequence—PDB files) to approximate the large conformational space. The scattering of each member of the pool is computed and a genetic algorithm is then run to obtain the subset of conformations that best fit the experimental SAXS data, i.e., minimizing the discrepancy χ^2:

$$\chi^2 = \frac{1}{K-1}\sum_{j=1}^{K}\left[\frac{\mu I(s_j) - I\ \exp(s_j)}{\sigma(s_j)}\right]^2 \tag{10}$$

where $I_{exp}(s)$ is the experimental scattering, K is the number of experimental points, $\sigma_{(sj)}$ are standard deviations and μ is a scaling factor.

The conformers selected by the genetic algorithm provide the low-resolution sample space used for the generation of statistical distributions of parameters. These distributions (R_g, D_{max}, end-to-end distance and conformer volume) describe the degree of flexibility and compactness of the sample measured, relative to the corresponding distributions of unrestricted conformational freedom derived from the initial pool (Fig. 8). As the size-distributions are essentially probability density functions and describe the degree of entropy of the system, $H_b(S)$ (Shannon & Weaver, 1949), a measure of the states available (i.e., flexibility of the system) can be derived (see Tria et al., 2015 for details) (Tria et al., 2015) such that:

$$R_{\text{flex}} = -H_b(S)$$

Where a broad Gaussian-like distribution of parameters corresponds to high-uncertainty, i.e., $H_b(S)$ tends to -1 (similar to that expected for a pool of random conformers), thus the metric of flexibility R_{flex} approaches 1 (fully flexible). A narrow size distribution provides low uncertainty, with $H_b(S)$ tending to 0, and thus R_{flex} approaches 0 (low flexibility). The information content, or entropy, can therefore be used as a quantitative measure of flexibility with the metric R_{flex}. For convenience, R_{flex} can be reported as a percentage in the range 0 to 100%, with $R_{flex}=100\%$ representing maximum flexibility. Using R_{flex}, the uncertainty of the selected ensemble distribution can be compared to that of the reference pool.

4.2.3 Quasi-atomistic modeling of membranes, detergents and membrane proteins

Membrane proteins solubilized in solution using lipids, lipidic scaffolds and detergents are readily measured in solution by SAXS. A significant issue to

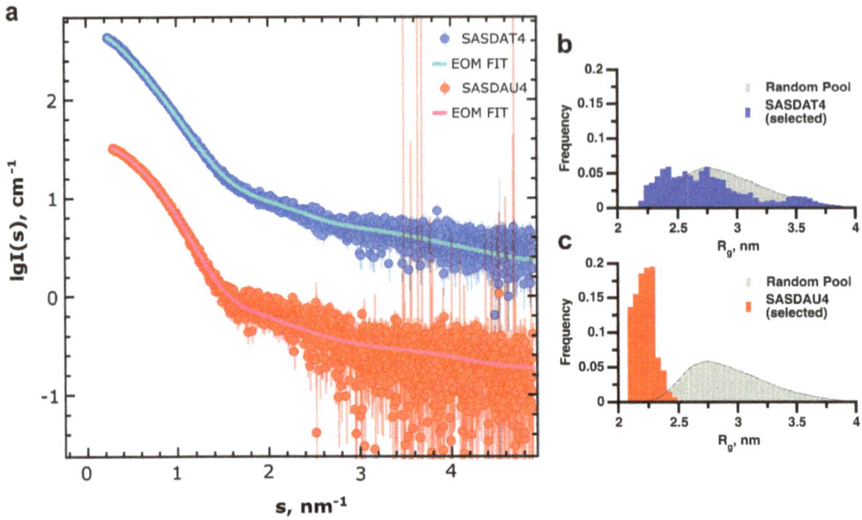

Fig. 8 EOM analysis of Urokinase Plasminogen Activator Receptor, uPAR wildtype (blue) and disulfide-bridged mutant (red). (A) Fits of the best ensembles selected by EOM are shown on the left for both constructs (data offset vertically for better visualization). R_g size-distributions for the selected ensembles and the initial random pool are shown on the right for wildtype (B) and mutant (C). The metric for flexibility is $R_{flex} = 81\%$ (wildtype) and 47% (mutant), relative to 85% for the pool. A broad distribution of compact and extended structures is selected for the wildtype case, whereas a narrow distribution of compact structures is selected for the mutant—indicating a stable compact structure in solution with little flexibility. Ensembles generated from data deposited at www.sasbdb.org for entries SASDAT4 (wildtype), and SASDAU4 (mutant).

consider for the measurement of membrane proteins in detergents is the separation of the scattered intensities of the protein-detergent-complex (PDC) from that of the free detergent in the sample. It is often not practical to match the detergent background of the buffer solution to that of the sample, as the equilibrium of free and bound detergent in the sample will lead to a different concentration of free detergent compared to the buffer containing the detergent (and no interacting PDC). Although extensive dialysis can be used to attempt to obtain a matched detergent concentration, the most robust approach is to measure the sample with in-line HPLC. Provided that the elution of the PDC and free detergent micelles is well separated, this facilitates extraction of an optimally buffer subtracted PDC SAXS curve.

The complex contrast situation mentioned in Section 4.1.2 for ab initio modeling also provides difficulties for hybrid modeling approaches, however, this can be overcome computationally again if appropriate chemical information is provided. In this case, structures of the detergent/lipid

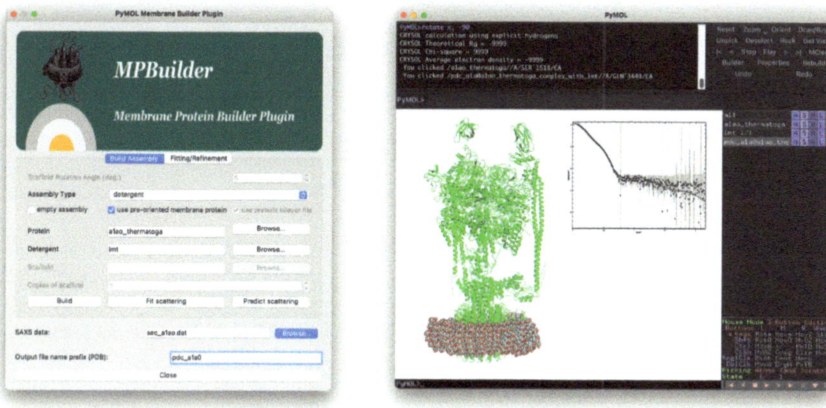

Fig. 9 MPBuilder interface to PYMOL. BUILD operation for the A1AO-ATP synthase from Therm. Thermophilus and a detergent corona of dodecylmaltoside (DDM) shown. Model generated from data deposited at www.sasbdb.org for entry SASDKK4.

components (PDB/CIF file formats) are used to accurately calculate the scattering intensities of the solubilizing components in solution. In the ATSAS package the PYMOL plugin MPBuilder (Molodenskiy, Svergun, & Mertens, 2021) provides a convenient interface for biological users to construct candidate membrane protein models from atomic structures in lipidic/detergent environments, and to calculate the fit of these models to experimental SAXS data (Fig. 9).

MPBuilder requires an input membrane protein structure file (PDB/CIF format), and for best results it is recommended that an alignment tool such as that provided by the PPM3.0 program at the OPM website (https://opm.phar.umich.edu/ppm_server) (Lomize, Pogozheva, Joo, Mosberg, & Lomize, 2012) be used to optimally position the transmembrane region with respect to the origin (center of the membrane or detergent corona to be modeled). A database of structures of commonly used detergents, lipids and protein scaffolds (nanodisc scaffold protein belts and saposin proteins) is provided in CIF format. The tool can be used in build or refinement mode, with the latter conducting a grid search of available parameters (e.g., number of detergent molecules—polar angles and density of detergent monomers, number of scaffold monomers, geometrical offset of scaffold monomers relative to the origin, and angle of scaffold monomers with respect to the membrane plane—e.g., XY).

In addition to MPBuilder, ATSAS provides a tool for the construction of liposomes and mixed micellar systems (e.g., bicelles). The program ELLLIP

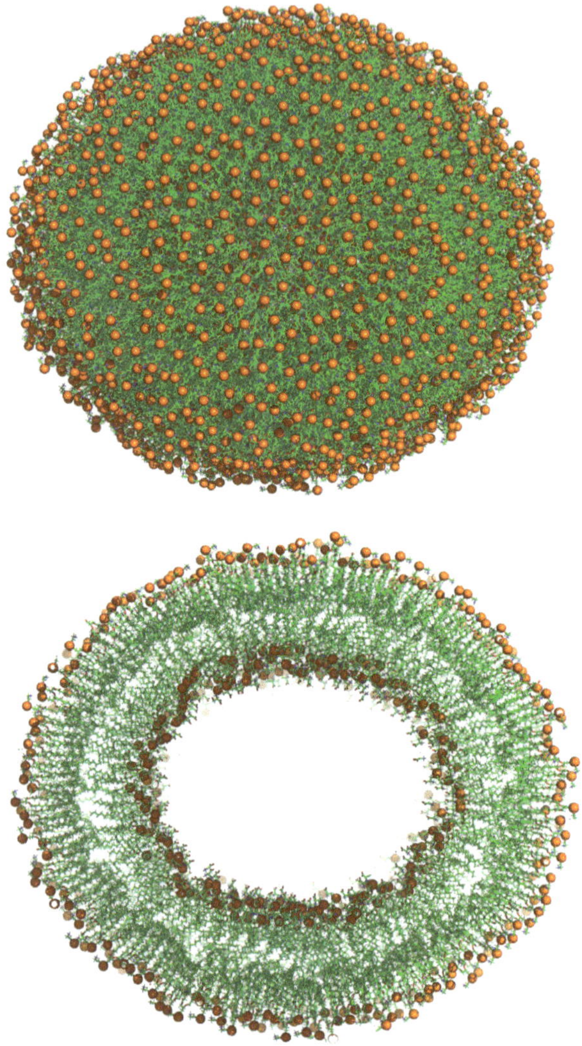

Fig. 10 Ellipsoidal POPE:POPS liposome generated with ELLLIP. Model constructed using the POPE (outer leaflet) and POPS (inner leaflet) PDB files extracted from the MPBuilder CIF database. Phosphorus atoms of the headgroups are shown as orange spheres.

(ELlipsoidal Lipid) (Petukhov et al., 2020) facilitates construction of ellipsoidal bilayered membrane particles (Fig. 10.), using an angular grid to optimize the positions of user defined input lipid molecules (PDB format). Further randomization can be conducted as required to reduce ideality and more closely resemble biophysically relevant lipid structures in solution. The model membrane particles constructed can then be checked for agreement with scattering data with CRYSOL.

4.3 Polydisperse systems

In monodisperse ideal solutions, it is assumed that all particles present are identical. This facilitates application of the ab initio and hybrid modeling procedures outlined above, with an additional assumption that inter-particle interactions are minimal (i.e., sample is significantly dilute). These assumptions cannot be made when the sample measured is polydisperse. Analysis of such samples requires approaches that account for the different particle sizes/ structures present and their corresponding volume fractions. For a mixture, similarly to that expressed in Eq. (9), the scattering profile can be represented as the volume-weighted sum of the scattering intensities of the individual components:

$$I(s) = \sum_{k=1}^{N} v_k I_k(s) \tag{11}$$

Where V_k and I_k correspond to the volume fraction and scattering profile (form-factor) of each distinct scattering component N. ATSAS provides the program OLIGOMER (Konarev et al., 2003) to analyze multiple sets of polydisperse scattering data and obtain an optimized solution that best describes the volume fractions of components present. Such an analysis can be conducted using pre-computed intensities (form-factors) from components with known three dimensional structure (i.e., PDB files), or individual scattering profiles previously measured, for each component. Association/dissociation processes and evolving systems are well-suited to this type of analysis.

4.3.1 Ab initio methods applied to polydisperse solutions

The ab initio methods described in Section 4.1 aim to reconstruct molecular shape information directly from scattering intensities. The success of this approach relies heavily on the assumption that samples are monodisperse and significantly dilute such that interparticle-interactions can be ignored. However, the ATSAS tool DAMMIX (Konarev & Svergun, 2018) has been developed specifically for the ab initio shape reconstruction of a transient component in an evolving polydisperse solution. In addition to the low-resolution structure the volume fraction of the intermediate is determined. From a set of k scattering curves representing time points, and knowledge of the structure of the initial and final components, the structure/shape of the unknown intermediate component and its volume fraction are determined from Eq. (4) and the following relationship:

$$V_{start}(k) + V_{intermediate}(k) + V_{final}(k) = 1 \tag{12}$$

Where V_{start} $V_{intermediate}$ and V_{final} are the volume fractions of the starting structure, intermediate/unknown and final state, respectively.

DAMMIX can be applied to two-component evolving systems such as oligomerization from an initial monomeric state through to the formation of a large aggregate, and systems with multiple assembly states such as viruses and nanoparticles (Fig. 11).

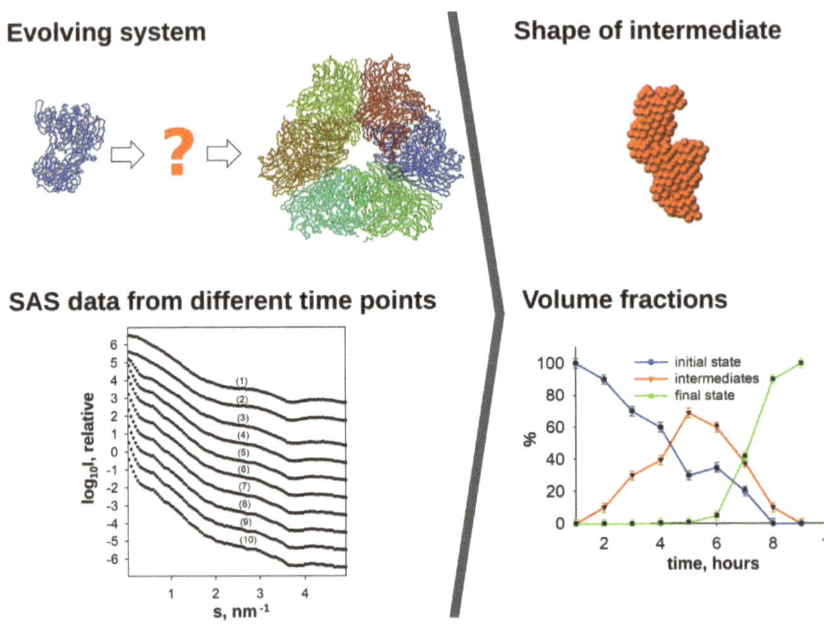

Fig. 11 DAMMIX facilitates low-resolution structure/shape reconstruction of an unknown intermediate in an evolving system. Knowledge of the initial and final state structure is required, and optimization is conducted using multiple SAXS data sets to determine the intermediate structure and the volume fractions of the initial, intermediate and final states at each time point. *Reused with permission from IUCRJ Manalastas-Cantos, K., Konarev, P.V., Hajizadeh, N.R., Kikhney, A.G., Petoukhov, M.V., Molodenskiy, D.S., Panjkovich, A., Mertens, H.D.T., Gruzinov, A., Borges, C., Jeffries, C.M., Svergun, D.I., Franke, D. (2021). ATSAS 3.0: Expanded functionality and new tools for small-angle scattering data analysis.* Journal of Applied Crystallography *54: 343–355. doi: 10.1107/S1600576720013412.*

4.3.2 Modeling polydisperse solutions of lipid vesicles

Lipid vesicles are often used as model biological membranes and scattering techniques are well suited to their analysis in solution (Heftberger et al., 2014; Konarev, Gruzinov, Mertens, & Svergun, 2021; Pabst, Rappolt, Amenitsch, & Laggner, 2000; Pencer & Hallett, 2000; Pérez & Koutsioubas, 2015). Several tools in ATSAS have been recently developed to facilitate the analysis of lipid vesicles, including methods for the reconstruction of electron density profiles across the lipid bilayer, and for the characterization of vesicle size distributions.

Scattering data from solutions of lipid vesicles can be used as input for the tools BILMIX (Konarev et al., 2020) and LIPMIX (Konarev et al., 2021) to generate profiles of the electron density, $\rho(z)$ across the membrane bilayer. An initial approximation of the SAXS intensities is made as the product of the form-factors of a thin spherical shell (FF_{TSS}) and a flat lipid bilayer (FF_{FLB}) (Kiselev, Lesieur, Kisselev, Lombardo, & Aksenov, 2002; Pencer, Krueger, Adams, & Katsaras, 2006). The former defines the size of the vesicles and the latter describes the electron density across the bilayer. The scattering profile of each vesicle k present in solution is then expressed as:

$$I_k(s) \cong \frac{1}{s^2}\left|\int FF_{\mathrm{TSS}}(s,\ r)_k D_{\mathrm{v}}(r)_k \mathrm{d}r\right|^2 \left|FF_{\mathrm{FLB}}(s)_k\right|^2 \sum_{i=1}^{M} w_i s_i^{\mathrm{FLB}}(s) \qquad (13)$$

Where $D_\nu(r)$ is the size distribution of the vesicles, and the last term (implemented only in LIPMIX) accounts for the presence of M distinct multilayers of occupancy w_i and inter-bilayer structure factor S_i^{FLB} (Zhang, Suter, & Nagle, 1994). The form factor of the flat lipid bilayer, $FF_{FLB}(s)$ is the Fourier transform of the electron-density profile $\rho(z)$:

$$\begin{aligned}\rho(z) = &\sum_{i=1}^{2} A_i\left\{ \exp\left[\frac{-(z - z_{\mathrm{H}i})^2}{2\sigma_{\mathrm{H}i}^2}\right] + \exp\left[\frac{-(z + z_{\mathrm{H}i})^2}{2\sigma_{\mathrm{H}i}^2}\right]\right\}\\ &- \rho_r \exp\left(\frac{-z^2}{2\sigma_{\mathrm{c}}^2}\right)\end{aligned} \qquad (14)$$

Where the Gaussian terms of width σ_{H1} and positioned at $\pm z_{H1}$ represent the polar lipid head groups, and the Gaussian term of width σ_c positioned at the center of the bilayer $z = 0$ represent the electron density of the apolar lipid core. Modeling of an asymmetric electron density profile is facilitated in BILMIX by the addition of two Gaussian terms of width σ_{H2} positioned at $\pm z_{H2}$.

LIPMIX and BILMIX can be used to model the size distribution of vesicles/liposomes and to determine bilayer electron-density profiles (Fig. 12).

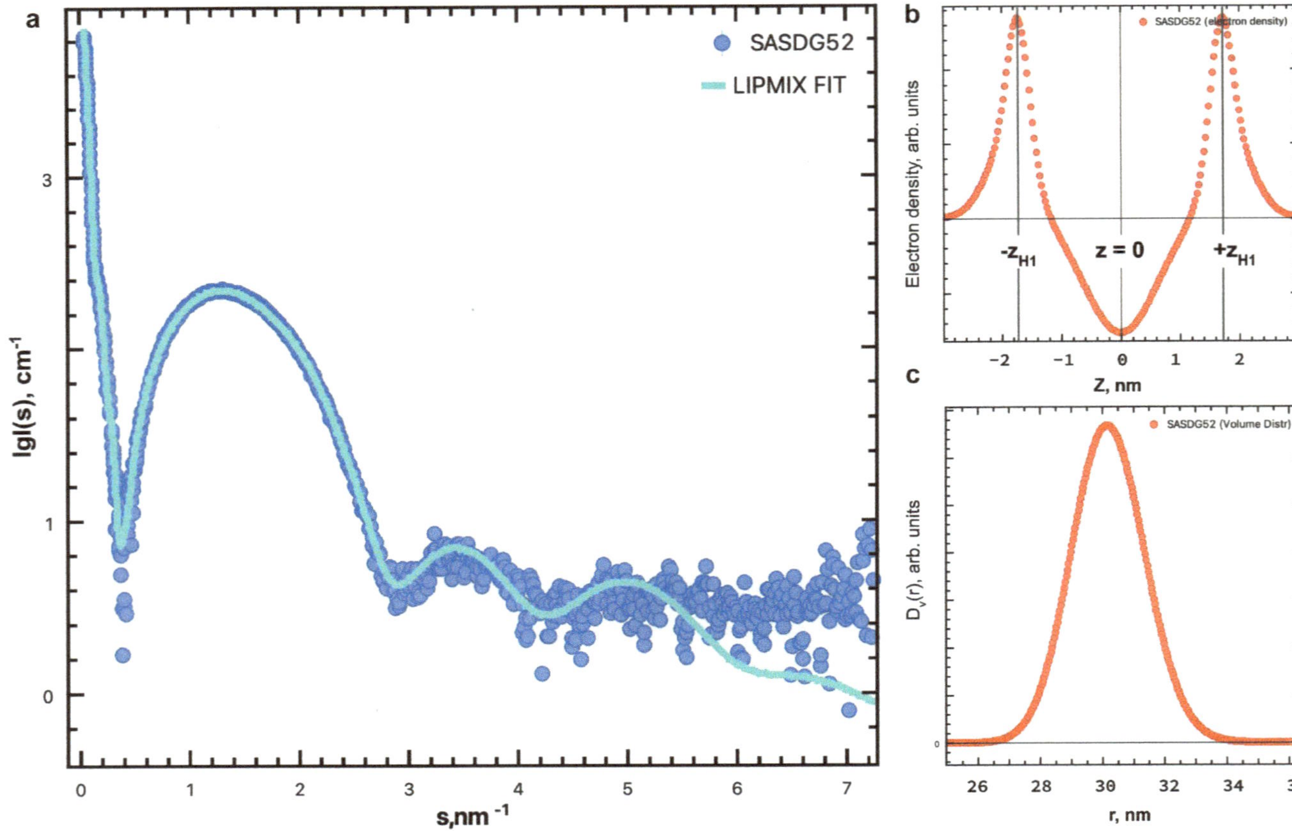

Fig. 12 LIPMIX modeling of the bilayer electron density profile and size-distribution of a 1,2-dimyristoyl-sn-glycero-3-phosphocholine (DMPC) liposome solution, extruded through a 30 nm membrane. The model fit to the data (cyan line) is shown (A). The electron density profile (B) and volume distribution (C) are shown on the right. Data deposited at www.sasbdb.org for entry SASDG52.

This facilitates analysis of the behavior of model membranes in solution and can also be employed to examine potential interactions and disruption or morphological changes caused by additional species.

5. ATSAS summary

The ATSAS suite provides an extensive set of tools for reduction, analysis and structural modeling from experimental SAXS data. Not all available programs and tools are covered in this short review, however, potential users are encouraged to download the ATSAS suite and explore possible applications in their own SAXS analyses. ATSAS has been updated to include support of the mmCIF format, a new standard adopted by the PDB for structural information. A database initiated by the ATSAS team in collaboration with the PDB, the Small-Angle Scattering Biological DataBase (SASBDBD), has been successfully launched and adapted as a standard repository of biological scattering data (Kikhney et al., 2020). SASBDBD provides an excellent resource for new users to investigate the use of available data analysis and modeling procedures with curated experimental data. ATSAS is under constant development and will continue to be maintained at EMBL with the introduction of new tools and further improvements.

The latest ATSAS package can be installed locally (Windows/Mac OS/ Linux) available at:

https://www.embl-hamburg.de/biosaxs/software.html.

Alternatively, many programs can be run on the EMBL Hamburg cluster via the ATSAS online interface:

https://www.embl-hamburg.de/biosaxs/atsas-online/

References

Alexandrov, V., Lehnert, U., Echols, N., Milburn, D., Engelman, D., & Gerstein, M. (2005). Normal modes for predicting protein motions: A comprehensive database assessment and associated web tool. *Protein Science*, *14*, 633–643. https://doi.org/10.1110/ps.04882105.

Alves, C., Pedersen, J. S., & Oliveira, C. L. P. (2017). Calculation of two-dimensional scattering patterns for oriented systems. *Journal of Applied Crystallography*, *50*, 840–850. https://doi.org/10.1107/S1600576717005179.

Anscombe, F. J. (1948). The transformation of poisson, binomial and negative-binomial data. *Biometrika*, *35*, 246–254. https://doi.org/10.2307/2332343.

Bernadó, P., Mylonas, E., Petoukhov, M. V., Blackledge, M., & Svergun, D. I. (2007). Structural characterization of flexible proteins using small-angle X-ray scattering. *Journal of the American Chemical Society*, *129*, 5656–5664.

Breßler, I., Kohlbrecher, J., & Thünemann, A. F. (2015). SASfit: A tool for small-angle scattering data analysis using a library of analytical expressions. *Journal of Applied Crystallography*, *48*, 1587–1598. https://doi.org/10.1107/S1600576715016544.

Bressler, I., Pauw, B. R., & Thünemann, A. F. (2015). McSAS: Software for the retrieval of model parameter distributions from scattering patterns. *Journal of Applied Crystallography*, *48*, 962–969. https://doi.org/10.1107/S1600576715007347.

Brookes, E., Vachette, P., Rocco, M., & Pérez, J. (2016). US-SOMO HPLC-SAXS module: Dealing with capillary fouling and extraction of pure component patterns from poorly resolved SEC-SAXS data. *Journal of Applied Crystallography*, *49*, 1827–1841. https://doi.org/10.1107/S1600576716011201.

Brosey, C. A., & Tainer, J. A. (2019). Evolving SAXS versatility: Solution X-ray scattering for macromolecular architecture, functional landscapes, and integrative structural biology. *Current Opinion in Structural Biology*, *58*, 197–213. https://doi.org/10.1016/j.sbi.2019.04.004.

Da Vela, S., & Svergun, D. I. (2020). Methods, development and applications of small-angle X-ray scattering to characterize biological macromolecules in solution. *Current Research in Structural Biology*, *2*, 164–170. https://doi.org/10.1016/j.crstbi.2020.08.004.

Denisov, I. G., Grinkova, Y. V., Lazarides, A. A., & Sligar, S. G. (2004). Directed self-assembly of monodisperse phospholipid bilayer nanodiscs with controlled size. *Journal of the American Chemical Society*, *126*, 3477–3487. https://doi.org/10.1021/ja0393574.

Dobbins, S. E., Lesk, V. I., & Sternberg, M. J. E. (2008). Insights into protein flexibility: The relationship between normal modes and conformational change upon protein–protein docking. *Proceedings of the National Academy of Sciences*, *105*, 10390–10395. https://doi.org/10.1073/pnas.0802496105.

Franke, D., Hajizadeh, N. R., & Svergun, D. I. (2020). Simulation of small-angle X-ray scattering data of biological macromolecules in solution. *Journal of Applied Crystallography*, *53*, 536–539. https://doi.org/10.1107/S1600576720000527.

Franke, D., Jeffries, C. M., & Svergun, D. I. (2018). Machine learning methods for X-ray scattering data analysis from biomacromolecular solutions. *Biophysical Journal*, *114*, 2485–2492. https://doi.org/10.1016/j.bpj.2018.04.018.

Franke, D., Kikhney, A. G., & Svergun, D. I. (2012). Automated acquisition and analysis of small angle X-ray scattering data. *Nuclear Instruments and Methods in Physics Research, Section A: Accelerators, Spectrometers, Detectors and Associated Equipment*, *689*, 52–59. https://doi.org/10.1016/j.nima.2012.06.008.

Franke, D., Petoukhov, M. V., Konarev, P. V., Panjkovich, A., Tuukkanen, A., Mertens, H. D. T., et al. (2017). ATSAS 2.8: A comprehensive data analysis suite for small-angle scattering from macromolecular solutions. *Journal of Applied Crystallography*, *50*, 1212–1225. https://doi.org/10.1107/S1600576717007786.

Franke, D., & Svergun, D. I. (2009). DAMMIF, a program for rapid ab-initio shape determination in small-angle scattering. *Journal of Applied Crystallography*, *42*, 342–346.

Glatter, O. (1977). A new method for the evaluation of small-angle scattering data. *Journal of Applied Crystallography*, *10*, 415–421. https://doi.org/10.1107/S0021889877013879.

Glover, K. J., Whiles, J. A., Wu, G., Yu, N., Deems, R., Struppe, J. O., et al. (2001). Structural evaluation of phospholipid Bicelles for solution-state studies of membrane-associated biomolecules. *Biophysical Journal*, *81*, 2163–2171. https://doi.org/10.1016/S0006-3495(01)75864-X.

Goldstein, H. (1980). Classical mechanics. *Addison-Wesley series in physics* (2nd ed.). Reading, Mass: Addison-Wesley Pub. Co.

Grant, T. D. (2018). Ab initio electron density determination directly from solution scattering data. *Nature Methods*, *15*, 191–193. https://doi.org/10.1038/nmeth.4581.

Gruzinov, A. Y., Schroer, M. A., Manalastas-Cantos, K., Kikhney, A. G., Hajizadeh, N. R., Schulz, F., et al. (2021). Anomalous SAXS at P12 beamline EMBL Hamburg: Instrumentation and applications. *Journal of Synchrotron Radiation*, *28*.

Guinier, A. (1939). La diffraction des rayons X aux tres petits angles; application a l'etude de phenomenes ultramicroscopiques. *Annales de Physique*, *12*, 161–237.
Hajizadeh, N. R., Franke, D., Jeffries, C. M., & Svergun, D. I. (2018). Consensus Bayesian assessment of protein molecular mass from solution X-ray scattering data. *Scientific Reports*, *8*, 7204. https://doi.org/10.1038/s41598-018-25355-2.
Hammel, M., Rosenberg, D. J., Bierma, J., Hura, G. L., Thapar, R., Lees-Miller, S. P., et al. (2021). Visualizing functional dynamicity in the DNA-dependent protein kinase holoenzyme DNA-PK complex by integrating SAXS with cryo-EM. *Progress in Biophysics and Molecular Biology*, *163*, 74–86. https://doi.org/10.1016/j.pbiomolbio.2020.09.003.
Hansen, S. (2012). Bayesian methods in SAXS and SANS structure determination. In T. Hamelryck, K. Mardia, & J. Ferkinghoff-Borg (Eds.), *Bayesian methods in structural bioinformatics* (pp. 313–342). Berlin, Heidelberg: Springer. https://doi.org/10.1007/978-3-642-27225-7_13.
Harris, F. J. (1978). On the use of windows for harmonic analysis with the discrete Fourier transform. *Proceedings of the IEEE*, *66*, 51–83. https://doi.org/10.1109/PROC.1978.10837.
Heftberger, P., Kollmitzer, B., Heberle, F. A., Pan, J., Rappolt, M., Amenitsch, H., et al. (2014). Global small-angle X-ray scattering data analysis for multilamellar vesicles: The evolution of the scattering density profile model. *Journal of Applied Crystallography*, *47*, 173–180. https://doi.org/10.1107/S1600576713029798.
Hendrickson, W. A. (2014). Anomalous diffraction in crystallographic phase evaluation. *Quarterly Reviews of Biophysics*, *47*, 49–93. https://doi.org/10.1017/S0033583514000018.
Hopkins, J. B., Gillilan, R. E., & Skou, S. (2017). BioXTAS RAW: Improvements to a free open-source program for small-angle X-ray scattering data reduction and analysis. *Journal of Applied Crystallography*, *50*, 1545–1553. https://doi.org/10.1107/S1600576717011438.
Iglewicz, B., & Hoaglin, D. C. (1993). *How to detect and handle outliers*. Milwaukee, Wis: ASQC Quality Press.
James, R. W. (1950). *The optical principles of the diffraction of X-rays*. Repr. Bell.
Jumper, J., Evans, R., Pritzel, A., Green, T., Figurnov, M., Ronneberger, O., et al. (2021). Highly accurate protein structure prediction with AlphaFold. *Nature*, *596*, 583–589. https://doi.org/10.1038/s41586-021-03819-2.
Kachala, M., Valentini, E., & Svergun, D. I. (2015). Application of SAXS for the structural characterization of IDPs. In I. C. Felli, & R. Pierattelli (Eds.), *Intrinsically disordered proteins studied by NMR spectroscopy* (pp. 261–289). Cham: Springer International Publishing. https://doi.org/10.1007/978-3-319-20164-1_8.
Kikhney, A. G., Borges, C. R., Molodenskiy, D. S., Jeffries, C. M., & Svergun, D. I. (2020). SASBDB: Towards an automatically curated and validated repository for biological scattering data. *Protein Science*, *29*, 66–75. https://doi.org/10.1002/pro.3731.
Kikhney, A. G., & Svergun, D. I. (2015). A practical guide to small angle X-ray scattering (SAXS) of flexible and intrinsically disordered proteins. *FEBS Letters*, *589*, 2570–2577.
Kiselev, M. A., Lesieur, P., Kisselev, A. M., Lombardo, D., & Aksenov, V. L. (2002). Model of separated form factors for unilamellar vesicles. *Applied Physics A: Materials Science & Processing*, *74*, s1654–s1656. https://doi.org/10.1007/s003390201837.
Konarev, P. V., Gruzinov, A. Y., Mertens, H. D., & Svergun, D. I. (2021). Restoring structural parameters of lipid mixtures from small-angle X-ray scattering data. *Journal of Applied Crystallography*, *54*(Pt 1), 169–179. https://doi.org/10.1107/S1600576720015368.
Konarev, P. V., Petoukhov, M. V., Dadinova, L. A., Fedorova, N. V., Volynsky, P. E., Svergun, D. I., et al. (2020). BILMIX: A new approach to restore the size polydispersity and electron density profiles of lipid bilayers from liposomes using small-angle X-ray scattering data. *Journal of Applied Crystallography*, *53*, 236–243. https://doi.org/10.1107/S1600576719015656.

Konarev, P. V., Petoukhov, M. V., Volkov, V. V., & Svergun, D. I. (2006). ATSAS 2.1, a program package for small-angle scattering data analysis. *Journal of Applied Crystallography*, *39*, 277–286. https://doi.org/10.1107/S0021889806004699.

Konarev, P. V., & Svergun, D. I. (2018). Direct shape determination of intermediates in evolving macromolecular solutions from small-angle scattering data. *IUCrJ*, *5*, 402–409. https://doi.org/10.1107/S2052252518005900.

Konarev, P. V., Volkov, V. V., Sokolova, A. V., Koch, M. H. J., & Svergun, D. I. (2003). PRIMUS: A windows PC-based system for small-angle scattering data analysis. *Journal of Applied Crystallography*, *36*, 1277–1282.

Krebs, W. G., Alexandrov, V., Wilson, C. A., Echols, N., Yu, H., & Gerstein, M. (2002). Normal mode analysis of macromolecular motions in a database framework: Developing mode concentration as a useful classifying statistic. *Proteins: Structure, Function, and Bioinformatics*, *48*, 682–695. https://doi.org/10.1002/prot.10168.

Lipfert, J., & Doniach, S. (2007). Small-angle X-ray scattering from RNA, proteins, and protein complexes. *Annual Review of Biophysics and Biomolecular Structure*, *36*, 307–327. https://doi.org/10.1146/annurev.biophys.36.040306.132655.

Liu, H., Morris, R. J., Hexemer, A., Grandison, S., & Zwart, P. H. (2012). Computation of small-angle scattering profiles with three-dimensional Zernike polynomials. *Acta Crystallographica. Section A*, *68*, 278–285. https://doi.org/10.1107/S010876731104788X.

Lomize, M. A., Pogozheva, I. D., Joo, H., Mosberg, H. I., & Lomize, A. L. (2012). OPM database and PPM web server: Resources for positioning of proteins in membranes. *Nucleic Acids Research*, *40*, D370–D376. https://doi.org/10.1093/nar/gkr703.

Manalastas-Cantos, K., Konarev, P. V., Hajizadeh, N. R., Kikhney, A. G., Petoukhov, M. V., Molodenskiy, D. S., et al. (2021). ATSAS 3.0: Expanded functionality and new tools for small-angle scattering data analysis. *Journal of Applied Crystallography*, *54*, 343–355. https://doi.org/10.1107/S1600576720013412.

Mertens, H. D., & Svergun, D. I. (2017). Combining NMR and small angle X-ray scattering for the study of biomolecular structure and dynamics. *Archives of Biochemistry and Biophysics*, *628*, 33–41.

Miake-Lye, R. C., Doniach, S., & Hodgson, K. O. (1983). Anomalous x-ray scattering from terbium-labeled parvalbumin in solution. *Biophysical Journal*, *41*, 287–292. https://doi.org/10.1016/S0006-3495(83)84440-3.

Molodenskiy, D. S., Svergun, D. I., & Mertens, H. D. T. (2021). MPBuilder: A PyMOL plugin for building and refinement of solubilized membrane proteins against small angle X-ray scattering data. *Journal of Molecular Biology*, *433*, 166888. https://doi.org/10.1016/j.jmb.2021.166888.

Mylonas, E., & Svergun, D. I. (2007). Accuracy of molecular mass determination of proteins in solution by small-angle X-ray scattering. *Journal of Applied Crystallography*, *40*, s245–s249.

Orthaber, D., Bergmann, A., & Glatter, O. (2000). SAXS experiments on absolute scale with Kratky systems using water as a secondary standard. *Journal of Applied Crystallography*, *33*, 218–225. https://doi.org/10.1107/S0021889899015216.

Pabst, G., Rappolt, M., Amenitsch, H., & Laggner, P. (2000). Structural information from multilamellar liposomes at full hydration: Full q-range fitting with high quality x-ray data. *Physical Review E*, *62*, 4000.

Panjkovich, A., & Svergun, D. I. (2016). Deciphering conformational transitions of proteins by small angle X-ray scattering and normal mode analysis. *Physical Chemistry Chemical Physics*, *18*, 5707–5719. https://doi.org/10.1039/C5CP04540A.

Pencer, J., & Hallett, F. R. (2000). Small-angle neutron scattering from large unilamellar vesicles: An improved method for membrane thickness determination. *Physical Review E*, *61*, 3003–3008. https://doi.org/10.1103/PhysRevE.61.3003.

Pencer, J., Krueger, S., Adams, C. P., & Katsaras, J. (2006). Method of separated form factors for polydisperse vesicles. *Journal of Applied Crystallography*, *39*, 293–303. https://doi.org/10.1107/S0021889806005255.

Pérez, J., & Koutsioubas, A. (2015). Memprot: A program to model the detergent corona around a membrane protein based on SEC–SAXS data. *Acta Crystallographica. Section D, Biological Crystallography*, *71*, 86–93. https://doi.org/10.1107/S1399004714016678.

Perkins, S. J., Wright, D. W., Zhang, H., Brookes, E. H., Chen, J., Irving, T. C., et al. (2016). Atomistic modelling of scattering data in the collaborative computational project for small angle scattering (CCP-SAS). *Journal of Applied Crystallography*, *49*, 1861–1875. https://doi.org/10.1107/S160057671601517X.

Petoukhov, M. V., Franke, D., Shkumatov, A. V., Tria, G., Kikhney, A. G., Gajda, M., et al. (2012). New developments in the ATSAS program package for small-angle scattering data analysis. *Journal of Applied Crystallography*, *45*, 342–350. https://doi.org/10.1107/S0021889812007662.

Petoukhov, M. V., Konarev, P. V., Kikhney, A. G., & Svergun, D. I. (2007). ATSAS 2.1—Towards automated and web-supported small-angle scattering data analysis. *Journal of Applied Crystallography*, *40*, s223–s228.

Petoukhov, M. V., & Svergun, D. I. (2005). Global rigid body modeling of macromolecular complexes against small-angle scattering data. *Biophysical Journal*, *89*, 1237–1250. https://doi.org/10.1529/biophysj.105.064154.

Petukhov, M. V., Konarev, P. V., Dadinova, L. A., Fedorova, N. V., Volynsky, P. E., Svergun, D. I., et al. (2020). Quasi-atomistic approach to modeling of liposomes. *Crystallography Reports*, *65*, 258–263. https://doi.org/10.1134/S1063774520020182.

Piiadov, V., Ares de Araújo, E., Oliveira Neto, M., Craievich, A. F., & Polikarpov, I. (2019). SAXSMoW 2.0: Online calculator of the molecular weight of proteins in dilute solution from experimental SAXS data measured on a relative scale. *Protein Science*, *28*, 454–463. https://doi.org/10.1002/pro.3528.

Porod, G. (1951). Die Röntgenkleinwinkelstreuung von dichtgepackten kolloiden Systemen. *Kolloid Zeitschrift*, *124*, 83–114. https://doi.org/10.1007/BF01512792.

Rambo, R. P., & Tainer, J. A. (2013). Accurate assessment of mass, models and resolution by small-angle scattering. *Nature*, *496*, 477–481. https://doi.org/10.1038/nature12070.

Schneidman-Duhovny, D., Hammel, M., & Sali, A. (2010). FoXS: A web server for rapid computation and fitting of SAXS profiles. *Nucleic Acids Research*, *38*, W540–W544. https://doi.org/10.1093/nar/gkq461.

Seddon, A. M., Curnow, P., & Booth, P. J. (2004). Membrane proteins, lipids and detergents: Not just a soap opera. *Biochimica et Biophysica Acta (BBA)—Biomembranes, Lipid-Protein Interactions*, *1666*, 105–117. https://doi.org/10.1016/j.bbamem.2004.04.011.

Semenyuk, A. V., & Svergun, D. I. (1991). GNOM—A program package for small-angle scattering data processing. *Journal of Applied Crystallography*, *24*, 537–540. https://doi.org/10.1107/S002188989100081X.

Shannon, C. E., & Weaver, W. (1949). *The mathematical theory of communication*. Illinois: Urbana EEUU.

Stuhrmann, H. B. (1970). Interpretation of small-angle scattering functions of dilute solutions and gases. A representation of the structures related to a one-particle scattering function. *Acta Crystallogr. Sect. A*, *26*, 297–306. https://doi.org/10.1107/S0567739470000748.

Stuhrmann, H. B., & Notbohm, H. (1981). Configuration of the four iron atoms in dissolved human hemoglobin as studied by anomalous dispersion. *Proceedings of the National Academy of Sciences*, *78*, 6216–6220.

Svergun, D. I. (1992). Determination of the regularization parameter in indirect-transform methods using perceptual criteria. *Journal of Applied Crystallography*, *25*, 495–503. https://doi.org/10.1107/S0021889892001663.

Svergun, D. I. (1999). Restoring low resolution structure of biological macromolecules from solution scattering using simulated annealing. *Biophysical Journal, 76*, 2879–2886.

Svergun, D., Barberato, C., & Koch, M. H. J. (1995). CRYSOL-a program to evaluate X-ray solution scattering of biological macromolecules from atomic coordinates. *Journal of Applied Crystallography, 28*, 768–773.

Svergun, D. I., Koch, M. H., Timmins, P. A., & May, R. P. (2013). *Small angle X-ray and neutron scattering from solutions of biological macromolecules*. Oxford University Press.

Svergun, D. I., & Nierhaus, K. H. (2000). A map of protein-rRNA distribution in the 70 SEscherichia coli ribosome. *The Journal of Biological Chemistry, 275*, 14432–14439.

Svergun, D. I., Petoukhov, M. V., & Koch, M. H. J. (2001). Determination of domain structure of proteins from X-ray solution scattering. *Biophysical Journal, 80*, 2946–2953. https://doi.org/10.1016/S0006-3495(01)76260-1.

Tama, F., & Sanejouand, Y.-H. (2001). Conformational change of proteins arising from normal mode calculations. *Protein Engineering, Design & Selection, 14*, 1–6. https://doi.org/10.1093/protein/14.1.1.

Tanford, C., & Reynolds, J. A. (1976). Characterization of membrane proteins in detergent solutions. *Biochimica et Biophysica Acta (BBA) - Reviews on Biomembranes, 457*, 133–170. https://doi.org/10.1016/0304-4157(76)90009-5.

Tobi, D., & Bahar, I. (2005). Structural changes involved in protein binding correlate with intrinsic motions of proteins in the unbound state. *Proceedings of the National Academy of Sciences, 102*, 18908–18913. https://doi.org/10.1073/pnas.0507603102.

Trewhella, J. (2022). Recent advances in small-angle scattering and its expanding impact in structural biology. *Structure, 30*, 15–23. https://doi.org/10.1016/j.str.2021.09.008.

Tria, G., Mertens, H. D., Kachala, M., & Svergun, D. I. (2015). Advanced ensemble modelling of flexible macromolecules using X-ray solution scattering. *IUCrJ, 2*, 207–217.

Wako, H., & Endo, S. (2017). Normal mode analysis as a method to derive protein dynamics information from the protein data Bank. *Biophysical Reviews, 9*, 877–893. https://doi.org/10.1007/s12551-017-0330-2.

Zhang, R., Suter, R. M., & Nagle, J. F. (1994). Theory of the structure factor of lipid bilayers. *Physical Review E, 50*, 5047–5060. https://doi.org/10.1103/PhysRevE.50.5047.

CHAPTER EIGHT

Multi-state modeling of antibody-antigen complexes with SAXS profiles and deep-learning models

Tomer Cohen[a], Matan Halfon[a], Lester Carter[b,†], Beth Sharkey[c], Tushar Jain[d], Arvind Sivasubramanian[d,*], and Dina Schneidman-Duhovny[a,*]

[a]The Rachel and Selim Benin School of Computer Science and Engineering, The Hebrew University of Jerusalem, Jerusalem, Israel
[b]Stanford Synchrotron Radiation Lightsource, SLAC National Accelerator Laboratory, Menlo Park, CA, United States
[c]High-Throughput Expression, Adimab LLC, Lebanon, NH, United States
[d]Computational Biology, Adimab LLC, Palo Alto, CA, United States
*Corresponding authors: e-mail address: arvind.sivasubramanian@adimab.com; dina.schneidman@mail.huji.ac.il

Contents

Abstract

Antibodies are an established class of human therapeutics. Epitope characterization is an important part of therapeutic antibody discovery. However, structural characterization of antibody-antigen complexes remains challenging. On the one hand, X-ray crystallography or cryo-electron microscopy provide atomic resolution characterization of the epitope, but the data collection process is typically long and the success rate is low. On the other hand, computational methods for modeling antibody-antigen structures

† Current address: Exscientia PLC., The Schrödinger Building, Oxford Science Park, Oxford OX4 4GE, United Kingdom.

Methods in Enzymology, Volume 678
ISSN 0076-6879
https://doi.org/10.1016/bs.mie.2022.11.003

from the individual components frequently suffer from a high false positive rate, rarely resulting in a unique solution. Recent deep learning models for structure prediction are also successful in predicting protein-protein complexes. However, they do not perform well for antibody-antigen complexes. Small Angle X-ray Scattering (SAXS) is a reliable technique for rapid structural characterization of protein samples in solution albeit at low resolution. Here, we present an integrative approach for modeling antigen-antibody complexes using the antibody sequence, antigen structure, and experimentally determined SAXS profiles of the antibody, antigen, and the complex. The method models antibody structures using a novel deep-learning approach, NanoNet. The structures of the antibodies and antigens are represented using multiple 3D conformations to account for compositional and conformational heterogeneity of the protein samples that are used to collect the SAXS data. The complexes are predicted by integrating the SAXS profiles with scoring functions for protein-protein interfaces that are based on statistical potentials and antibody-specific deep-learning models. We validated the method via application to four Fab:EGFR and one Fab:PCSK9 antibody:antigen complexes with experimentally available SAXS datasets. The integrative approach returns accurate predictions (interface RMSD < 4 Å) in the top five predictions for four out of five complexes (respective interface RMSD values of 1.95, 2.18, 2.66 and 3.87 Å), providing support for the utility of such a computational pipeline for epitope characterization during therapeutic antibody discovery.

1. Introduction

Antibodies are an established class of human therapeutics. Therapeutic antibody discovery is typically performed using animal immunization, or in vitro selections using display technologies, with a typical campaign resulting in multiple antibodies against a variety of target epitopes (Laustsen, Greiff, Karatt-Vellatt, Muyldermans, & Jenkins, 2021). Neutralizing antibodies can also be isolated from convalescent individuals, as in SARS-CoV-2 (Zost et al., 2020) or Ebola (Bornholdt et al., 2016). Selection of the lead antibody candidate for further development may be prioritized based on considerations such as developability (Jain et al., 2017) and the targeted epitope. Therefore, methods that can rapidly and accurately predict antibody conformations, antibody-antigen structures and epitopes would contribute significantly to the successful development of therapeutic antibodies.

X-ray crystallography and Cryo-electron microscopy (cryo-EM) provide atomic resolution structures but the data collection process is typically long and the success rate and throughput are low (Lin, 2018; Yip, Fischer, Paknia, Chari, & Stark, 2020). Alternatively, structures of the antibody-antigen complexes can be modeled with computational docking tools using the 3D structures of the unbound proteins. While there has been a significant advance in modeling antibodies (Norn, Lapidoth, & Fleishman, 2017; Ruffolo, Guerra, Mahajan, Sulam, & Gray, 2020; Ruffolo, Sulam, & Gray, 2022; Sivasubramanian, Sircar, Chaudhury, & Gray, 2009; Weitzner & Gray, 2017) and antibody-antigen complexes (Krawczyk, Liu, Baker, Shi, & Deane, 2014; Sela-Culang et al., 2014; Sircar & Gray, 2010; Weitzner et al., 2017), the methods are computationally demanding and suffer from high false positive rate (Kilambi & Gray, 2017). There are several challenges to address for accurate modeling of antibody-antigen complexes. First, experimental and structural studies have found that antibodies, similarly to other proteins, can populate multiple structural states (Sela-Culang, Alon, & Ofran, 2012). This conformational plasticity is responsible for efficient antigen recognition, multi-specificity, (Keskin, 2007) and increase in the effective size of the functional antibody repertoire, but may also lead to autoimmunity and allergy (James, Roversi, & Tawfik, 2003). To obtain an accurate docking model, structural changes in the antibody and the antigen upon binding need to be considered (Lexa & Carlson, 2012). Second, the absence of a reliable scoring function (Barradas-Bautista, Moal, & Fernandez-Recio, 2017; Moal, Moretti, Baker, & Fernandez-Recio, 2013; Moal, Torchala, Bates, & Fernandez-Recio, 2013; Ritchie, 2008) that can distinguish the correct binding mode from multiple incorrect ones, further affects the accuracy of docking methods. Third, paired multiple sequence alignments are typically not available for antibody:antigen interactions; thus, novel deep-learning models that rely on co-evolutionary signal for structure prediction cannot be applied in a straightforward manner (Baek et al., 2021; Evans et al., 2021; Jumper et al., 2021). Therefore, new techniques are necessary for this task.

The integrative structure modeling approach has been successful in rapid and accurate structural characterization of multi-subunit complexes by computational integration of data from a variety of fast and accessible experimental techniques (Alber et al., 2007; Bonomi et al., 2019; Drake, Seffernick, & Lindert, 2022; Masrati et al., 2021; Rantos, Karius, & Kosinski, 2022; Russel et al., 2012; Schneidman-Duhovny, Pellarin, & Sali, 2014). One such technique, Small Angle X-ray scattering (SAXS) is a high-throughput technique for low-resolution structural characterization of molecules in solution (Hura et al., 2009; Petoukhov & Svergun, 2013; Putnam, Hammel, Hura, & Tainer, 2007; Rambo & Tainer, 2013a, 2013b; Tiede, Mardis, & Zuo, 2009; Trewhella, 2016; Whitten & Trewhella, 2009). A key strength of SAXS is that it is informative about the shapes of macromolecules as large as 1000 Å in the 10–50 Å resolution range at near-physiological conditions. The experiment is performed with ~1.0 mg/mL of a macromolecular sample in a ~15 μL volume, and usually takes only a few minutes on a well-equipped synchrotron beam line (Hura et al., 2009). Moreover, SAXS profiles can be rapidly collected for a variety of experimental conditions, such as ligand-bound and unbound protein samples, ligand titration series, different temperatures, or pH values. The SAXS profile of a macromolecule, $I(q)$, is computed by subtracting the SAXS profile of the buffer from the SAXS profile of the macromolecule in the buffer. The profile can be converted into an approximate distribution of pairwise atomic distances of the macromolecule (i.e., the pair-distribution function) via a Fourier transform (Putnam et al., 2007). In contrast to static atomic resolution structures solved by X-ray crystallography, SAXS profiles, while limited in resolution, contain information about the conformational and compositional states of the system in solution (Bernado, Mylonas, Petoukhov, Blackledge, & Svergun, 2007; Hura et al., 2013; Rambo & Tainer, 2011).

Several software packages and webservers are available for protein-protein docking with SAXS profile (Jimenez-Garcia, Pons, Svergun, Bernado, & Fernandez-Recio, 2015; Karaca & Bonvin, 2013; Petoukhov & Svergun, 2005; Pons et al., 2010; Xia et al., 2015). We have also developed such an approach, FoXSDock (Carter et al., 2015; Schneidman-Duhovny, Hammel, & Sali, 2011), that was validated on an unbound docking benchmark of protein-protein complexes, including ~30 antibody-antigen complexes (Schneidman-Duhovny et al., 2011). These methods generate a large pool of complex models by global rigid docking, followed by filtering of the complex models based on the SAXS profile, and re-scoring by a composite score that consists of the interface and the SAXS score terms. These methods are validated using simulated SAXS profiles with some noise, because SAXS data is not available for most of the docking benchmark. In contrast to simulated data, experimental samples can contain a mixture of antibody, antigen, and antibody-antigen complexes, with antigens and antibodies populating multiple 3D conformations. To account for such sample heterogeneity, which reduces the modeling accuracy, the data can be fitted using multi-state models that contain several structural states and their population weights (Schneidman-Duhovny et al., 2014).

Here we develop, validate, and apply a protocol for antibody epitope prediction based on the SAXS datasets. To address the challenges in interpretation of data coming from heterogeneous samples, our method explicitly considers the conformational and compositional heterogeneity of antibodies, antigens and antibody-antigen samples subjected to SAXS analysis. We rely on the SAXS profiles of an antibody, an antigen, and their complex. The SAXS profiles of the antibody and the antigen enable us to model their conformational states in solution. Our protocol combines FoXSDock (Schneidman-Duhovny et al., 2011) and MultiFoXS (Schneidman-Duhovny, Hammel, Tainer, & Sali, 2016) that we have previously developed for protein-protein docking and ensemble selection, respectively. In addition, we model the antibody structures using an antibody-specific end-to-end deep learning model we have recently developed (Cohen, Halfon, & Schneidman-Duhovny, 2022). The SAXS fit scores are integrated with scores for assessing protein-protein interactions, including SOAP (Dong, Fan, Schneidman-Duhovny, Webb, & Sali, 2013) and ContactNet (in preparation), a novel deep learning-based scoring function for antibody-antigen interfaces. The method is validated using five antibody: antigen interactions with experimental SAXS profiles and published crystal structures of the complexes.

2. Materials and methods

2.1 Software

The following software packages are used in the protocols described below:

1. Integrative Modeling Package (IMP)—a software package that includes a SAXS module can be downloaded from `http://salilab.org/imp/download.html`. It is available in binary form for most common machine types and operating systems, or it can be built from the source code; either the stable 2.7.0 release of IMP, or a recent development version, should be used. The code related to the protocols described here are mainly in the `saxs`, `foxs`, `multi_state`, and `integrative_docking` IMP modules
2. PatchDock—a software for protein-protein docking can be downloaded from `http://bioinfo3d.cs.tau.ac.il/PatchDock/`
3. MODELER—a software for comparative modeling of protein structures can be downloaded from `https://salilab.org/modeller/download_installation.html`
4. NanoNet—a deep-learning model for antibody modeling can be downloaded from `https://github.com/dina-lab3D/NanoNet`
5. ContactNet—a deep-learning model for classifying antibody-antigen complexes, will soon be available from `https://github.com/dina-lab3D/ContactNet`

6. Gnuplot (http://www.gnuplot.info/) is used for plotting by the scripts provided with the examples used here
7. Colab notebooks for training NanoNet and for antigen-nanobody docking with a SAXS profile are available from https://github.com/dina-lab3D/tutorials

2.2 Integrative modeling

Integrative structure determination typically proceeds in four stages (Fig. 1): (i) gathering information, (ii) model representation, (iii) sampling models with good scores, and (iv) analyzing models and information (Alber et al., 2007; Russel et al., 2012; Schneidman-Duhovny et al., 2014). In the information-gathering stage, we collect SAXS profiles for the antigen, antibody, and their complexes. In the second stage, the SAXS data are used to select one or more conformational states of the antibody and antigen for protein-protein docking. Specifically, we use a deep-learning model to accurately predict the conformation of the Fv, followed by sampling of the elbow angles to arrive at the conformation of the Fab. SAXS data-fitting is used to decide if the antigen and the antibody will be represented by one or more structural states. In the third stage, the models of the Fab:antigen complex are sampled using protein-protein docking. In the final and fourth stage, we score the models based on the SAXS profiles of the complex, and interface complementarity using SOAP (Dong et al., 2013) and ContactNet.

2.2.1 Gathering information

Sample preparation. The ECD (Extra Cellular Domain) of EGFR (UniProt ID: P00533, UniProt residue numbering 25–642) was expressed in Sf21 insect cells (BlueSky Bioservices, Worcester, MA). Conditioned and frozen media from the insect cell production was thawed at 4C prior to primary capture by Ni Sepharose resin. The protein was then polished in 10 mg batches by size exclusion chromatography using a GE Healthcare HighLoad 16/600 Superdex 200 column and a 10 mM sodium phosphate, 230 mM sodium chloride, pH 7.3 running buffer. Active peak fractions were pooled and normalized to a target protein concentration of 10 mg/mL.

Fab constructs encoding the heavy and light chain variable domain sequences from the respective PDB entries (cetuximab: 1yy9; necitumumab: 3b2u; matuzumab: 3c09; duligotuzumab: 3p0y) linked to human WT CH1 and CK sequences were cloned, followed by Fab expression in HEK293 cells. Transiently transfected HEK293 cultures were harvested by centrifugation for 5 min at 2400 × *g*. The clarified supernatant was decanted from the

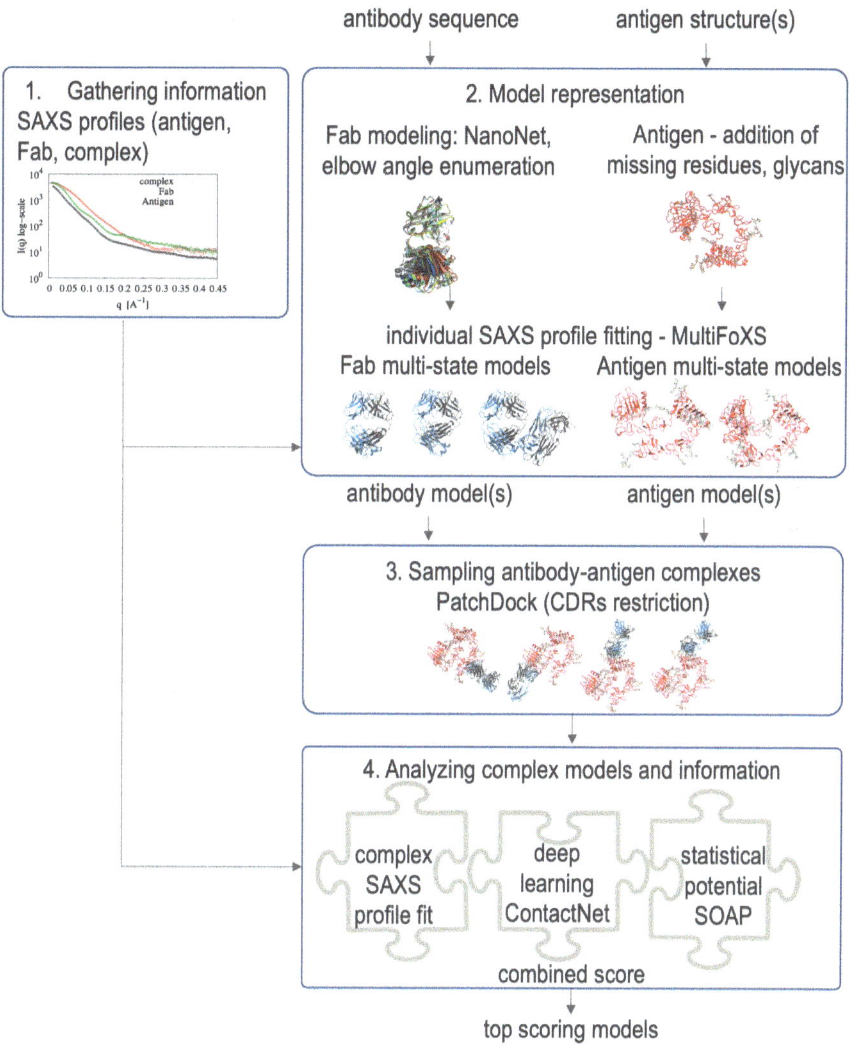

Fig. 1 A protocol for integrative modeling of antibody-antigen complexes with SAXS profiles.

pellet and centrifuged a second time prior to primary capture by either ThermoFisher CaptureSelect CH1 or GE Healthcare KappaSelect resin. The resulting protein was buffer-exchanged into 25 mM HEPES, 150 mM sodium chloride, pH 7.3 and normalized to a target protein concentration of 10 mg/mL.

PCSK9 sample and data were collected as previously described (Schneidman-Duhovny et al., 2012).

SAXS Data Collection. SAXS data were collected at Beamline 4–2 at the Stanford Synchrotron Radiation Lightsource (SSRL) in the SLAC National Accelerator Laboratory. All data were collected using a Mar225 CCD detector. SAXS profiles of each sample were recorded at protein concentrations of 0.5, 1.0, and 5.0 mg/mL in a protein storage buffer (Table S1 in Supplementary Material in the online version at https://doi.org/10.1016/bs.mie.2022.11.003). The intensities from the buffer data were subtracted from the sample data to produce the final scattering profile. Data were initially analyzed (Table S2 in Supplementary Material in the online version at https://doi.org/10.1016/bs.mie.2022.11.003) using the ATSAS package PRIMUS program (Petoukhov & Svergun, 2007) and SCÅTTER (Tully, Tarbouriech, Rambo, & Hutin, 2021).

2.2.2 Model representation

SAXS profile fitting. We computed the profiles using the Debye formula with the FoXS program (Schneidman-Duhovny, Hammel, Tainer, & Sali, 2013). For a given structural model there are three adjustable parameters: excluded volume (c_1), hydration layer density (c_2), and the scaling factor (c) that are optimized to fit the experimental profile as measured by the χ^2 score:

$$\chi^2 = \frac{1}{S}\sum_{i=1}^{S}\left(\frac{I_{exp}(q_i) - cI(q_i, c_1, c_2)}{\sigma(q_i)}\right)^2$$

where $I_{exp}(q)$ is the experimental profile, which is a function of the momentum transfer $q = (4\pi \, sin \, \theta)/\lambda$, where 2θ is the scattering angle and λ is the wavelength of the incident X-ray beam, $\sigma(q)$ is the experimental error of the measured profile, and S is the number of points in the profile. $I(q, c_1, c_2)$ is the computed profile, given by the Debye formula (Schneidman-Duhovny et al., 2013), with c_1 and c_2 optimized to minimize the score.

Multi-state SAXS profile fitting. It is possible that the sample includes compositional heterogeneity. In this case, the score of a multi-state model is given by:

$$\chi^2 = \frac{1}{S}\sum_{i=1}^{S}\left(\frac{I_{exp}(q_i) - c\Sigma_n w_n I_n(q_i, c_1, c_2)}{\sigma(q_i)}\right)^2$$

where $I_n(q, c_1, c_2)$ and w_n are the computed profile and the corresponding weight, respectively, for each considered structural state (Schneidman-Duhovny et al., 2016). When fitting multi-state models, the weights can be optimized using non-negative least-square fitting (Schneidman-Duhovny et al., 2016; Schneidman-Duhovny & Hammel, 2018). The enumeration of multi-state models of size N (typically, $1 < N < 5$), such that the corresponding sum of weighted SAXS profiles fits the experimental SAXS data, is performed iteratively using a branch-and-bound method.

Fab modeling. The structure of the antibody variable domains were modeled by NanoNet (Cohen et al., 2022). While NanoNet was originally designed for modeling heavy-chain only antibodies, the latest version simultaneously models both light and heavy chain variable domains along with optimization of the inter-domain orientation. This was achieved by training the network to predict the transformation (rotation and translation) of the light chain variable domain with respect to the heavy chain variable domain. The constant domain of the Fab was modeled based on the most sequence-similar PDB structure. To model Fab elbow angle variability, 28 Fab structures from the PDB with elbow angles in the range of 125°–190° (Stanfield, Zemla, Wilson, & Rupp, 2006) were used as templates for MODELER 9.7 (Sali & Blundell, 1993) and 10 models were generated for each elbow angle, resulting in 280 Fab models. In principle, any antibody modeling tool can be used, including IgFold (Ruffolo, Chu, Mahajan, & Gray, 2022) and AlphaFold2 (Jumper et al., 2021).

Fab SAXS profile fitting. Next, SAXS profiles were calculated for the 280 Fab models. We observed that in some cases, the fit to the SAXS profile is significantly better with a small fraction of Fab dimers in addition to monomers. To account for a fraction of Fab dimers, we extracted Fab structures from the PDB with such dimeric assemblies in their asymmetric unit (Fig. 2). This dataset was used along with all Fab structures to select single- and multi-state models that best fit the experimental SAXS profile.

Antigen modeling. It is important to complete all the missing fragments in the antigen, so that SAXS profiles can be accurately calculated from the structures. Missing fragments, such as loops, or N-, and C-termini can be added by MODELLER or similar tools (Sali & Blundell, 1993). Post-translational modifications, such as glycosylation, can be added using specialized force fields (Kirschner et al., 2008). Single- and multi-state modeling is used to select one or more structures that best fit the SAXS profile of the antigen.

2.2.3 Sampling good-scoring models

Given n antibody and m antigen conformations from their respective multi-state models we used a rigid docking approach, PatchDock, to perform nm cross-docking runs of all antibody conformations against all antigen

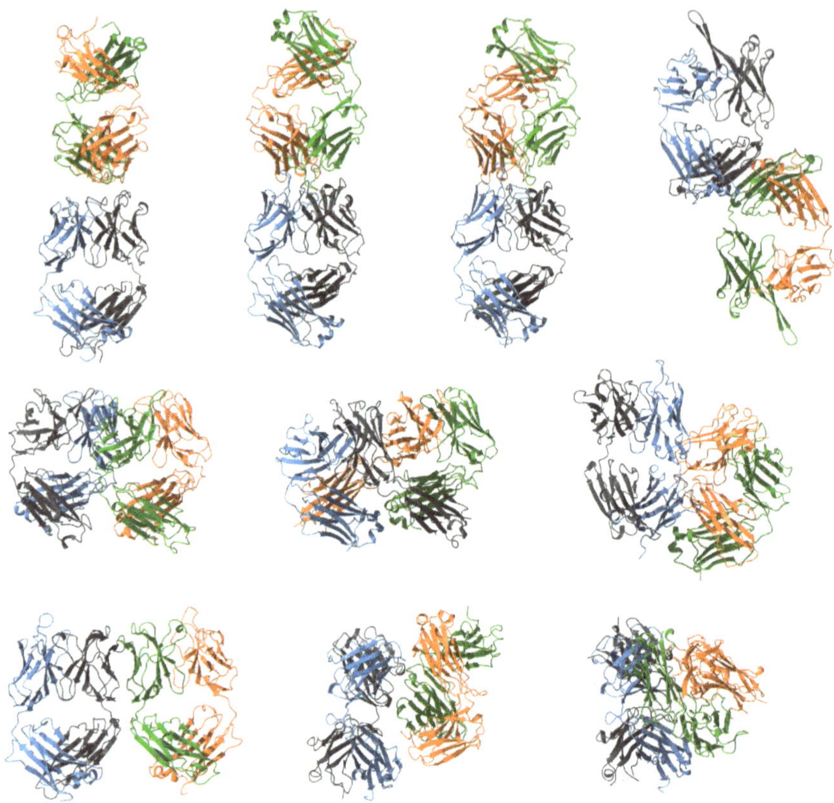

Fig. 2 Fab dimerization structures from PDB (1pg7,6mxr, 6my4,1q1j, 2a6j, 3oay,1igf, 6n32, 2a6j, 6u8c).

conformations. PatchDock is an efficient geometric rigid docking that maximizes geometric shape complementarity (Duhovny, Nussinov, & Wolfson, 2002; Schneidman-Duhovny, Inbar, Nussinov, & Wolfson, 2005). The sampling precision can be controlled by the resolution of the surface representation (i.e., the minimal distance between surface points used to generate docking models) and clustering parameters. The antibody-antigen docking protocol automatically detects the Complementarity Determining Regions (CDRs) and focuses the search space accordingly. The output docked structures are required to have some of the heavy chain CDR3 residues in contact with an antigen. For each of the top scoring Fab models from SAXS profile fitting, docking was performed with increased sampling precision. Usually, docking methods balance configurational sampling precision against the accuracy and efficiency of the scoring function, with the goal of retaining a sufficiently accurate model within a sufficiently small fraction of the best scoring models. For SAXS-based scoring function a higher sampling precision was needed to guarantee that the interface and the global shape of the complex are well sampled, typically generating ~30,000 models.

2.2.4 Analyzing models and information

The docking models were analyzed using three scoring functions that assess the quality of fit to the SAXS profile of the complex, as well as the interface complementarity scores.

SAXS profile fit. The SAXS profile of each complex was calculated and compared to the experimental profile. To account for sample compositional and conformational heterogeneity, in addition to the complex, we calculated the score of a multi-state model that contains the free antibody and antigen states from stage 2.

Interface scoring with statistical potential (SOAP). We used a statistical potential for scoring protein-protein interfaces to assess each generated antibody-antigen interface (Dong et al., 2013). Statistically Optimized Atomic Potential (SOAP) is a distance-dependent potential optimized on protein-protein docking benchmark 4.0 using a Bayesian approach.

Interface scoring with a deep-learning model (ContactNet). ContactNet is an attention-based deep network architecture, to classify protein-protein interaction distance maps obtained from docking into near-native and non-native ones (in preparation). We trained the ContactNet model to classify antibody-antigen interactions using ~1000 complexes from a non-redundant AbDb dataset (Ferdous & Martin, 2018) with modeled antibodies. To generalize to the typical scenario when the antibody crystal structure is unavailable, the Fv regions were modeled using AlphaFold2 and docked to the antigens to produce a set of docked complexes for classification. Docked complexes with modeled antibodies enable training of the network to recognize and classify correct complexes even if CDR conformations and the antibody orientation relative to an antigen are not highly accurate. The success rate of ContactNet (number of docking cases with acceptable accuracy models (Mendez, Leplae, Lensink, & Wodak, 2005) among 10 best scoring ones) on the validation set complexes is ~45% compared to ~25% for SOAP. Here, the average score of two trained models was used as a final ContactNet score.

Final combined score. The individual scores (SAXS and interface score) for each antibody-antigen model were normalized into *Z*-scores using the average and standard deviation of all docking model scores. This normalization enables combination and comparison of scores for different data types.

3. Results

Dataset. Our benchmark dataset consists of five complexes with experimental SAXS profiles: four EGFR-Fab and one PCSK9-Fab interaction (Table 1). EGFR (epidermal growth factor receptor) is a transmembrane tyrosine kinase receptor involved in the proliferation and survival of cancer cells. It is one of the first molecular targets of anti-cancer monoclonal antibodies (Martinelli, De Palma, Orditura, De Vita, & Ciardiello, 2009). PCSK9 (serine protease proprotein convertase subtilisin/kexin type 9) is an important target for lowering LDL-cholesterol levels. PCSK9 antibodies decrease LDL-cholesterol concentrations by inhibiting PCSK9 interaction with the low-density lipoprotein receptor (Gouni-Berthold, 2015). We have previously modeled this PCSK9-J16 Fab complex using SAXS (Schneidman-Duhovny et al., 2012). Here, we apply multi-state modeling for better SAXS profile fits and integrate SAXS scores with statistical potentials and deep learning-based scoring. Crystal structures are available for these complexes and were used for method validation (Li et al., 2005; Li, Kussie, & Ferguson, 2008; Liang et al., 2012; Schaefer et al., 2011; Schmiedel, Blaukat, Li, Knochel, & Ferguson, 2008).

Fab modeling and fitting to SAXS profiles. The variable domains of Fabs were modeled with NanoNet (Fig. 3). Accurate models (CDR H3 RMSD ~1 Å) were obtained for all Fabs except PDB 3p0y where the CDR H3 RMSD is 2.33 Å (Table 1). Subsequently, we found that the latter was the only one not included in the NanoNet training set. Retraining of NanoNet with exclusion of all the Fabs used here, resulted in a slightly higher CDR H3 RMSDs of 1.4 Å (1yy9), 2.8 Å (3b2u), 2.5 Å (3c09), 1.8 Å (3p0y), and 1.7 Å (3sqo). This CDR H3 accuracy (1.0–3.0 Å) leads

Table 1 NanoNet prediction accuracy for the variable domains of the five Fabs in the dataset as measured by backbone atoms RMSD for VL and VH (light and heavy chain variable domains), as well as for CDR3 loops only.

Fab	PDB	RMSD VL	RMSD L3	RMSD VH	RMSD H3	RMSD all
Cetuximab	1yy9	0.82	0.97	0.88	0.69	1.57
Necitumumab	3b2u	1.16	0.95	0.66	1.06	1.22
Matuzumab	3c09	0.86	0.98	0.49	0.83	1.03
Duligotuzumab	3p0y	1.00	1.62	1.01	2.33	1.29
J16	3sqo	0.72	0.75	0.61	0.77	0.94

The modeled and experimental (complex) Fab structures were superimposed using the residues in the antibody framework regions (Cohen et al., 2022).

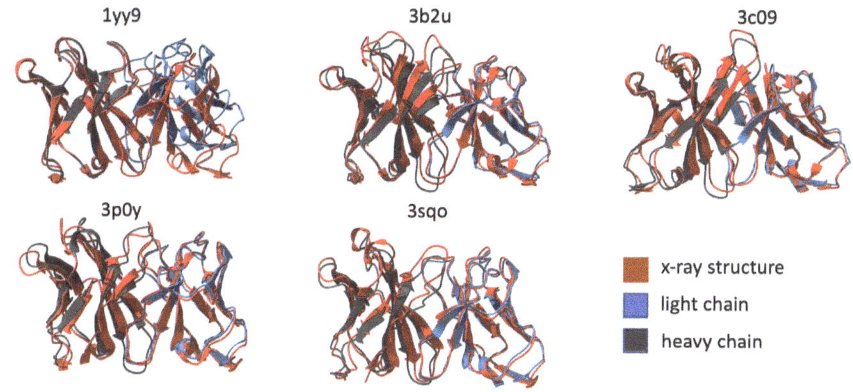

Fig. 3 NanoNet models of variable domains superimposed to x-ray structures.

Table 2 Multi-state modeling results of Fabs. The second and the third columns report the χ^2 values for the x-ray Fab structure and our best scoring Fab model, respectively.

PDB	χ^2 X-ray Fab	χ^2 Single-state	Min χ^2 without Fab dimers	# States in the final multi-state model	% Fab dimers	χ^2 Multiple states
1yy9	27.3	9.6	9.6	3	3.4	2.3
3b2u	20.4	11.3	11.3	4	2.4	5.4
3c09	19.0	4.9	4.9	4	5.5	2.1
3p0y	9.4	3.5	3.4	3	3.8	2.8
3sqo	69.0	6.5	3.6	4	2.9	1.6

The fourth column contains the number of states that best fit the SAXS profile (lowest χ^2 value—column 6). The fifth column contains the percentage of Fab dimers in the multi-state models.

to a sufficient number of correct models that can be further identified with SAXS profiles. The orientation of the VH with respect to the VL was accurate, as indicated by overall RMSD of the variable domain, except for PDB 1yy9, where there was a shift in the orientation (Fig. 3).

The constant domains were added to obtain an ensemble of Fab models with varying elbow angles (Methods). The models were compared against the experimental SAXS profile. The minimal χ^2 values across all single-state models were in the range of 3.5–11.3 indicating that the fit is suboptimal (Table 2). The fit of the Fab crystal structures was even worse, indicating that the problem is not related only to the NanoNet model accuracy. Given previous observations of Fab dimerization, we obtained PDB Fab structures with dimers (Fig. 2) in asymmetric units and attempted to fit the SAXS profiles using multiple states. The fit was improved significantly with 3- to 4-state models to χ^2 values in the range of 1.6–5.4 (Table 2 and Fig. 4A–D). These multi-state models usually contained two Fab

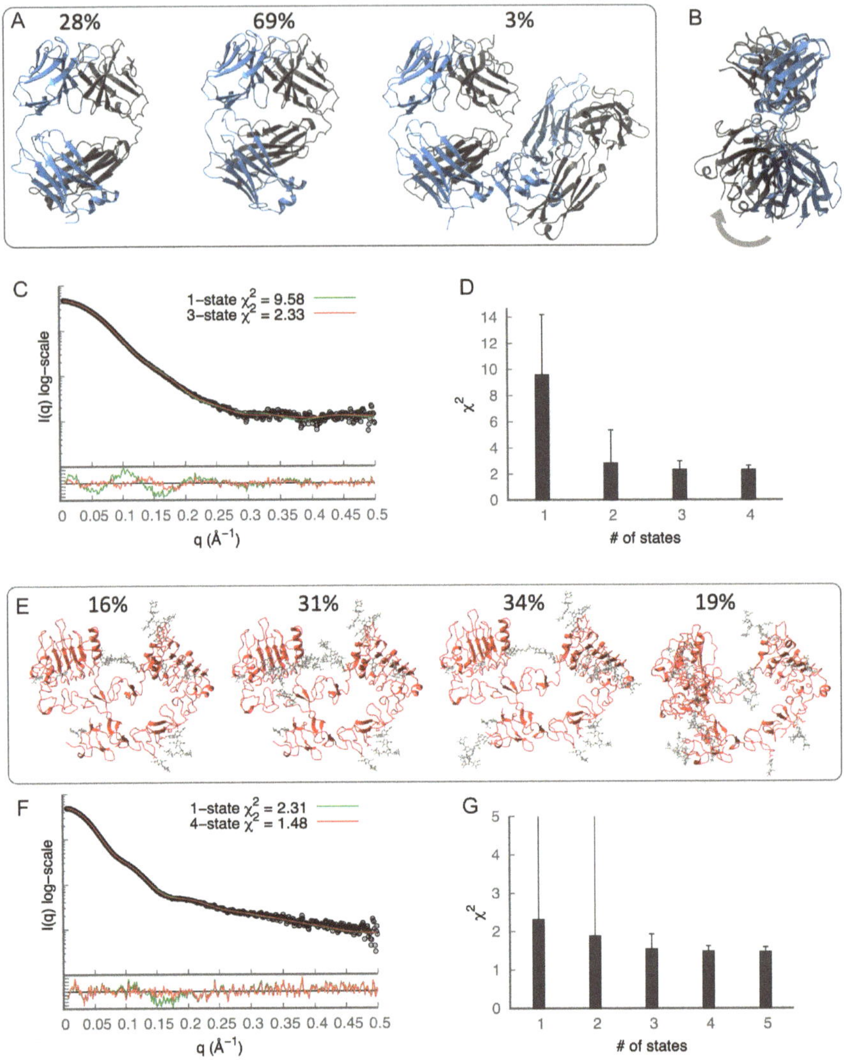

Fig. 4 SAXS-based multi-state models for the Cetuximab antibody and the EGFR antigen. (A) 3-state Cetuximab model that contains the Fab with two different elbow angles (as shown in B) and 3% of Fab dimers. (C) SAXS profile fits to experimental data (black dots) for 1-state model (green) vs. 3-state model (red). (D) Number of states vs. the quality of the fit to the SAXS profile as measured by χ^2. The error bar is calculated based on the average score of top scoring multi-state models. (E) 4-state EGFR model and the corresponding weights. (F) SAXS profile fits to experimental data (black dots) for 1-state model (green) vs. 4-state model (red). (G) Number of states vs. the quality of the fit to the SAXS profile (as in D).

conformations with two different elbow angles, representing the flexibility of the Fab, as well as 2.4–5.5% of dimeric Fab states (Fig. 4A–D). The final number of states was selected as a minimal number after which there is no significant drop in χ^2 values. These multi-state models were included in SAXS-based scoring of complexes after docking.

Antigen modeling and fitting to SAXS profiles. Missing residues were added to the PCSK9 structure using MODELER 9.7 and a good fit to SAXS profile was obtained (Table 3). For EGFR antigen, there are three structures in the PDB that show slight conformational changes and have variable glycosylation patterns. We have used information from mass spectrometry (Wu et al., 2006) to model glycans according to the identified pattern. Because glycosylation is dynamic and heterogeneous, a multi-state model was required to fit the SAXS profile (Fig. 4E–G and Table 3).

Sampling antibody-antigen complexes. The modeled Fabs were docked to the antigen using PatchDock (Methods) with final clustering RMSD of 2 Å. The two antigens, EGFR and PCSK9, present a sampling challenge due to their size of over 600 amino acids spanning multiple domains. Due to elbow angle variability and the large effect that it has on the Fab RMSD, we rely on interface RMSD (IRMSD) as a measure of docking accuracy (Mendez et al., 2005; Mendez, Leplae, De Maria, & Wodak, 2003). Overall, acceptable quality models (IRMSD < 4 Å) were obtained for all Fab:antigen complexes (Table 4) and comprised between 0.23% and 0.9% of all models (hit rate). For 4/5 Fab:antigen complexes, medium accuracy models (IRMSD < 2 Å) were found. No medium-accuracy docking models were obtained for the duligotuzumab:EGFR case (PDB 3p0y), presumably on account of the lower accuracy CDR H3 prediction. This emphasizes one of the main bottlenecks in modeling antibody-antigen complexes: accurate complex models can be obtained only when the Fab, especially the CDR H3 conformation, has been modeled accurately.

Table 3 SAXS profile fitting for EGFR and PCSK9 antigens.

Antigen	PDB	χ^2 X-ray	χ^2 Single-state	# States in a final multi-state model	χ^2 Multiple states
EGFR	3qwq	30.7	2.3	4	1.5
PCSK9	2p4e	54.5	2.6	1	–

The third and the fourth columns report the χ^2 values for the x-ray antigen structure and our best model, respectively. The fifth column contains the number of states that best fit the SAXS profile (lowest χ^2 value—column 6).

Table 4 Accuracy of sampling and scoring functions.

PDB	# of Docked models	# of medium accuracy models (IRMSD <2 Å)	# of acceptable accuracy models (IRMSD <4 Å)	Rank (IRMSD) by SAXS c^2	Rank (IRMSD) by SOAP	Rank (IRMSD) by ContactNet	Rank (IRMSD) by combined score
1yy9	18,671	6 (0.03%)	144 (0.77%)	1 (1.8 Å)	30 (2.49 Å)	1 (1.95 Å)	1 (1.95 Å)
3b2u	34,186	43 (0.13%)	281 (0.82%)	1 (3.9 Å)	2 (1.07 Å)	13 (2.18 Å)	4 (2.18 Å)
3c09	27,254	18 (0.07%)	64 (0.23%)	776 (1.5 Å)	192 (2.67 Å)	2122 (3.77 Å)	178 (2.67 Å)
3p0y	37,461	0	89 (0.24%)	139 (3.78 Å)	1228 (3.50 Å)	45 (3.87 Å)	2 (3.87 Å)
3sqo	17,346	12 (0.07%)	160 (0.9%)	1022 (2.57 Å)	38 (2.61 Å)	1 (2.66 Å)	1 (2.66 Å)

Analyzing models and information. We score the docked complexes using SAXS, SOAP, and ContactNet to analyze the discrimination ability of each function separately and combined. Our main criteria are the rank of the near-native model with IRMSD < 4 Å (Table 4). We also plot the "funnels" (score vs. IRMSD scatterplot) to visualize the score distributions vs. IRMSD (Fig. 5). We find that the discrimination ability of each score varies significantly between the complexes. The most consistent

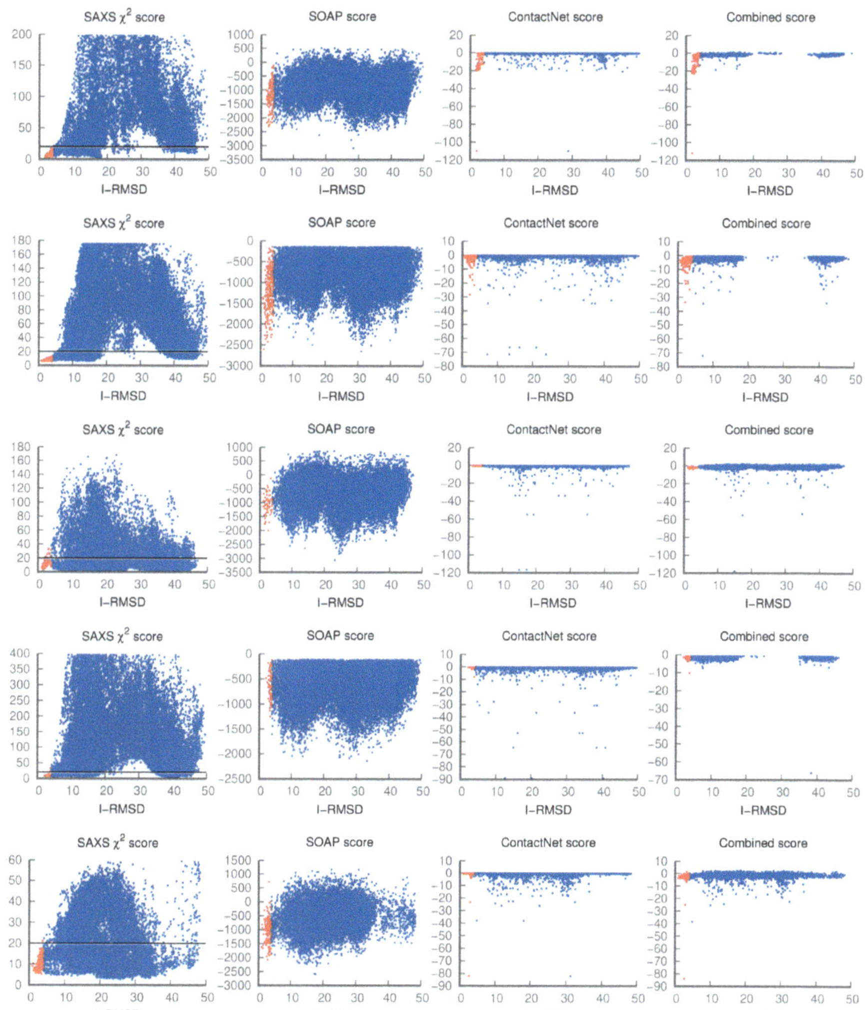

Fig. 5 Funnels for SAXS, SOAP, ContactNet, and combined scores for 1yy9, 3b2u, 3c09, 3p0y, and 3sqo, respectively.

is the SAXS score, where near-native models (Fig. 5, left column, red) have χ^2 values below 20, enabling filtering of a large fraction of models. For two complexes (1yy9, 3b2u), the top scoring model based on SAXS is a near-native model (Table 4, Fig. 6). The SOAP score has varying success with a near-native model ranked 2nd for 3b2u and a pronounced funnel (Fig. 5), but limited success for other complexes. The main feature of ContactNet scores is to classify as "incorrect" a large fraction of complexes, discarding a large fraction of false positive models. This enables the high ranking of near-natives that were classified correctly. However, in some cases all the near-native models are discarded (3c09). Unlike the SAXS-based score, where most near-natives have low χ^2 values that enable filtering, SOAP and ContactNet have a wide range of values for near-natives.

The combination of scores (after filtering all models with $\chi^2 > 20$) shows the true value of data integration. The near-native model is among best five predictions according to the combined score in 4 out of 5 complexes (Table 4 and Fig. 5). Two of them have the correct model as top scoring (1yy9, 3sqo). Only in the case of 3c09, the correct model is not ranked high (178), probably due to insufficient signal from the three scoring functions. Notably, the integrative approach succeeds on the 3p0y complex, where the CDR3 loop is of low accuracy, no medium accuracy models were found by docking, and none of the scores was successful in ranking on its own (Fig. 6).

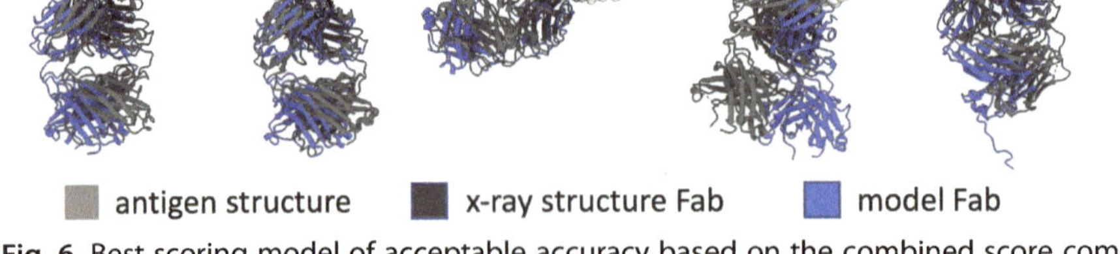

Fig. 6 Best scoring model of acceptable accuracy based on the combined score compared to X-ray structures. The antigens (surface representation) were superimposed to show the difference in the Fab position.

4. Protocol

4.1 SAXS profile fitting with FoXS. We compare the SAXS profile of an antibody (`ab.dat`) to a single structure (`ab.pdb`) by running FoXS (IMP package)

```
> foxs ab.pdb ab.dat
```

4.2 Multi-state SAXS profile fitting with MultiFoXS. To compare multiple structures to a single profile run multi-state enumeration (IMP package) as follows

```
> multi_foxs ab.dat ab1.pdb ab2.pdb ...
```

4.3 NanoNet antibody modeling. Run NanoNet by providing a fasta file with the sequences of fab variable domains

```
> python NanoNet.py <fasta file>
```

4.4 Docking with PatchDock. To run PatchDock, prepare PDB files for an antibody (`ab.pdb`) and an antigen (`ag.pdb`)

```
# prepare parameter file
 > buildParams.pl Ag.pdb Fab.pdb 2.0 AA
 # run PatchDock
 > patch_dock.Linux params.txt docking.res
 # prepare transformation file for further scoring
 > PatchDockOut2Trans.pl docking.res > trans
```

4.5 Scoring with SAXS profile of the complex. Use IMP `saxs_score` program a for faster scoring of complexes that includes multiple states

```
> saxs_score Ag.pdb Fab.pdb trans complex.dat -t --no_filtering_by_rg -p profile_filenames -o saxs_score.res
```

where `Ag.pdb`, `Fab.pdb` and `trans` are from the docking stage above, `complex.dat` is the SAXS profile of the complex. `-t` indicates that multi-state witting is performed that includes antibody and antigen in addition to complex. Additional conformations can be added by calculating their SAXS profiles and including profile filenames in the `profile_filenames` file. The output is written into the `saxs_score.res` file.

4.6 Scoring with SOAP statistical potential. Similarly, to SAXS scoring, run the `soap_score` program from IMP

```
> soap_score Ag.pdb Fab.pdb trans -o soap_score.res
```

4.7 Scoring with ContactNet. The scoring function will soon be available with format similar to SOAP score.

4.8 Combining different scores. To combines the scores, use `combines_scores` program from IMP by providing score files and their weights. Here, same weight is used:

```
> combine_scores saxs_score.res 1 soap_score.res 1 contactnet_
  score.res 1
```

5. Discussion

We present a protocol for accurate integrative modeling of antibody-antigen complexes using SAXS profiles, protein-protein docking, deep-learning models, and statistical potentials. An experimentally determined input antibody structure is not required, since this is modeled using the NanoNet deep-learning approach and the experimental SAXS profile of the Fab. The protocol considers the compositional and conformational heterogeneity of the sample using multiple structures and conformations in SAXS fitting. High-throughput SAXS data collection enables efficient and timely prioritization of lead antibody candidates targeting the desired antigen epitope. We expect the protocol presented here will be used for epitope mapping and validation of structural hypotheses in antibody design and development projects involving macromolecular antigens. Moreover, it can aid in the characterization of the immune repertoire antibodies (Georgiou et al., 2014).

Using SAXS data greatly reduces the number of possible complexes when compared to docking with interface scores only. Near-native models could be obtained among the top-scoring predictions even for antigens with over 600 amino acids. However, sometimes additional information is needed to narrow down to the correct cluster. In these cases, the approach can be complemented using additional data (Schneidman-Duhovny et al., 2012), such as cross-linking mass spectrometry (Xiang et al., 2020; Xiang et al., 2021), mutagenesis, or hydrogen-deuterium exchange (Puchades et al., 2019). Additional improvement in model accuracy can be obtained by refinement of top scoring models using AlphaFold2, by providing them as template structures (Ghani et al., 2021).

Acknowledgments

We thank the Beamline 4-2 at the Stanford Synchrotron Radiation Lightsource (SSRL) in the SLAC National Accelerator Laboratory team for enabling data collection for this project. For assistance with Fab:EGFR sample preparation and characterization, we thank Ngarewa Houston, Heather Lynaugh, Urszula Sharples, Christin Strong, Yuan Cao, Beata Bobrowicz, Stephanie Durand, Isabelle Caffry, Yohann Misquita, Tingwan Sun and, Yingda Xu at Adimab. This work was supported by ISF 1466/18 (D.S.).

References

Alber, F., Dokudovskaya, S., Veenhoff, L. M., Zhang, W., Kipper, J., Devos, D., et al. (2007). Determining the architectures of macromolecular assemblies. *Nature*, *450*(7170), 683–694 (pii) https://doi.org/10.1038/nature06404.

Baek, M., DiMaio, F., Anishchenko, I., Dauparas, J., Ovchinnikov, S., Lee, G. R., et al. (2021). Accurate prediction of protein structures and interactions using a three-track neural network. *Science*, eabj8754. https://doi.org/10.1126/science.abj8754.

Barradas-Bautista, D., Moal, I. H., & Fernandez-Recio, J. (2017). A systematic analysis of scoring functions in rigid-body protein docking: The delicate balance between the predictive rate improvement and the risk of overtraining. *Proteins*, *85*(7), 1287–1297. https://doi.org/10.1002/prot.25289.

Bernado, P., Mylonas, E., Petoukhov, M. V., Blackledge, M., & Svergun, D. I. (2007). Structural characterization of flexible proteins using small-angle X-ray scattering. *Journal of the American Chemical Society*, *129*(17), 5656–5664. https://doi.org/10.1021/ja069124n.

Bonomi, M., Hanot, S., Greenberg, C. H., Sali, A., Nilges, M., Vendruscolo, M., et al. (2019). Bayesian weighing of electron cryo-microscopy data for integrative structural modeling. *Structure*, *27*(1), 175–188 (e176).

Bornholdt, Z. A., Turner, H. L., Murin, C. D., Li, W., Sok, D., Souders, C. A., et al. (2016). Isolation of potent neutralizing antibodies from a survivor of the 2014 Ebola virus outbreak. *Science*, *351*(6277), 1078–1083. https://doi.org/10.1126/science.aad5788.

Carter, L., Kim, S. J., Schneidman-Duhovny, D., Stohr, J., Poncet-Montange, G., Weiss, T. M., et al. (2015). Prion protein-antibody complexes characterized by chromatography-coupled small-angle X-ray scattering. *Biophysical Journal*, *109*(4), 793–805. https://doi.org/10.1016/j.bpj.2015.06.065.

Cohen, T., Halfon, M., & Schneidman-Duhovny, D. (2022). NanoNet: Rapid and accurate end-to-end nanobody modeling by deep learning. *Frontiers in Immunology*, *13*, 958584. https://doi.org/10.3389/fimmu.2022.958584.

Dong, G. Q., Fan, H., Schneidman-Duhovny, D., Webb, B., & Sali, A. (2013). Optimized atomic statistical potentials: Assessment of protein interfaces and loops. *Bioinformatics*, *29*(24), 3158–3166. https://doi.org/10.1093/bioinformatics/btt560.

Drake, Z. C., Seffernick, J. T., & Lindert, S. (2022). Protein complex prediction using Rosetta, AlphaFold, and mass spectrometry covalent labeling. *bioRxiv*. https://doi.org/10.1101/2022.04.30.490108.

Duhovny, D., Nussinov, R., & Wolfson, H. J. (2002). Efficient unbound docking of rigid molecules. In *Lecture notes in computer science second international workshop, WABI 2002, Rome, Italy*.

Evans, R., O'Neill, M., Pritzel, A., Antropova, N., Senior, A., Green, T., et al. (2021). Protein complex prediction with AlphaFold-Multimer. *bioRxiv*. https://doi.org/10.1101/2021.10.04.463034.

Ferdous, S., & Martin, A. C. (2018). AbDb: Antibody structure database—A database of PDB-derived antibody structures. *Database*, *2018*, 1–9. https://doi.org/10.1093/database/bay040.

Georgiou, G., Ippolito, G. C., Beausang, J., Busse, C. E., Wardemann, H., & Quake, S. R. (2014). The promise and challenge of high-throughput sequencing of the antibody repertoire. *Nature Biotechnology*, *32*(2), 158–168. https://doi.org/10.1038/nbt.2782.

Ghani, U., Desta, I., Jindal, A., Khan, O., Jones, G., Kotelnikov, S., et al. (2021). Improved docking of protein models by a combination of alphafold2 and cluspro. *bioRxiv*.

Gouni-Berthold, I. (2015). PCSK9 antibodies: A new class of lipid-lowering drugs. *Atherosclerosis. Supplements*, *18*, 21–27. https://doi.org/10.1016/j.atherosclerosissup.2015.02.003.

Hura, G. L., Budworth, H., Dyer, K. N., Rambo, R. P., Hammel, M., McMurray, C. T., et al. (2013). Comprehensive macromolecular conformations mapped by quantitative SAXS analyses. *Nature Methods*, *10*(6), 453–454. https://doi.org/10.1038/nmeth.2453.

Hura, G. L., Menon, A. L., Hammel, M., Rambo, R. P., Poole, F. L., 2nd, Tsutakawa, S. E., et al. (2009). Robust, high-throughput solution structural analyses by small angle X-ray scattering (SAXS). *Nature Methods*, *6*(8), 606–612 (pii) https://doi.org/10.1038/nmeth.1353.

Jain, T., Sun, T., Durand, S., Hall, A., Houston, N. R., Nett, J. H., et al. (2017). Biophysical properties of the clinical-stage antibody landscape. *Proceedings of the National Academy of Sciences of the United States of America*, *114*(5), 944–949. https://doi.org/10.1073/pnas.1616408114.

James, L. C., Roversi, P., & Tawfik, D. S. (2003). Antibody multispecificity mediated by conformational diversity. *Science*, *299*(5611), 1362–1367. https://doi.org/10.1126/science.1079731.

Jimenez-Garcia, B., Pons, C., Svergun, D. I., Bernado, P., & Fernandez-Recio, J. (2015). pyDockSAXS: Protein-protein complex structure by SAXS and computational docking. *Nucleic Acids Research*, *43*(W1), W356–W361. https://doi.org/10.1093/nar/gkv368.

Jumper, J., Evans, R., Pritzel, A., Green, T., Figurnov, M., Ronneberger, O., et al. (2021). Highly accurate protein structure prediction with AlphaFold. *Nature*. https://doi.org/10.1038/s41586-021-03819-2.

Karaca, E., & Bonvin, A. M. (2013). On the usefulness of ion-mobility mass spectrometry and SAXS data in scoring docking decoys. *Acta Crystallographica. Section D, Biological Crystallography*, *69*(Pt. 5), 683–694. https://doi.org/10.1107/S0907444913007063.

Keskin, O. (2007). Binding induced conformational changes of proteins correlate with their intrinsic fluctuations: A case study of antibodies. *BMC Structural Biology*, 7, 31. https://doi.org/10.1186/1472-6807-7-31.

Kilambi, K. P., & Gray, J. J. (2017). Structure-based cross-docking analysis of antibody-antigen interactions. *Scientific Reports*, 7(1), 8145. https://doi.org/10.1038/s41598-017-08414-y.

Kirschner, K. N., Yongye, A. B., Tschampel, S. M., Gonzalez-Outeirino, J., Daniels, C. R., Foley, B. L., et al. (2008). GLYCAM06: A generalizable biomolecular force field. Carbohydrates. *Journal of Computational Chemistry*, *29*(4), 622–655. https://doi.org/10.1002/jcc.20820.

Krawczyk, K., Liu, X., Baker, T., Shi, J., & Deane, C. M. (2014). Improving B-cell epitope prediction and its application to global antibody-antigen docking. *Bioinformatics*, *30*(16), 2288–2294. https://doi.org/10.1093/bioinformatics/btu190.

Laustsen, A. H., Greiff, V., Karatt-Vellatt, A., Muyldermans, S., & Jenkins, T. P. (2021). Animal immunization, in vitro display technologies, and machine learning for antibody discovery. *Trends in Biotechnology*, *39*(12), 1263–1273. https://doi.org/10.1016/j.tibtech.2021.03.003.

Lexa, K. W., & Carlson, H. A. (2012). Protein flexibility in docking and surface mapping. *Quarterly Reviews of Biophysics*, *45*(3), 301–343. https://doi.org/10.1017/S0033583512000066.

Li, S., Kussie, P., & Ferguson, K. M. (2008). Structural basis for EGF receptor inhibition by the therapeutic antibody IMC-11F8. *Structure*, *16*(2), 216–227. https://doi.org/10.1016/j.str.2007.11.009.

Li, S., Schmitz, K. R., Jeffrey, P. D., Wiltzius, J. J., Kussie, P., & Ferguson, K. M. (2005). Structural basis for inhibition of the epidermal growth factor receptor by cetuximab. *Cancer Cell*, 7(4), 301–311. https://doi.org/10.1016/j.ccr.2005.03.003.

Liang, H., Chaparro-Riggers, J., Strop, P., Geng, T., Sutton, J. E., Tsai, D., et al. (2012). Proprotein convertase substilisin/kexin type 9 antagonism reduces low-density lipoprotein cholesterol in statin-treated hypercholesterolemic nonhuman primates. *The Journal of Pharmacology and Experimental Therapeutics*, *340*(2), 228–236. https://doi.org/10.1124/jpet.111.187419.

Lin, Y. (2018). What's happened over the last five years with high-throughput protein crystallization screening? *Expert Opinion on Drug Discovery*, *13*(8), 691–695. https://doi.org/10.1080/17460441.2018.1465924.

Martinelli, E., De Palma, R., Orditura, M., De Vita, F., & Ciardiello, F. (2009). Anti-epidermal growth factor receptor monoclonal antibodies in cancer therapy. *Clinical and Experimental Immunology*, *158*(1), 1–9. https://doi.org/10.1111/j.1365-2249.2009.03992.x.

Masrati, G., Landau, M., Ben-Tal, N., Lupas, A., Kosloff, M., & Kosinski, J. (2021). Integrative structural biology in the era of accurate structure prediction. *Journal of Molecular Biology*, *433*(20), 167127.

Mendez, R., Leplae, R., De Maria, L., & Wodak, S. J. (2003). Assessment of blind predictions of protein-protein interactions: Current status of docking methods. *Proteins*, *52*(1), 51–67. https://doi.org/10.1002/prot.10393.

Mendez, R., Leplae, R., Lensink, M. F., & Wodak, S. J. (2005). Assessment of CAPRI predictions in rounds 3–5 shows progress in docking procedures. *Proteins*, *60*(2), 150–169. https://doi.org/10.1002/prot.20551.

Moal, I. H., Moretti, R., Baker, D., & Fernandez-Recio, J. (2013). Scoring functions for protein-protein interactions. *Current Opinion in Structural Biology*, *23*(6), 862–867. https://doi.org/10.1016/j.sbi.2013.06.017.

Moal, I. H., Torchala, M., Bates, P. A., & Fernandez-Recio, J. (2013). The scoring of poses in protein-protein docking: Current capabilities and future directions. *BMC Bioinformatics*, *14*, 286. https://doi.org/10.1186/1471-2105-14-286.

Norn, C. H., Lapidoth, G., & Fleishman, S. J. (2017). High-accuracy modeling of antibody structures by a search for minimum-energy recombination of backbone fragments. *Proteins*, *85*(1), 30–38. https://doi.org/10.1002/prot.25185.

Petoukhov, M. V., & Svergun, D. I. (2005). Global rigid body modeling of macromolecular complexes against small-angle scattering data. *Biophysical Journal*, *89*(2), 1237–1250 (pii) https://doi.org/10.1529/biophysj.105.064154.

Petoukhov, M. V., & Svergun, D. I. (2007). Analysis of X-ray and neutron scattering from biomacromolecular solutions. *Current Opinion in Structural Biology*, *17*(5), 562–571 (pii) https://doi.org/10.1016/j.sbi.2007.06.009.

Petoukhov, M. V., & Svergun, D. I. (2013). Applications of small-angle X-ray scattering to biomacromolecular solutions. *The International Journal of Biochemistry & Cell Biology*, *45*(2), 429–437. https://doi.org/10.1016/j.biocel.2012.10.017.

Pons, C., D'Abramo, M., Svergun, D. I., Orozco, M., Bernado, P., & Fernandez-Recio, J. (2010). Structural characterization of protein-protein complexes by integrating computational docking with small-angle scattering data. *Journal of Molecular Biology*, *403*(2), 217–230 (pii) https://doi.org/10.1016/j.jmb.2010.08.029.

Puchades, C., Kűkrer, B., Diefenbach, O., Sneekes-Vriese, E., Juraszek, J., Koudstaal, W., et al. (2019). Epitope mapping of diverse influenza hemagglutinin drug candidates using HDX-MS. *Scientific Reports*, *9*(1), 1–10.

Putnam, C. D., Hammel, M., Hura, G. L., & Tainer, J. A. (2007). X-ray solution scattering (SAXS) combined with crystallography and computation: Defining accurate macromolecular structures, conformations and assemblies in solution. *Quarterly Reviews of Biophysics*, *40*(3), 191–285 (pii) https://doi.org/10.1017/S0033583507004635.

Rambo, R. P., & Tainer, J. A. (2011). *Characterizing flexible and intrinsically unstructured biological macromolecules by SAS using the Porod-Debye law [Research Support]*. N.I.H. Extramural.

Rambo, R. P., & Tainer, J. A. (2013a). *Accurate assessment of mass, models and resolution by small-angle scattering [Research Support]*. N.I.H., Extramural.

Rambo, R. P., & Tainer, J. A. (2013b). Super-resolution in solution x-ray scattering and its applications to structural systems biology. *Annual Review of Biophysics*, *42*, 415–441. https://doi.org/10.1146/annurev-biophys-083012-130301.

Rantos, V., Karius, K., & Kosinski, J. (2022). Integrative structural modeling of macromolecular complexes using Assembline. *Nature Protocols*, *17*(1), 152–176.

Ritchie, D. W. (2008). Recent progress and future directions in protein-protein docking. *Current Protein & Peptide Science*, *9*(1), 1–15. http://www.ncbi.nlm.nih.gov/entrez/query.fcgi?cmd=Retrieve&db=PubMed&dopt=Citation&list_uids=18336319.

Ruffolo, J. A., Chu, L.-S., Mahajan, S. P., & Gray, J. J. (2022). Fast, accurate antibody structure prediction from deep learning on massive set of natural antibodies. *bioRxiv*. https://doi.org/10.1101/2022.04.20.488972.

Ruffolo, J. A., Guerra, C., Mahajan, S. P., Sulam, J., & Gray, J. J. (2020). Geometric potentials from deep learning improve prediction of CDR H3 loop structures. *Bioinformatics*, *36*(Suppl. 1), i268–i275. https://doi.org/10.1093/bioinformatics/btaa457.

Ruffolo, J. A., Sulam, J., & Gray, J. J. (2022). Antibody structure prediction using interpretable deep learning. *Patterns*, *3*(2), 100406. https://doi.org/10.1016/j.patter.2021.100406.

Russel, D., Lasker, K., Webb, B., Velazquez-Muriel, J., Tjioe, E., Schneidman-Duhovny, D., et al. (2012). Putting the pieces together: Integrative modeling platform software for structure determination of macromolecular assemblies. *PLoS Biology*, *10*(1), e1001244. https://doi.org/10.1371/journal.pbio.1001244.

Sali, A., & Blundell, T. L. (1993). Comparative protein modelling by satisfaction of spatial restraints. *Journal of Molecular Biology*, *234*(3), 779–815 (pii) https://doi.org/10.1006/jmbi.1993.1626.

Schaefer, G., Haber, L., Crocker, L. M., Shia, S., Shao, L., Dowbenko, D., et al. (2011). A two-in-one antibody against HER3 and EGFR has superior inhibitory activity compared with monospecific antibodies. *Cancer Cell*, *20*(4), 472–486. https://doi.org/10.1016/j.ccr.2011.09.003.

Schmiedel, J., Blaukat, A., Li, S., Knochel, T., & Ferguson, K. M. (2008). Matuzumab binding to EGFR prevents the conformational rearrangement required for dimerization. *Cancer Cell*, *13*(4), 365–373. https://doi.org/10.1016/j.ccr.2008.02.019.

Schneidman-Duhovny, D., & Hammel, M. (2018). Modeling structure and dynamics of protein complexes with SAXS profiles. In *Protein complex assembly* (pp. 449–473). New York, NY: Humana Press.

Schneidman-Duhovny, D., Hammel, M., & Sali, A. (2011). Macromolecular docking restrained by a small angle X-ray scattering profile. *Journal of Structural Biology*, *173*(3), 461–471. https://doi.org/10.1016/j.jsb.2010.09.023 (pii).

Schneidman-Duhovny, D., Hammel, M., Tainer, J. A., & Sali, A. (2013). Accurate SAXS profile computation and its assessment by contrast variation experiments. *Biophysical Journal*, *105*(4), 962–974. https://doi.org/10.1016/j.bpj.2013.07.020.

Schneidman-Duhovny, D., Hammel, M., Tainer, J. A., & Sali, A. (2016). FoXS, FoXSDock and MultiFoXS: Single-state and multi-state structural modeling of proteins and their complexes based on SAXS profiles. *Nucleic Acids Research*, *44*(W1), W424–W429. https://doi.org/10.1093/nar/gkw389.

Schneidman-Duhovny, D., Inbar, Y., Nussinov, R., & Wolfson, H. J. (2005). PatchDock and SymmDock: Servers for rigid and symmetric docking. *Nucleic Acids Research*, *33*(Web Server issue), W363–W367 (pii) https://doi.org/10.1093/nar/gki481.

Schneidman-Duhovny, D., Pellarin, R., & Sali, A. (2014). Uncertainty in integrative structural modeling. *Current Opinion in Structural Biology*, *28C*, 96–104. https://doi.org/10.1016/j.sbi.2014.08.001.

Schneidman-Duhovny, D., Rossi, A., Avila-Sakar, A., Kim, S. J., Velazquez-Muriel, J., Strop, P., et al. (2012). A method for integrative structure determination of protein-protein complexes. *Bioinformatics*, *28*(24), 3282–3289. https://doi.org/10.1093/bioinformatics/bts628.

Sela-Culang, I., Alon, S., & Ofran, Y. (2012). A systematic comparison of free and bound antibodies reveals binding-related conformational changes. *Journal of Immunology*, *189*(10), 4890–4899. https://doi.org/10.4049/jimmunol.1201493.

Sela-Culang, I., Benhnia, M. R., Matho, M. H., Kaever, T., Maybeno, M., Schlossman, A., et al. (2014). Using a combined computational-experimental approach to predict antibody-specific B cell epitopes. *Structure*, *22*(4), 646–657. https://doi.org/10.1016/j.str.2014.02.003.

Sircar, A., & Gray, J. J. (2010). SnugDock: Paratope structural optimization during antibody-antigen docking compensates for errors in antibody homology models. *PLoS Computational Biology*, *6*(1), e1000644. https://doi.org/10.1371/journal.pcbi.1000644.

Sivasubramanian, A., Sircar, A., Chaudhury, S., & Gray, J. J. (2009). Toward high-resolution homology modeling of antibody Fv regions and application to antibody-antigen docking. *Proteins*, *74*(2), 497–514. https://doi.org/10.1002/prot.22309.

Stanfield, R. L., Zemla, A., Wilson, I. A., & Rupp, B. (2006). Antibody elbow angles are influenced by their light chain class. *Journal of Molecular Biology*, *357*(5), 1566–1574 (pii) https://doi.org/10.1016/j.jmb.2006.01.023.

Tiede, D. M., Mardis, K. L., & Zuo, X. (2009). X-ray scattering combined with coordinate-based analyses for applications in natural and artificial photosynthesis. *Photosynthesis Research*. https://doi.org/10.1007/s11120-009-9475-6.

Trewhella, J. (2016). Small-angle scattering and 3D structure interpretation. *Current Opinion in Structural Biology*, *40*, 1–7. https://doi.org/10.1016/j.sbi.2016.05.003.

Tully, M. D., Tarbouriech, N., Rambo, R. P., & Hutin, S. (2021). Analysis of SEC-SAXS data via EFA deconvolution and scatter. *JoVE (Journal of Visualized Experiments)*, *167*, e61578.

Weitzner, B. D., & Gray, J. J. (2017). Accurate structure prediction of CDR H3 loops enabled by a novel structure-based C-terminal constraint. *Journal of Immunology*, *198*(1), 505–515. https://doi.org/10.4049/jimmunol.1601137.

Weitzner, B. D., Jeliazkov, J. R., Lyskov, S., Marze, N., Kuroda, D., Frick, R., et al. (2017). Modeling and docking of antibody structures with Rosetta. *Nature Protocols*, *12*(2), 401–416. https://doi.org/10.1038/nprot.2016.180.

Whitten, A. E., & Trewhella, J. (2009). Small-angle scattering and neutron contrast variation for studying bio-molecular complexes. *Methods in Molecular Biology*, *544*, 307–323. https://doi.org/10.1007/978-1-59745-483-4_20.

Wu, S. L., Kim, J., Bandle, R. W., Liotta, L., Petricoin, E., & Karger, B. L. (2006). Dynamic profiling of the post-translational modifications and interaction partners of epidermal growth factor receptor signaling after stimulation by epidermal growth factor using extended range proteomic analysis (ERPA). *Molecular & Cellular Proteomics*, *5*(9), 1610–1627. https://doi.org/10.1074/mcp.M600105-MCP200.

Xia, B., Mamonov, A., Leysen, S., Allen, K. N., Strelkov, S. V., Paschalidis, I., et al. (2015). Accounting for observed small angle X-ray scattering profile in the protein-protein docking server ClusPro. *Journal of Computational Chemistry*, *36*(20), 1568–1572. https://doi.org/10.1002/jcc.23952.

Xiang, Y., Nambulli, S., Xiao, Z., Liu, H., Sang, Z., Duprex, W. P., et al. (2020). Versatile and multivalent nanobodies efficiently neutralize SARS-CoV-2. *Science*, *370*(6523), 1479–1484. https://doi.org/10.1126/science.abe4747.

Xiang, Y., Sang, Z., Bitton, L., Xu, J., Liu, Y., Schneidman-Duhovny, D., et al. (2021). Integrative proteomics identifies thousands of distinct, multi-epitope, and high-affinity nanobodies. *Cell Systems*. https://doi.org/10.1016/j.cels.2021.01.003.

Yip, K. M., Fischer, N., Paknia, E., Chari, A., & Stark, H. (2020). Atomic-resolution protein structure determination by cryo-EM. *Nature*, *587*(7832), 157–161. https://doi.org/10.1038/s41586-020-2833-4.

Zost, S. J., Gilchuk, P., Chen, R. E., Case, J. B., Reidy, J. X., Trivette, A., et al. (2020). Rapid isolation and profiling of a diverse panel of human monoclonal antibodies targeting the SARS-CoV-2 spike protein. *Nature Medicine*, *26*(9), 1422–1427.

CHAPTER NINE

Combining NMR, SAXS and SANS to characterize the structure and dynamics of protein complexes

Florent Delhommel[a,b], Santiago Martínez-Lumbreras[a,b], and Michael Sattler[a,b,*]
[a]Institute of Structural Biology, Helmholtz Zentrum München, Neuherberg, Germany
[b]Bavarian NMR Center, Department of Chemistry, Technical University of Munich, Garching, Germany
*Corresponding author: e-mail address: sattler@helmholtz-muenchen.de

Contents

Abstract

Understanding the structure and dynamics of biological macromolecules is essential to decipher the molecular mechanisms that underlie cellular functions. The description of structure and conformational dynamics often requires the integration of complementary techniques. In this review, we highlight the utility of combining nuclear magnetic

Methods in Enzymology, Volume 678
ISSN 0076-6879
https://doi.org/10.1016/bs.mie.2022.09.020

resonance (NMR) spectroscopy with small angle scattering (SAS) to characterize these challenging biomolecular systems. NMR can assess the structure and conformational dynamics of multidomain proteins, RNAs and biomolecular complexes. It can efficiently provide information on interaction surfaces, long-distance restraints and relative domain orientations at residue-level resolution. Such information can be readily combined with high-resolution structural data available on subcomponents of biomolecular assemblies. Moreover, NMR is a powerful tool to characterize the dynamics of biomolecules on a wide range of timescales, from nanoseconds to seconds. On the other hand, SAS approaches provide global information on the size and shape of biomolecules and on the ensemble of all conformations present in solution. Therefore, NMR and SAS provide complementary data that are uniquely suited to investigate dynamic biomolecular assemblies. Here, we briefly review the type of data that can be obtained by both techniques and describe different approaches that can be used to combine them to characterize biomolecular assemblies. We then provide guidelines on which experiments are best suited depending on the type of system studied, ranging from fully rigid complexes, dynamic structures that interconvert between defined conformations and systems with very high structural heterogeneity.

1. Introduction

Biological processes rely on intricate networks of interacting macromolecules such as proteins and nucleic acids. Structural biology aims at uncovering these processes by investigating the three-dimensional structure of macromolecules at atomic resolution, and by characterizing their interactions. Within the last 30 years, outstanding advances in biochemical methods and structural biology technologies, i.e. synchrotron radiation, ultra-high-field NMR spectrometers and the "resolution revolution" in cryo-electron microscopy, enabled the structural analysis of proteins and their complexes, covering about 30% of the human proteome. This wealth of data coupled with the increasing availability of full genome sequences from varied species enabled the development of machine learning tools for structure prediction (Baek et al., 2021; Jumper et al., 2021), that now extend the predictive structural coverage of human proteome to about 50% (Porta-Pardo, Ruiz-Serra, Valentini, & Valencia, 2022).

However, in humans; about a third of the proteome does not adopt a defined globular structure, and the majority of human proteins contain intrinsically disordered regions (IDRs), that play essential roles for biological functions and that are linked to human diseases. Eukaryotic proteins often encompass multiple domains connected by flexible regions that lack defined ternary structures and allow sampling of a large conformational space

(Davey, 2019). This allows to combine and synergize the functions of individual domains and greatly expands the possibility of regulation of molecular interactions in spite of the limited number of domain structures (Lunde, Moore, & Varani, 2007). This is further tuned by multidomain organization and by interdomain flexibly, transient interactions of linkers with globular domains and conformational switches (Mackereth & Sattler, 2012). Finally, it is now well established that intrinsically disordered regions are essential for the regulation of many proteins, as they contain conserved motifs, post translation modification (PTM) sites and can regulate protein localization and function by liquid-liquid phase separation (Banani, Lee, Hyman, & Rosen, 2017; Borcherds, Bremer, Borgia, & Mittag, 2021; Davey, 2019).

The study of dynamic biomolecular assemblies is challenging and individual techniques are often not sufficient to accurately describe their behavior in solution: X-ray crystallography requires a high degree of rigidity for crystal growth and is limited to highly structured regions of proteins. Cryogenic electron microscopy (cryo-EM) provides an environment closer to solution but still requires high sample homogeneity. In addition, class averaging can lead to the selection of a specific conformation to the detriment of others. The characterization of conformational ensembles from cryo-EM is possible, but remains highly challenging (Bock et al., 2013). Deep-learning methods for structure prediction provide single conformations of proteins based on coevolution but do not account for conformational dynamics or the presence of stabilizing cofactors and PTMs and consequently often predict erroneously the conformations of isolated, nonmodified proteins (Perrakis & Sixma, 2021). Finally, nuclear magnetic resonance spectroscopy (NMR) can characterize conformational dynamics in solution using a wide range of experimental approaches (Alderson & Kay, 2021), but is challenged by the molecular weight of the object studied in solution. Consequently, the faithful description of the dynamics and flexibility of proteins in solution requires the integration of data acquired from several techniques that provide information at multiple time scales and variable spatial resolution (Rout & Sali, 2019).

The integration of NMR with small angle scattering (SAS) methods, such as small angle X-ray (SAXS) or neutron scattering (SANS) has been particularly powerful for the study of dynamic multidomain proteins and complexes that comprise structured and disordered regions and for which the interactions are transient (Delhommel, Gabel, & Sattler, 2020). In this chapter, we first summarize the information that can be gained from NMR experiments. We then show how it can be used to generate or validate structural models, how to obtain long-range distance and orientational information on larger complexes and how it provides information on dynamics from picosecond to second timescales. We then present the information on global structure that can be extracted from SAXS and the benefits of using SANS for the study of hetero-oligomeric macromolecular complexes. With recent examples we demonstrate and highlight the utility of integrative structural biology of these solution techniques, and discuss different computational approaches to describe conformational ensembles of such dynamic systems.

2. Sample preparation

2.1 Sample requirements

NMR and SAS share various features: first, they can be applied to samples in aqueous solution, where molecules are free to tumble and second, their observed signals average existing conformational states in solution. Both methods require rather concentrated samples, typically in the range of a few mg/mL, with volumes of 50 μL for SAXS or 200 μL for SANS and >150 μL for NMR.

The efficient production and purification of recombinant proteins in high yield is therefore a prerequisite for studies involving these techniques, particularly for NMR and SANS where isotope labeling is required. The presence of rare codons in the recombinant gene can strongly affect expression level, particularly for longer constructs; thus it is advisable to use insert sequences that are codon optimized for the specific expression system. The choice of expression vector is also important to avoid potential leaking expression of the recombinant protein; the series of pETM vectors developed at the EMBL (Dümmler, Lawrence, & de Marco, 2005) are particularly well suited for high protein expression in bacterial systems. The use of specific bacterial strains can also drastically improve the expression of unoptimized sequences (Codon+, Rosetta cells), toxic sequences (pLysS) or disulfide brides containing proteins (Origami). Bacterial expression is not always possible for all human proteins, particularly for large, multidomain proteins and complexes. High yield can also be achieved in eukaryote expression systems and most notably using insect cells (Smith, Summers, & Fraser, 1983) infected with baculovirus containing FastBac vectors for single protein expression or MacroBac vectors for multiprotein complexes (Gradia et al., 2017). Finally, the use of a SUMO-tag can drastically improve protein production yields, in addition to simplifying later steps of purification protocols (Brondyk, 2009).

Recombinant proteins intended for any structural biology technique require high purity and homogeneity. It is usually recommended to purify proteins using three different types of chromatography: first an affinity chromatography would separate the protein of interest from the expression system background. The affinity tag is then removed using a specific protease and the protein of interests is isolated by ion-exchange purification, to remove potential contaminants. Finally, the size exclusion chromatography (SEC) step will remove potential aggregates and separate different oligomeric states. Ideally, the homogeneity of the final sample should be evaluated using SDS-PAGE, SLS and/or DLS. Proteins stored in long-term at 4 or −80 °C can also be affected by oligomerization and/or degradation of the protein of interest, particularly when their C-terminus is unstructured. Therefore, the quality of samples should also be evaluated after conservation and just before performing the experiments.

In the case of nucleic acids, production of pure samples at high yield is also required, although SAXS benefits from about fourfold higher scattering of nucleic acids compared to proteins due to the electron-rich phosphate backbone. Chemical synthesis is typically used for short oligonucleotides, while longer RNA sequences can be readily obtained by *in vitro* transcription using recombinant T7 polymerase (Baronti, Karlsson, Marušič, & Petzold, 2018; Edelmann, Niedner, & Niessing, 2014). It is important to analyze sample homogeneity after the refolding process; the presence of different folding pathways depending on buffer and experimental conditions needs to be considered when analyzing data (Jones & Sattler, 2019; Rambo & Tainer, 2010). A general concern with smaller RNA oligonucleotides is that monomeric hairpin structures tend to dimerize in solution at high concentrations and long incubation times, as for example required for crystallization. Therefore, thermal melting followed by snap cooling is an important aspect of sample preparation to avoid the presence of dimeric species (Duszczyk, Zanier, & Sattler, 2008).

Finally, for protein-protein or protein-nucleic acid complexes, it is critical to ensure complex homogeneity by size exclusion chromatography, and ideally an extensive biophysical characterization of the complex stoichiometry, binding affinities, and kinetic interconversion rates.

2.2 Isotope labeling

Biomolecular NMR requires the isotopic labeling of atoms of interests. Most commonly, ^{15}N and/or ^{13}C are uniformly introduced to proteins by using minimum media, such as M9 medium for *E. coli* expression, in which the sole source of these atoms is already labeled. This approach results in slightly lower yields than expression in rich media and is also more expensive due to the cost of precursors, thus expression optimization is critical. ^{2}H labeling is beneficial and required for NMR studies of larger proteins (above 30 kDa), as it improves sensitivity by reducing NMR relaxation that causes line-broadening (Gardner & Kay, 1998). Deuteration requires protein expression in D_2O solution that profoundly affects the metabolism of bacterial host cells, which need to be adapted to heavy water by successive precultures with increasing concentrations of D_2O. Deuterated proteins are also of particular interest in SANS, to perform contrast matching and specifically observe the scattering curve of a certain component of a complex while masking others. It is possible to recycle deuterated culture medium multiple times to reduce costs, although this leads to dilution of ^{2}H. Isotope labeling is also possible in eukaryotic cells, i.e. insect and mammalian cells, albeit with more complex protocols and higher costs (Franke et al., 2018). Nevertheless, this is an active area of research given that some proteins cannot be expressed and purified from bacterial cells.

Protocols have been established to segmentally label proteins. In this approach, two sections of a protein are expressed and purified independently with the desired labeling scheme, typically with one labeled and the other not. The final construct is then reconstituted using transpeptidases, such as Sortase A, or protein trans splicing, which will ligate the two segments into one (Clark, Sievers, & Debelouchina, 2022). This approach introduces additional non-native residues in between segments that will remain after the reaction. It is thus essential to ensure that the ligation region is well accessible and that the introduction of the ligation motif does not affect the structure and function of the protein. Typically, the ligation site is introduced in a nonconserved region of a solvent-exposed linker between domains of interest.

Lastly, to optimize specific NMR applications, for example for measuring dynamics in large protein complexes, sophisticated site-specific labeling approaches have been developed. These include amino acid and site-specific labeling (Lacabanne, Meier, & Böckmann, 2018), but most prominently the selective and stereoselective methyl ^{1}H,^{13}C-labeling in an otherwise deuterated protein (Kerfah, Plevin, Sounier, Gans, & Boisbouvier, 2015). The specific spectroscopic properties of methyl groups allow to perform NMR experiments with proteins with much higher molecular weights than what is permitted by uniform labeling approaches (Schütz & Sprangers, 2020).

This has broadly extended the use of NMR to megadalton complexes, that could not be studied prior to this development. The major drawback is the loss of residue-level resolution to up to about a third of the protein sequence, comprising methyl-containing amino acids.

Isotope labeling of RNA can be achieved by chemical synthesis with the possibility to design site-specific labeling (Kawahara et al., 2012). By *in vitro* transcription, it is possible to do uniformly or nucleotide-selective labeling, while segmental isotope labeling is readily available through enzymatic ligation of oligonucleotides, for example using T4 ligases (Duss, Maris, Von Schroetter, & Allain, 2010).

3. NMR spectroscopy

NMR is a broadly applicable and versatile technique for the study of biomolecules. While it allows high-resolution *de novo* structure determination of proteins and protein complex of up to 30 kDa, it can efficiently provide unique residue-level information on binding interfaces and conformational dynamics even for large protein complexes, up to hundreds of kilodalton molecular weight (Alderson & Kay, 2021). There is a wide range of NMR tools to study the conformation, interactions and dynamics of biological macromolecules from picosecond to second timescales. In the following, the most commonly used methods in NMR are briefly described (Fig. 1), while for further details we refer to dedicated relevant reviews.

3.1 Structural analysis

High-resolution structure determination by NMR has been extensively used for domains of protein, nucleic acids and small complexes, albeit to a lesser extent than X-ray crystallography, due to its challenging procedures and size limitations. It involves the combination of different data-derived restraints into computational calculations to produce a structural ensemble that fulfills all applied restraints.

The chemical shift (CS) is the most direct and sensitive NMR observable. It senses the magnetic environment of the observed nuclear spin and is therefore affected by its surrounding atoms. It is very useful for characterizing the local conformation of a biomolecule: in proteins, the secondary chemical shift, i.e. the deviation of a given chemical shift from values observed in a random coil conformation, provides unique information on backbone torsion angles, defining the secondary structure of the polypeptide

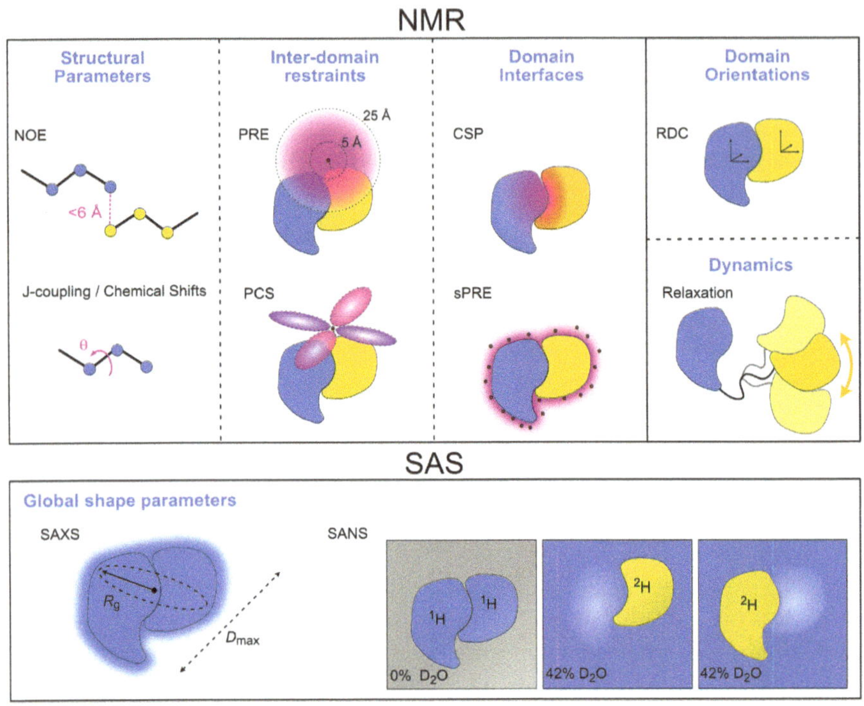

Fig. 1 Different structural parameters that can be measured by NMR and SAS, details are described in the text.

chains (Mielke & Krishnan, 2009; Wishart & Sykes, 1994). Similarly, in nucleic acids, the secondary structure and base paring can be determined by the presence of imino-proton signals and the conformation of the sugar pucker and the phosphodiester bond can be obtained based on the CS values of the involved nuclei (Fürtig, Richter, Wöhnert, & Schwalbe, 2003). Information about torsion angles is also available from scalar coupling constants across the three bonds (^3J), the size of which is modulated as described by Karplus curves. For high-resolution structure calculations, the nuclear Overhauser effect (NOE), which results from cross-relaxation between dipolar coupled spins, provides essential distance-dependent information. The magnitude of the NOE is directly related to the distance between atoms, providing useful proton-proton distance restraints up to 6–8 Å. The measurement of NOE is challenged by signal overlap and line-broadening with increasing molecular weight, i.e. total number of proton spins. Methyl-specific isotope-labeling schemes can be applied for larger molecules to reduce signal overlap and line-broadening but at the cost of a reduced number of restraints that can be obtained, limited to methyl-containing side chains.

Complementary to the NOEs, residual dipolar couplings (RDC), that can be observed in anisotropic solution by weakly aligning a biomolecule, can provide orientational restraints. Various options are available for partial alignment of the biomolecule with respect to the external magnetic field, such as dilute solutions of filamentous phages, polyacrylamide gels or liquid crystalline media (Bax, Kontaxis, & Tjandra, 2001; Blackledge, 2005). The RDC is most sensitively measured between directly bonded backbone atoms in proteins (^{1}H-^{15}N) or base and sugar atoms (^{1}H-^{13}C) in nucleic acids (Latham, Brown, McCallum, & Pardi, 2005) and provides information about the orientation of the magnetic dipole-dipole interaction vector between the pair of nuclei within a common reference frame, related to the axis of the external magnetic field. The analysis of multiple RDCs for a given rigid structural module (i.e. peptide bond, or a domain within a multidomain protein or protein complex) provides information on the relative orientation (Blackledge, 2005) and can also be used to determine the relative orientational arrangement of folded regions, i.e. domains (Simon, Madl, Mackereth, Nilges, & Sattler, 2010).

Distances, torsion angles and orientational restraints are routinely used in restrained molecular dynamics structure calculations. Available software, such as CYANA, Xplor-NIH, ARIA2 or CNS (Güntert & Buchner, 2015; Rieping et al., 2007; Schwieters, Kuszewski, & Marius Clore, 2006), implement the NMR-derived restraints into torsion and/or cartesian dynamics and simulated annealing protocols that yield a structural ensemble that fulfills the experimental restraints. Current efforts aim to the automation of the process and thus minimize user input and computation time (Allain, Mareuil, Ménager, Nilges, & Bardiaux, 2020; Pritišanac, Alderson, & Güntert, 2020). High-resolution structures of complexes are obtained through a combination of torsion angle, orientational and short distance restraints of each component and additional intermolecular distance restraints. Unambiguous intermolecular NOEs can be used to solve structures of complexes and normally are obtained via differential labeling of the samples and the use of isotope-edited and filtered experiments (Gross, Gelev, & Wagner, 2003; Sattler, Schleucher, & Griesinger, 1999). For larger protein complexes, NOE measurements can be quite challenging and focus on NOEs involving methyl groups, therefore requiring the use of additional experimental restraints, i.e. from RDCs or paramagnetic NMR experiments.

Finally, it is also possible to use NMR observables for the validation of structural models based on other techniques or on computational predictions. As we will see later, efficient structural validation is possible by combining NMR with SAXS data.

3.2 Binding studies and domain interactions

Upon addition of the binding partner, a biomolecule experiences changes in its chemical environment, either exclusively affecting the binding interface regions or extending to other regions of the molecule by conformational changes or allosteric effects. In any case, these changes can be readily detected by NMR. Upon addition of increasing amounts of a titrant, changes in the chemical environment of a given nuclear spin yield chemical shift perturbations (CSPs), i.e. changes in NMR resonance frequencies, which can be monitored and quantified, typically in ^{1}H-^{15}N or ^{1}H-^{13}C correlation experiments. These titration experiments are commonly used for the analysis of the binding sites and interfaces in biomolecules but also to qualitatively determine the affinity of the interaction (Williamson, 2013). Two extreme situations can be observed during a titration depending on the kinetic exchange rate (k_{ex}) of the binding ($k_{ex} = k_{on}[L] + k_{off}$, [L] being the ligand concentration) and the difference of the NMR resonance frequencies ($\delta\nu$) of the free and bound state of the protein: fast and slow exchange regimes, which produce different types of titration spectra. In the first case, a single signal is observed at the population-weighted average chemical shift; in the slow exchange, the signals of both states are observed with intensities reflecting the respective population. The quantitative analysis of the interaction and dissociation constant derivation can be done for the fast exchange scenario following both the chemical shift changes and the line shape of the signals (Waudby, Ramos, Cabrita, & Christodoulou, 2016). In addition, NMR CSP analysis can also be applied to identify intramolecular interactions between individual domains or regions of a protein. The comparison of spectra of the isolated domains with the multidomain constructs in the same experimental conditions allows to determine interfaces, even for very transient interactions. This is a unique feature of NMR respect to other structural biology methods, therefore several approaches have been implemented for the study of domain arrangements and conformational dynamics in multidomain proteins and their complexes (Göbl, Madl, Simon, & Sattler, 2014).

NMR can be used for structural analysis of larger complexes or multi-domain proteins, also in situations where it is difficult to identify intermolecular or interdomain NOEs. In these cases, RDCs, paramagnetic relaxation enhancement (PRE) and pseudo contact shift (PCS) data can be measured to determine long-range distances and further orientational restraints (Hass & Ubbink, 2014). For this, the biomolecule is modified to introduce the paramagnetic specie; typically by introducing cysteines in proteins or thiouridines in nucleic acids at specific sites (Hennig et al., 2015). Several paramagnetic tags are available to conjugate with the thio groups to incorporate a site-specific paramagnetic center: nitroxyl radicals with an unpaired electron are commonly used for PRE measurement, while lanthanide containing tags, such DOTA cages, allow PCS measurement, but also produce PRE and RDC data (Joss, Walliser, Zimmermann, & Häussinger, 2018). In both cases a reference diamagnetic sample has to be produced for comparison.

PRE effects result from spin-spin relaxation between the unpaired electron of the paramagnetic center and the nuclear spin observed by NMR. The PRE can be extracted by comparison of the proton relaxation rates between the paramagnetic and the reference diamagnetic sample and is a function of the distance between the unpaired electron and the nucleus, up to 25 Å for nitroxyl radicals (Clore, 2015). When the paramagnetic center possesses anisotropic magnetic susceptibility tensor, i.e. for lanthanides, a partial magnetic alignment of the particle in the magnetic field is produced, allowing RDC measurements. Moreover, dipolar interaction between the electron and the nuclear spins also results in changes of the NMR chemical shifts, known as PCSs. These can be very efficiently measured by comparing the spectra of the paramagnetic and diamagnetic versions of the tag and depend on the distance and orientation of the vector between the electron and the observed nucleus with respect to the external magnetic field. Therefore, the long-range distance and orientational information obtained from PREs, PCSs and RDCs is extremely useful to locate and orient the different components of a complex, when their individual structures are known (Kuenze, Bonneau, Leman, & Meiler, 2019; Simon et al., 2010). In addition, paramagnetic restraints have unique potential for drug discovery (Softley, Bostock, Popowicz, & Sattler, 2020). It is important to note that PRE and PCS are extremely sensitive to dynamics and the presence of multiple conformations (Clore, 2015), and their use for structural analysis is therefore only advised for rigid complexes.

Solvent PRE (sPRE) data, i.e. PRE effects induced by soluble, inert, paramagnetic molecules designed not to interact with the biomolecule (Madl, Bermel, & Zangger, 2009), allow to map the solvent accessible surface of a biomolecule, and can thus be used for structure validation of a model, for direct refinement during structure calculations (Hartlmüller et al., 2017) and for structural determination of complexes, typically in combination with additional NMR data (Gong, Gu, Guo, Wang, & Tang, 2017; Madl, Güttler, Görlich, & Sattler, 2011).

3.3 Conformational dynamics

Biomolecules in solution are often dynamic and experience changes in their conformations. Solution techniques, such as NMR and SAS, are directly sensitive to such dynamics. In fact, there is a vast repertoire of NMR tools to analyze the dynamics of biomolecules, covering the different requirements of each system. Standard NMR relaxation measurements of longitudinal (R_1) and transversal (R_2) relaxation rates together with the heteronuclear cross-relaxation, i.e. $\{^1H\}$-^{15}N NOEs, provide information about the rotational diffusion of the molecule, and internal motions in the ps to ns range at residue-level resolution (Chao & Byrd, 2018; Mittermaier & Kay, 2006). These experiments provide direct information about conformational flexibility, for example, flexible terminal regions, linkers connecting structural domains and intrinsically disordered proteins (IDPs).

NMR relaxation dispersion measurements directly measure conformational dynamics at μs-ms time scales in proteins and nucleic acids (Fig. 2) (Palmer, 2014). Relaxation dispersion approaches can detect lowly populated conformational states that cannot be directly detected and may exist in dynamic equilibrium with a major ground state conformation that is observed by NMR or other structural biology techniques. Two experimental approaches have been developed to measure these hidden states, namely the Carr-Purcell-Meiboom-Gill (CPMG) experiment and the $R_{1\rho}$ experiments. Fitting of the NMR relaxation data to typically two-state models provide information on the chemical shift difference between the two states

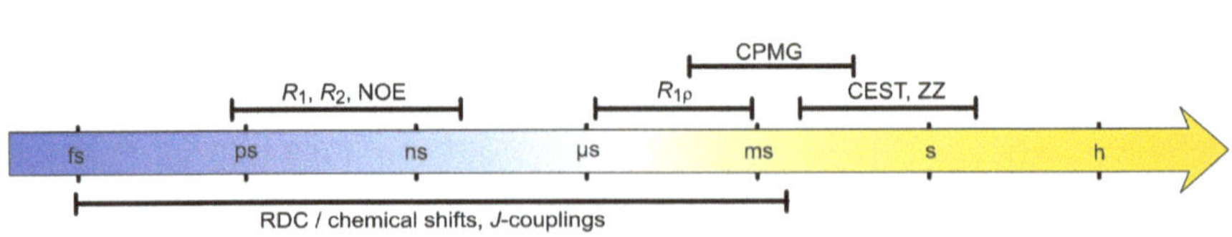

Fig. 2 NMR experiments and parameters to analyze conformational dynamics in biomolecules over a wide range of timescales.

and their exchange rates (Bothe et al., 2011; Loria, Rance, & Palmer, 1999). In addition, RDCs, but also scalar J-couplings, are averaged by internal motions up to millisecond timescales and thus are sensitive to structural dynamics at μs-ms time scales which are related to biological function (Al-Hashimi, 2013; Bailor et al., 2007; Igumenova, Brath, Akke, & Palmer, 2007). For slower timescales, in the millisecond regime, Chemical Exchange Saturation Transfer (CEST) experiments can provide information about the presence and population of a minor state by scanning saturating frequencies across the spectral range (Vallurupalli, Bouvignies, & Kay, 2012). Slower exchange processes, when both states are sufficient populated, can be characterized by ZZ-exchange experiments to directly fit the kinetic exchange rates between the two states (Jeener, Meier, Bachmann, & Ernst, 1979; Li & Palmer, 2009). These approaches highlight the strength of NMR in the study of dynamical systems, as they allow the complete characterization of exchange with even extremely minor states, down to a few percents.

4. Small angle X-ray and neutron scattering (SAXS/SANS)

In the last 20 years, small angle scattering approaches have become increasingly popular for the study of biological macromolecules (Brosey & Tainer, 2019; Mertens & Svergun, 2010). This was particularly driven by the general improvement of detection systems, the better automatization of data acquisition, a larger availability of beamline and bench-top SAXS apparatuses and the apparition of new software for the analysis of scattering profiles of macromolecules.

SAS relies on the scattering of an incident beam of X-rays or neutrons on a solution of isotropically oriented macromolecules. The scattering pattern is integrated into a one-dimensional plot *I(Q)* of the intensity with respect to the scattering vector, related to the angle to the incoming beam. The bulk solvent itself is by far the main component of the macromolecular solution, consequently its scattering has to be subtracted to obtain the curve resulting from the macromolecule and its hydration shell. SAS measurements observe the average over all conformations present in solution, over several seconds and are theoretically not limited by the size of the observed macromolecule. Consequently, SAS methods are particularly suited to investigate flexibility and conformational changes. However, SAS is particularly sensitive to sample inhomogeneity and aggregates. It is thus essential to ensure that the solution is monodisperse and does not show aggregation in the concentration required for SAS. To avoid these issues, size exclusion chromatography coupled to SAXS (SEC-SAXS) has been developed and is now available at most beamlines (David & Pérez, 2009). This approach has many advantages: (1) removal of aggregates and/or different oligomeric states that would affect scattering, (2) separation of unbound components of a complex, (3) ensure that the buffer used as a reference is identical to the one containing the macromolecule in the selected frames.

4.1 Structural information from SAXS

SAXS experiments can be performed in a few seconds per sample using high brilliance synchrotron sources or alternatively with bench-top sources and longer exposure of a few minutes to half an hour. With a single measurement of SAXS, low-resolution structural information about conformations existing in solution can be obtained from various parameters (Fig. 1). First, the Guinier approximation at very small angle (Guinier, 1939), provides the radius of gyration (R_g) of the macromolecule. In practice, the R_g is a measure of the overall compactness and size of the observed macromolecule. The $I(0)$ is also extrapolated from this technique and can be used to estimate the molecular weight of globular proteins, if the exact concentration of the sample is known. A different method of molecular weight determination has been proposed, based on the R_g and the ratio of the $I(0)$ with the total scattered intensity. This method has the advantage of not relying on any assumption on the biomolecule and of being independent from sample concentration, making it applicable in a SEC-SAXS setup (Rambo & Tainer, 2013a). The conversion to real-space of the scattering profile by Fourier transform shows the pairwise distance distribution of the macromolecule, $P(r)$. Its profile is directly linked to the averaged conformations of the macromolecule. The maximum observable interatomic distance (D_{max}) of the $P(r)$ reports on the extension of the largest observable conformation (Bernado, 2010). The dimensionless Kratky representation of the scattering data is particularly useful to identify and compare the presence of unstructured regions by enhancing the scattering intensities at larger angles (Durand et al., 2010). For globular proteins the Kratky representation is bell shaped, while for completely unstructured proteins increasing values are observed at larger angles. This can be used to determine qualitatively the compactness and flexibility of a macromolecule, using only the raw data. Finally, the direct relation between interatomic distances and the scattering profile, allows to calculate directly the theoretical diffraction curve from a 3D structure. The hydration shell also has a non-negligible contribution to the scattering and has to be taken into account explicitly or implicitly during the back-calculation of the structure (Svergun et al., 1998). This is now routinely done in most modern software. Different types of biomolecules have different X-ray scattering contrast with regard to the solvent. Therefore, for a mixture composed of more than one type of macromolecules, such as protein/nucleic acid complex or proteins embedded in membrane, the contribution of the individual components to the overall scattering has to be taken into account during the analysis.

Various software tools have been developed to use SAXS data to model the structure of macromolecular complexes in solution. There are two main applications depending on the conformational homogeneity of the sample. Therefore, an initial evaluation of the Kratky transformation is essential to determine which type of approach should be applied to a given sample. First, for homogeneous samples with low flexibility and for which a unique conformation of the macromolecule is responsible for the scattering, an *ab initio* shape consistent with the scattering profile of macromolecules can be generated. These approaches will either reduce a large sphere to a smaller volume with a specific shape, for which the back-calculated profile fit best to the measured data (DAMMIN) (Svergun, 1999), or fold a residue-like bead chain into a shape fitting best the data (GASBOR) (Svergun, Petoukhov, & Koch, 2001). All these approaches will provide a general volume that is consistent with scattering data and that represents the conformation average of a protein and its hydration shell in solution. However, the resulting models are qualitative and ambiguous (Petoukhov & Svergun, 2015). Alternatively, if high-resolution structures or models of individual domains and or subcomplexes are available, rigid-body docking supplemented by bead-chains can be used to fit the data (CORAL). This is particularly useful for complexes and multidomain proteins, with interacting domains. In this approach too, a single model is used to describe the data and thus requires a homogeneous sample with a single conformation (Manalastas-Cantos et al., 2021). For dynamic assemblies, complexes with significant flexible regions or for fully flexible molecules (IDPs), specific tools have been developed to describe the macromolecules as ensemble of conformations. These approaches can be used starting without prior structural information, or starting from the structure of folded regions complemented by flexible models (EOM) (Bernadó, Mylonas, Petoukhov, Blackledge, & Svergun, 2007; Kikhney & Svergun, 2015). The ensemble approach has to be used carefully: first, the ensemble is selected based on the quality of fit of its average and does not guarantee that any of the conformation are realistic. Secondly, ensemble fitting adds more freedom than with single structure, and is thus more likely to provide a better fit. As we will see later in this chapter, ensemble analysis often requires the addition of other restraints coming from additional experiments.

4.2 Contrast matching and structural analysis of subdomains from SANS

SANS is similar to SAXS in principle but relies on the scattering of neutrons rather than X-rays (Mahieu & Gabel, 2018). A drawback of neutron scattering is the much lower flux compared to X-rays that requires longer exposure for comparable sensitivity. In addition, a relatively small number of facilities are available for such experiments. However, the use of neutron scattering offers unique information, complementary to SAXS. First of all, the requirement for longer exposure is compensated by the lack of radiation damage that neutrons induce. Secondly, neutron scattering in H_2O is much less dependent on the hydration shell of proteins compared to SAXS and renders the back-calculation from 3D structures more accurate (Kim & Gabel, 2015). And finally, by far the most used characteristic of SANS is the possibility to contrast-match specific macromolecules. Contrast matching uses the large difference of scattering properties of ^1H and ^2H (deuterium, D) isotopes that encompass the average of protonated biomolecules composed of (C, O, N, S, P). Thus, specific H_2O/D_2O ratios can be used to mask specific components of a heterogeneous complex while obtaining the scattering profile of the others by matching exactly its scattering length density. For example, the signal of DNA or RNA can be match with values of $60\% \pm 3\%$ and $66\% \pm 3\%$ D_2O, respectively and with values of $42\% \pm 2\%$ D_2O for proteins (Mahieu & Gabel, 2018). The value depends on the exact atomic composition of the biomolecules and can be calculated online (http://psldc.isis.rl.ac.uk/Psldc/). It is also possible to increase the contrast of a biomolecule by its full or partial deuteration. In more advance methods, the domains of a given protein can be differentially labeled in order to mask only a portion of the protein (Sonntag et al., 2017). Thus, it is also possible to mask certain components of a homotypic complex by having some of them protonated and some deuterated. Often, one given complex is measured at different H_2O/D_2O ratio of the buffer and the shape of the complex is extracted using multiphase modeling.

5. Integration of NMR and SAS

As we have outlined above, NMR and SAS experiments provide complementary data that are uniquely suited to investigate the conformation and dynamics of protein complexes in solution: NMR is particularly efficient at providing local restraints, such as binding interfaces, local interatomic distances, dihedral angles and relative domain orientations. However, NMR spectroscopy becomes challenging with increasing molecular weight of the assembly studied. On the other hand, SAS has no size restriction and on the contrary; becomes more sensitive for larger objects. It provides global size and shape parameters that can be used to delineate the overall shape of an object. However, the capacity of SAS to sense local conformations is limited to low resolution. In this section of the chapter, we present how these two techniques can be combined to determine accurately the behavior of proteins in solution. We first show how these techniques can be used to validate and expand structures and models of proteins, an application that will surely become more prominent given the recent breakthroughs in deep learning approaches for structure prediction. We then exemplify how NMR and SAS can be used to describe the dynamics of proteins, an essential step to select which approaches should be chosen to model a given macromolecule. We describe how mainly rigid assemblies can be studied with NMR, SAXS and SANS to finally move on to the more complex description of various dynamic assemblies induced either by local flexibility or transient interactions.

5.1 Validation of structural models in solution

X-ray crystallography is still by far the largest source of structural information. In most cases, crystal structures are precise and accurate, however, the conformation of biomolecules can be affected by crystal packing, potentially producing erroneous conformations or selecting states that are not representative of the variability of the protein in solution. The same is also true for cryo-EM, in which only some conformations may survive the freezing process or in which the class selection can favor certain conformations. The combined use of NMR and SAXS is a fast and reliable tool to validate and refine structures that were not obtained in solution state. From the NMR point of view, RDCs in a single alignment medium can provide a fast and accurate validation of the local conformation of the backbone while

SAXS and SANS can validate the overall shape of the entity studied (Hennig et al., 2014). As an example, this protocol was used on the tandem RRM domains of hnRNP A1, which interact with RNA. The protein crystalized in a 2:2 complex where the RNA molecules orthogonally cross the side-by-side arrangement of two tandem RRMs. The authors showed that the RRM/RNA interfaces from the crystal structures are correct, but the intermolecular cross-bridge interactions with the RNA, which result in a 2:2 complex, is not present in solution. Therefore, a 1:1 complex was modeled based on the crystallographic domain structures and experimentally validated and refined using NMR RDC and SAXS data measured for the complex in solution (Kooshapur et al., 2018).

The recent breakthroughs in structure prediction by AlphaFold2 and RoseTTAFold are profoundly changing the field of structural biology (Baek et al., 2021; Jumper et al., 2021). However, even with the impressive increase in prediction accuracy, the models derived require validation from experimental data. Moreover, given that these algorithms have learned from the PDB, which is dominated by static crystal structures, representing proteins that have been optimized for crystallization, information about regulatory flanking regions and flexible linkers connecting domains are not well considered, requiring experimental analysis. The combination of NMR with SAXS can be a fast and reliable way to assess the validity of structural models, and also is an efficient method to determine the structure of complex of interest from available substructures (Mirdita et al., 2022). We expect that such approaches will be increasingly important and applied in the upcoming years.

5.2 Characterization of rigid and dynamic complexes

Before attempting to model a protein or protein complex, it is important to perform an initial characterization and, most notably, to evaluate the amount of flexibility of the system. This will be essential to help choosing a modeling strategy that is applicable and suited to the system and experimental data available (Fig. 3). The combination of NMR and SAXS is particularly efficient for determining global and local flexibility. Initially, NMR fingerprint spectra, i.e. $^{1}H,^{15}N$ correlations of a protein can give a general idea of the presence of a globular fold and rigidity. Spectra with a good peak dispersion in the proton dimension (roughly from 6 to 10 ppm) would indicate residues within a folded domain. On the contrary, a spectrum with a narrow amide protein chemical shift spread (8.2 ± 0.3 ppm) indicates a lack of globular fold

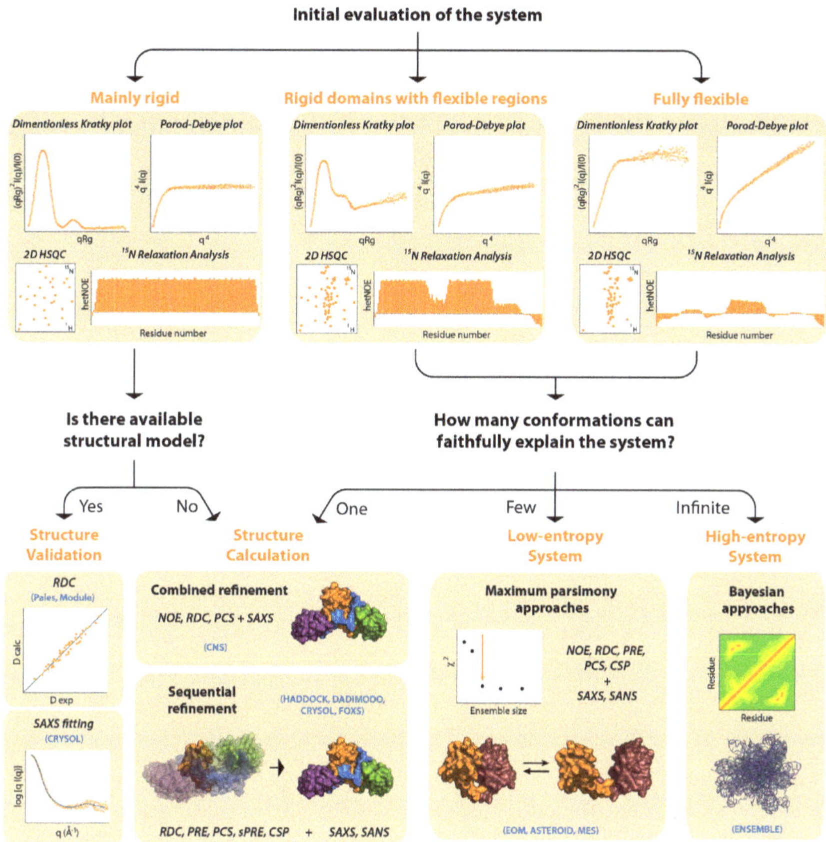

Fig. 3 Scheme showing the different applications of combining and NMR and SAXS analyses, depending on the dynamic behavior of a biomolecular system. Experimental data and (computational) approaches mentioned in the text are indicated.

and that all peptide bonds are exposed to solvent and not hydrogen-bonded. Homogenous line-widths of the signals in the $^1H,^{15}N$ correlation spectra are indicative of a well behaved globular folded protein, while presence of additional line-broadening is indicative of regions undergoing conformational exchange (Alderson & Kay, 2021). NMR relaxation experiments provide information on internal motion and flexibility at residue-level resolution. This is particularly important in the case of complexes of multidomain proteins to analyze the behaviors of the linkers. These relatively simple NMR approaches are efficient at characterizing local flexibility and dynamics but remain relatively blind to large domain motions.

SAXS can directly detect large scale flexibility from a number of techniques. The representation of SAXS data by the dimensionless Kratky plot will qualitatively indicate the amount of flexibility present in a macromolecule (Durand et al., 2010) and is particularly efficient for detecting highly flexible segments even within the presence of globular folded domains in a protein. Similarly, the Porod-Debye exponent analysis indicates the flexibility of macromolecule and highlights particularly well flexibility between folded regions (Rambo & Tainer, 2011). It is advised to combine these two approaches to obtain a comprehensive analysis of the conformational flexibility of a biomolecular system studied. Finally, a pairwise distance distribution, $P(r)$, "tailing" for larger distances can also be a sign of flexibility; however, this can also be the result of a strong anisotropy of the object shape, for example, in the case of double stranded nucleic acids (Brosey & Tainer, 2019; Kikhney & Svergun, 2015).

Depending on the amount of flexibility of a macromolecule, different approaches should be chosen to model their structure. In general, the simplest modeling approach that best explains the data should be favored, to reduce the risk of overfitting. Therefore, for rigid protein assemblies a single conformation would be sufficient to faithfully describe the system. In the case of rigid modular organizations of multidomain proteins, multiple conformations of the linkers can be generated to reflect more accurately their dynamics. For systems with expected multiple conformations and for which some structural information is already available, the generation of a minimum conformational ensemble that explains best the data is required. Finally, for highly flexible systems such as IDPs, an ensemble of many conformers is required to accurately represent and explain the experimental data, which are averaged over the ensemble (Bonomi, Heller, Camilloni, & Vendruscolo, 2017). The description of conformational ensembles requires a careful assessment of the conformational degrees of freedom with the available experimental data. We will describe these different approaches in the following with several examples.

5.3 Rigid complexes

Once flexibility has been evaluated and the macromolecular complex is found to be rigid with a well-defined globular structure, NMR and SAXS data can be combined with structural information of subdomains to assemble a model. Two main strategies can be employed: simultaneous use of data from both techniques, either by direct refinement or by grid search-based methods, and a posteriori scoring and filtering of NMR-based models with the scattering data.

The direct refinement against NMR- and SAXS-derived experimental restraints during structure calculation can be done with a specific target function to include SAXS data (Grishaev, Wu, Trewhella, & Bax, 2005). This allows the direct addition of the SAXS data during an NMR structure calculation. To reduce the computational cost of an all-atom fit to the SAXS data, the structures are transformed into an accurate coarse-grained approximation of its overall shape. The authors could show that this approach significantly improves the accuracy of an NMR structure calculated based on RDCs and NOE datasets of varying size but also with RDCs only (Grishaev, Tugarinov, Kay, Trewhella, & Bax, 2008). The addition of the SAXS data reduces the conformational space of potential structures with information about the overall global dimension and therefore greatly complements the local distance information provided by NOEs (Rambo & Tainer, 2013b).

Many protocols have been developed specifically to obtain the structure of protein-protein or protein-RNA complexes using the combination of NMR RDC and SAXS data. This combination is particularly attractive as NMR RDCs provide "global" structural information to define the relative orientations of domains in a complex, while SAXS is sensitive to the relative domain positions but is less efficient to discriminate between orientations, in particular for globular spherical-shaped domains. However, RDCs do not provide any information on the distance between these domains or on the overall shape of the complex. This is thus perfectly complemented by the SAXS data. Structure calculation protocols rely on scoring of the assembly, in which the individual domains/structures are treated as (semi) rigid bodies. Hence, a necessary initial step is to validate the rigid-body assumption and show that the individual domains do not change conformation upon complex formation. Alternatively, it is possible and recommended to minimize the individual domain structure based on the RDCs measured in the complex (Simon et al., 2010). A grid-search approach can be used for combining these data. The relative orientation of rigid-body domains can be defined based on the RDC data. To explore the domain positions (i.e. center-of-mass distance of the domains) a grid search can be performed, where for each structure the agreement to SAXS and RDC data is tested and the model best fitting the data is selected

(Wang et al., 2009). An alternative, computationally efficient protocol has been proposed to combine these methods to calculate the model of complexes (Gabel, Simon, & Sattler, 2006).

By far the most commonly used approach to combine these techniques is to initially generate an ensemble of models of the assembly that satisfies NMR data (which can include different types and be rather sparse), and then use the SAXS scattering curve to filter these models and select the most representative objects (Madl, Gabel, & Sattler, 2011). It is also possible to generate ensembles by randomization, i.e. by molecular dynamics or Monte-Carlo simulations, within constraints of a force field and filter the best conformers on the SAXS and NMR data, in a repeated process that improves the starting ensemble at each iteration. The main advantages of this approach are its ease of use and versatility. Various experiments and techniques are available to generate restraints from NMR. Multiple, complementary sets of RDC data can provide orientation restraints, while chemical shift perturbations and solvent PREs will provide ambiguous restraints to define binding interfaces. Finally PRE or PCS data provide interatomic distances restraints of varying precisions (Lapinaite et al., 2013). For the latter, it is strongly recommended to use data from multiple positions of the paramagnetic probe and for all components of the complex, to reduce ambiguity and to detect and avoid possible structural effects from conjugation of the paramagnetic probe. In most cases, the pool of conformations will be generated using rigid-body docking, followed by minimization. In that case, as we have seen before, it is important that the conformation of the individual domains is minimized using data (RDC, PREs) collected on the individual domains. Software, such as ARIA (Simon et al., 2010), HADDOCK (Dominguez, Boelens, & Bonvin, 2003) or DADIMODO (Evrard, Mareuil, Bontems, Sizun, & Perez, 2011) support these approaches. Alternatively, the selection can also be done independently using back-calculation tools such as CRYSOL (Manalastas-Cantos et al., 2021) or FOXS (Schneidman-duhovny et al., 2016). In addition to NMR and SAXS, other experimental data can easily be considered in this integrative approach. For example, restraints obtained from cross-linking combined with mass spectrometry can be used in the generation of the initial pool of conformers. It also allows to set a second independent filter with SANS data, which provide information about subunit structures using contrast matching (Hennig, Wang, Sonntag, Gabel, & Sattler, 2013). As with any modeling approach, the final model proposed must be validated using independent data not used in the initial calculation of the model. For example, keeping some of the NMR data out of the calculation to be used for cross-validation. Alternatively, mutations can be generated based on the model expected to drastically reduce the interactions of the protein and experimentally tested.

The modeling of the complex formed by the multidomain RNA-binding protein TIA-1 and RNA combined SAS and NMR data. TIA-1 is comprised of three RNA recognition motif (RRM) domains. The authors first determined high-resolution structures of the individual RRM domains bound to RNA (Wang et al., 2014). The dynamic of the protein with and without nucleic acid determine that the RRM domains of TIA-1 are independent in solution. After addition of RNA, the D_{max} and overall flexibility is reduced and the domains tumble together, indicating a general compaction of the complex. The authors then model the structure of the three-domain construct bound to a single RNA: first they generate a structural ensemble with CNS using distance restraints obtained from the structure of the individual domains and bound RNA fragments to keep their fold rigid, and then randomized the linkers, which were shown to be flexible based on NMR relaxation data. Then a first filter is applied using SAXS data of the complex to select an ensemble of structures consistent with the overall shape determined by SAXS data. Finally, contrast-matched SANS data obtained with segmentally labeled proteins and RNA are used to further filter on the previously selected SAXS ensemble. The final ensemble shows good convergence, with a well-defined relative position of the three domains. However, as not further NMR data were included (such as RDC data) the relative domain orientations in the final remain poorly defined (Sonntag et al., 2017).

5.4 Ensembles of dynamic assemblies

The combination of SAXS with NMR can accurately describe the conformational dynamics of flexible proteins in solution (Chan-Yao-Chong, Durand, & Ha-Duong, 2019). This is particularly useful to study proteins that contain intrinsically disordered regions or for nucleic acids that are, by nature, more flexible than proteins. Dynamic systems can be separated into two main categories: low-entropy systems that are characterized by a finite and defined number of stable states, and high-entropy systems that adopt a continuum of conformations. Depending on the type of system studied, different approaches have been developed (Fig. 3).

Low-entropy systems include protein complexes that alternate between multiple conformations. To study this type of system, maximum parsimony approaches are usually favored. Here, a large initial pool of conformers is generated and the smallest pool of conformations, for which the weighted average best explains the data, is selected. This approach only assesses the overall ensemble and does not guarantee that each of the individual conformations are relevant and actually present in solution. It is therefore essential to validate the conformers independently before analyzing individual structures composing the ensemble (Delhommel et al., 2017). Note, that the final selected ensemble strongly depends on the quality of the initial ensemble generated. Therefore, the pool of conformers should ideally represent the complete conformational landscape of the system studied (Thomasen & Lindorff-Larsen, 2022). Many approaches employ a randomization of linkers connecting rigid-body domains. Alternatively molecular dynamics (MD) can be employed to sample dynamic conformations starting from a given initial model, however, in this case conformational space is likely not sampled exhaustively. Software are available for these applications, such as EOM (Bernadó et al., 2007) for SAXS only, ASTEROID (Salmon et al., 2010) or MES (Ihms & Foster, 2015). These mainly differ in the way initial conformers are generated and how ensembles are selected. One of the main difficulties of this approach is to find optimal weights for different experimental data used for the selection of the ensembles. Ideally, the level of noise and resolution of each dataset should be evaluated to define appropriate weights.

Mackereth et al. investigated the solution structure of a construct of U2AF2 that contains two RRM domains connected by a 20-residue flexible linker, free and in the presence of a native RNA ligand (Mackereth et al., 2011). To do so, the authors combined rigid-body calculations using a large number of interdomain distances restraints derived from PRE data, orientation restraints from RDCs and structures of the individual domains. The authors identified two distinct conformations, a major "closed state" in the apo form and an "open state" after addition of RNA. Notably, PRE data indicated that the free tandem RRMs also sample the open state, even in absence of nucleic acid, that is consistent with the open domain arrangement in the RNA bound structure. However, SAXS data of the free tandem RRMs clearly indicated that neither the major closed state nor the lowly populated open state, which are both compact with the two domains being in close proximity, could explain the scattering profile. Therefore, an ensemble was generated using independent rigid bodies of the two domains and a full randomization of the linker using *flexible-meccano* (Ozenne et al., 2012). The ensemble was supplemented with structures representing the closed and open conformations that are defined by the NMR PRE data. From these, a final ensemble was selected by simultaneously considering

NMR RDC and PRE together with SAXS data. It is crucial to avoid overfitting and perform appropriate cross-validation in these calculations, as in general the number of experimental restraints is limited. In the analysis, 10% of the experimental data were not included in the calculation and used for cross-validation to avoid overfitting. Also, the information and contribution of the different NMR (PRE, RDCs) and SAXS data was carefully assessed by analyzing conformational ensembles obtained by calculations that include subsets of the experimental data. The resulting ensemble showed that the two domains not only sample the previous PRE-defined compact tandem domain arrangements, but also a wide range of "detached" conformations where the domains sample large distances from each other enabled by the flexible linker (Huang et al., 2014). These detached conformations may play a role in finding RNA targets where local diffusion of the individual domains can help binding the target RNA motifs which can be spread over a larger region in the RNA ligands that serve as substrates of U2AF2. In this example, the use of NMR was essential in characterizing the compact (closed and open) states but was blind to the extended states for which no PREs are observable and RDCs are dynamically averaged. On the other hand, SAXS provides an accurate representation of the overall dimensions of the ensemble, but cannot discriminate the individual, compact domain arrangements.

When structures for distinct conformations of a protein are available, NMR and SAXS can be used to quantify their respective populations in solution. This approach has recently been employed to evaluate the changes of local domain orientations in the molecular chaperone Hsp90 upon addition of ATP (Lopez et al., 2021). Hsp90 is a large and dimeric chaperone protein that adopts a conformational ensemble of open resting states that are driven toward closed, more compact states upon interaction with ATP. Crystal structures available for Hsp90 in different species show distinct overall opening of the Hsp90 dimer, and imply a large domain rotation from one state to another. However, these structures provide snap shots that are likely affected by crystal packing. Using solution NMR upon addition of ATP, the methyl-TROSY spectrum of Hsp90 shows the appearance of a minor population of NMR signals of about 20% population. The major and minor state signals could be assigned to different relative rotation of the N-terminal ATP-binding and middle domains based on experimental PREs data. Interestingly, none of the two conformations can explain the experimental SAXS data, which suggests that the overall dimer is not fully closed and remains dynamic. Structural ensembles were generated based on the conformations identified by NMR and SAXS was used to estimate their respective populations. The analysis of the SAXS data shows that the Hsp90 dimer still samples a large range of open to partially closed conformations in presence of ATP and validates the populations observed by NMR. This illustrates the complementarity of NMR and SAXS, in reporting on local and global conformations, respectively, which is required to describe the dynamic ensemble in solution.

For highly flexible systems that do not particularly favor any specific conformation, the maximum parsimony approaches are insufficient to characterize their conformational continuum. Instead, strategies aiming at maximizing entropy should be considered. In this approach, the behavior of the protein is represented by the averaging of hundreds of conformers. Multiple approaches can be developed to generate such a large pool. The most common approaches consist in generating large datasets only based on chemistry to then reweight and select large ensembles of conformers to describe the data. This approach is relatively costly in term of computation but is easy to set up and flexible. Another more recent approach, consists in the direct restriction the ensemble generation based on the available data. The major difficulty here is that the error of precision for each type of dataset has to be known and introduced before the generation of the ensemble. Bayesian approaches have been developed to tackle this kind of problems (Rieping, Habeck, & Nilges, 2005). The Bayesian approach takes into account the quality of various dataset and estimate the uncertainty of the final model. This has been developed to analyze SAXS data only (Hummer & Köfinger, 2015) and more recently to combine NMR and SAXS with many different parameters (Lincoff et al., 2020).

6. Conclusions and future perspectives

We have outlined the utility of combining information about local structure from NMR with the low-resolution global structural information provided by SAS. The combination of these two techniques is broadly applicable to a variety of systems, including multidomain proteins and large assemblies. The unique strength in combining the two methods lies in the capacity to study the dynamics of biological macromolecules in solution and probe conformational changes, for systems that sample a few defined conformations, but also for fully flexible systems.

Cryo-EM is now a major pilar of structural biology and has permitted to solve numerous structures of large proteins and assemblies that were previously not tractable by X-ray crystallography. While truly impressive structural insight is provided by cryo-EM single-particle reconstructions, information about inherent flexible regions that many biological assemblies require for their function remains invisible, due to conformational averaging. Conformational variability can prevent high-resolution structural analysis due to dynamic averaging. In that case, NMR structural data (but also crystallographic or predicted structures) for these regions can be combined with low-resolution density maps from EM to describe the complete structure (Gauto et al., 2019). EM fails to observe highly flexible regions connected to rigid complexes, that are averaged-out in the selected classes. Because of their dynamics, such regions remain visible by NMR even for very high-molecular weight complexes. It is thus advantageous to combine NMR, EM and SAXS, to determine a high-resolution structure of the rigid part of the complex, while characterizing flexible regions using NMR and SAXS (Kumar et al., 2021).

Finally, the development of AlphaFold and RoseTTAFold is changing the way we are considering structural biology. However, in many cases, it has already been observed that the predicted structures do not reflect the main conformation of a protein in solution, but instead, might predict a conformation only observable in a complex or stabilized by cofactors. It thus remains essential to experimentally validate these models using solution data. As we have shown in this chapter, the validation of structures by NMR coupled with SAS is already well established and is particularly efficient. In addition, AlphaFold can now be used to predict complexes and oligomers. Combined with NMR binding studies and SAXS data this could become a very efficient approach to model and validate assemblies formed in solution. We expect in the future this integrative approach may facilitate early steps of structural analysis and will allow structural biology research to focus on the elucidation of the molecular mechanisms.

The examples highlighted in this review show how unique information is obtained by combining NMR and SAS data, where typically NMR provides higher-resolution information about local structure, while SAS contributes complementary information about global structure. The combination of these two solution techniques will greatly help to uncover dynamic conformations and conformational ensembles, involving intrinsically disordered regions and their role in multidomain proteins and complexes. This information is not provided by crystallographic structures and cryo-EM approaches and also not by AlphaFold predictions as the deep learning has been trained on globular and rigid protein structures.

Acknowledgments

We are grateful to Frank Gabel and Annika Elimelech for critical reading of the manuscript and all members of the Sattler group for discussions. We apologize for not being able to discuss many other studies due to space constraints.

M.S. acknowledges funding by the Deutsche Forschungsgemeinschaft (grants GRK1721, SFB1035, SFB1309 and SPP1935). F.D. is supported by an EMBO Longterm Fellowship (ALTF 243-2018). S.M.-L. is supported by an EU HORIZON 2020 research and innovation program under the Marie Skłodowska-Curie grant agreement No. 792692. We acknowledge support by the Helmholtz Association Initiative and Networking Fund under project number ZT-I-0003 (to M.S.)

References

Alderson, T. R., & Kay, L. E. (2021). NMR spectroscopy captures the essential role of dynamics in regulating biomolecular function. *Cell*, *184*(3), 577–595. https://doi.org/10.1016/j.cell.2020.12.034.

Al-Hashimi, H. M. (2013). NMR studies of nucleic acid dynamics. *Journal of Magnetic Resonance*, *237*, 191–204. https://doi.org/10.1016/j.jmr.2013.08.014.

Allain, F., Mareuil, F., Ménager, H., Nilges, M., & Bardiaux, B. (2020). ARIAweb: A server for automated NMR structure calculation. *Nucleic Acids Research*, *48*(1), W41–W47. https://doi.org/10.1093/nar/gkaa362.

Baek, M., Dimaio, F., Anishchenko, I., Dauparas, J., Lee, G. R., Wang, J., et al. (2021). Accurate prediction of protein structures and interactions using a 3-track neural network. *Science*, *373*(6557), 871–876. https://doi.org/10.1126/science.abj8754.Accurate.

Bailor, M. H., Musselman, C., Hansen, A. L., Gulati, K., Patel, D. J., & Al-Hashimi, H. M. (2007). Characterizing the relative orientation and dynamics of RNA A-form helices using NMR residual dipolar couplings. *Nature Protocols*, *2*(6), 1536–1546. https://doi.org/10.1038/nprot.2007.221.

Banani, S. F., Lee, H. O., Hyman, A. A., & Rosen, M. K. (2017). Biomolecular condensates: Organizers of cellular biochemistry. *Nature Reviews. Molecular Cell Biology*, *18*(5), 285–298. https://doi.org/10.1038/nrm.2017.7.

Baronti, L., Karlsson, H., Marušič, M., & Petzold, K. (2018). A guide to large-scale RNA sample preparation. *Analytical and Bioanalytical Chemistry*, *410*(14), 3239–3252. https://doi.org/10.1007/s00216-018-0943-8.

Bax, A., Kontaxis, G., & Tjandra, N. (2001). Dipolar couplings in macromolecular structure determination. *Methods in Enzymology*, *339*(1997), 127–174. https://doi.org/10.1016/S0076-6879(01)39313-8.

Bernado, P. (2010). Effect of interdomain dynamics on the structure determination of modular proteins by small-angle scattering. *European Biophysics Journal*, *39*(5), 769–780. https://doi.org/10.1007/s00249-009-0549-3.

Bernadó, P., Mylonas, E., Petoukhov, M. V., Blackledge, M., & Svergun, D. I. (2007). Structural characterization of flexible proteins using SAXS. *Journal of the American Chemical Society*, *129*(ww), 5656–5664.

Blackledge, M. (2005). Recent progress in the study of biomolecular structure and dynamics in solution from residual dipolar couplings. *Progress in Nuclear Magnetic Resonance Spectroscopy*, *46*(1), 23–61. https://doi.org/10.1016/j.pnmrs.2004.11.002.

Bock, L. V., Blau, C., Schröder, G. F., Davydov, I. I., Fischer, N., Stark, H., et al. (2013). Energy barriers and driving forces in tRNA translocation through the ribosome. *Nature Structural and Molecular Biology*, *20*(12), 1390–1396. https://doi.org/10.1038/nsmb.2690.

Bonomi, M., Heller, G. T., Camilloni, C., & Vendruscolo, M. (2017). Principles of protein structural ensemble determination. *Current Opinion in Structural Biology*, *42*, 106–116. https://doi.org/10.1016/j.sbi.2016.12.004.

Borcherds, W., Bremer, A., Borgia, M. B., & Mittag, T. (2021). How do intrinsically disordered protein regions encode a driving force for liquid–liquid phase separation? *Current Opinion in Structural Biology*, *67*, 41–50. https://doi.org/10.1016/j.sbi.2020.09.004.

Bothe, J. R., Nikolova, E. N., Eichhorn, C. D., Chugh, J., Hansen, A. L., & Al-Hashimi, H. M. (2011). Characterizing RNA dynamics at atomic resolution using solution-state NMR spectroscopy. *Nature Methods*, *8*(11). https://doi.org/10.1038/nmeth.1735.

Brondyk, W. H. (2009). Selecting an appropriate method for expressing a recombinant protein. *Methods in Enzymology*, *463*, 131–147. https://doi.org/10.1016/S0076-6879(09)63011-1.

Brosey, C. A., & Tainer, J. A. (2019). Evolving SAXS versatility: Solution X-ray scattering for macromolecular architecture, functional landscapes, and integrative structural biology. *Current Opinion in Structural Biology*, *52*, 1–17. https://doi.org/10.1016/j.sbi.2019.04.004.

Chan-Yao-Chong, M., Durand, D., & Ha-Duong, T. (2019). Molecular dynamics simulations combined with nuclear magnetic resonance and/or small-angle X-ray scattering data for characterizing intrinsically disordered protein conformational ensembles. *Journal of Chemical Information and Modeling*, *59*(5), 1743–1758. https://doi.org/10.1021/acs.jcim.8b00928.

Chao, F., & Byrd, R. A. (2018). Protein dynamics revealed by NMR relaxation methods. *Emerging Topics in Life Sciences*, *2*, 93–105.

Clark, E. T., Sievers, E. E., & Debelouchina, G. T. (2022). A chemical biology primer for NMR spectroscopists. *Journal of Magnetic Resonance Open*, *10–11*, 100044. https://doi.org/10.1016/j.jmro.2022.100044.

Clore, G. M. (2015). Practical aspects of paramagnetic relaxation enhancement in biological macromolecules. *Methods in Enzymology*, *564*, 485–497. https://doi.org/10.1016/bs.mie.2015.06.032.

Davey, N. E. (2019). The functional importance of structure in unstructured protein regions. *Current Opinion in Structural Biology*, *56*, 155–163. https://doi.org/10.1016/j.sbi.2019.03.009.

David, G., & Pérez, J. (2009). Combined sampler robot and high-performance liquid chromatography: A fully automated system for biological small-angle X-ray scattering experiments at the Synchrotron SOLEIL SWING beamline. *Journal of Applied Crystallography*, *42*, 892–900. https://doi.org/10.1107/S0021889809029288.

Delhommel, F., Cordier, F., Bardiaux, B., Bouvier, G., Colcombet-Cazenave, B., Brier, S., et al. (2017). Structural and dynamical properties of the PDZ tandem of Whirlin reveal an original interaction. *Structure*, *25*(11), 1645–1656. https://doi.org/10.1016/j.str.2017.08.013.

Delhommel, F., Gabel, F., & Sattler, M. (2020). Current approaches for integrating solution NMR spectroscopy and small angle scattering to study the structure and dynamics of biomolecular complexes. *Journal of Molecular Biology*. https://doi.org/10.1016/j.jmb.2020.03.014.

Dominguez, C., Boelens, R., & Bonvin, A. M. J. J. (2003). HADDOCK: A protein-protein docking approach based on biochemical or biophysical information. *Journal of the American Chemical Society*, *125*(7), 1731–1737. https://doi.org/10.1021/ja026939x.

Dümmler, A., Lawrence, A. M., & de Marco, A. (2005). Simplified screening for the detection of soluble fusion constructs expressed in E. coli using a modular set of vectors. *Microbial Cell Factories*, *4*, 1–10. https://doi.org/10.1186/1475-2859-4-34.

Durand, D., Vivès, C., Cannella, D., Pérez, J., Pebay-Peyroula, E., Vachette, P., et al. (2010). NADPH oxidase activator p67phox behaves in solution as a multidomain protein with semi-flexible linkers. *Journal of Structural Biology*, *169*(1), 45–53. https://doi.org/10.1016/j.jsb.2009.08.009.

Duss, O., Maris, C., Von Schroetter, C., & Allain, F. H. T. (2010). A fast, efficient and sequence-independent method for flexible multiple segmental isotope labeling of RNA using ribozyme and RNase H cleavage. *Nucleic Acids Research*, *38*(20). https://doi.org/10.1093/nar/gkq756.

Duszczyk, M. M., Zanier, K., & Sattler, M. (2008). A NMR strategy to unambiguously distinguish nucleic acid hairpin and duplex conformations applied to a Xist RNA A-repeat. *Nucleic Acids Research*, *36*(22), 7068–7077. https://doi.org/10.1093/nar/gkn776.

Edelmann, F. T., Niedner, A., & Niessing, D. (2014). Production of pure and functional RNA for in vitro reconstitution experiments. *Methods*, *65*(3), 333–341. https://doi.org/10.1016/j.ymeth.2013.08.034.

Evrard, G., Mareuil, F., Bontems, F., Sizun, C., & Perez, J. (2011). DADIMODO: A program for refining the structure of multidomain proteins and complexes against small-angle scattering data and NMR-derived restraints. *Journal of Applied Crystallography*, *44*(6), 1264–1271. https://doi.org/10.1107/S0021889811035758.

Franke, B., Opitz, C., Isogai, S., Grahl, A., Delgado, L., Gossert, A. D., et al. (2018). Production of isotope-labeled proteins in insect cells for NMR. *Journal of Biomolecular NMR*, *71*(3), 173–184. https://doi.org/10.1007/s10858-018-0172-7.

Fürtig, B., Richter, C., Wöhnert, J., & Schwalbe, H. (2003). NMR spectroscopy of RNA. *Chembiochem*, *4*(10), 936–962. https://doi.org/10.1002/cbic.200300700.

Gabel, F., Simon, B., & Sattler, M. (2006). A target function for quaternary structural refinement from small angle scattering and NMR orientational restraints. *European Biophysics Journal*, *35*(4), 313–327. https://doi.org/10.1007/s00249-005-0037-3.

Gardner, K. H., & Kay, L. E. (1998). The use of 2H, 13C, 15N multidimensional NMR to study the structure and dynamics of proteins. *Annual Review of Biophysics and Biomolecular Structure*, *27*, 357–406. https://doi.org/10.1146/annurev.biophys.27.1.357.

Gauto, D. F., Estrozi, L. F., Schwieters, C. D., Effantin, G., Macek, P., Sounier, R., et al. (2019). Integrated NMR and cryo-EM atomic-resolution structure determination of a half-megadalton enzyme complex. *Nature Communications*, *10*(1), 1–12. https://doi.org/10.1038/s41467-019-10490-9.

Göbl, C., Madl, T., Simon, B., & Sattler, M. (2014). NMR approaches for structural analysis of multidomain proteins and complexes in solution. *Progress in Nuclear Magnetic Resonance Spectroscopy*, *80*, 26–63. https://doi.org/10.1016/j.pnmrs.2014.05.003.

Gong, Z., Gu, X. H., Guo, D. C., Wang, J., & Tang, C. (2017). Protein structural ensembles visualized by solvent paramagnetic relaxation enhancement. *Angewandte Chemie (International Ed. in English)*, *56*(4), 1002–1006. https://doi.org/10.1002/anie.201609830.

Gradia, S. D., Ishida, J. P., Tsai, M. S., Jeans, C., Tainer, J. A., & Fuss, J. O. (2017). MacroBac: New technologies for robust and efficient large-scale production of recombinant multiprotein complexes. *Methods in Enzymology*, *592*, 1–26. https://doi.org/10.1016/bs.mie.2017.03.008.

Grishaev, A., Tugarinov, V., Kay, L. E., Trewhella, J., & Bax, A. (2008). Refined solution structure of the 82-kDa enzyme malate synthase G from joint NMR and synchrotron SAXS restraints. *Journal of Biomolecular NMR*, *40*(2), 95–106. https://doi.org/10.1007/s10858-007-9211-5.

Grishaev, A., Wu, J., Trewhella, J., & Bax, A. (2005). Refinement of multidomain protein structures by combination of solution small-angle X-ray scattering and NMR data. *Journal of the American Chemical Society*, *127*(47), 16621–16628. https://doi.org/10.1021/ja054342m.

Gross, J. D., Gelev, V. M., & Wagner, G. (2003). A sensitive and robust method for obtaining intermolecular NOEs between side chains in large protein complexes. *Journal of Biomolecular NMR*, *25*(3), 235–242. https://doi.org/10.1023/A:1022890112109.

Guinier, A. (1939). Diffraction of X-rays of very small angles—Application to the study of ultramicroscopic phenomenon. *Annals of Physics*, *12*, 161–237.

Güntert, P., & Buchner, L. (2015). Combined automated NOE assignment and structure calculation with CYANA. *Journal of Biomolecular NMR*, *62*(4), 453–471. https://doi.org/10.1007/s10858-015-9924-9.

Hartlmüller, C., Günther, J. C., Wolter, A. C., Wöhnert, J., Sattler, M., & Madl, T. (2017). RNA structure refinement using NMR solvent accessibility data. *Scientific Reports*, 7(1), 1–10. https://doi.org/10.1038/s41598-017-05821-z.

Hass, M. A. S., & Ubbink, M. (2014). Structure determination of protein-protein complexes with long-range anisotropic paramagnetic NMR restraints. *Current Opinion in Structural Biology*, *24*(1), 45–53. https://doi.org/10.1016/j.sbi.2013.11.010.

Hennig, J., Militti, C., Popowicz, G. M., Wang, I., Sonntag, M., Geerlof, A., et al. (2014). Structural basis for the assembly of the Sxl-Unr translation regulatory complex. *Nature*, *515*(7526), 287–290. https://doi.org/10.1038/nature13693.

Hennig, J., Wang, I., Sonntag, M., Gabel, F., & Sattler, M. (2013). Combining NMR and small angle X-ray and neutron scattering in the structural analysis of a ternary protein-RNA complex. *Journal of Biomolecular NMR*, *56*(1), 17–30. https://doi.org/10.1007/s10858-013-9719-9.

Hennig, J., Warner, L. R., Simon, B., Geerlof, A., Mackereth, C. D., & Sattler, M. (2015). Structural analysis of protein-RNA complexes in solution using NMR paramagnetic relaxation enhancements. *Methods in Enzymology*, *558*(1), 333–362. https://doi.org/10.1016/bs.mie.2015.02.006.

Huang, J. R., Warner, L. R., Sanchez, C., Gabel, F., Madl, T., Mackereth, C. D., et al. (2014). Transient electrostatic interactions dominate the conformational equilibrium sampled by multidomain splicing factor U2AF65: A combined NMR and SAXS study. *Journal of the American Chemical Society*, *136*(19), 7068–7076. https://doi.org/10.1021/ja502030n.

Hummer, G., & Köfinger, J. (2015). Bayesian ensemble refinement by replica simulations and reweighting. *Journal of Chemical Physics*, *143*(24). https://doi.org/10.1063/1.4937786.

Igumenova, T. I., Brath, U., Akke, M., & Palmer, A. G. (2007). Characterization of chemical exchange using residual dipolar coupling. *Journal of the American Chemical Society*, *129*(44), 13396–13397. https://doi.org/10.1021/ja0761636.

Ihms, E. C., & Foster, M. P. (2015). MESMER: Minimal ensemble solutions to multiple experimental restraints. *Bioinformatics*, *31*(12), 1951–1958. https://doi.org/10.1093/bioinformatics/btv079.

Jeener, J., Meier, B. H., Bachmann, P., & Ernst, R. R. (1979). Investigation of exchange processes by two-dimensional NMR spectroscopy. *The Journal of Chemical Physics*, *71*(11), 4546–4553. https://doi.org/10.1063/1.438208.

Jones, A. N., & Sattler, M. (2019). Challenges and perspectives for structural biology of lncRNAs—The example of the Xist lncRNA A-repeats. *Journal of Molecular Cell Biology*, *11*(10), 845–859. https://doi.org/10.1093/jmcb/mjz086.

Joss, D., Walliser, R. M., Zimmermann, K., & Häussinger, D. (2018). Conformationally locked lanthanide chelating tags for convenient pseudocontact shift protein nuclear magnetic resonance spectroscopy. *Journal of Biomolecular NMR*, *72*(1–2), 29–38. https://doi.org/10.1007/s10858-018-0203-4.

Jumper, J., Evans, R., Pritzel, A., Green, T., Figurnov, M., Ronneberger, O., et al. (2021). Highly accurate protein structure prediction with AlphaFold. *Nature*, *596*(7873), 583–589. https://doi.org/10.1038/s41586-021-03819-2.

Kawahara, I., Haruta, K., Ashihara, Y., Yamanaka, D., Kuriyama, M., Toki, N., et al. (2012). Site-specific isotope labeling of long RNA for structural and mechanistic studies. *Nucleic Acids Research*, *40*(1), 1–8. https://doi.org/10.1093/nar/gkr951.

Kerfah, R., Plevin, M. J., Sounier, R., Gans, P., & Boisbouvier, J. (2015). Methyl-specific isotopic labeling: A molecular tool box for solution NMR studies of large proteins. *Current Opinion in Structural Biology*, *32*, 113–122. https://doi.org/10.1016/j.sbi.2015.03.009.

Kikhney, A. G., & Svergun, D. I. (2015). A practical guide to small angle X-ray scattering (SAXS) of flexible and intrinsically disordered proteins. *FEBS Letters*, *589*, 2570–2577.

Kim, H. S., & Gabel, F. (2015). Uniqueness of models from small-angle scattering data: The impact of a hydration shell and complementary NMR restraints. *Acta Crystallographica. Section D: Biological Crystallography*, *71*, 57–66. https://doi.org/10.1107/S1399004714013923.

Kooshapur, H., Choudhury, N. R., Simon, B., Mühlbauer, M., Jussupow, A., Fernandez, N., et al. (2018). Structural basis for terminal loop recognition and stimulation of pri-miRNA-18a processing by hnRNP A1. *Nature Communications*, *9*(1). https://doi.org/10.1038/s41467-018-04871-9.

Kuenze, G., Bonneau, R., Leman, J. K., & Meiler, J. (2019). Integrative protein modeling in RosettaNMR from sparse paramagnetic restraints. *Structure*, *27*(11). https://doi.org/10.1016/j.str.2019.08.012. 1721–1734.e5.

Kumar, A., Yu, C. W. H., Rodríguez-Molina, J. B., Li, X. H., Freund, S. M. V., & Passmore, L. A. (2021). Dynamics in Fip1 regulate eukaryotic mRNA 3′ end processing. *Genes and Development*, *35*(21–22), 1510–1526. https://doi.org/10.1101/GAD.348671.121.

Lacabanne, D., Meier, B. H., & Böckmann, A. (2018). Selective labeling and unlabeling strategies in protein solid-state NMR spectroscopy. *Journal of Biomolecular NMR*, *71*(3), 141–150. https://doi.org/10.1007/s10858-017-0156-z.

Lapinaite, A., Simon, B., Skjaerven, L., Rakwalska-Bange, M., Gabel, F., & Carlomagno, T. (2013). The structure of the box C/D enzyme reveals regulation of RNA methylation. *Nature*, *502*(7472), 519–523. https://doi.org/10.1038/nature12581.

Latham, M. P., Brown, D. J., McCallum, S. A., & Pardi, A. (2005). NMR methods for studying the structure and dynamics of RNA. *Chembiochem*, *6*(9), 1492–1505. https://doi.org/10.1002/cbic.200500123.

Li, Y., & Palmer, A. G. (2009). TROSY-selected ZZ-exchange experiment for characterizing slow chemical exchange in large proteins. *Journal of Biomolecular NMR*, *45*(4), 357–360. https://doi.org/10.1007/s10858-009-9385-0.

Lincoff, J., Haghighatlari, M., Krzeminski, M., Teixeira, J. M. C., Gomes, G. N. W., Gradinaru, C. C., et al. (2020). Extended experimental inferential structure determination method in determining the structural ensembles of disordered protein states. *Communications Chemistry*, *3*(1), 1–12. https://doi.org/10.1038/s42004-020-0323-0.

Lopez, A., Dahiya, V., Delhommel, F., Freiburger, L., Stehle, R., Asami, S., et al. (2021). Client binding shifts the populations of dynamic Hsp90 conformations through an allosteric network. *Science Advances*, *7295*(December).

Loria, J. P., Rance, M., & Palmer, A. G. (1999). A relaxation-compensated Carr-Purcell-Meiboom-Gill sequence for characterizing chemical exchange by NMR spectroscopy. *Journal of the American Chemical Society*, *16*, 2331–2332. https://doi.org/10.1021/ja983961a.

Lunde, B. M., Moore, C., & Varani, G. (2007). RNA-binding proteins: Modular design for efficient function. *Nature Reviews. Molecular Cell Biology*, *8*(6), 479–490. https://doi.org/10.1038/nrm2178.

Mackereth, C. D., Madl, T., Bonnal, S., Simon, B., Zanier, K., Gasch, A., et al. (2011). Multi-domain conformational selection underlies pre-mRNA splicing regulation by U2AF. *Nature*, *475*(7356), 408–411. https://doi.org/10.1038/nature10171.

Mackereth, C. D., & Sattler, M. (2012). Dynamics in multi-domain protein recognition of RNA. *Current Opinion in Structural Biology*, *22*(3), 287–296. https://doi.org/10.1016/j.sbi.2012.03.013.

Madl, T., Bermel, W., & Zangger, K. (2009). Use of relaxation enhancements in a paramagnetic environment for the structure determination of proteins using NMR spectroscopy. *Angewandte Chemie (International Ed. in English)*, *48*(44), 8259–8262. https://doi.org/10.1002/anie.200902561.

Madl, T., Gabel, F., & Sattler, M. (2011). NMR and small-angle scattering-based structural analysis of protein complexes in solution. *Journal of Structural Biology, 173*(3), 472–482. https://doi.org/10.1016/j.jsb.2010.11.004.

Madl, T., Güttler, T., Görlich, D., & Sattler, M. (2011). Structural analysis of large protein complexes using solvent paramagnetic relaxation enhancements. *Angewandte Chemie (International Ed. in English), 50*(17), 3993–3997. https://doi.org/10.1002/anie.201007168.

Mahieu, E., & Gabel, F. (2018). Biological small-angle neutron scattering: Recent results and development. *Acta Crystallographica. Section D: Biological Crystallography, 74*, 715–726. https://doi.org/10.1107/S2059798318005016.

Manalastas-Cantos, K., Konarev, P. V., Hajizadeh, N. R., Kikhney, A. G., Petoukhov, M. V., Molodenskiy, D. S., et al. (2021). ATSAS 3.0: Expanded functionality and new tools for small-angle scattering data analysis. *Journal of Applied Crystallography, 54*, 343–355. https://doi.org/10.1107/S1600576720013412.

Mertens, H. D. T., & Svergun, D. I. (2010). Structural characterization of proteins and complexes using small-angle X-ray solution scattering. *Journal of Structural Biology, 172*(1), 128–141. https://doi.org/10.1016/j.jsb.2010.06.012.

Mielke, S. P., & Krishnan, V. V. (2009). Characterization of protein secondary structure from NMR chemical shifts. *Progress in Nuclear Magnetic Resonance Spectroscopy, 54*(2–3), 141–165. https://doi.org/10.1016/b978-1-4832-2943-0.50008-x.

Mirdita, M., Schütze, K., Moriwaki, Y., Heo, L., Ovchinnikov, S., & Steinegger, M. (2022). ColabFold: Making protein folding accessible to all. *Nature Methods, 19*(6), 679–682. https://doi.org/10.1038/s41592-022-01488-1.

Mittermaier, A., & Kay, L. E. (2006). New tools provide new insights in NMR studies of protein dynamics. *Science, 312*(5771), 224–228. https://doi.org/10.1126/science.1124964.

Ozenne, V., Bauer, F., Salmon, L., Huang, J. R., Jensen, M. R., Segard, S., et al. (2012). Flexible-meccano: A tool for the generation of explicit ensemble descriptions of intrinsically disordered proteins and their associated experimental observables. *Bioinformatics, 28*(11), 1463–1470. https://doi.org/10.1093/bioinformatics/bts172.

Palmer, A. G. (2014). Chemical exchange in biomacromolecules: Past, present, and future. *Journal of Magnetic Resonance, 241*(1), 3–17. https://doi.org/10.1016/j.jmr.2014.01.008.

Perrakis, A., & Sixma, T. K. (2021). AI revolutions in biology. *EMBO Reports, 22*(11), 1–6. https://doi.org/10.15252/embr.202154046.

Petoukhov, M. V., & Svergun, D. I. (2015). Ambiguity assessment of small-angle scattering curves from monodisperse systems. *Acta Crystallographica. Section A: Biological Crystallography, 71*, 1051–1058. https://doi.org/10.1107/S1399004715002576.

Porta-Pardo, E., Ruiz-Serra, V., Valentini, S., & Valencia, A. (2022). The structural coverage of the human proteome before and after AlphaFold. *PLoS Computational Biology, 18*(1), e1009818. https://doi.org/10.1371/journal.pcbi.1009818.

Pritišanac, I., Alderson, T. R., & Güntert, P. (2020). Automated assignment of methyl NMR spectra from large proteins. *Progress in Nuclear Magnetic Resonance Spectroscopy, 118–119*, 54–73. https://doi.org/10.1016/j.pnmrs.2020.04.001.

Rambo, R. P., & Tainer, J. A. (2010). Improving small-angle X-ray scattering data for structural analyses of the RNA world. *RNA, 16*(3), 638–646. https://doi.org/10.1261/rna.1946310.

Rambo, R. P., & Tainer, J. A. (2011). Characterizing flexible and intrinsically unstructured biological macromolecules by SAS using the Porod-Debye law. *Biopolymers, 95*(8), 559–571. https://doi.org/10.1002/bip.21638.

Rambo, R. P., & Tainer, J. A. (2013a). Accurate assessment of mass, models and resolution by small-angle scattering. *Nature, 496*(7446), 477–481. https://doi.org/10.1038/nature12070.

Rambo, R. P., & Tainer, J. A. (2013b). Super-resolution in solution x-ray scattering and its applications to structural systems biology. *Annual Review of Biophysics*, *42*(1), 415–441. https://doi.org/10.1146/annurev-biophys-083012-130301.

Rieping, W., Habeck, M., Bardiaux, B., Bernard, A., Malliavin, T. E., & Nilges, M. (2007). ARIA2: Automated NOE assignment and data integration in NMR structure calculation. *Bioinformatics (Oxford, England)*, *23*(3), 381–382. https://doi.org/10.1093/bioinformatics/btl589.

Rieping, W., Habeck, M., & Nilges, M. (2005). Inferential structure determination. *Science*, *309*(5732), 303–306. https://doi.org/10.1126/science.1110428.

Rout, M. P., & Sali, A. (2019). Principles for integrative structural biology studies. *Cell*, *177*(6), 1384–1403. https://doi.org/10.1016/j.cell.2019.05.016.

Salmon, L., Nodet, G., Ozenne, V., Yin, G., Jensen, M. R., Zweckstetter, M., et al. (2010). NMR characterization of long-range order in intrinsically disordered proteins. *Journal of the American Chemical Society*, *132*(24), 8407–8418. https://doi.org/10.1021/ja101645g.

Sattler, M., Schleucher, J., & Griesinger, C. (1999). Heteronuclear multidimensional NMR experiments for the structure determination of proteins in solution employing pulsed field gradients. *Progress in Nuclear Magnetic Resonance Spectroscopy*, *34*(2), 93–158. https://doi.org/10.1016/S0079-6565(98)00025-9.

Schneidman-duhovny, D., Hammel, M., Tainer, J. A., Sali, A., Angle, S., & Saxs, X. S. (2016). FoXS, FoXSDock and MultiFoXS: Single-state and multi-state structural modeling of proteins and their complexes based on SAXS profiles. *Nucleic Acids Research*, *44*(May), 424–429. https://doi.org/10.1093/nar/gkw389.

Schütz, S., & Sprangers, R. (2020). Methyl TROSY spectroscopy: A versatile NMR approach to study challenging biological systems. *Progress in Nuclear Magnetic Resonance Spectroscopy*, *116*, 56–84. https://doi.org/10.1016/j.pnmrs.2019.09.004.

Schwieters, C. D., Kuszewski, J. J., & Marius Clore, G. (2006). Using Xplor-NIH for NMR molecular structure determination. *Progress in Nuclear Magnetic Resonance Spectroscopy*, *48*(1), 47–62. https://doi.org/10.1016/j.pnmrs.2005.10.001.

Simon, B., Madl, T., Mackereth, C. D., Nilges, M., & Sattler, M. (2010). An efficient protocol for NMR-spectroscopy-based structure determination of protein complexes in solution. *Angewandte Chemie (International Ed. in English)*, *49*(11), 1967–1970. https://doi.org/10.1002/anie.200906147.

Smith, G. E., Summers, M. D., & Fraser, M. J. (1983). Production of human beta interferon in insect cells infected with a baculovirus expression vector. *Molecular and Cellular Biology*, *3*(12), 2156–2165. https://doi.org/10.1128/mcb.3.12.2156-2165.1983.

Softley, C. A., Bostock, M. J., Popowicz, G. M., & Sattler, M. (2020). Paramagnetic NMR in drug discovery. *Journal of Biomolecular NMR*, *74*(6–7), 287–309. https://doi.org/10.1007/s10858-020-00322-0.

Sonntag, M., Jagtap, P. K. A., Simon, B., Appavou, M. S., Geerlof, A., Stehle, R., et al. (2017). Segmental, domain-selective perdeuteration and small-angle neutron scattering for structural analysis of multi-domain proteins. *Angewandte Chemie (International Ed. in English)*, *56*(32), 9322–9325. https://doi.org/10.1002/anie.201702904.

Svergun, D. I. (1999). Restoring low resolution structure of biological macromolecules from solution scattering using simulated annealing. *Biophysical Journal*, *76*(6), 2879–2886. https://doi.org/10.1016/S0006-3495(99)77443-6.

Svergun, D. I., Petoukhov, M. V., & Koch, M. H. J. (2001). Determination of domain structure of proteins from X-ray solution scattering. *Biophysical Journal*, *80*(January), 2946–2953.

Svergun, D. I., Richard, S., Koch, M. H. J., Sayers, Z., Kuprin, S., & Zaccai, G. (1998). Protein hydration in solution: Experimental observation by x-ray. *Proceedings of the National Academy of Sciences of the United States of America*, *95*(March), 2267–2272.

Thomasen, F. E., & Lindorff-Larsen, K. (2022). Conformational ensembles of intrinsically disordered proteins and flexible multidomain proteins. *Biochemical Society Transactions*, *50*(1), 541–554. https://doi.org/10.1042/bst20210499.

Vallurupalli, P., Bouvignies, G., & Kay, L. E. (2012). Studying "invisible" excited protein states in slow exchange with a major state conformation. *Journal of the American Chemical Society*, *134*, 8148–8161. https://doi.org/10.1021/ja3001419.

Wang, I., Hennig, J., Jagtap, P. K. A., Sonntag, M., Valcárcel, J., & Sattler, M. (2014). Structure, dynamics and RNA binding of the multi-domain splicing factor TIA-1. *Nucleic Acids Research*, *42*(9), 5949–5966. https://doi.org/10.1093/nar/gku193.

Wang, J., Zuo, X., Yu, P., Byeon, I. J. L., Jung, J., Wang, X., et al. (2009). Determination of multicomponent protein structures in solution using global orientation and shape restraints. *Journal of the American Chemical Society*, *131*(30), 10507–10515. https://doi.org/10.1021/ja902528f.

Waudby, C. A., Ramos, A., Cabrita, L. D., & Christodoulou, J. (2016). Two-dimensional NMR lineshape analysis. *Scientific Reports*, *6*(April), 1–8. https://doi.org/10.1038/srep24826.

Williamson, M. P. (2013). Using chemical shift perturbation to characterise ligand binding. *Progress in Nuclear Magnetic Resonance Spectroscopy*, *73*, 1–16. https://doi.org/10.1016/j.pnmrs.2013.02.001.

Wishart, D. S., & Sykes, B. D. (1994). The 13C chemical-shift index: A simple method for the identification of protein secondary structure using 13C chemical-shift data. *Journal of Biomolecular NMR*, *4*(2), 171–180. https://doi.org/10.1007/BF00175245.

CHAPTER TEN

From dilute to concentrated solutions of intrinsically disordered proteins: Interpretation and analysis of collected data

Samuel Lenton[a,b], Eric Fagerberg[b], Mark Tully[d], and Marie Skepö[b,c,]*

[a]Drug Delivery and Biophysics of Biopharmaceuticals, Department of Pharmacy, University of Copenhagen, Copenhagen, Denmark
[b]Division of Theoretical Chemistry, Lund University, Lund, Sweden
[c]LINXS—Institute of Advanced Neutron and X-ray Science, Lund, Sweden
[d]BioSAXS beamline, BM29, European Synchrotron Radiation Facility, ESRF, Grenoble, France
*Corresponding author: e-mail address: marie.skepo@teokem.lu.se

Contents

Methods in Enzymology, Volume 678
ISSN 0076-6879
https://doi.org/10.1016/bs.mie.2022.09.021

Abstract

Intrinsically disordered proteins (IDPs) have a broad energy landscape and consequently sample many different conformations in solution. The innate flexibility of IDPs is exploited in their biological function, and in many instances allows a single IDP to regulate a range of processes *in vivo*. Due to their highly flexible nature, characterizing the structural properties of IDPs is not straightforward. Often solution-based methods such as Nuclear Magnetic Resonance (NMR), Förster Resonance Energy Transfer (FRET), and Small-Angle X-ray Scattering (SAXS) are used. SAXS is indeed a powerful technique to study the structural and conformational properties of IDPs in solution, and from the obtained SAXS spectra, information about the average size, shape, and extent of oligomerization can be determined. In this chapter, we will introduce model-free methods that can be used to interpret SAXS data and introduce methods that can be used to interpret SAXS data beyond analytical models, for example, by using atomistic and different levels of coarse-grained models in combination with molecular dynamics (MD) and Monte Carlo simulations.

1. Introduction

In contrast to globular proteins, intrinsically disordered proteins (IDPs) adopt many different conformations in solution (Dunker et al., 2013; Oldfield & Dunker, 2014; Tompa, 2002). This transient structural nature means that IDPs are well suited to adopting a range of different functions however, it makes them challenging to study with many traditional biophysical techniques (Eliezer, 2009; Fisher & Stultz, 2011; Uversky, 2013). Small-Angle X-ray Scattering (SAXS) is a fast and powerful technique that can be used to characterize IDPs in solution at a broad range of concentrations (Bernado, Mylonas, Petoukhov, Blackledge, & Svergun, 2007; Kachala, Valentini, & Svergun, 2015; Martin, Hopkins, & Mittag, 2021), yielding information on the structural properties of the protein, the oligomerization state, and protein-protein interactions within the solution (Bernado & Blackledge, 2009; Cragnell, Staby, Lenton, Kragelund, & Skepö, 2019; Johansen, Jeffries, Hammouda, Trewhella, & Goldenberg, 2011; Kikhney & Svergun, 2015). Several of the analysis techniques discussed in this chapter are also applicable to globular proteins that contain intrinsically disordered regions; however, in this chapter we focus on IDPs, solely. During a SAXS experiment, a solution containing the protein is exposed to an incident X-ray beam, which is scattered by all the components of the solution. A two-dimensional detector measures the intensity of the scattered X-rays as a function of the scattering angle, q, according to:

$$q = 4\pi \sin(\theta)/\lambda \tag{1}$$

where λ is the wavelength of the incident X-ray beam, and 2θ is the angle between the incident and scattered beam. In solution, the protein molecules are randomly oriented, resulting in an isotropic scattering pattern, which is recorded, and radially averaged, giving a one-dimensional scattering curve of intensity, I(q), against q. During the experiment both the protein and the buffer scatter the incident X-rays, hence, to investigate the species of interest, the contributions of the buffer must be removed through subtraction. The resulting final scattering curve is a product of the form factor, P(q), the structure factor, S(q), a contrast factor, K, the concentration, c, and the molecular weight, M_w, of the sample:

$$I(q) = c \bullet k \bullet M_w \bullet P(q) \bullet S(q). \tag{2}$$

In Eq. 2, S(q) originates from the interactions between neighboring particles, and P(q) is caused by the overall shape of the scattering sample. Under dilute conditions, defined by the absence of inter-particle interactions, $S(q) = 1$, hence:

$$I(q) = P(q) \bullet S(q) => I(q) = P(q) \bullet 1, \tag{3}$$

where, I(q) depends only on P(q), and the contributions with respect to contrast, concentration, and size. P(q) is a Fourier transform of all interatomic distances, pair distribution function, P(r) within the measured protein:

$$I(q) = 4\pi \int_0^{D_{max}} P(r) \frac{\sin(qr)}{qr} dr, \tag{4}$$

where D_{max} is the maximum distance between any two electron pairs and r the distance. For a monodisperse solution of particles that contain a well-defined shape, for example, globular protein solutions, the Fourier transform of I(q) yields the real-space P(r), which can act as the basis for developing low-resolution models of the protein (Svergun, 1992; Svergun, Semenyuk, & Feigin, 1988). However, because IDPs do not adopt a single well-defined conformation, I(q) is the volume-weighted sum of all scattering profiles, or conformations, in the solution:

$$I(q) = \sum_{n=1}^{k} v_n I_n(q), \tag{5}$$

where v_n is the volume of each conformation, n, and I_n the corresponding scattering profile of each conformation for k distinct conformations. This characteristic is displayed graphically in Fig. 1, where each individual conformation contributes to the measured averaged scattering curve in a volume-weighted manner, resulting in a rather featureless scattering curve due to the averaging out of structural features.

The volume-weighted averaged scattering curve, collected during a SAXS experiment can be considered advantageous, as a single SAXS experiment captures contributions from the whole structural ensemble. However, because the contribution from each conformation is weighted by its respective volume, SAXS is less sensitive to sparse populations. One of the main goals of the analysis of IDP SAXS data is to obtain the weighted structural ensemble from the experimental data. This is typically achieved by either refining the data against structural ensembles or by performing simulations and either using the experimental data as a restraint, or comparing the scattering curve calculated from the simulation against the experimental scattering curve. This chapter will start by discussing methods that can be used to determine whether the measured sample behaves as an IDP and if so, how to analyze the ensemble properties of IDP scattering data. In the latter part of this chapter the focus will shift to deconvoluting the structural ensemble that gives rise to the experimental scattering spectra, including atomistic and coarse-grained modeling and computer simulations as a tool.

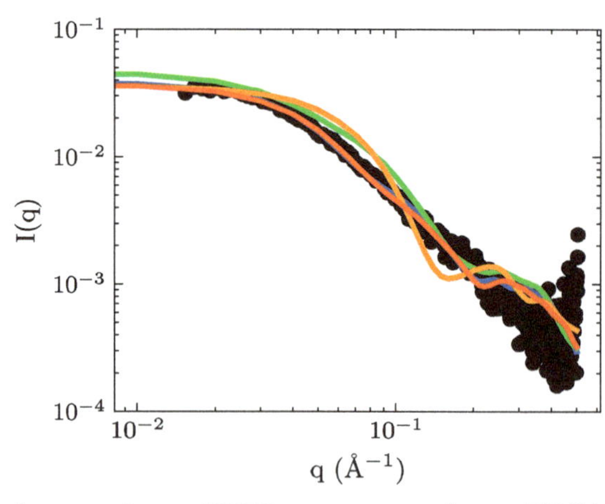

Fig. 1 Example of an experimental SAXS curve measured on an IDP (black dots) and the corresponding conformations depicted with colored lines that contribute to the average measured SAXS curve.

2. How to tell if a protein is an IDP?

Prior to performing a biological SAXS experiment, it is generally recommended to characterize the protein of interest using, for example, Dynamic Light Scattering (DLS) and Size Exclusion Chromatography (SEC) (Rambo, 2017), to obtain information about the stability, the average size, the mass distribution, and wether the sample is monodisperse or if there is any oligomerization or aggregation of the sample. Through Circular Dichroism (CD), it is possible to achieve qualitative information about the secondary structure, and if it behaves as an IDP (Chemes, Alonso, Noval, & Prat-Gay, 2012).

SAXS data provides several quick checks that can confirm the folded or unfolded state of the protein. In the first instance, some insight can be gained from a log-log plot of I(q) vs. q. As previously mentioned, the scattering profiles of IDPs are typically quite featureless. Due to the volume-weighted ensemble averaging of SAXS data, this translates into a smooth scattering profile, see Fig. 2A, in comparison to globular proteins, which are more feature-rich as shown in Fig. 2B.

Another control that can be easily performed is to plot the data as $I(q) \cdot q^{2.}$ vs. q, known as the Kratky plot, from which any distinct structural properties in the sample can be obtained, see Fig. 2C and D for an example of a typical Kratky plot observed for an IDP and a globular protein, respectively. For the former, the Kratky plot is, in a similar manner to the scattering curve itself, featureless with either a plateau or an increase in intensity at higher q, while for the latter, the Kratky plot is characterized by a clear maximum followed by a downturn at higher q. Determining the globular or disordered nature of a protein via the Kratky plot is straightforward if the protein is either completely folded or disordered, however, the interpretation is more problematic if the protein is globular with some disordered regions. In this case, when there are globular and flexible regions present, an alternative method could use complementary power transforms of the scattering curve by q^3, and q^4 to help detect any flexibility present (Brosey & Tainer, 2019). A plateau in the q^4 plot denotes a globular protein while a plateau in the q^3 plot denotes a degree of flexibility in the system.

A more detailed analysis can be performed through a dimensionless Kratky plot, where the data is normalized by the contributions due to the size, through the radius of gyration, R_g, and the mass/concentration, I(0),

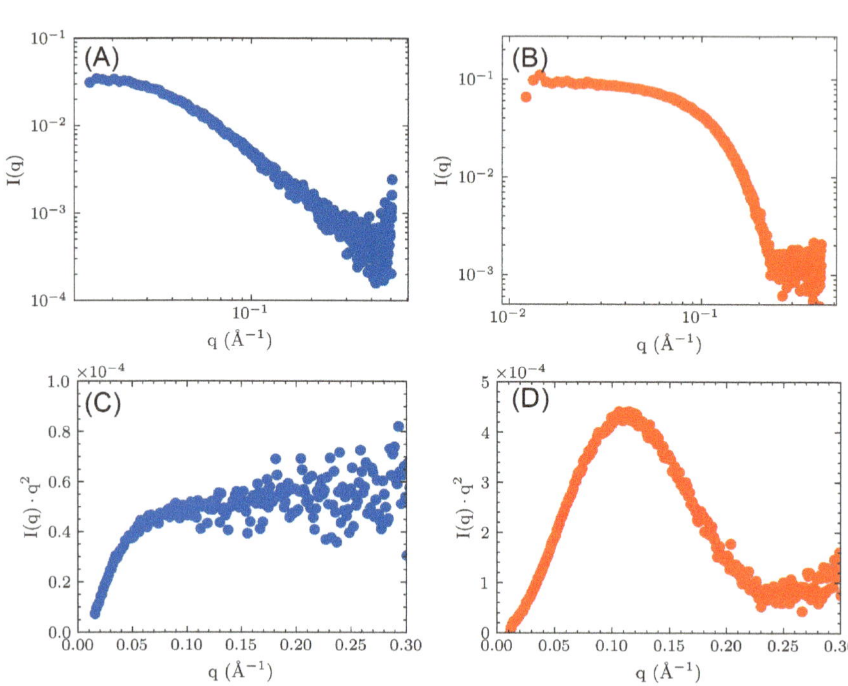

Fig. 2 (A) and (B) show I(q) of an IDP (blue) and a globular protein (red), respectively. It is noted that the IDP displays fewer features due to the averaging of all conformations. In (C) and (D) the corresponding Kratky plots are shown. A disordered protein is characterized by either a plateau or increase in intensity with increasing q, whereas globular proteins are defined by maximum followed by decay at higher q.

depicted as $q \cdot R_g^2 \cdot I(q)/I(0)$ vs. qR_g (Durand et al., 2010; Receveur-Brechot & Durand, 2012). This plot highlights features caused by the overall shape of the scattering sample. An example of a normalized Kratky plot for a globular protein and an IDP is displayed in Fig. 3. In the dimensionless Kratky plot, a globular protein display, independent of its size, a characteristic peak at $q \cdot R_g = \sqrt{(3)}$ with an intensity of 1.104. Absence of this characteristic peak, is indicative of either IDPs or globular proteins, that contain considerable disordered regions, where samples containing a mixture of disordered and globular regions will show intermediate behavior. By comparison of dimensionless Kratky plots, the extent of the disordered nature of different proteins can therefore be compared, irrespective of the concentration and the size of the protein.

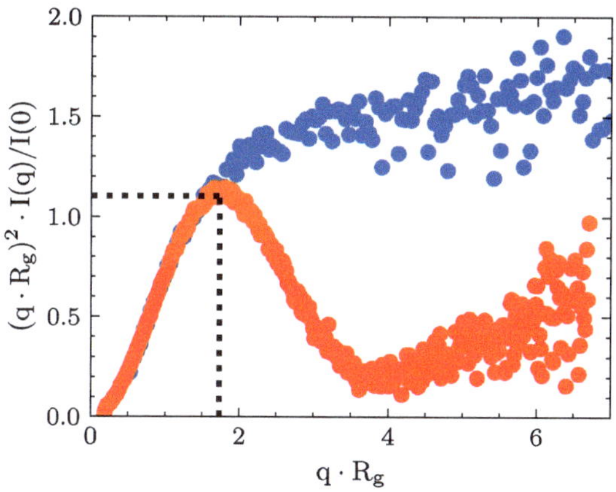

Fig. 3 Comparison of a dimensionless Kratky plot obtained for an IDP (blue) and a globular protein (red), respectively. Globular proteins are characterized by a peak at $q = \sqrt{(3)}$, which has an intensity of 1.104. This peak is not present in the case of disordered proteins.

3. The conformational ensemble

In general, a SAXS curve can be divided into three regimes that contain information on the size, shape, and surface of the scattering sample, see Fig. 4. Note that the precise location of these regimes is dependent upon the properties of the scattering sample. If the SAXS data is obtained for an IDP, analysis of these regimes yields information on the average properties, determined by the weighted contribution of all different conformations of the protein thus the conformational ensemble.

3.1 Determination of the size through the radius of gyration

According to the Guinier approximation, the scattered intensity is dependent only upon two parameters at small angles: R_g and the forward scattering angle, I(0), according to (Guinier, 1939):

$$I(q) = I(0) \exp\left(-\frac{qR_g^2}{3}\right). \tag{6}$$

Hence, by depicting SAXS data as $\ln(I(q))$ vs. q^2, a linear plot should be revealed in the low-q region, from which both R_g and I(0) can be determined.

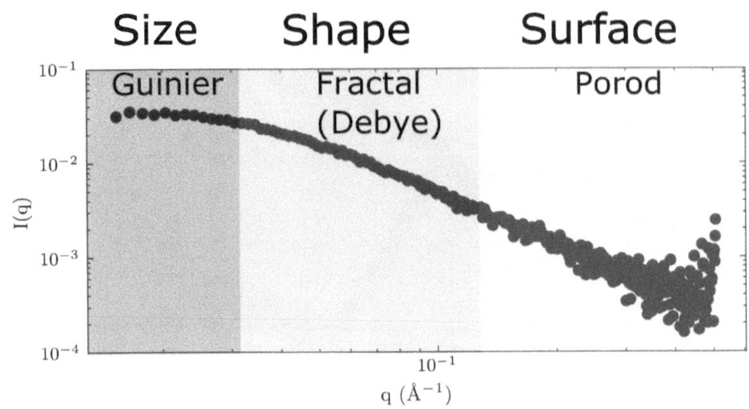

Fig. 4 Different regions of the SAXS curves provide information on different scales. Above three major regions and the information they contain are highlighted. Note that the boundaries are only examples and may vary in their specific q-range depending upon the properties of the measured protein.

For globular proteins, the Guinier regime is identified by the relationship $q \cdot R_g \leq 1.3$ (Putnam, Hammel, Hura, & Tainer, 2007). The absence of linearity in the Guinier plot indicates the presence of interactions between proteins in the sample, where an upturn indicates attractive interactions and a downturn repulsive interactions between the proteins. If so, changes to the experimental method can be made to overcome the contribution caused by inter-particle interactions, for example, by decreasing the protein concentration, changing the buffer conditions, or performing SEC-SAXS experiments. If the presence of inter-particle interactions has been detected, the Guinier approximation is no longer valid. If, in these instances, the Guinier approximation is still applied the obtained R_g will either be over- or underestimated for attractive and repulsive interactions, respectively (Borgia et al., 2016). For IDPs the Guinier analysis is known to underestimate the R_g when using the fitting range of $q \cdot R_g \leq 1.3$. Hence, a narrower q-range is used of $q \cdot R_g \leq 1.1$ or 0.8 (Borgia et al., 2016; Receveur-Brechot & Durand, 2012; Zheng & Best, 2018). The maximum q-range depends both on the compactness and length of the IDP studied, and in general, the more the protein shape deviates from a sphere, the smaller q-range can be used. Since q-max is dependent on the R_g, longer IDPs have a narrower Guinier region, hence fewer data points are available. In these cases, other methods can be used to estimate R_g, for example, by fitting the Debye function or using the extended Guinier approximation (Zheng & Best, 2018). Compared to globular proteins of the same M_w,

IDPs have significantly larger R_g's due to their less compact nature. Indeed, the Flory-scaling law can estimate the expected R_g according to:

$$R_g = R_0 N^v, \tag{7}$$

where N is the number of amino acids, and the exponent v is 0.5–0.6 for IDPs, dependent upon solvent conditions, and about 0.3 for globular proteins (Hofmann et al., 2012). Determining R_g from the Guinier regime is a fast and model-free method. Furthermore, the value of R_g combined with the number of amino acids in the protein provides insight into the compactness of the protein (Ahmed, Crehuet, & Lindorff-Larsen, 2019). Finally, the I(0) value determined from the Guinier analysis can be used as an estimate of M_w, as will be discussed in the following section.

3.2 Determination of size through the molecular mass

Calculation of M_w from the SAXS data can be used to determine the monomeric or oligomeric nature of the IDP in solution. If the collected data is on an absolute scale, then I(0), determined from the Guinier plot, is directly proportional to the number of electrons in the sample, the contrast of the sample, and the concentration of the sample according to the equation (Orthaber, Bergmann, & Glatter, 2000):

$$M_w = \frac{N_a I(0)}{c \Delta p_m^2}, \tag{8}$$

where, c is the concentration, N_a is Avogadro's number, and Δp_m^2 is the scattering contrast per mass. If both the concentration and the contrast of the sample are known, then M_w can be calculated directly using this relationship. If the scattering data was not collected on an absolute scale but the concentration of the measured protein is known, M_w can alternatively be estimated by comparing I(0) with I(0) of a standard of known M_w measured at a known concentration using the relationship:

$$M_{w,\,exp} = \left(\frac{I(0)_{exp}}{c_{exp}}\right)\left(\frac{M_{w,standard}}{\left(\frac{I(0)_{standard}}{c_{standard}}\right)}\right), \tag{9}$$

where I(0) is the forward scattering angle, c, is the concentration of the protein, and the subscripts represent the measured protein (exp) and the standard of known concentration, and M_w (standard). This method assumes that both the partial specific volume and contrast of the known standard are the same as the sample. Usually, this assumption provides a reasonable estimate if the known standard is also a protein, with the typical error being on the order of 10% (Mylonas & Svergun, 2007). The accuracy of this method can assist in determining the oligomeric state of the protein. Notice, that if only a small percentage of the protein is oligomeric then it may not be observable based solely on M_w calculation due to the inherent error. Other methods have also been developed to estimate M_w from the final scattering curve, without requiring knowledge of the precise protein concentration or further experimental data from known standards. These methods are particularly useful in situations where it is difficult to determine the precise protein concentration of the sample, e.g., SEC-SAXS experiments, and are based upon various parameters extracted from the scattering curve, for example, estimating the protein volume (de Neto et al., 2022; Hajizadeh, Franke, Jeffries, & Svergun, 2018) or through the calculation of the model-independent parameter, volume-of-correlation, V_c (Rambo & Tainer, 2013). Typically, well-established relationships between the volume and M_w are used as a reference if the protein has the same relationship between the volume and mass. This assumption is quite accurate for globular proteins, however, IDPs typically occupy a larger volume in solution resulting in these calculations often yielding an overestimate of M_w, thus caution should be used with these calculative methods in the case of IDP samples.

3.3 Determination of the shape through the pair distribution function

The second regime contains information about the overall shape of the protein in solution and is often combined with the information on the overall size. A common method used to analyze scattering data from SAXS data of monodisperse globular proteins is to determine a low-resolution structural envelope that represents the shape of the protein, which has a low resolution compared to, for example, crystallography experiments (Chacon, Moran, Diaz, Pantos, & Andreu, 1998; Franke & Svergun, 2009). Nonetheless, it can be used to determine oligomer- and fibril formation or mechanisms of proteins that cannot be crystallized (Hallin, Hasan, Guo, & Akerlund, 2016; Sagar, Herranz-Trillo, Langkilde, Vestergaard, & Bernado, 2021). Prior to performing shape analysis, it is important to confirm, by the Guinier plot, the absence of inter-particle interactions as the analysis should only be performed on the P(q). The first step in this process is to calculate P(r) from the scattering data, which is a Fourier transform of P(q):

$$P(r) = \frac{r^2}{2\pi^2} \int_0^\infty \frac{q^2 I(q) \sin(qr)}{qr} dq, \tag{10}$$

where r is the distance between two points. Once determined, both R_g and the volume of the scattering particle can be estimated from P(r). This method considers a larger part of the scattering curve and does not rely only on the few points that may be present in the narrower Guinier range of IDPs. Hence, it may provide a more realistic estimate of R_g for IDPs. Frequently used programs to calculate P(r) from SAXS curves are GNOM from the ATSAS package (Manalastas-Cantos et al., 2021) or ScÅtterIV (Tully, Tarbouriech, Rambo, & Hutin, 2021), where GNOM calculates P(r) via indirect Fourier transform. Visual investigation of P(r) can therefore provide insight into the protein structure. At longer distances the P(r) function should always drop to zero, the point at which this happens defines the longest distance between atoms within the protein and is known as D_{max}. Fig. 5 shows P(r) for a globular protein and an IDP. Globular proteins depict a clear maximum followed by decay, whereas IDPs show a much slower decay to D_{max}, due to the longer intra-particle distances present in IDPs compared to globular proteins of similar M_w. Thus, P(r) presents another method to evaluate the extent of disorder from SAXS data. P(r) from monodisperse solutions of globular proteins can be used as the basis for the determination of a

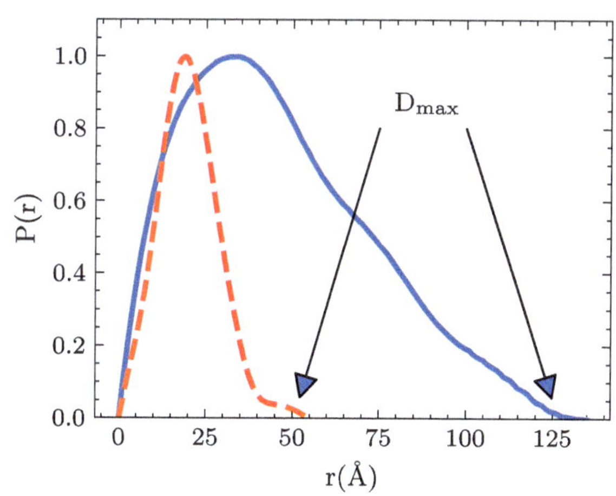

Fig. 5 Examples of P(r) for a globular (red) and an IDP (blue), respectively. As the scattering spectra, P(r) of IDPs is less feature-rich than obtained for a globular protein and is characterized by a slow decay with increasing r, caused by the extended period distances present within IDPs.

low-resolution structural model, whereas in the case of IDPs, this analysis is less meaningful as P(r) represents the averaged structural properties of the many conformations present in the solution.

3.4 Determination of the shape through the Porod regime

In the Porod regime, at intermediate-to-high q, length scales smaller than the overall scattering object is probed. Porod's law states that at high-q, I(q) decays according to:

$$I(q) \propto q^{-D}, \tag{11}$$

where D is the Porod exponent, which describes the degree of foldedness vs. flexibility in a sample (Rambo & Tainer, 2011). The exponent can, for example, be determined by fitting a straight line to the mid-high-q values of the log-log plot of the data or by using the flexibility plot in the software ScÅtterIV (Tully et al., 2021). The precise value of the Porod exponent can yield information on the three-dimensional properties of the protein, examples of Porod exponents for one-dimensional (1D) rods, a two-dimensional (2D) flat sheet, and a three-dimensional (3D) sphere are shown in Fig. 6.

3.5 Notes on the average conformational properties

Several methods have been presented that can be used to analyze the average properties of an IDP from SAXS data. The methods presented each focus on slightly different physical properties of the protein measured by SAXS, i.e., the size, shape, or M_w. The physical properties determined using these methods can act as indicators of disorder/order and provide insight into the flexible nature of the studied protein. Ideally several of the methods

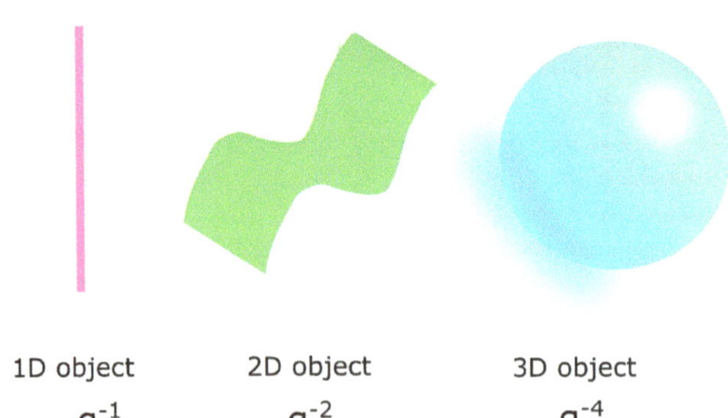

Fig. 6 Examples of shapes and their corresponding Porod exponent.

should be combined. For example, a plateau or increase at high-q in the Kratky plot may be caused by poor background subtraction or a narrow Guinier range may be due to aggregation. Although these methods only yield physical properties of the ensemble average, they can provide insight into whether the protein studied has properties consistent with an IDP or a globular protein and yield useful average properties of the structural ensemble represented by the measured I(q).

Furthermore, they have the advantage of not applying models to the data and thus can be rapidly applied to SAXS data, requiring only minimal changes between different proteins as determining the precise q-range required for meaningful analysis. When considered together, the results from this initial analysis may also influence the choice of subsequent, in-depth analysis.

4. Ensemble optimization methods

In the previous section, several model-free methods that can be used to obtain information from the average of the ensemble of conformations measured by SAXS were introduced. In this section, the focus will shift to more advanced methods that can yield not only information on the average properties of the full conformational ensemble but also information about the structural ensemble itself.

The use of ensemble methods to characterize disordered proteins was initially developed for NMR studies and later adapted to the analysis of SAXS data of IDPs (Bernado et al., 2007; Tria, Mertens, Kachala, & Svergun, 2015). The first developed method to analyze SAXS data by fitting structural ensembles was the ensemble optimization method, EOM (Sibille & Bernado, 2012). This method is currently still one of the most widely used. There are also other ensemble-based methods each of which takes a slightly different approach, for example, MultiFoXS uses a population-weighted ensemble modeling (Schneidman-Duhovny, Hammel, Tainer, & Sali, 2016) that explores the space of the φ and ψ main chain dihedral angles of a user-defined input (Schneidman-Duhovny et al., 2016). The underlying principles of the different ensemble methods to fit SAXS data of IDPs typically follow similar steps. The first step of most ensemble methods is to generate a pool of protein conformations that realistically samples the proteins' conformational space while providing conformations that are physically realistic. In the case of EOM, the random pool can be generated by the ATSAS program RanCh, which generates models based on the sequence of the protein given by the user, combined with an algorithm that takes into consideration bond vs. dihedral angle distributions (Bernado et al., 2007). For completely unfolded proteins the pool can be generated based on this latter parameter for either chemically denatured proteins or natively unfolded proteins which differ slightly in the allowable α-carbon dihedral angles of the backbone. A range of other options is also available with the EOM method, such as modeling flexible linkers between globular domains and introducing symmetry within the generated pool. Typically, several tens of thousands of models are generated using this method, which can be investigated by plotting distributions of key structural parameters such as R_g or D_{max}. EOM does not only accept ensembles generated by the ATSAS software RanCh, but external ensembles can also be used as input. Examples of externally generated ensembles include those generated from experimental data, for example, by NMR or ensembles determined from computational simulations (Cheng, Peng, & Zhang, 2017).

The second step in the EOM process is to select a subset of structures from the generated pool that agree well with the scattering data. This is typically achieved by calculating the theoretical scattering curve of each conformation and then fitting different ensembles of conformations to the experimental SAXS data. EOM achieves this step via the use of a genetic algorithm implemented in the GAJOE program, which is divided in five steps:

1. Several sub-ensembles of structures from the random pool are selected, each consisting of 5–50 structures. By using the nomenclature of genetic algorithms these subsets are referred to as chromosomes. The precise number of structures present in each chromosome varies based on the flexibility of the protein, with highly flexible proteins being more accurately represented by larger ensembles, while SAXS of less-flexible proteins can be represented by smaller ensembles. Using EOM in the default mode generates 50 sub-ensembles. In the latest version of EOM, the ensemble size can be automatically optimized during the selection procedure. In this mode each conformation is also assigned a weighting, in the case of a highly rigid protein fewer conformations will be present in each ensemble and thus overfitting of the data is reduced.
2. In the second step the chromosomes, generated in Step 1 are modified, by either replacing a selected percentage of structures, default is set to 20%, with others from the original pool, referred to as mutations, or by exchanging structures between chromosomes, denoted crossing. In EOM the default settings result in the generation of 100 new chromosomes, 50 for each operation.

3. Next, the average theoretical scattering curves are calculated for each chromosome, both those produced in Step 1 and the modified chromosomes produced in Step 2, and the average fit of its individual SAXS profiles are compared with the experimental scattering curve, through the fitness parameter χ^2.
4. The 50 best fitting chromosomes are selected and used to repeat the process starting from the first step.
5. The whole process is then repeated several times, in the default mode it corresponds to 100 times.
6. The final best-fitting chromosomes are then combined, and various physical parameters including R_g and D_{max} are calculated. The program also gives information about several models that best represent protein conformations.

The physical parameters obtained are then compared with those in the initial random pool, an example is given in Fig. 7, which covers the whole conformational landscape of the protein. By comparing the selected ensemble with the original ensemble, comparisons can be made with respect to the compact/extended nature of the protein. For example, in Fig. 7, the selected pool favors more extended conformations than the random pool. The flexibility of the protein can be judged qualitatively by the width of the pool thus

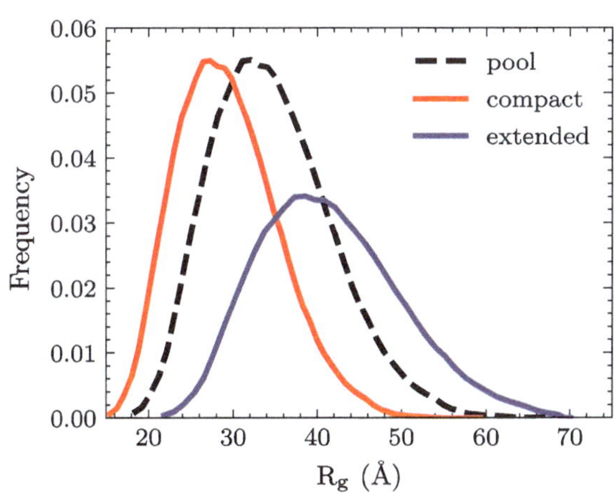

Fig. 7 Example output of the EOM analysis applied to an IDP SAXS curve. The black dashed line represents the different conformations present in the generated pool. The red and blue lines represent the selected ensembles obtained by fitting to the SAXS data. By comparing the selected ensemble with the random pool, information can be gained about the flexibility and judgments about the IDPs extended or collapsed nature in the solution.

does the protein adopt a broad range of conformations or not. In Fig. 7 the extended conformation covers a much wider range of sizes, indicating that this protein is more flexible. The latest version of EOM also offers a quantitative measure of protein flexibility through the R_{flex} parameter, where a R_{flex} of 0% is fully rigid and conversely a R_{flex} 100% is completely flexible. In some instances, EOM may yield a bimodal distribution with peaks appearing at low and high R_g compared to the pool, indicating that the protein can adopt both compact or more extended conformations. Ensemble methods are a powerful tool where more detailed analysis can reveal the existence and characterize the properties of a wide range of conformations that would be diminished if only applying analysis of the volume-weighted ensemble average data depicted in the I(q) spectra. Hence, ensemble methods can provide a more qualitative description of SAXS data from IDPs.

Apart from the EOM software offered by the ATSAS package, other ensemble-based methods have been successfully applied to the analysis of SAXS data from IDPs. In general, each method offers different pros and cons compared to the EOM method. For example, the program ENSEMBLE is not able to generate or modify structures on the fly, but instead requires the user to input all structures. Generated by other methods, it also offers the possibility to utilize a range of experimental data during the ensemble refinement process, e.g., SAXS, NMR, and the hydrodynamic radius (Gallat et al., 2012; Manalastas-Cantos et al., 2021). A more recent example is Bayesian Ensemble SAXS (BE-SAXS) which uses a generative model and does not impose restrictions on the ensemble size, thereby allowing sampling of the full conformational space of the protein, i.e., not limited by the number of models in the pool (Antonov, Olsson, Boomsma, & Hamelryck, 2016). Many other methods for determining ensemble structures based on IDP SAXS data exist including MES (Pelikan, Hura, & Hammel, 2009), BSS-SAXS (Yang, Blachowicz, Makowski, & Roux, 2010), MultiFoXS (Schneidman-Duhovny et al., 2016), and more advanced, computer simulation methods based on, for example, Monte Carlo or MD. The latter methods will be discussed later in this chapter, see Section 6.

5. Special considerations for crowded IDP solutions

Crowding experiments can be broadly classified as either self-crowding or secondary crowding (Ellis, 2001; Zimmerman & Minton, 1993). In the case of self-crowding, the concentration of the protein of interest is raised to high levels whereas secondary crowding means an increase of the concentration of a secondary excipient (Fagerberg, Lenton, & Skepö, 2019; Fagerberg, Månsson, Lenton, & Skepö, 2020; Grimaldo et al., 2015). Typically, polymers such as PEG or Ficoll are used for crowding studies but other molecules, such as, for example, cell lysate have also been utilized for this purpose (Bonucci et al., 2021; Grimaldo et al., 2019). Crowding can result in several effects, in some instances, IDPs have been shown to change their conformational ensemble upon crowding, although other studies have shown no effect (Fagerberg et al., 2019; Johansen et al., 2011). Certain IDPs can also participate in liquid-liquid phase separation, driven by protein-protein interactions, in the presence of a high concentration of excipients (Ray et al., 2020; Sawner et al., 2021). These changes observed under crowded conditions, i.e., modification of the structural ensemble and the onset of inter-protein interactions are ideally suited to be studied by small-angle scattering. Small-angle neutron scattering (SANS) can be particularly useful to study the protein behavior in the case of two-component systems (Goldenberg & Argyle, 2014; Johansen et al., 2011; Miller, Kim, & Mittal, 2016). By taking advantage of the difference in the scattering length density of D_2O to neutrons compared with X-rays, the contrast of deuterated proteins or excipients can be matched out, effectively making them invisible. For example, a deuterated PEG crowding agent allows the investigation of the protein properties ignoring contributions from the excipient (Goldenberg & Argyle, 2014; Johansen et al.; 2011; Miller et al., 2016). In both instances of crowding, one should be careful when interpreting the resulting SAXS data since high-concentration protein solutions are, by their nature, more likely to be affected by inter-protein interactions and/or aggregation, which can affect the Rg as mentioned earlier and may disturb the experimental data's consequent analysis.

Inter-protein interactions can also be probed by SAXS, if a concentration series of measurements are performed in a suitable concentration range. Such an analysis relates the osmotic pressure virial expansion:

$$\Pi/\rho kT = 1 + B_2\rho + B_3\rho^2 + \ldots \tag{12}$$

where Π is the osmotic pressure, B_n is the nth virial coefficient and ρ is the particle density, with the structure factor at zero angle (after converting particle density to concentration):

$$S(0\ c) = \frac{RT}{M_w}\left(\frac{\partial \Pi}{\partial c}\right)^{-1}. \tag{13}$$

Limiting the expansion to only the second virial coefficient, and writing the second virial coefficient as $A_2 = B_2 N_a / M_w^2$, a function:

$$I(0\ c) = cK \frac{1}{1 + 2cA_2 M_w} \tag{14}$$

is attained, which can be fitted to the experimental scattering data. The second virial coefficient is sensitive to solvent conditions and is positive for repulsive interactions and negative for attractive interactions.

6. Beyond analytical models

6.1 Models for the simulation of IDPs

Rather than generating a pool of structures from which a selection needs to be chosen, an ensemble that is directly representing a real system can be generated using simulation techniques. These rely on a description of the underlying physical driving forces, known by many names such as the intermolecular potential, the energy function, or the force field, partly depending on the technique used. The system itself, including the environment in which the IDP exists in, also needs to have a topological description. The interaction potential is specific to the topology used. While the added description of the surrounding environment may add complexity, it also offers a range of added capabilities, such as taking into consideration other types of solvents, the addition of ligands or interaction partners, different concentrations of salt, etc.

Probably, the most intuitive topological description is formed by considering each atom in the system. Usually, one specifically denotes the protein to be all-atom and the solvent explicitly described. Models using all-atom, explicitly solvent, often termed "force fields" have a history longer than the concept of IDPs, where proteins, in general, have been considered. Well-known examples of such models include the AMBER (Cornell et al., 1996), CHARMM (Brooks et al., 1983), OPLS (Jorgensen & Tiradorives, 1988), and GROMOS force fields (http://www.gromos.net).

Simulations of all-atom, explicit solvent models are highly computationally demanding and may require high-performance computing resources. To reduce the computational complexity, the overall environment surrounding the IDP can be considered implicitly, while keeping the description of the IDP on an atomistic level. Examples of such models are the ABSINTH (Choi & Pappu, 2019) or ProFasi (Irbäck & Mohanty, 2006) models. Models that further decrease the computational cost may use a reduction in the resolution of the topological description of an IDP. This decreases the number of particles that need to be considered, and in general shape the interaction potential to be smoother, making it easier to reach convergence. Such models are referred to as coarse-grained models. The level of coarse-graining can vary and may not be strict. A widely used model is the MARTINI force field (de Jong et al., 2013), which maps approximately four heavy atoms into one effective "interaction center", other terms used are "bead" or "dummy particle," except for ring structures. Other mappings use one bead for the protein side chain and one for the protein backbone, while an even simpler mapping models the amino acids as one bead, examples of different levels of coarse-graining are shown in Fig. 8 (Cragnell, Durand, Cabane, & Skepö, 2016; Dignon, Zheng, Kim, Best, & Mittal, 2018). Coarse-graining may still use explicit solvent, as in the MARTINI model, or use implicit solvent.

Coarse-graining has the drawback of not being able to consider more high-resolution features of IDPs. For instance, secondary structures are usually not considered in coarse-grained models, although there are exceptions, such as the PLUM model (Bereau & Deserno, 2009). The most coarse-grained models may not even distinguish between all types of amino acids, while such features are probably more dependent on the interaction potential used rather than the topological resolution. Even with such restrictions, they may still provide good performance (Cragnell, Rieloff, & Skepö, 2019), why knowledge of precision requirements for the problem at hand is pertinent.

While coarse-grained models are obviously more approximate in nature, and therefore in principle may not offer the same level of accuracy or ability to model the same range of IDP systems as atomistic models, the choice of

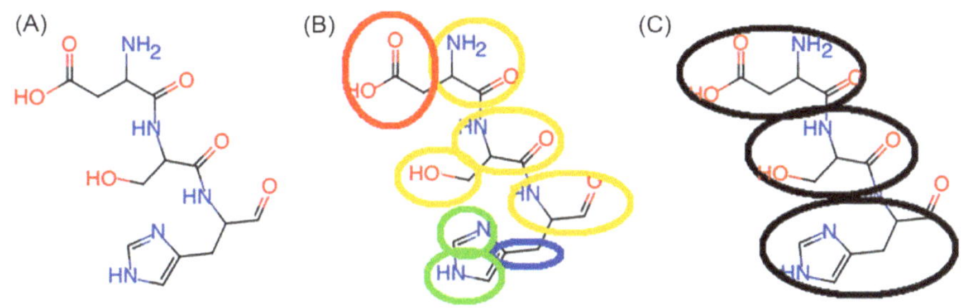

Fig. 8 Examples of coarse-graining schemes. (A) The original all-atom depiction of aspartic acid-serine-histidine peptide, (B) The MARTINI mapping of the peptide, and (C) One bead per amino acid mapping of the peptide.

the model must be carefully considered. Even in the case of the more computationally complex all-atom explicit solvent models, early application of these to IDPs showed them to produce more compact structures than what the experiment indicated (Henriques, Cragnell, & Skepö, 2015). This prompted further refinement of these models, as well as the construction of new models specifically for the class of IDPs. Even with this general development in mind, a newer model may not necessarily be better performing than an older model for a given system (Mu, Liu, Zhang, Luo, & Chen, 2021). A literature investigation should be performed to determine which model may be best suited for the system at hand, and a comparison of several models of a reference system.

6.2 Simulation techniques

To generate structures from a given model, several schemes are available. The historically first of these is the Monte Carlo method, a general method finding use far and wide beyond protein simulations. In this method, after system setup, trial moves are performed, which are accepted if the interaction potential is lower in the new configuration, or it is accepted with a probability related to the Boltzmann distribution, where configurations with higher interaction potential have a lower probability to be accepted. The moves can be unphysical, which is one reason why a standard Monte Carlo simulation will not yield time-dependent information, such as diffusion. On the other hand, the moves can be tailored to fit the system of consideration.

Another very popular method used for simulations is MD. Here, Newton's equations of motion are solved to evolve the system forward in time. This yields very intuitive simulations, where, in a sense, observation of the system can be analogous to watching processes through a "molecular microscope." However, one should be aware that due to numerical considerations, no trajectory produced from MD is "realistic," but this is not an issue as the goal is to attain statistical predictions. Additional care needs to be taken regarding having the correct temperature and pressure, however, most software have algorithms implemented for these considerations, and interaction potentials usually specify which algorithms they were developed together with. More generally,MDs has the drawback of limitations in which interaction potentials that can be used—discontinuous interaction potentials can in general not be used.

A third method for performing a simulation is through a stochastic simulation method, such as Brownian dynamics. In this case, evolving the system forward is done via the Langevin equation, which uses a stochastic term for the impact of solvent on the protein of interest. This type of simulation is usually performed when longer time scales are considered.

Simulations have in general two major limitations: The accuracy of the interaction potential and achieving accurate sampling. Sampling is about the ability to run simulations for a long enough time to assure that properties are converged, and simulate all relevant configurations of the system, not being limited to a subpopulation. Sampling is coupled with the limits of computation; hence why coarse-grained methods are one way to manage this problem. Another method is to use so-called enhanced sampling. These include replica-exchange MD, REMD (Sugita & Okamoto, 1999), and metadynamics (Laio & Parrinello, 2002). REMD runs several copies of the simulation at different temperatures in parallel, swapping between systems according to criteria similar to what is used in a Monte Carlo simulation. The partial simulation of the system at higher temperatures enables energy barriers to be crossed that could not be overcome at lower temperatures. Metadynamics on the other hand biases the simulation to progress in accordance with chosen collective variables.

6.3 Simulations of the single IDP and SAXS-related analysis

A simulation can be compared with SAXS data either directly through the calculation of intensity profiles or indirectly, by comparing extracted parameters, for example, the R_g. From simulation configurations (or "frames"), R_g can be computed by:

$$R_g{}^2 = \frac{\sum_{i=1}^{p} |r_i - r_{cM}|^2}{N_p} \tag{15}$$

where N_p is the number of monomers, r_i is the position of bead i, and r_{cM} is the position of the center of mass. P(r) can be computed as well for comparison with the experimentally determined data as mentioned above.

For the calculation of intensity profiles, several algorithms/software are available, including CRYSOL (Svergun, Barberato, & Koch, 1995), FoXS (Schneidman-Duhovny, Hammel, & Sali, 2010; Schneidman-Duhovny, Hammel, Tainer, & Sali, 2013), and Pepsi-SAXS (Grudinin, Garkavenko, & Kazennov, 2017). These are commonly referred to as "forward models". In general, the calculation of an intensity curve follows Eq. 16:

$$I(q) = \left|\sum_{j}^{N} b_j\, e^{-iqr_j}\right|^2 \tag{16}$$

with b_j being the generalized scattering length of particle j (implicitly dependent on q in SAXS), N is the total number of particles, and r_j the position of particle j. However, this equation is specific for a particular scattering orientation. By assuming no preference in molecular orientation, one can integrate over all orientations to yield the Debye equation:

$$I(q) = \sum_{j=1}^{N_j} \sum_{i=1}^{N_i} f_i(q) f_j(q) \frac{\sin qr_{ij}}{qr_{ij}} \tag{17}$$

where f is the atomistic scattering factor for particle i, j, and r_{ij} is the distance between them. This equation is used in the FoXS software (Schneidman-Duhovny et al., 2010), while other algorithms may use different approximations of the Debye equation. One advantage is that it is possible to compute partial structure factors, where the structure factor for a specific group of particles in a simulation is determined. If several partial structure factors are produced, covering all particles, they can be combined to yield the full structure factor.

In the case of coarse-grained modeling, forward models have been constructed to parameterize form factors for the dummy particles used in a simulation (Niebling, Bjoerling, & Westenhoff, 2014; Stovgaard, Andreetta, Ferkinghoff-Borg, & Hamelryck, 2010). The most coarse-grained models, which do not differentiate between all amino acids may assume an "average" form factor for all dummy particles. Another alternative is to fine-grain a coarse-grained simulation (Rotkiewicz & Skolnick, 2008), back-mapping the dummy particles into fully atomistic resolution, and thereafter perform the SAXS computation with an all-atom forward model. There are mainly two optimization targets for these algorithms: accuracy, in terms of replicating the SAXS curve for structures that are well-known, and speed of computation. Even if the calculation of an intensity profile for one system configuration may only take a few seconds, a simulation may generate hundreds of thousands, if not millions, of configurations, hence why the speed can indeed be an important aspect when considering which software to use.

A given force field can be found to be inadequate when comparing simulation results and experimental data. A combination of simulation with experimental data can be used, either by reweighting the ensemble produced after a simulation (Rozycki, Kim, & Hummer, 2011) or by directly incorporating the SAXS data into the simulation (Hermann & Hub, 2019). The latter approach uses the SAXS data as bias, or a restraint, along the progression of the simulation, which can be favorable because an ensemble generated by a simulation may lack important configurations. This can be due to configurations that erroneously have high energies according to the force field, which would not be sampled unless the simulation is very long.

6.4 Simulations of concentrated solutions of IDPs and SAXS-related analysis

For systems of concentrated IDP solutions, the simulation time may increase due to two reasons. Firstly, if an implicit solvent was used, the increase of IDP chains in the system will increase the number of particles needed to be considered. In the case of explicit solvent, solvent molecules are replaced by IDP chains, which is why the overall number of particles is roughly the same. Secondly, having many chains in a system may cause entanglement, during which some subpopulations of the ensemble may become more difficult to reach. This inevitably leads to increasing the time of convergence when concentrated systems of chains are considered. If liquid-liquid phase separation is to be studied by simulation, a slab-method simulation setup can be used (Dignon et al., 2018). This method reduces the length of the simulation required to reach convergence.

In the concentrated regime, the SAXS curve will have a non-negligible contribution from S(q). While several analytical models exist that incorporate such effects (Pedersen, 1997), a general treatment would assume that changes in P(q) cannot be separated from the contribution of S(q) at high concentration. Consequently, R_g cannot be determined and indirect comparison between simulation and experiment cannot be used. Direct comparison through a calculated SAXS intensity curve would thus be advised, but also in this regard there are possible pitfalls. For example, changes in the hydration layer (discussed in Section 7) due to high concentration should also be considered (Balu et al., 2016).

7. General notes on the treatment of hydration layers of IDPs

As mentioned, calculation of SAXS intensity profiles through computer simulations can be separated into two groups, depending on whether the solvent is modeled implicitly, as a continuous electron density, which is the case in many coarse-grained models, or considered explicitly, calculating all parameters, which is commonly used in all-atomistic models. The former offers reduced computational cost and the advantage that it is possible to simulate more complex and concentrated systems, but on the other hand, it requires the definition of several free parameters, for example, the excess density of the solvation layer. A common approach is to include the size of the hydration layer in the size of the amino acid, if, for example, the coarse-graining is on the amino acid level. The question is then how to determine the optimal values to use and their impact. For example, for SAXS experiments, many forward models include parameters that describe the contribution of the hydration layer to the background-subtracted experimental data. In an earlier study, it was shown that small variations of the contrast of the solvation shell, $\delta\rho$, severely affect the outcome, analysis, and interpretation of computed SAXS profiles for folded and disordered proteins (Henriques, Arleth, Lindorff-Larsen, & Skepö, 2018). An example of this behavior is shown in Fig. 9, and additionally demonstrated in Table 1, where R_g is determined from the Guinier approximation from the calculated curves.

Forward models are used to relate the simulated structures to the experimental data, see Section 6.3 for a list of commonly used programs. The programs can either predict the solution scattering curve from a structure or fit an experimental scattering curve to the structure. Fitting is achieved by varying one or more of three parameters describing: (i) the average displaced solvent volume per atomic group, (ii) the contrast (excess scattering) of the hydration shell, and (iii) the relative background. For globular proteins, the default value for the thickness of the hydration shell is typically chosen to be 3 Å, and the average electron densities (ρ_o and ρ_b) of bulk water and a typical globular protein are 0.334 e/Å^{3}and 0.43 e/Å^3, respectively. The latter number works well for the example Lysozyme, which is a globular

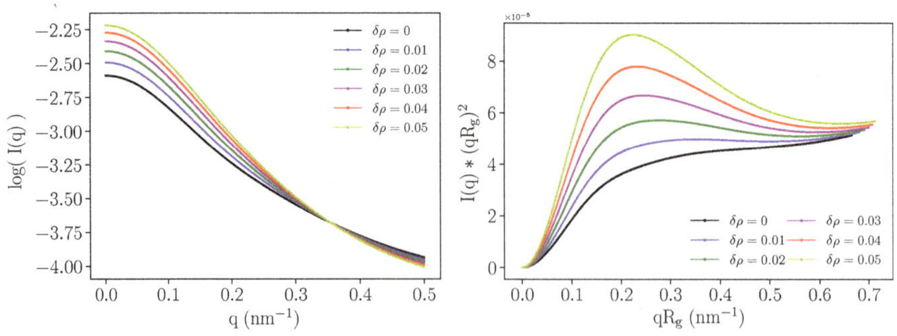

Fig. 9 SAXS curves calculated via Crysol using different values of $\delta\rho$. From a simulation of Histatin 5, as found in Henriques et al., where the AMBER99-SBN-ILDN force field was used.

Table 1 The radius of gyration, R_g, *determined* via Guinier approximation (using different limits of qR_g) as a function of the contrast hydration shell used in the calculation of scattering curve by Crysol.

$\delta\rho$	0	0.01	0.02	0.03	0.04	0.05
R_g (Å) ($qR_g \leq 0.8$)	13.31	13.60	13.82	1.400	14.15	14.27
R_g (Å) ($qR_g \leq 0.9$)	13.25	13.55	13.77	13.95	14.10	14.22
R_g (Å) ($qR_g \leq 1.0$)	13.17	13.47	13.70	13.89	14.04	14.17
R_g (Å) ($qR_g \leq 1.1$)	13.09	13.39	13.64	13.82	13.98	14.10
R_g (Å) ($qR_g \leq 1.2$)	12.99	13.31	13.55	13.74	13.90	14.03
R_g (Å) ($qR_g \leq 1.3$)	12.88	13.21	13.46	13.65	13.81	13.95

protein, but it is not evident what to use for IDPs, due to their flexible conformation, and because the surface properties of IDPs are very different from those of folded proteins (Gallat et al., 2012).

SAXS predictions based on explicit-solvent MD are computationally more expensive but hold several advantages since they naturally reproduce the increased R_g due to the hydration layer (Chen & Hub, 2014). Thereby R_g is not fitted against the experimental data, whereas another advantage is that no reduced P(q)'s are used, eliminating the uncertainties and fitting parameters due to the atomic volumes. The fact that SAXS data contains contributions from the hydration layer, an integrated investigation provides novel and untapped opportunities to gain insight into hydration and macromolecular function by combining SAXS with computer simulations.

Both implicit and explicit modeling provide useful data when determining hydration layers for IDPs. Implicit forward models are especially useful during or just after SAXS data collection or to test new models against SAXS data as they are computationally very quick to perform. While the more computationally heavy explicit MD provides greater accuracy and refinement but with a cost of time etc.

8. Summary and conclusions

SAXS provides a powerful solution-based technique to characterize the structural properties of IDPs in solution. The intrinsic flexibility present within IDPs can be observed visually by investigation of I(q), which will be featureless compared to those of globular proteins. Dimensionless Kratky format further determines the extent of disorder in the protein. Methods have also been highlighted to determine R_g from the SAXS data, without relying on the fitting of models. Since the Guinier regime often is narrow and contains a few data points, one should be careful when applying these methods to the SAXS data of IDPs. Also, obtaining P(r) for IDPs is more challenging in comparison to globular proteins. Moreover, both the extent of the flexibility and R_g determined using these non-model-based methods are ensemble averages, weighted by the population of each distinct conformation within the structural ensemble. A more detailed analysis can be performed to obtain the distribution of distinct conformations within the ensemble and a popular and well-renowned method is the ensemble optimization method available within the ATSAS package (Manalastas-Cantos et al., 2021). Using this method, it is possible to obtain a pool of structural conformations that fit the scattering data, revealing structural properties hidden within the ensemble averages.

It is also possible to obtain structural ensembles using more complex, simulation-based methods. Here the user needs to make a few decisions, for example, on which level of detail the protein shall be described, such as atomistic or coarse-grained, and on whether the solvent is modeled implicitly as continuous electron density or considered explicitly. The former offers reduced computational cost but require the definition of several free parameters to account for, such as the excess density of the solvation layer. Through modeling and computer simulations, it is possible to achieve a molecular understanding of the underlying physics of the macroscopic properties of the sample, as obtained from SAXS. Simulations can also be used for a predictive purpose, for example, the effect of increasing the protein concentration, varying the physicochemical properties such as pH, ionic strength, the addition of multivalent ions, and amino-acid specific mutations and variants. Another possibility is to use the scattering data as a restraint in the simulations. Hence, SAXS is a very powerful tool indeed.

Acknowledgments

We acknowledge financial support from the Crafoord Foundation, Sweden. The authors would like to thank the European Synchrotron Radiation Facility (ESRF) in Grenoble, France, for the use of beamline BM29 and the excellent support over the years.

References

Ahmed, M. C., Crehuet, R., & Lindorff-Larsen, K. (2019). Computing, analyzing and comparing the radius of gyration and hydrodynamic radius in conformational ensembles of intrinsically disordered proteins. *bioRxiv*, 679373. https://doi.org/10.1101/679373.

Antonov, L. D., Olsson, S., Boomsma, W., & Hamelryck, T. (2016). Bayesian inference of protein ensembles from SAXS data. *Physical Chemistry Chemical Physics*, *18*(8), 5832–5838. https://doi.org/10.1039/c5cp04886a.

Balu, R., Mata, J. P., Knott, R., Elvin, C. M., Hill, A. J., Choudhury, N. R., et al. (2016). Effects of crowding and environment on the evolution of conformational ensembles of the multi-stimuli-responsive intrinsically disordered protein, Rec1-Resilin: A small-angle scattering investigation. *Journal of Physical Chemistry B, 120*(27), 6490–6503. https://doi.org/10.1021/acs.jpcb.6b02475.

Bereau, T., & Deserno, M. (2009). Generic coarse-grained model for protein folding and aggregation. *Journal of Chemical Physics, 130*(23), 235106. https://doi.org/10.1063/1.3152842.

Bernado, P., & Blackledge, M. (2009). A self-consistent description of the conformational behavior of chemically denatured proteins from NMR and small angle scattering. *Biophysical Journal, 97*(10), 2839–2845. https://doi.org/10.1016/j.bpj.2009.08.044.

Bernado, P., Mylonas, E., Petoukhov, M. V., Blackledge, M., & Svergun, D. I. (2007). Structural characterization of flexible proteins using small-angle X-ray scattering. *Journal of the American Chemical Society, 129*(17), 5656–5664. https://doi.org/10.1021/ja069124n.

Bonucci, A., Palomino-Schatzlein, M., Malo de Molina, P., Arbe, A., Pierattelli, R., Rizzuti, B., et al. (2021). Crowding effects on the structure and dynamics of the intrinsically disordered nuclear chromatin protein NUPR1. *Frontiers in Molecular Biosciences, 8*, 643. https://doi.org/10.3389/fmolb.2021.684622.

Borgia, A., Zheng, W., Buholzer, K., Borgia, M. B., Schueler, A., Hofmann, H., et al. (2016). Consistent view of polypeptide chain expansion in chemical denaturants from multiple experimental methods. *Journal of the American Chemical Society, 138*(36), 11714–11726. https://doi.org/10.1021/jacs.6b05917.

Brooks, B. R., Bruccoleri, R. E., Olafson, B. D., States, D. J., Swaminathan, S., & Karplus, M. (1983). CHARMM—A program for macromolecular energy, minimization, and dynamics calculations. *Journal of Computational Chemistry, 4*(2), 187–217. https://doi.org/10.1002/jcc.540040211.

Brosey, C. A., & Tainer, J. A. (2019). Evolving SAXS versatility: Solution X-ray scattering for macromolecular architecture, functional landscapes, and integrative structural biology. *Current Opinion in Structural Biology, 58*, 197–213. https://doi.org/10.1016/j.sbi.2019.04.004.

Chacon, P., Moran, F., Diaz, J. F., Pantos, E., & Andreu, J. M. (1998). Low-resolution structures of proteins in solution retrieved from X-ray scattering with a genetic algorithm. *Biophysical Journal, 74*(6), 2760–2775. https://doi.org/10.1016/s0006-3495(98)77984-6.

Chemes, L. B., Alonso, L. G., Noval, M. G., & Prat-Gay, D. G. (2012). Circular dichroism techniques for the analysis of intrinsically disordered proteins and domains. In V. Uversky, & K. Dunker (Eds.), *Vol. 895. Intrinsically disordered protein analysis: Volume 1, methods and experimental tools* (pp. 387–404). Totowa, NJ: Humana Press.

Chen, P.-C., & Hub, J. S. (2014). Validating solution ensembles from molecular dynamics simulation by wide-angle X-ray scattering data. *Biophysical Journal, 107*(2), 435–447. https://doi.org/10.1016/j.bpj.2014.06.006.

Cheng, P., Peng, J., & Zhang, Z. (2017). SAXS-oriented ensemble refinement of flexible biomolecules. *Biophysical Journal, 112*(7), 1295–1301. https://doi.org/10.1016/j.bpj.2017.02.024.

Choi, J.-M., & Pappu, R. V. (2019). Improvements to the ABSINTH force field for proteins based on experimentally derived amino acid specific backbone conformational statistics. *Journal of Chemical Theory and Computation, 15*(2), 1367–1382. https://doi.org/10.1021/acs.jctc.8b00573.

Cornell, W. D., Cieplak, P., Bayly, C. I., Gould, I. R., Merz, K. M., Ferguson, D. M., et al. (1996). A second generation force field for the simulation of proteins, nucleic acids, and organic molecules (vol 117, pg 5179, 1995). *Journal of the American Chemical Society, 118*(9), 2309. https://doi.org/10.1021/ja955032e.

Cragnell, C., Durand, D., Cabane, B., & Skepö, M. (2016). Coarse-grained modeling of the intrinsically disordered protein Histatin 5 in solution: Monte Carlo simulations in combination with SAXS. *Proteins*, *84*(6), 777–791. https://doi.org/10.1002/prot.25025.

Cragnell, C., Rieloff, E., & Skepö, M. (2019). Utilizing coarse-grained modelling and Monte Carlo simulations to evaluate the conformational ensemble of intrinsically disordered proteins and regions. *Journal of Molecular Biology*, *430*(16), 2478–2492. https://doi.org/10.1016/j.jmb.2018.03.006.

Cragnell, C., Staby, L., Lenton, S., Kragelund, B. B., & Skepö, M. (2019). Dynamical oligomerisation of histidine rich intrinsically disordered proteins is regulated through zinc-histidine interactions. *Biomolecules*, *9*(168), 1–20. https://doi.org/10.3390/biom9050168.

de Jong, D. H., Singh, G., Bennett, W. F. D., Arnarez, C., Wassenaar, T. A., Schafer, L. V., et al. (2013). Improved parameters for the Martini coarse-grained protein force field. *Journal of Chemical Theory and Computation*, *9*(1), 687–697. https://doi.org/10.1021/ct300646g.

de Neto, M. O., de Fernandes, A. F., Piiadov, V., Craievich, A. F., de Araujo, E. A., & Polikarpov, I. (2022). SAXSMoW 3.0: New advances in the determination of the molecular weight of proteins in dilute solutions from SAXS intensity data on a relative scale. *Protein Science*, *31*(1), 251–258. https://doi.org/10.1002/pro.4227.

Dignon, G. L., Zheng, W., Kim, Y. C., Best, R. B., & Mittal, J. (2018). Sequence determinants of protein phase behavior from a coarse-grained model. *PLOS Computational Biology*, *14*(1), e1005941. https://doi.org/10.1371/journal.pcbi.1005941.

Dunker, A. K., Babu, M. M., Barbar, E., Blackledge, M., Bondos, S. E., Dosztanyi, Z., et al. (2013). What's in a name? Why these proteins are intrinsically disordered: Why these proteins are intrinsically disordered. *Intrinsically Disordered Proteins*, *1*(1), e24157. https://doi.org/10.4161/idp.24157.

Durand, D., Vives, C., Cannella, D., Perez, J., Pebay-Peyroula, E., Vachette, P., et al. (2010). NADPH oxidase activator P67(phox) behaves in solution as a multidomain protein with semi-flexible linkers. *Journal of Structural Biology*, *169*(1), 45–53. https://doi.org/10.1016/j.jsb.2009.08.009.

Eliezer, D. (2009). Biophysical characterization of intrinsically disordered proteins. *Current Opinion in Structural Biology*, *19*(1), 23–30. https://doi.org/10.1016/j.sbi.2008.12.004.

Ellis, R. J. (2001). Macromolecular crowding: Obvious but underappreciated. *Trends in Biochemical Sciences*, *26*(10), 597–604. https://doi.org/10.1016/s0968-0004(01)01938-7.

Fagerberg, E., Lenton, S., & Skepö, M. (2019). Evaluating models of varying complexity of crowded intrinsically disordered protein solutions against SAXS. *Journal of Chemical Theory and Computation*, *15*(12), 6968–6983. https://doi.org/10.1021/acs.jctc.9b00723.

Fagerberg, E., Månsson, L. K., Lenton, S., & Skepö, M. (2020). The effects of chain length on the structural properties of intrinsically disordered proteins in concentrated solutions. *The Journal of Physical Chemistry. B*, *124*(52), 11843–11853. https://doi.org/10.1021/acs.jpcb.0c09635.

Fisher, C. K., & Stultz, C. M. (2011). Constructing ensembles for intrinsically disordered proteins. *Current Opinion in Structural Biology*, *21*(3), 426–431. https://doi.org/10.1016/j.sbi.2011.04.001.

Franke, D., & Svergun, D. I. (2009). DAMMIF, a program for rapid ab-initio shape determination in small-angle scattering. *Journal of Applied Crystallography*, *42*, 342–346. https://doi.org/10.1107/s0021889809000338.

Gallat, F. X., Laganowsky, A., Wood, K., Gabel, F., van Eijck, L., Wuttke, J., et al. (2012). Dynamical coupling of intrinsically disordered proteins and their hydration water: Comparison with folded soluble and membrane proteins. *Biophysical Journal*, *103*(1), 129–136. https://doi.org/10.1016/j.bpj.2012.05.027.

Goldenberg, D. P., & Argyle, B. (2014). Minimal effects of macromolecular crowding on an intrinsically disordered protein: A small-angle neutron scattering study. *Biophysical Journal, 106*(4), 905–914. https://doi.org/10.1016/j.bpj.2013.12.003.

Grimaldo, M., Lopez, H., Beck, C., Roosen-Runge, F., Moulin, M., Devos, J. M., et al. (2019). Protein short-time diffusion in a naturally crowded environment. *Journal of Physical Chemistry Letters, 10*(8), 1709–1715. https://doi.org/10.1021/acs.jpclett.9b00345.

Grimaldo, M., Roosen-Runge, F., Hennig, M., Zanini, F., Zhang, F., Jalarvo, N., et al. (2015). Hierarchical molecular dynamics of bovine serum albumin in concentrated aqueous solution below and above thermal denaturation. *Physical Chemistry Chemical Physics, 17*(6), 4645–4655. https://doi.org/10.1039/c4cp04944f.

Grudinin, S., Garkavenko, M., & Kazennov, A. (2017). Pepsi-SAXS: An adaptive method for rapid and accurate computation of small-angle X-ray scattering profiles. *Acta Crystallographica. Section A, 73*, 449–464. https://doi.org/10.1107/s2059798317005745.

Guinier, A. (1939). La diffraction des rayons X aux tres petits angles; application a l'etude de phenomenes ultramicroscopiques. *Annals of Physics (Paris), 12*, 161–237.

Hajizadeh, N. R., Franke, D., Jeffries, C. M., & Svergun, D. I. (2018). Consensus Bayesian assessment of protein molecular mass from solution X-ray scattering data. *Scientific Reports, 8*, 7204. https://doi.org/10.1038/s41598-018-25355-2.

Hallin, E. I., Hasan, M., Guo, K., & Akerlund, H.-E. (2016). Molecular studies on structural changes and oligomerisation of violaxanthin de-epoxidase associated with the pH-dependent activation. *Photosynthesis Research, 129*(1), 29–41. https://doi.org/10.1007/s11120-016-0261-y.

Henriques, J., Arleth, L., Lindorff-Larsen, K., & Skepö, M. (2018). On the calculation of SAXS profiles of folded and intrinsically disordered proteins from computer simulations. *Journal of Molecular Biology, 430*, 2521–2539. https://doi.org/10.1016/j.jmb.2018.03.002.

Henriques, J., Cragnell, C., & Skepö, M. (2015). Molecular dynamics simulations of intrinsically disordered proteins: Force field evaluation and comparison with experiment. *Journal of Chemical Theory and Computation, 11*(7), 3420–3431.

Hermann, M. R., & Hub, J. S. (2019). SAXS-restrained ensemble simulations of intrinsically disordered proteins with commitment to the principle of maximum entropy. *Journal of Chemical Theory and Computation, 15*(9), 5103–5115. https://doi.org/10.1021/acs.jctc.9b00338.

Hofmann, H., Soranno, A., Borgia, A., Gast, K., Nettels, D., & Schuler, B. (2012). Polymer scaling laws of unfolded and intrinsically disordered proteins quantified with single-molecule spectroscopy. *Proceedings of the National Academy of Sciences of the United States of America, 109*(40), 16155–16160. https://doi.org/10.1073/pnas.1207719109.

Irbäck, A., & Mohanty, S. (2006). PROFASI: A Monte Carlo simulation package for protein folding and aggregation. *Journal of Computational Chemistry, 27*(13), 1548–1555. https://doi.org/10.1002/jcc.20452.

Johansen, D., Jeffries, C. M. J., Hammouda, B., Trewhella, J., & Goldenberg, D. P. (2011). Effects of macromolecular crowding on an intrinsically disordered protein characterized by small-angle neutron scattering with contrast matching. *Biophysical Journal, 100*(4), 1120–1128. https://doi.org/10.1016/j.bpj.2011.01.020.

Jorgensen, W. L., & Tiradorives, J. (1988). The opls potential functions for proteins—Energy minimizations for crystals of cyclic-peptides and crambin. *Journal of the American Chemical Society, 110*(6), 1657–1666. https://doi.org/10.1021/ja00214a001.

Kachala, M., Valentini, E., & Svergun, D. I. (2015). Application of SAXS for the structural characterization of IDPs. In I. C. Felli, & R. Pierattelli (Eds.), *Vol. 870. Intrinsically disordered proteins studied by Nmr spectroscopy* (pp. 261–289). Springer, Cham.

Kikhney, A. G., & Svergun, D. I. (2015). A practical guide to small angle X-ray scattering (SAXS) of flexible and intrinsically disordered proteins. *FEBS Letters*, *589*(19), 2570–2577. https://doi.org/10.1016/j.febslet.2015.08.027.

Laio, A., & Parrinello, M. (2002). Escaping free-energy minima. *Proceedings of the National Academy of Sciences of the United States of America*, *99*(20), 12562–12566. https://doi.org/10.1073/pnas.202427399.

Manalastas-Cantos, K., Konarev, P. V., Hajizadeh, N. R., Kikhney, A. G., Petoukhov, M. V., Molodenskiy, D. S., et al. (2021). ATSAS 3.0: Expanded functionality and new tools for small-angle scattering data analysis. *Journal of Applied Crystallography*, *54*, 343–355. https://doi.org/10.1107/s1600576720013412.

Martin, E. W., Hopkins, J. B., & Mittag, T. (2021). Small-angle X-ray scattering experiments of monodisperse intrinsically disordered protein samples close to the solubility limit. In C. D. Keating (Ed.), *Vol. 646. Liquid-liquid phase coexistence and Membraneless organelles* (pp. 185–222). Academic Press. https://doi.org/10.1016/bs.mie.2020.07.002.

Miller, C. M., Kim, Y. C., & Mittal, J. (2016). Protein composition determines the effect of crowding on the properties of disordered proteins. *Biophysical Journal*, *111*(1), 28–37. https://doi.org/10.1016/j.bpj.2016.05.033.

Mu, J., Liu, H., Zhang, J., Luo, R., & Chen, H.-F. (2021). Recent force field strategies for intrinsically disordered proteins. *Journal of Chemical Information and Modeling*, *61*(3), 1037–1047. https://doi.org/10.1021/acs.jcim.0c01175.

Mylonas, E., & Svergun, D. I. (2007). Accuracy of molecular mass determination of proteins in solution by small-angle X-ray scattering. *Journal of Applied Crystallography*, *40*, S245–S249. https://doi.org/10.1107/s002188980700252x.

Niebling, S., Bjoerling, A., & Westenhoff, S. (2014). MARTINI bead form factors for the analysis of time-resolved X-ray scattering of proteins. *Journal of Applied Crystallography*, *47*, 1190–1198. https://doi.org/10.1107/s1600576714009959.

Oldfield, C. J., & Dunker, A. K. (2014). Intrinsically disordered proteins and intrinsically disordered protein regions. In R. D. Kornberg (Ed.), *Vol. 83. Annual review of biochemistry* (pp. 553–584). Annual Reviews.

Orthaber, D., Bergmann, A., & Glatter, O. (2000). SAXS experiments on absolute scale with Kratky systems using water as a secondary standard. *Journal of Applied Crystallography*, *33*, 218–225. https://doi.org/10.1107/s0021889899015216.

Pedersen, J. S. (1997). Analysis of small-angle scattering data from colloids and polymer solutions: Modeling and least-squares fitting. *Advances in Colloid and Interface Science*, *70*, 171–210. https://doi.org/10.1016/s0001-8686(97)00312-6.

Pelikan, M., Hura, G. L., & Hammel, M. (2009). Structure and flexibility within proteins as identified through small angle X-ray scattering. *General Physiology and Biophysics*, *28*(2), 174–189. https://doi.org/10.4149/gpb_2009_02_174.

Putnam, C. D., Hammel, M., Hura, G. L., & Tainer, J. A. (2007). X-ray solution scattering (SAXS) combined with crystallography and computation: Defining accurate macromolecular structures, conformations and assemblies in solution. *Quarterly Reviews of Biophysics*, *40*(3), 191–285. https://doi.org/10.1017/s0033583507004635.

Rambo, R. P. (2017). Considerations for sample preparation using size-exclusion chromatography for home and synchrotron sources. In B. Chaudhuri, I. G. Munoz, S. Qian, & V. S. Urban (Eds.), *Vol. 1009. Biological small angle scattering: Techniques, strategies and tips* (pp. 31–45). Springer.

Rambo, R. P., & Tainer, J. A. (2011). Characterizing flexible and intrinsically unstructured biological macromolecules by SAS using the Porod-Debye law. *Biopolymers*, *95*(8), 559–571. https://doi.org/10.1002/bip.21638.

Rambo, R. P., & Tainer, J. A. (2013). Accurate assessment of mass, models and resolution by small-angle scattering. *Nature*, *496*(7446), 477. https://doi.org/10.1038/nature12070.

Ray, S., Singh, N., Kumar, R., Patel, K., Pandey, S., Datta, D., et al. (2020). Alpha-Synuclein aggregation nucleates through liquid-liquid phase separation. *Nature Chemistry*, *12*(8), 705–716. https://doi.org/10.1038/s41557-020-0465-9.

Receveur-Brechot, V., & Durand, D. (2012). How random are intrinsically disordered proteins? A small angle scattering perspective. *Current Protein & Peptide Science*, *13*(1), 55–75. https://doi.org/10.2174/138920312799277901.

Rotkiewicz, P., & Skolnick, J. (2008). Fast procedure for reconstruction of full-atom protein models from reduced representations. *Journal of Computational Chemistry*, *29*(9), 1460–1465. https://doi.org/10.1002/jcc.20906.

Rozycki, B., Kim, Y. C., & Hummer, G. (2011). SAXS ensemble refinement of ESCRT-III CHMP3 conformational transitions. *Structure*, *19*(1), 109–116. https://doi.org/10.1016/j.str.2010.10.006.

Sagar, A., Herranz-Trillo, F., Langkilde, A. E., Vestergaard, B., & Bernado, P. (2021). Structure and thermodynamics of transient protein-protein complexes by chemometric decomposition of SAXS datasets. *Structure*, *29*(9), 1074. https://doi.org/10.1016/j.str.2021.03.017.

Sawner, A. S., Ray, S., Yadav, P., Mukherjee, S., Panigrahi, R., Poudyal, M., et al. (2021). Modulating alpha-synuclein liquid-liquid phase separation. *Biochemistry*, *60*(48), 3676–3696. https://doi.org/10.1021/acs.biochem.1c00434.

Schneidman-Duhovny, D., Hammel, M., & Sali, A. (2010). FoXS: A web server for rapid computation and fitting of SAXS profiles. *Nucleic Acids Research*, *38*, W540–W544. https://doi.org/10.1093/nar/gkq461.

Schneidman-Duhovny, D., Hammel, M., Tainer, J. A., & Sali, A. (2013). Accurate SAXS profile computation and its assessment by contrast variation experiments. *Biophysical Journal*, *105*(4), 962–974. https://doi.org/10.1016/j.bpj.2013.07.020.

Schneidman-Duhovny, D., Hammel, M., Tainer, J. A., & Sali, A. (2016). FoXS, FoXSDock and MultiFoXS: Single-state and multi-state structural modeling of proteins and their complexes based on SAXS profiles. *Nucleic Acids Research*, *44*(W1), W424–W429. https://doi.org/10.1093/nar/gkw389.

Sibille, N., & Bernado, P. (2012). Structural characterization of intrinsically disordered proteins by the combined use of NMR and SAXS. *Biochemical Society Transactions*, *40*, 955–962. https://doi.org/10.1042/bst20120149.

Stovgaard, K., Andreetta, C., Ferkinghoff-Borg, J., & Hamelryck, T. (2010). Calculation of accurate small angle X-ray scattering curves from coarse-grained protein models. *BMC Bioinformatics*, *11*, 429. https://doi.org/10.1186/1471-2105-11-429.

Sugita, Y., & Okamoto, Y. (1999). Replica-exchange molecular dynamics method for protein folding. *Chemical Physics Letters*, *314*(1–2), 141–151. https://doi.org/10.1016/s0009-2614(99)01123-9.

Svergun, D. I. (1992). Determination of the regularization parameter in indirect-transform methods using perceptual criteria. *Journal of Applied Crystallography*, *25*, 495–503. https://doi.org/10.1107/s0021889892001663.

Svergun, D., Barberato, C., & Koch, M. H. J. (1995). CRYSOL—A program to evaluate x-ray solution scattering of biological macromolecules from atomic coordinates. *Journal of Applied Crystallography*, *28*, 768–773. https://doi.org/10.1107/s0021889895007047.

Svergun, D. I., Semenyuk, A. V., & Feigin, L. A. (1988). Small-angle scattering—Data treatment by the regularization method. *Acta Crystallographica Section A*, *44*, 244–250. https://doi.org/10.1107/s0108767387011255.

Tompa, P. (2002). Intrinsically unstructured proteins. *Trends in Biochemical Sciences*, *27*(10), 527–533. https://doi.org/10.1016/s0968-0004(02)02169-2.

Tria, G., Mertens, H. D. T., Kachala, M., & Svergun, D. I. (2015). Advanced ensemble modelling of flexible macromolecules using X-ray solution scattering. *IUCrJ*, *2*, 207–217. https://doi.org/10.1107/s205225251500202x.

Tully, M. D., Tarbouriech, N., Rambo, R. P., & Hutin, S. (2021). Analysis of SEC-SAXS data via EFA deconvolution and scatter. *Journal of Visualized Experiments: JoVE, 167*, e61578. https://doi.org/10.3791/61578.

Uversky, V. N. (2013). Unusual biophysics of intrinsically disordered proteins. *Biochimica Et Biophysica Acta-Proteins and Proteomics, 1834*(5), 932–951. https://doi.org/10.1016/j.bbapap.2012.12.008.

Yang, S., Blachowicz, L., Makowski, L., & Roux, B. (2010). Multidomain assembled states of Hck tyrosine kinase in solution. *Proceedings of the National Academy of Sciences of the United States of America, 107*(36), 15757–15762. https://doi.org/10.1073/pnas.1004569107.

Zheng, W., & Best, R. B. (2018). An extended Guinier analysis for intrinsically disordered proteins. *Journal of Molecular Biology, 430*(16), 2540–2553. https://doi.org/10.1016/j.jmb.2018.03.007.

Zimmerman, S. B., & Minton, A. P. (1993). Macromolecular crowding—Biochemical, biophysical, and physiological consequences. *Annual Review of Biophysics and Biomolecular Structure, 22*, 27–65. https://doi.org/10.1146/annurev.bb.22.060193.000331.

CHAPTER ELEVEN

Applying HT-SAXS to chemical ligand screening

Chris A. Brosey[a,*], Runze Shen[a], Davide Moiani[a], Darin E. Jones[c], Kathryn Burnett[d], Greg L. Hura[d,e], and John A. Tainer[a,b,d,*]

[a]Department of Molecular and Cellular Oncology, The University of Texas MD Anderson Cancer Center, Houston, TX, United States
[b]Department of Cancer Biology, University of Texas MD Anderson Cancer Center, Houston, TX, United States
[c]Department of Pharmaceutical Sciences, University of Arkansas for Medical Sciences, Little Rock, AR, United States
[d]Molecular Biophysics and Integrated Bioimaging Division, Lawrence Berkeley National Laboratory, Berkeley, CA, United States
[e]Department of Chemistry and Biochemistry, University of California Santa Cruz, Santa Cruz, CA, United States
*Corresponding authors: e-mail address: CABrosey@mdanderson.org; JTainer@mdanderson.org

Contents

Abstract

Chemical probes are invaluable tools for investigating essential biological processes. Understanding how small-molecule probes engage biomolecular conformations is critical to developing their functional selectivity. High-throughput solution X-ray scattering is well-positioned to profile target-ligand complexes during probe development, bringing conformational insight and selection to traditional ligand binding assays. Access to high-quality synchrotron SAXS datasets and high-throughput data analysis now allows routine academic users to incorporate conformational information into small-molecule development pipelines. Here we describe a general approach for benchmarking and preparing HT-SAXS chemical screens from small fragment libraries. Using the allosteric oxidoreductase Apoptosis-Inducing Factor (AIF) as an exemplary system, we illustrate how HT-SAXS efficiently identifies an allosteric candidate among hits of a microscale thermophoresis ligand screen. We discuss considerations for pursuing HT-SAXS chemical screening with other systems of interest and reflect on advances to extend screening throughput and sensitivity.

Methods in Enzymology, Volume 678
ISSN 0076-6879
https://doi.org/10.1016/bs.mie.2022.09.022

Abbreviations

AIF Apoptosis-Inducing Factor
ALS Advanced Light Source
FAD flavin adenine dinucleotide
FMN flavin adenine mononucleotide
HT high-throughput
IDP intrinsically disordered protein
MST microscale thermophoresis
NADH nicotinamide adenine dinucleotide
PPI protein-protein interface
R_g radius-of-gyration
RNA ribonucleic acid
SAXS small-angle X-ray scattering
SEC size-exclusion chromatography
SSM structural similarity map
V_R volatility-of-ratio

1. Introduction

As tools of research and the clinic, small-molecule compounds probe critical biological processes by modulating activities and architectures of their molecular targets (Arrowsmith et al., 2015). Notable examples from the field of DNA damage signaling and repair include Mre11 nuclease (Moiani et al., 2018; Shibata et al., 2014), DNA polymerase theta (Zhou et al., 2021), poly(ADP-ribose) glycohydrolase (PARG) (Houl et al., 2019), and O6-methylguanine DNA methyltransferase (MGMT) (Daniels et al., 2004). Understanding how these chemical probes impact biomolecular conformation, assembly formation, or allostery can lead to improved selectivity and specificity and advance new biology. High-throughput small-angle X-ray scattering (HT-SAXS) (Brosey & Tainer, 2019; Dyer et al., 2014; Hura et al., 2009; Putnam, Hammel, Hura, & Tainer, 2007; Rambo & Tainer, 2013a) is well positioned to report on global target-ligand conformations relevant to assembly and allostery and to complement information from other high-throughput binding and functional assays.

A key strength of SAXS is its translatability. Since SAXS is not limited by sample labeling or molecular weight requirements, optimized binding conditions and purified reagents can be applied to the solution format of SAXS assays. HT-SAXS sample preparation can also leverage established protocols to dispense and assemble target-ligand complexes in multi-well plates suitable for synchrotron data collection (Dyer et al., 2014). Ongoing development of HT-SAXS screening methods continues to expand sample throughput at synchrotron SAXS beamlines (Chen & Hennig, 2018; Chen, Masiewicz, Perez, & Hennig, 2020; Chen, Masiewicz, Rybin, Svergun, & Hennig, 2018; Dyer et al., 2014; Hura et al., 2013), allowing novice and experienced users to acquire high-quality target-ligand datasets. Importantly, SAXS enables access to conformational targets at the frontiers of drug discovery, including dynamically flexible systems (Rambo & Tainer, 2011), intrinsically disordered proteins (IDPs) (Joshi & Vendruscolo, 2015), nucleic acids (Warner, Hajdin, & Weeks, 2018), protein-protein interfaces (PPIs) (Scott, Bayly, Abell, & Skidmore, 2016) and allosteric assemblies (Brosey & Tainer, 2019).

Here, we present a methodological approach for screening target-ligand complexes with HT-SAXS. Our method is directed toward discovery and optimization of chemical probes but is relevant for investigating the conformational impact of any small-molecule panel. We demonstrate our approach with Apoptosis-Inducing Factor (AIF), a mitochondrial redox switch that allosterically dimerizes upon binding and reduction by NADH (Brosey et al., 2016; Sevrioukova, 2009). Dimeric AIF supports the mitochondrial disulfide import pathway and regulates biogenesis of respiratory complexes performing oxidative phosphorylation (OXPHOS) (Hangen et al., 2015), raising interest in its therapeutic potential for cancer (Ashton, McKenna, Kunz-Schughart, & Higgins, 2018; Lemberg, Gori, Tsukamoto, Rais, & Slusher, 2022). We describe parallel screening of AIF with microscale thermophoresis (MST) and HT-SAXS to identify chemical ligands and allosteric effectors from a specialized 96-compound library. Our HT-SAXS assay design leverages the accessible mail-in service offered by the SIBYLS SAXS (12.3.1) beamline at the Advanced Light Source (ALS) synchrotron (https://sibyls.als.lbl.gov/) (Classen et al., 2013) and utilizes standard analytic tools to assess and summarize target-ligand datasets. We also outline important considerations and concerns for applying and benchmarking HT-SAXS screening with other biomolecular systems and discuss priority areas for future development.

2. Considerations for SAXS target and library selection

An informative HT-SAXS assay relies on strategic selection of conformational targets and chemical libraries that are well suited for solution X-ray scattering (Fig. 1A). SAXS optimally detects global changes in biomolecular conformation; thus, monomer-dimer equilibria, protein assembly/disassembly, intra- or inter-domain rearrangements, and significant changes to molecular

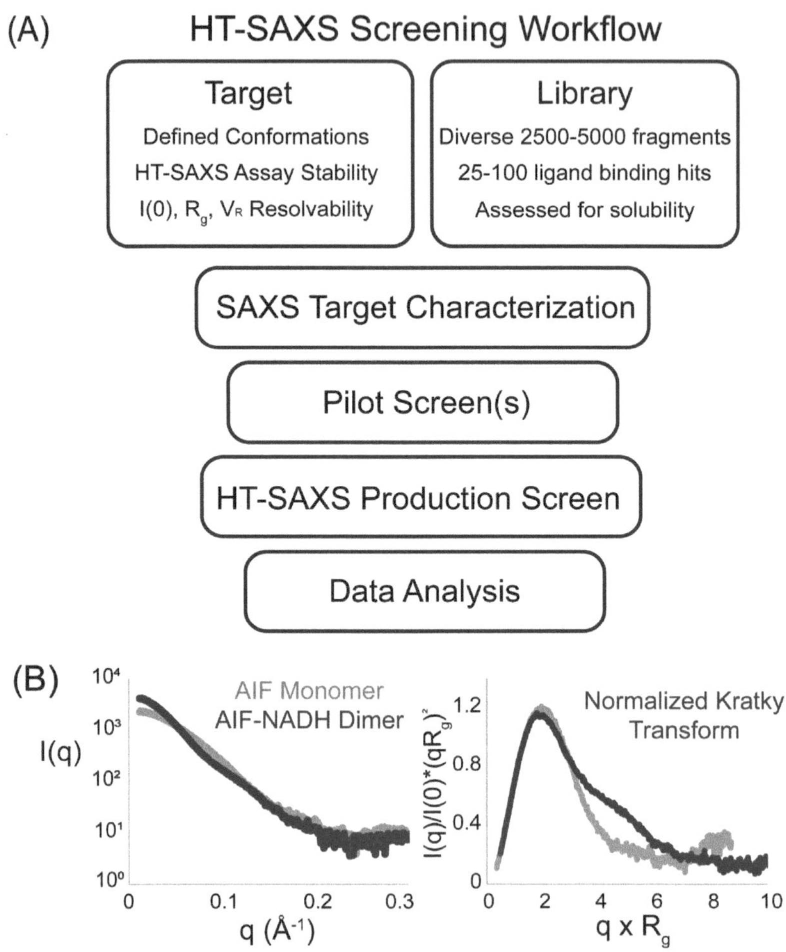

(C)

AIF	0% DMSO		5% DMSO		
	I(0)	Rg	I(0)	Rg	V_R (0%/5%)
Monomer	2545 (4)	28.4 (0.2)	2302 (4)	28.8 (0.3)	0.45
Dimer	5885 (14)	41.3 (0.5)	5325 (10)	42.2 (0.4)	0.95
I(0) Ratio	2.3		2.3		
V_R (M/D)	7.4		7.5		

Fig. 1 HT-SAXS Screening Work Flow. (A) Workflow of a prototypical HT-SAXS chemical screen with criteria for targets and libraries. (B) SAXS analysis of monomeric AIF and the NADH-stimulated dimer (AIF + 10-fold molar excess NADH) reveals well-resolved scattering curves, I(q). Kratky transformation of I(q) highlights distinct, folded monomer and dimer architectures. (C) Titration of 5% DMSO does not significantly alter AIF architecture or interfere with NADH-induced dimerization, as indicated by comparable I(0) and R_g values and $V_R < 1$. The I(0) ratio and volatility-of-ratio (0.015–0.15 Å^{-1}) between AIF monomer and dimer are unchanged by DMSO. Values in parentheses are errors of fitting.

flexibility are well captured by X-ray scattering analysis (Brosey & Tainer, 2019; Nishimura et al., 2009; Rambo & Tainer, 2011). Improvements to the sensitivity and interpretation of high-resolution SAXS data hold the potential to monitor subtler changes in biomolecular architecture as the field continues to progress (Rambo & Tainer, 2013b).

To assess whether SAXS analysis can resolve known conformational endpoints of a potential target, scattering experiments on high-quality protein or nucleic acid reagents are necessary (Dyer et al., 2014; Graewert & Jeffries, 2017; Rambo, 2017). Such experiments will also establish whether a target is sufficiently stable during sample shipment and data acquisition to produce usable SAXS data. Key parameters to test at this stage are sample concentration (1–10 mg/mL), sample buffer (pH, salt, additives, see also https://bl1231.als.lbl.gov/saxs_protocols/saxs_sample_prep.php), the ability to match buffers containing native ligands, and shipping format (chilled or frozen). For AIF, HT-SAXS data were previously acquired on concentration series of the ligand-free monomeric (60 kDa) and NADH-bound dimeric (120 kDa) proteins (Brosey et al., 2016). These experiments established a target concentration (4 mg/mL), standard buffer (25 mM HEPES, pH 7.5, 150 mM NaCl), and shipping condition (dry ice) for SAXS analysis.

Visual comparison of AIF monomer and dimer X-ray scattering curves and their Kratky transforms highlights distinct scattering profiles for each allosteric state (Fig. 1B), while linear Guinier transforms support aggregation-free samples (data not shown). Our HT-SAXS analysis relies upon the extrapolated zero-angle scattering intensity, I(0), radius-of-gyration, R_g, and the volatility-of-ratio (V_R) metric (Hura et al., 2013) (https://sibyls.als.lbl.gov/saxs-similarity/) to quantify and rank ligand-induced differences relative to a baseline state. V_R captures differences between two SAXS curves by quantifying fluctuations in their normalized I(q) ratio or R(q). Perfect sample identity results in R(q) of 1 for the entire q-range and V_R of 0; differences between samples cause R(q) deviations from unity and increase V_R. To assess the resolving power of V_R when applied to AIF's allosteric states, one can calculate a *Z*-factor (Z) from replicate monomer and dimer V_R values, referencing the AIF monomer as a baseline state.

$$Z = 1 - \frac{3\sigma_1 + 3\sigma_2}{|\mu_1 - \mu_2|} \tag{1}$$

where μ is the average V_R of 2–3 replicate samples and σ is the associated standard deviation. We chose to focus V_R calculation over the low- and mid-q region of the R(q) ratio (0.015–0.15 Å^{-1}), the region of maximum monomer-dimer signal divergence and highest signal-to-noise within the I(q) curve. This yields AIF monomer and dimer V_R values of 2.3 ± 0.2 and 8.0 ± 0.1, respectively. The resulting *Z*-factor is 0.83, where Z-factors within 0.5–1.0 are considered to provide good discrimination between baseline and maximum signal. Similar calculations for I(0) (22.6 ± 1.1 Å vs 51.8 ± 1.7 Å) and R_g (28.9 ± 0.9 Å vs 43.2 ± 0.8 Å) yield Z-factors of 0.71 and 0.64, respectively.

For certain biomolecular systems, it may be desirable to estimate SAXS V_R resolvability before investing in production of SAXS-quality protein or nucleic acid samples. If experimental structures of target conformational states are available, theoretical X-ray scattering curves can be calculated from three-dimensional coordinates using FoXS (Schneidman-Duhovny, Hammel, & Sali, 2010; Schneidman-Duhovny, Hammel, Tainer, & Sali, 2013, 2016) or CRYSOL (Franke et al., 2017; Svergun, Barberato, & Koch, 1995) webservers. These theoretical curves can be supplemented with a 1–5% Gaussian error to simulate replicate measurements, then used in Z-factor calculations to quantify I(0), R_g, and V_R resolvability by SAXS. This approach, however, requires that known experimental structures accurately represent biomolecular conformations in solution.

Choice of chemical library also determines the effectiveness of HT-SAXS screening assays. Our screening efforts have relied on small, custom fragment libraries (Moiani et al., 2021) designed for chemical diversity, absence of problematic pan-assay interference (PAINS) (Baell & Holloway, 2010; Baell & Nissink, 2018), predicted favorable physicochemical properties, and tractability of chemical elaboration. These libraries are built from commercial fragments purchased from Life Chemicals and contain 2500–5000 fragments prepared as 10 mg/mL DMSO stocks (~20–70 mM). A ligand binding screen returning a 1–2% hit rate from this library yields 25–100 representative compounds. A single 96-well HT-SAXS screening plate can accommodate 28 compounds with controls (vide infra); thus, conformational characterization of projected library hits can be completed with 1–3 HT-SAXS plates, typically within 2 synchrotron SAXS shifts. Fragment pooling has the potential to increase throughput for other library formats while keeping HT-SAXS screening plates to a manageable number.

It is important to assess aqueous solubility of any candidate compounds before HT-SAXS screening, since highly insoluble material may clog SAXS sample loading and fluidics systems and/or produce unusable SAXS data. Complementary ligand binding assays should provide opportunity to assess and potentially optimize compound solubility (e.g., increasing DMSO).

Preparing small test samples of compounds mixed with protein or nucleic acid at chosen SAXS concentrations can also flag potential problems with precipitation at room temperature and upon freeze-thaw treatment.

3. HT-SAXS sample preparation

There are many excellent resources describing preparation of high-quality protein reagents for X-ray scattering analysis (Dyer et al., 2014; Graewert & Jeffries, 2017; Jeffries et al., 2016; Rambo, 2017; Rambo & Tainer, 2010). Here, we focus on specific preparation of AIF samples with matched buffer for HT-SAXS using size-exclusion chromatography. Expression and purification of AIF are previously described (Brosey et al., 2016). Freshly purified or thawed aliquots of AIF protein are concentrated to ~300–700 μL at a final concentration of 30–40 mg/mL. Following a 5-min high-speed spin to pellet stray aggregation, the concentrate is loaded onto a Superdex 200 10/300 Increase column equilibrated in 25 mM HEPES, pH 7.5, 150 mM NaCl and eluted at 0.4 mL/min with 0.5 mL fractions. SEC fractions are scanned by UV–Vis spectroscopy to measure concentration and confirm the presence of AIF's FAD cofactor (identified by $A_{270}/A_{450} \sim 7.5$ and $A_{260}/A_{280} \sim 0.8$). Fractions are also visualized by SDS-PAGE to confirm sample purity. Peak fractions free of contaminants are selected and pooled to 4.1 mg/mL (typically 1–2 mL) to provide the AIF protein stock to assemble the screen. Buffer fractions preceding the void volume (usually 2–4 mL into the elution) are also pooled for protein dilution and to prepare matched buffer samples. Using SEC-matched buffer for preparing AIF SAXS samples is crucial in order to ensure an accurate buffer subtraction between protein and buffer scattering curves. Prepped AIF protein and buffers can be kept on ice prior to assembling the plate or stored at 4 °C overnight if prepped the day before.

4. Benchmarking a pilot HT-SAXS screen

Performing at least two small trial screens is highly recommended prior to commencing a production HT-SAXS screen containing library compounds. A pilot screen serves several purposes. The first is to determine if titrated DMSO impacts target conformation or interferes with the binding of native ligands, providing a control for the library solvent. Our fragment libraries are usually screened with final concentrations of 3–10% DMSO. A second objective is testing the accuracy of buffer subtraction for a given sample assembly protocol. Because library compounds are often limited in material, the protocol reported here describes mixing DMSO controls and compounds into proteins and buffers manually. Since this may increase the margin of error for a poor background subtraction, it is critical to ensure the sample-buffer match is sufficient to produce usable data. Finally, pilot screens are useful for identifying and remedying any weaknesses in the logistics of preparing, shipping, and collecting SAXS screening plates at synchrotron beamlines.

A sample layout for a pilot SAXS screen in a 96-well PCR plate is presented in Fig. 2A, based upon SIBYLS recommendations for interleaving matched buffers with sample concentration series. DMSO is tested at 1–10% for three sample concentrations, where matched buffer wells contain identical DMSO concentrations. This layout can be repeated to include positive controls for known ligands or other conformational endpoints. All samples are prepared at 30 μL per well. Thus, a 10% DMSO AIF titration would involve combining 3 μL DMSO with 27 μL SEC-purified AIF (4.4 mg/mL stock) or 27 μL SEC-matched buffer. When manually adding titrant directly to samples, we recommend preparing the titrant concentration so that at least 1-μL volume is being added to the sample (i.e., adding 1 μL 30% DMSO to achieve 1% DMSO). If buffer subtraction becomes an issue, adjusting titrant concentration to allow larger, more accurate titrant volumes may help.

For AIF, manual titration of DMSO into ligand-free monomer or NADH-reduced dimer does not cause major disruptions in the buffer subtraction. Importantly, DMSO neither alters protein conformation nor interferes with NADH-induced dimerization (Fig. 1C). This is initially assessed by overlay and visual inspection of scattering curves, then quantitatively established by comparing extrapolated zero-angle scattering intensity I(0), radius-of-gyration (R_g), and V_R with reference to DMSO-free controls.

5. Ligand screen design and assembly

Once pilot HT-SAXS screens have identified reliable DMSO concentrations and benchmarked the sample shipping/data collection pipeline, preparing the production HT-SAXS screen should be a streamlined process. Before setting up a production screen, we recommend reaching out to your local SAXS beamline scientist to confirm shipment plans and to ensure available shift times will cover the planned data collection.

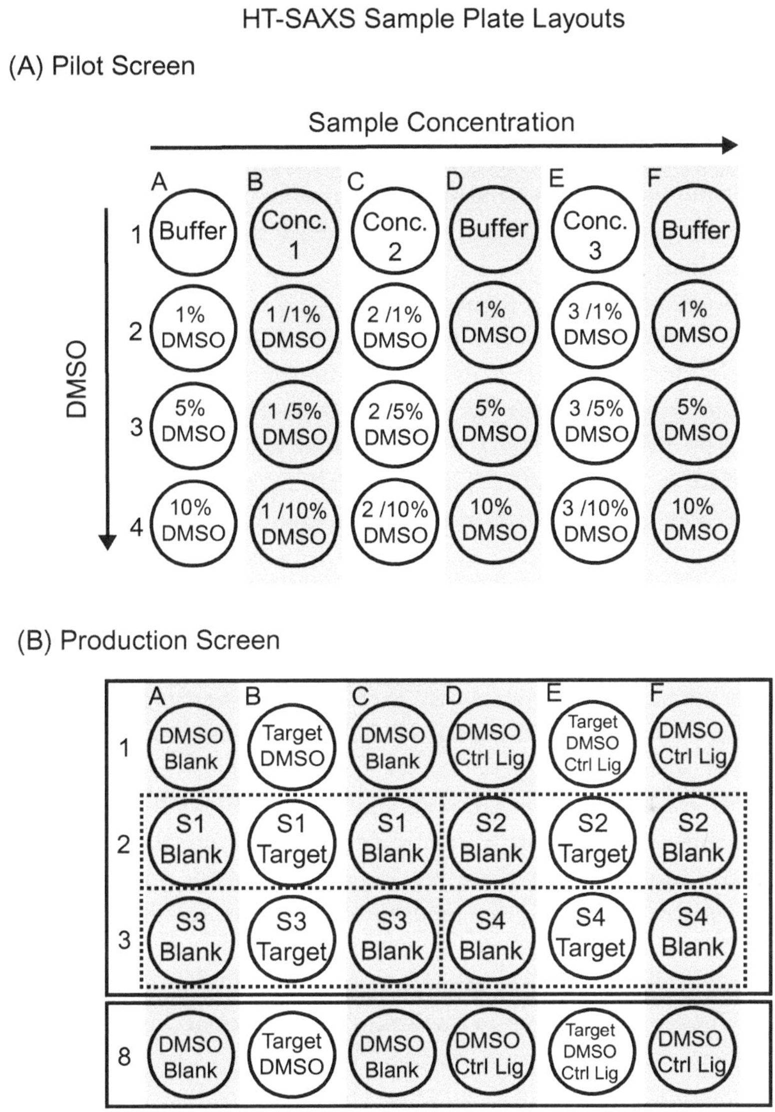

Fig. 2 HT-SAXS Sample Plate Layouts. (A) Sample 96-well plate layout for a HT-SAXS pilot screen. (B) Sample 96-well plate layout for a HT-SAXS production screen with control samples (rows 1 and 8) and 4 screening compounds (S1-S4, rows 2 and 3).

An exemplary layout of an HT-SAXS production screen is shown in Fig. 2B. Internal controls (DMSO, native ligand with DMSO) are placed at the beginning and end of each plate. Target-ligand samples are prepared with flanking ligand-matched buffers collected before and after the target-ligand complex. If a plate contains 4 control sections (12 wells), the remaining wells can accommodate 28 target-ligand complexes with matched-buffers. Overall, this plate design will require 3–9 μL compound for each target-ligand complex (one target, two matched buffers). If library materials are not limited, this design can be expanded to include duplicate target-ligand wells (two targets flanked by two matched buffers) to provide redundancy should bubbles or poor sample transfer result in unusable scattering profiles.

In this example, we screened AIF with a focused 96-fragment library, created by similarity search of high-ranking scaffolds from a previous discovery screen (Brosey et al., 2020). Ligand binding was initially assessed by microscale thermophoresis (MST) with Atto488-labeled AIF (Fig. 3A). Briefly, 100 nM Atto488-AIF was incubated with 0.3 mg/mL compound or DMSO in MST buffer (25 mM HEPES, pH 7.5, 150 mM NaCl, 0.01% Tween-20) for 15 min at room temperature. Thermophoresis curves were measured in triplicate at 25 °C using a Monolith NT.115 system (NanoTemper) with 30% LED power and 40% infrared excitation for 20 s (3-s equilibration and 1-s recovery periods). Time-averaged amplitudes were calculated over a 1-s window using NanoTemper's MO.Affinity Analysis software and exported to Excel for analysis. Screening hits were defined as fragments increasing Atto488-AIF MST amplitude three standard deviations above averaged DMSO controls (Fig. 3A). Twenty-five fragments qualified for this cutoff and were verified by MST titration experiments, which report micromolar affinity among the top candidates (Fig. 3B). Inspection of these top fragments reveals an enrichment in benzothiazolamine scaffolds, lending confidence to the MST results (Fig. 3C).

Two HT-SAXS screens were prepared in parallel to the MST experiments. The first focused upon assessing AIF conformational impact among the highest affinity fragment binders (14 samples). The second, broader screen examined the entire 96-fragment library in order to probe for cryptic ligands undetectable by MST. SAXS samples were prepared by combining library fragments or DMSO with SEC-purified AIF stock or SEC-matched buffer. Final concentrations were 4 mg/mL AIF, 1–5 mM library fragments, and 3–6.7% DMSO. For the broad 96-fragment screen, AIF-ligand samples were prepared in duplicate with flanking ligand-matched buffers. In order to conserve material and extend screening boundaries, the 96-fragment screen was prepared with a 15-μL sample volume; however, the prevalence of interfering bubbles during data collection supports the recommended

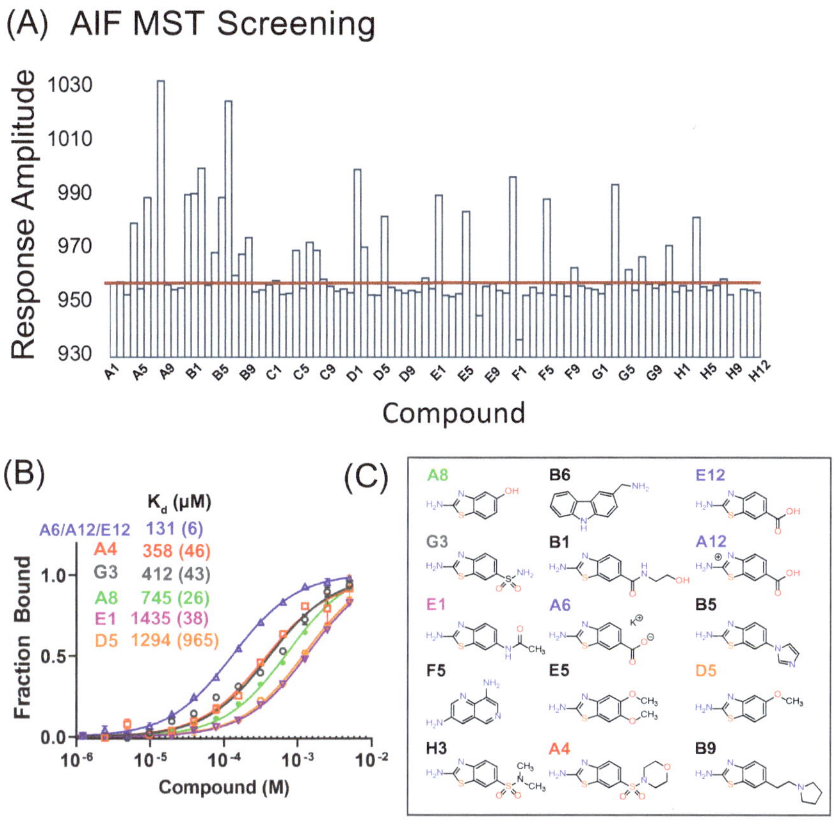

Fig. 3 AIF MST Screen. (A) AIF MST screening results from the focused 96-fragment library. Red line indicates three standard deviations from the average DMSO response amplitude. (B) MST titration curves and affinities for the top fragment ligands. Response amplitudes were exported and fit for one-site binding in GraphPad Prism 9. Values in parentheses are fitting errors. (C) The top MST candidates reveal an enrichment of benzothiazolamine scaffolds.

30-μL sample volume used in the focused 14-fragment screen. With DMSO controls, the focused and broad HT-SAXS screens required 1 and 4.5 96-well sample plates, respectively. Once assembled, SAXS plates were sealed, flash frozen by setting the plate in a shallow layer of liquid nitrogen and shipped on dry ice to the ALS SIBYLS beamline, where they were stored at −80 °C prior to data collection. For the user-provided data collection instructions (https://bl1231.als.lbl.gov/htsaxs/instructions/htsaxs), we specified a needle wash after collection of the target-ligand complex and also after the final ligand-matched buffer for each compound.

Ideally, problematic, precipitating compounds would be flagged and mitigated prior to preparing the production screen. Should ligand-induced precipitation occur once the plate is at the beamline, a high-speed centrifuge spin can pellet insoluble material and allow loading and data collection of the soluble remainder. This also is a point to discuss with a local beamline contact when planning the screening shipment.

6. Analysis of HT-SAXS screening datasets

The SIBYLS mail-in HT-SAXS program provides a package of processed, buffer-subtracted I(q) scattering curves with annotations on aggregated samples or problematic background subtractions. Data are collected as 200–300-millisecond frames over a 10-s exposure; thus each sample contains 30–50 time-resolved I(q) curves. We recommend an initial visual inspection of the I(q) data to sort out any samples with unusable curves—those with poor buffer subtraction or missing data (Fig. 4A). This can be achieved efficiently for the 25–100 compound screens described here using the drag-and-drop feature of ScÅtter (https://bl1231.als.lbl.gov/scatter/) to load and graph each sample time series.

Once unusable I(q) curves have been excluded from the screening dataset, a second triage step is applied to remove aggregated samples exhibiting dramatic upward deflection in the low-angle I(q) signal or non-linear Guinier regions. Here, initial I(q) curves ($t = 0.2$ or 0.3 s) from each sample are copied and organized into a directory, then uploaded into ScÅtter for visualization and comparison to DMSO or other controls. ScÅtter automatically calculates I(0) and R_g values, allowing for rapid quantitative comparison to a DMSO baseline. Guinier plots with residuals are also automatically calculated and can be assessed for linearity.

Once problematic and aggregated datasets have been excluded, one must decide whether (1) to continue analysis with all initial frames (higher throughput), (2) to average all or part of each time-series to produce a single I(q) curve (higher signal-to-noise), or (3) to analyze the SAXS data as time-resolved I(q) series (time-dependent or kinetic effects). Because AIF is a redox-regulated flavoprotein, X-ray-induced reduction unlocks its response to certain classes of allosteric fragment effectors (Brosey et al., 2020). Thus, some ligand impacts on AIF conformation are most efficiently revealed by time-resolved HT-SAXS analysis. The expectation is that X-ray-induced redox effects will primarily be relevant for redox-sensitive

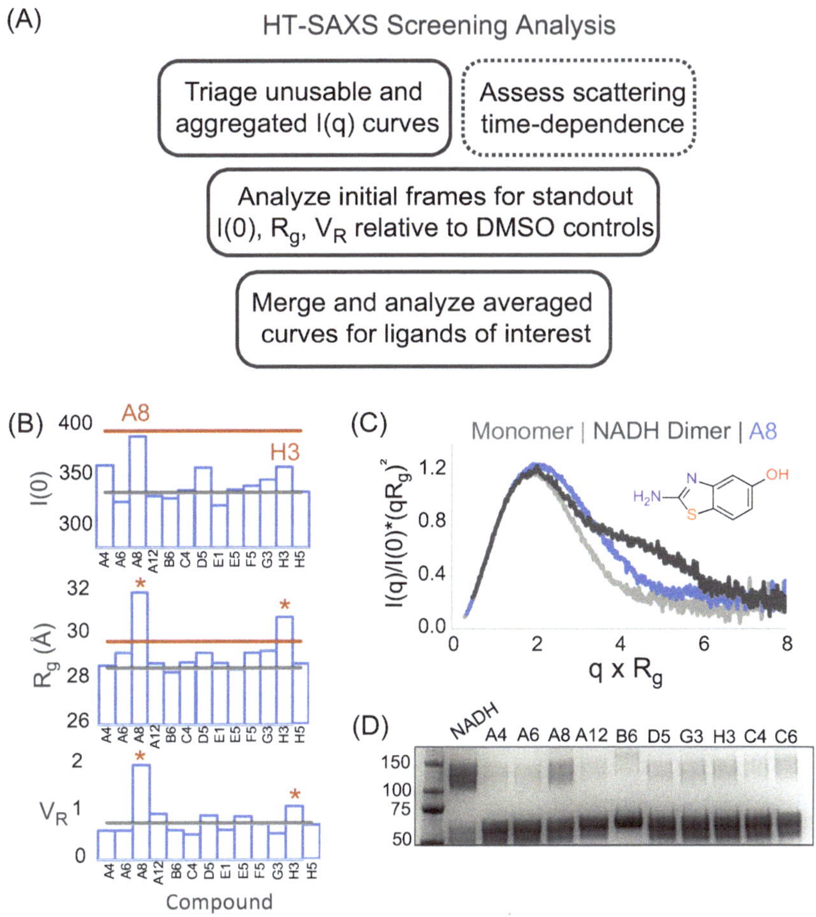

Fig. 4 AIF HT-SAXS Screening Analysis of 14 Focused Hits. (A) Workflow for triaging and analyzing large HT-SAXS screening datasets. (B) I(0), R_g, and V_R results from the focused 14-fragment AIF SAXS screen. Gray and red lines indicate average and three-standard deviations from AIF-DMSO controls. (C) Overlay of normalized Kratky transforms from A8 (blue) and AIF monomer (gray) and dimer (black) benchmarks. (D) SDS-PAGE of BS^3 amine-crosslinking reactions that capture AIF dimers in the presence of NADH among top candidates from MST screening. D, dimer; M, monomer.

protein targets, e.g., those containing metal centers, functional disulfides, or redox-related cofactors such as FAD or FMN.

If a target system has a likelihood of containing time-dependent information, one can quantitatively assess this by comparing time-evolved I(0) and R_g values relative to time-invariant controls using ScÅtter. Calculations of V_R relative to the initial frame or a control series may also be useful for detecting time-evolution within low- or mid-q intensities (0.015–0.15 Å). The SIBYLS SAXS similarity web application offers a drag-and-drop interface for rapidly calculating and ranking V_R values in structural similarity maps (Hura et al., 2013) (https://sibyls.als.lbl.gov/saxs-similarity/). As a caveat, radiation damage can also cause time-dependent aggregation and should be ruled out before kinetic analysis. This can usually be diagnosed by appearance of upward deflection in low-angle I(q) intensities, non-linear Guinier plots, and a continuous dose-dependent increase in I(0) and R_g values in later I(q) frames.

Time-resolved assessment of the AIF HT-SAXS datasets reveals radiation-induced aggregation in several samples. Thus, we chose to focus upon high-throughput analyses of the initial, aggregation-free I(q) snapshots. I(0) and R_g values were calculated with ScÅtter, then exported to Excel and GraphPad Prism for visualization and statistical analysis. V_R values were calculated using the SIBYLS SAXS similarity web application over q-ranges matched to exposure time (0.02–0.15 $Å^{-1}$ for 300-ms frames, 14-compound screen; 0.02–0.13 $Å^{-1}$ for 200-ms frames, 96-compound screen) to minimize high-q noise. V_R values were copied into Excel using the "Metric Matrix" option under the Operations menu.

Initial analysis focused upon assessing the consistency of replicate DMSO controls from each screen. For the focused 14-compound screen, the coefficient-of-variation among replicate DMSO samples was 6% and 1% for I(0) and R_g, respectively, while V_R was 0.76. For the broader 96-compound HT-SAXS screen, plotting AIF-DMSO I(0) and R_g indicates plate-to-plate consistency with the exception of I(0) outliers in Plates 2 and 3 (Fig. 5A). These screening plates required a longer collection period; we surmise that the outlier samples, which were located at the end of each plate, experienced evaporative concentration and corresponding elevation of concentration-dependent I(0). Coefficients-of-variation for I(0) and R_g excluding I(0) outliers are 1–5%, reflecting robust internal consistency of the screening assay. V_R, as referenced to the first AIF-DMSO sample of Plate 1, aligns with the pilot DMSO measurements (V_R 1.9 (0.6)).

Inspection of the 14-compound focused screening results highlights AIF-A8 as consistently increased across all three SAXS metrics (Fig. 4B). AIF-H3 exhibits an elevated R_g but does not differ significantly among the other SAXS metrics. In contrast, fragments with higher micromolar affinities for AIF (A6/A12/E12, A4, G3) do not cause significant deviation from AIF-DMSO controls. Normalized Kratky plots, which detect AIF dimerization as a shoulder of the primary maximum, suggest that A8 is

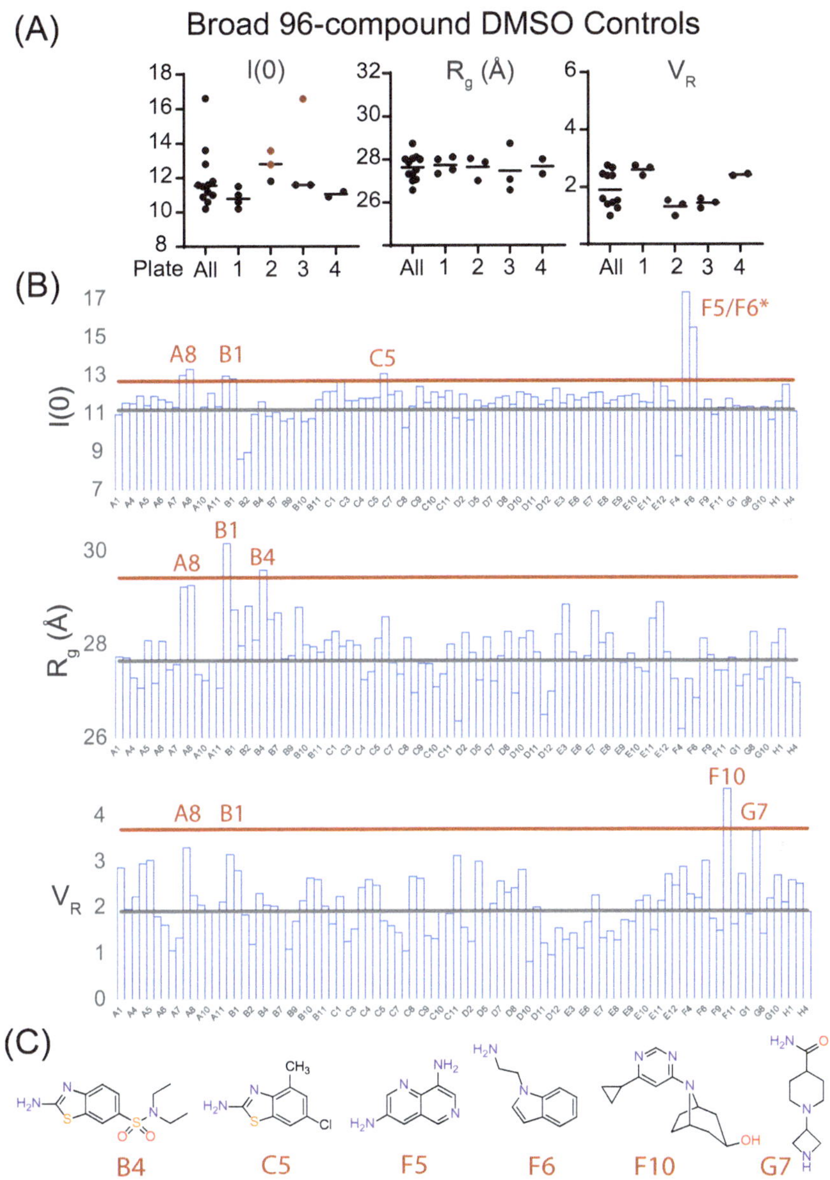

Fig. 5 AIF HT-SAXS Screening Analysis of 96-Fragment Library. (A) I(0), R_g, and V_R values from AIF-DMSO controls demonstrating consistency across screening plates. (B) I(0), R_g, and V_R analysis from curated 64-compound dataset. Gray and red lines indicate average and three-standard deviations from AIF-DMSO controls. (C) Novel library fragments exhibiting significant change in at least one SAXS metric.

stimulating weak dimerization (Fig. 4C). This is confirmed by amine-crosslinking, which captures a minor dimeric species when AIF is saturated with A8 (Fig. 4D). Fragments with highest affinities do not induce detectable dimerization. Thus, the SAXS counter screen usefully selects for fragments with allosteric properties among AIF ligands identified by MST.

After elimination of unusable and aggregated curves, the curated HT-SAXS dataset from the broad 96-compound screen resulted in 64 unique AIF-ligand scattering curves (Fig. 5B). Sample bubbles and AIF aggregation accounted for the majority of eliminated curves. Among I(0), R_g, and V_R values, fragments A8 and B1 are consistently elevated. Additional fragments are highlighted among individual metrics (B4, C5, D5/D6, F10, G7) and include additional benzothiazolamine scaffolds (Fig. 5C); however, these do not appear as consistently as the MST-identified fragments. B1 was originally excluded from focused screening, due to signs of aggregation in the original MST data. The aggregation-free SAXS data suggests that this candidate is well behaved and may provide a second allosteric lead for optimization with A8.

7. Summary and future perspectives

Screening approaches using HT-SAXS continue to advance and promise to provide unique, complementary insights to traditional ligand binding assays. In this application, we have paired HT-SAXS chemical screening with MST to identify potential allosteric effectors among ligands of Apoptosis-Inducing Factor. This protocol and data collection strategy are readily accessible to SAXS academic users. Importantly, our approach can be customized toward any biological system experiencing well-resolved conformational changes. Efforts continue to reduce sample volume and concentration through optimization of HT-SAXS sample loading systems and hold the potential to reduce sample consumption by half. Future improvements in detector hardware and analysis algorithms are expected to extend the throughput, sensitivity, and resolution of HT-SAXS chemical screening as a robust and practical approach.

Key Benchmarks for SAXS Chemical Screens

Target

- Are structural states of the target resolvable with SAXS parameters—V_R, R_g, I(0)? This can be assessed with…
  - Simulated SAXS data from available atomic structures
  - Experimental SAXS data

- Can the target be produced at sufficient quantities and purity for SAXS experiments?
- Is the target resilient during shipment to SAXS synchrotron beamlines?
- Does the target produce high-quality, aggregation-free HT-SAXS data?
- What are optimal SAXS sample concentrations and buffer conditions for screening?

Library

- Is the library soluble in aqueous screening buffer (as assessed by complementary ligand binding experiments)?
- Does the target remain soluble when mixed with candidate library hits at SAXS concentrations?
- Do target-ligand samples remain soluble after a freeze-thaw cycle (if required for shipment)?

Pilot Screen

- Does DMSO affect sample conformation or stability?
- Is DMSO buffer matching optimal?
- Are there weaknesses in the pipeline of screen preparation, shipping, and collection?

Production Screen

- Are shipment plans and shift scheduling confirmed with SAXS beamline staff?
- Are sufficient shifts available to collect the production screen?

Simple Scattering deposition

Curated, aggregation-free I(q) scattering curves of AIF-ligand complexes from the focused screen of Fig. 4 are deposited under XSLGMJSW in the Simple Scattering database (https://simplescattering.com/).

Acknowledgments

This work was supported by the National Institutes of Health (NIH) (P01 CA92584, R35 CA220430), the Cancer Prevention and Research Institute of Texas (RP180813), and a Robert A. Welch Chemistry Chair. SAXS experiments were conducted at the SIBYLS beamline (12.3.1) at the Advanced Light Source (ALS), a national user facility operated by Lawrence Berkeley National Laboratory on behalf of the Department of Energy, Office of Basic Energy Sciences, through the Integrated Diffraction Analysis Technologies (IDAT) program, supported by DOE Office of Biological and Environmental Research. Additional support comes from the National Institute of Health project ALS-ENABLE (P30 GM124169) and a High-End Instrumentation Grant S10OD018483.

References

Arrowsmith, C. H., Audia, J. E., Austin, C., Baell, J., Bennett, J., Blagg, J., et al. (2015). The promise and peril of chemical probes. *Nature Chemical Biology*, *11*(8), 536–541. https://doi.org/10.1038/nchembio.1867.

Ashton, T. M., McKenna, W. G., Kunz-Schughart, L. A., & Higgins, G. S. (2018). Oxidative phosphorylation as an emerging target in cancer therapy. *Clinical Cancer Research*, *24*(11), 2482–2490. https://doi.org/10.1158/1078-0432.CCR-17-3070.

Baell, J. B., & Holloway, G. A. (2010). New substructure filters for removal of pan assay interference compounds (PAINS) from screening libraries and for their exclusion in bioassays. *Journal of Medicinal Chemistry*, *53*(7), 2719–2740. https://doi.org/10.1021/jm901137j.

Baell, J. B., & Nissink, J. W. M. (2018). Seven year itch: Pan-assay interference compounds (PAINS) in 2017-utility and limitations. *ACS Chemical Biology*, *13*(1), 36–44. https://doi.org/10.1021/acschembio.7b00903.

Brosey, C. A., Ho, C., Long, W. Z., Singh, S., Burnett, K., Hura, G. L., et al. (2016). Defining NADH-driven allostery regulating apoptosis-inducing factor. *Structure*, *24*(12), 2067–2079. https://doi.org/10.1016/j.str.2016.09.012.

Brosey, C. A., Shen, R., Burnett, K., Hura, G., Moiani, D., Jones, D., et al. (2020). Incorporating HT-SAXS into drug-discovery pipelines. *Acta Crystallogr. Sect. A*, *76*(a1), a139. https://doi.org/10.1107/S010876732009861X.

Brosey, C. A., & Tainer, J. A. (2019). Evolving SAXS versatility: Solution X-ray scattering for macromolecular architecture, functional landscapes, and integrative structural biology. *Current Opinion in Structural Biology*, *58*, 197–213. https://doi.org/10.1016/j.sbi.2019.04.004.

Chen, P. C., & Hennig, J. (2018). The role of small-angle scattering in structure-based screening applications. *Biophysical Reviews*, *10*, 1295–1310. https://doi.org/10.1007/s12551-018-0464-x.

Chen, P. C., Masiewicz, P., Perez, K., & Hennig, J. (2020). Structure-based screening of binding affinities via small-angle X-ray scattering. *IUCrJ*, 7(Pt. 4), 644–655. https://doi.org/10.1107/S2052252520004169.

Chen, P. C., Masiewicz, P., Rybin, V., Svergun, D., & Hennig, J. (2018). A general small-angle X-ray scattering-based screening protocol validated for protein-RNA interactions. *ACS Combinatorial Science*, *20*(4), 197–202. https://doi.org/10.1021/acscombsci.8b00007.

Classen, S., Hura, G. L., Holton, J. M., Rambo, R. P., Rodic, I., McGuire, P. J., et al. (2013). Implementation and performance of SIBYLS: A dual endstation small-angle X-ray scattering and macromolecular crystallography beamline at the advanced light source. *Journal of Applied Crystallography*, *46*, 1–13. https://doi.org/10.1107/S0021889812048698.

Daniels, D. S., Woo, T. T., Luu, K. X., Noll, D. M., Clarke, N. D., Pegg, A. E., et al. (2004). DNA binding and nucleotide flipping by the human DNA repair protein AGT. *Nature Structural & Molecular Biology*, *11*(8), 714–720. https://doi.org/10.1038/nsmb791.

Dyer, K. N., Hammel, M., Rambo, R. P., Tsutakawa, S. E., Rodic, I., Classen, S., et al. (2014). High-throughput SAXS for the characterization of biomolecules in solution: A practical approach. *Methods in Molecular Biology*, *1091*, 245–258. https://doi.org/10.1007/978-1-62703-691-7_18.

Franke, D., Petoukhov, M. V., Konarev, P. V., Panjkovich, A., Tuukkanen, A., Mertens, H. D. T., et al. (2017). ATSAS 2.8: A comprehensive data analysis suite for small-angle scattering from macromolecular solutions. *Journal of Applied Crystallography*, *50*(Pt 4), 1212–1225. https://doi.org/10.1107/S1600576717007786.

Graewert, M. A., & Jeffries, C. M. (2017). Sample and buffer preparation for SAXS. *Advances in Experimental Medicine and Biology*, *1009*, 11–30. https://doi.org/10.1007/978-981-10-6038-0_2.

Hangen, E., Feraud, O., Lachkar, S., Mou, H., Doti, N., Fimia, G. M., et al. (2015). Interaction between AIF and CHCHD4 regulates respiratory chain biogenesis. *Molecular Cell*, *58*(6), 1001–1014. https://doi.org/10.1016/j.molcel.2015.04.020.

Houl, J. H., Ye, Z., Brosey, C. A., Balapiti-Modarage, L. P. F., Namjoshi, S., Bacolla, A., et al. (2019). Selective small molecule PARG inhibitor causes replication fork stalling and cancer cell death. *Nature Communications*, *10*(1), 5654. https://doi.org/10.1038/s41467-019-13508-4.

Hura, G. L., Budworth, H., Dyer, K. N., Rambo, R. P., Hammel, M., McMurray, C. T., et al. (2013). Comprehensive macromolecular conformations mapped by quantitative SAXS analyses. *Nature Methods*, *10*(6), 453–454. https://doi.org/10.1038/nmeth.2453.

Hura, G. L., Menon, A. L., Hammel, M., Rambo, R. P., Poole, F. L., 2nd, Tsutakawa, S. E., et al. (2009). Robust, high-throughput solution structural analyses by small angle X-ray scattering (SAXS). *Nature Methods*, *6*(8), 606–612. https://doi.org/10.1038/nmeth.1353.

Jeffries, C. M., Graewert, M. A., Blanchet, C. E., Langley, D. B., Whitten, A. E., & Svergun, D. I. (2016). Preparing monodisperse macromolecular samples for successful biological small-angle X-ray and neutron-scattering experiments. *Nature Protocols*, *11*(11), 2122–2153. https://doi.org/10.1038/nprot.2016.113.

Joshi, P., & Vendruscolo, M. (2015). Druggability of intrinsically disordered proteins. *Advances in Experimental Medicine and Biology*, *870*, 383–400. https://doi.org/10.1007/978-3-319-20164-1_13.

Lemberg, K. M., Gori, S. S., Tsukamoto, T., Rais, R., & Slusher, B. S. (2022). Clinical development of metabolic inhibitors for oncology. *The Journal of Clinical Investigation*, *132*(1), e148550. https://doi.org/10.1172/JCI148550.

Moiani, D., Link, T. M., Brosey, C. A., Katsonis, P., Lichtarge, O., Kim, Y., et al. (2021). An efficient chemical screening method for structure-based inhibitors to nucleic acid enzymes targeting the DNA repair-replication interface and SARS CoV-2. *Methods in Enzymology*, *661*, 407–431. https://doi.org/10.1016/bs.mie.2021.09.003.

Moiani, D., Ronato, D. A., Brosey, C. A., Arvai, A. S., Syed, A., Masson, J. Y., et al. (2018). Targeting allostery with avatars to design inhibitors assessed by cell activity: Dissecting MRE11 endo- and exonuclease activities. *Methods in Enzymology*, *601*, 205–241. https://doi.org/10.1016/bs.mie.2017.11.030.

Nishimura, N., Hitomi, K., Arvai, A. S., Rambo, R. P., Hitomi, C., Cutler, S. R., et al. (2009). Structural mechanism of abscisic acid binding and signaling by dimeric PYR1. *Science*, *326*(5958), 1373–1379. https://doi.org/10.1126/science.1181829.

Putnam, C. D., Hammel, M., Hura, G. L., & Tainer, J. A. (2007). X-ray solution scattering (SAXS) combined with crystallography and computation: Defining accurate macromolecular structures, conformations and assemblies in solution. *Quarterly Reviews of Biophysics*, *40*(3), 191–285. https://doi.org/10.1017/S0033583507004635.

Rambo, R. P. (2017). Considerations for sample preparation using size-exclusion chromatography for home and synchrotron sources. *Advances in Experimental Medicine and Biology*, *1009*, 31–45. https://doi.org/10.1007/978-981-10-6038-0_3.

Rambo, R. P., & Tainer, J. A. (2010). Improving small-angle X-ray scattering data for structural analyses of the RNA world. *RNA*, *16*(3), 638–646. https://doi.org/10.1261/rna.1946310.

Rambo, R. P., & Tainer, J. A. (2011). Characterizing flexible and intrinsically unstructured biological macromolecules by SAS using the Porod-Debye law. *Biopolymers*, *95*(8), 559–571. https://doi.org/10.1002/bip.21638.

Rambo, R. P., & Tainer, J. A. (2013a). Accurate assessment of mass, models and resolution by small-angle scattering. *Nature*, *496*(7446), 477–481. https://doi.org/10.1038/nature 12070.

Rambo, R. P., & Tainer, J. A. (2013b). Super-resolution in solution X-ray scattering and its applications to structural systems biology. *Annual Review of Biophysics*, *42*, 415–441. https://doi.org/10.1146/annurev-biophys-083012-130301.

Schneidman-Duhovny, D., Hammel, M., & Sali, A. (2010). FoXS: A web server for rapid computation and fitting of SAXS profiles. *Nucleic Acids Research*, *38*(Web Server issue), W540–W544. https://doi.org/10.1093/nar/gkq461.

Schneidman-Duhovny, D., Hammel, M., Tainer, J. A., & Sali, A. (2013). Accurate SAXS profile computation and its assessment by contrast variation experiments. *Biophysical Journal*, *105*(4), 962–974. https://doi.org/10.1016/j.bpj.2013.07.020.

Schneidman-Duhovny, D., Hammel, M., Tainer, J. A., & Sali, A. (2016). FoXS, FoXSDock and MultiFoXS: Single-state and multi-state structural modeling of proteins and their complexes based on SAXS profiles. *Nucleic Acids Research*, *44*(W1), W424–W429. https://doi.org/10.1093/nar/gkw389.

Scott, D. E., Bayly, A. R., Abell, C., & Skidmore, J. (2016). Small molecules, big targets: Drug discovery faces the protein-protein interaction challenge. *Nature Reviews. Drug Discovery*, *15*(8), 533–550. https://doi.org/10.1038/nrd.2016.29.

Sevrioukova, I. F. (2009). Redox-linked conformational dynamics in apoptosis-inducing factor. *Journal of Molecular Biology*, *390*(5), 924–938. https://doi.org/10.1016/j.jmb.2009.05.013.

Shibata, A., Moiani, D., Arvai, A. S., Perry, J., Harding, S. M., Genois, M. M., et al. (2014). DNA double-strand break repair pathway choice is directed by distinct MRE11 nuclease activities. *Molecular Cell*, *53*(1), 7–18. https://doi.org/10.1016/j.molcel.2013.11.003.

Svergun, D., Barberato, C., & Koch, M. H. J. (1995). CRYSOL—A program to evaluate X-ray solution scattering of biological macromolecules from atomic coordinates. *Journal of Applied Crystallography*, *28*(6), 768–773. https://doi.org/10.1107/S0021889895007047.

Warner, K. D., Hajdin, C. E., & Weeks, K. M. (2018). Principles for targeting RNA with drug-like small molecules. *Nature Reviews. Drug Discovery*, *17*(8), 547–558. https://doi.org/10.1038/nrd.2018.93.

Zhou, J., Gelot, C., Pantelidou, C., Li, A., Yücel, H., Davis, R. E., et al. (2021). A first-in-class polymerase theta inhibitor selectively targets homologous-recombination-deficient tumors. *Nature Cancer*, *2*(6), 598–610. https://doi.org/10.1038/s43018-021-00203-x.

CHAPTER TWELVE

Combining small angle X-ray scattering (SAXS) with protein structure predictions to characterize conformations in solution

Naga Babu Chinnam[a], Aleem Syed[a], Greg L. Hura[b,c], Michal Hammel[b], John A. Tainer[a,d], and Susan E. Tsutakawa[b,*]

[a]Department of Molecular and Cellular Oncology, The University of Texas MD Anderson Cancer Center, Houston, TX, United States
[b]Molecular Biophysics and Integrated Bioimaging Division, Lawrence Berkeley National Laboratory, Berkeley, CA, United States
[c]Department of Chemistry and Biochemistry, University of California Santa Cruz, Santa Cruz, CA, United States
[d]Department of Cancer Biology, University of Texas MD Anderson Cancer Center, Houston, TX, United States
*Corresponding author: e-mail address: setsutakawa@lbl.gov

Contents

Methods in Enzymology, Volume 678
ISSN 0076-6879
https://doi.org/10.1016/bs.mie.2022.09.023

Abstract

Accurate protein structure predictions, enabled by recent advances in machine learning algorithms, provide an entry point to probing structural mechanisms and to integrating and querying many types of biochemical and biophysical results. Limitations in such protein structure predictions can be reduced and addressed through comparison to experimental Small Angle X-ray Scattering (SAXS) data that provides protein structural information in solution. SAXS data can not only validate computational predictions, but can improve conformational and assembly prediction to produce atomic models that are consistent with solution data and biologically relevant states. Here, we describe how to obtain protein structure predictions, compare them to experimental SAXS data and improve models to reflect experimental information from SAXS data. Furthermore, we consider the potential for such experimentally-validated protein structure predictions to broadly improve functional annotation in proteins identified in metagenomics and to identify functional clustering on conserved sites despite low sequence homology.

1. Introduction

Deciphering a protein's functional mechanisms is enabled by accurate protein structure(s) and knowledge of the active site and interaction interfaces. In the past, the availability of structures was a major bottleneck. Scientists were limited by proteins whose structures could be experimentally determined at high resolution by X-ray crystallography, nuclear magnetic resonance (NMR), and, more recently, cryo-electron microscopy (Cryo-EM). Protein structure predictions, particularly for proteins >100 residues without a known orthologous structure, were not reliable, even for obtaining the correct topology (Moult, Fidelis, Kryshtafovych, Schwede, & Tramontano, 2018). After the 2020 Critical Assessment of Protein structure predictions (CASP14), it was clear that machine learning algorithms improved the accuracy of protein structure predictions to the level that predictions could be reliably used (Baek, DiMaio, et al., 2021; Jumper et al., 2021b,

2021a; Jumper & Hassabis, 2022; Kryshtafovych, Schwede, Topf, Fidelis, & Moult, 2021; Tunyasuvunakool et al., 2021; Varadi et al., 2022). Scientists can input the amino acid sequence into a machine learning server and receive a protein structure prediction that is likely to be highly accurate. Among many factors that contribute to the structure prediction accuracy, we highlight several factors in the top protein structure prediction algorithm, AlphaFold2. (1) Use of evolutionary co-variance, which is the concept that residues that vary together in evolution are more likely than not to be close in tertiary structure, first implemented by Deborah Marks and Chris Sanders (Marks et al., 2011; Morcos et al., 2011). By default, evolutionary covariance depends on the quality and depth (how many sequences) in the multiple sequence alignments of the target. (2) Use of inter-residue probability distribution distograms that are a spreadsheet of probabilities of any one residue being within distance to other residues in the protein. First applied in CASP13 by the Deepmind team, this opened up the algorithms to better incorporate long-distance relationships, such as N- and C-capping of alpha helices. (3) Application of attention machine learning algorithms that prioritize parts of the input as being more significant, was an impactful change in CASP14 from the traditional convolutional neural networks applied in CASP13. (4) Inclusion of the 3D model within the machine learning algorithm, allowed a confidence assessment in the B-factor column of the coordinates. In our and others' experience, low confidence regions often coincided with loop and other flexible regions and may be a means to efficiently identify such regions, as we show in our example below.

Yet, current protein structure prediction algorithms have limitations. (1) Exposed regions. Although globular regions with substantial contacts and strong evolutionary conservation are likely to be highly accurate in predicting, regions with fewer contacts or flexibility are likely to be mispredicted. (2) Proteins with few family members. The number of homologous sequences matters. Accuracy depends partly on the depth of the multiple sequence alignment and derived evolutionary covariances from sets of residues that mutate simultaneously. Thus, proteins with low sequence homology to other proteins will have a lower probability of obtaining an accurate structure. (3) Protein-protein, protein-DNA, and protein-RNA complexes. The latest machine learning algorithms have not been tested in CASP for the accurate prediction of complexes. Also, given that computational time is limited for many publicly available servers, the calculation for complexes is more complicated than for individual proteins. For complexes with RNA or DNA, machine learning algorithms cannot predict RNA and DNA conformations sculpted by protein interactions. (4) Bias toward deposited structures. Accuracy is a relative term as it depends on the standard. In its assessment of protein structure prediction algorithms, the CASP committee uses high resolution structures determined mainly by X-ray crystallography. However, we and others have shown that proteins in solution can be conformationally distinct from proteins trapped in crystal lattices and that sometimes may not adopt biologically-relevant conformations (Andrec et al., 2007; Ogorzalek et al., 2018). As the machine learning algorithms use the PDB as their training database, their predictions are influenced by the presence of co-factors, nucleic acids, or ligands, in addition to the salt and aggregating conditions to form ordered crystallographic protein arrays. With the target being biologically relevant conformations, we and others anecdotally have found predictions to be closer to crystal structures than to the solution structures. (5) Small molecules and ligands. Current machine learning algorithms cannot predict the structural impact of bound ligands, which can be functionally important for allosteric binders and induced fits. (6) Impact of mutations. Machine learning algorithms minimize disagreements between multiple inputs but, by necessity, remove outliers. Thus, current algorithms are challenged to predict the structural consequences of a mutation.

Although there are limitations in protein structure predictions, comparison to experimental data, even if sparse, has the potential to test, validate, and guide protein structure predictions. In particular, Small Angle X-ray Scattering (SAXS) can provide global structural restraints. SAXS data contains information on the distances between all electron pairs. Unlike NMR, these distances are not assigned, and it is currently not sufficient to derive an atomic model. However, all distances are included, and it is possible to calculate the SAXS curves from atomic models—allowing a quantitative comparison between atomic models and experimental data (Rambo & Tainer, 2013a; Schneidman-Duhovny, Hammel, & Sali, 2010; Schneidman-Duhovny, Hammel, Tainer, & Sali, 2013, 2016; Schneidman-Duhovny, Kim, & Sali, 2012; Svergun, Barberato, & Koch, 1995). One step further, atomic models can be altered and/or or used as an ensemble of multiple atomic models, to test against the SAXS data and gain information on the conformation(s) and dynamics in solution (Schneidman-Duhovny et al., 2010, 2012, 2013, 2016; Schneidman-Duhovny & Hammel, 2018). Here, we focus on how to use atomic models from protein structure algorithms with experimental SAXS data to validate prediction and/or to characterize the structure in solution. A match between the prediction and the SAXS data will support the validity of the protein structure prediction. We will show how to compare atomic models to experimental SAXS data. Similarity but with slight differences between calculated and SAXS data suggest that the solution structure is different, either in conformation, dynamics, or stoichiometry. We will show servers that can alter the predicted model structure and that can test ensembles of these altered atomic models.

The rationale for using SAXS data to validate structure predictions is that it has a major advantage compared to other structural methods such as X-ray crystallography. SAXS data collection is straightforward with minimal sample preparation and is essentially available without cost through synchrotron SAXS beamlines to any academic scientist with purified protein (Classen et al., 2010, 2013; Dyer et al., 2014; Hura et al., 2009). SAXS synchrotron data collection for research to be published is typically free. No labeling is required. The particle size limitation at our SIBYLS beamline in the Advanced Light Source Synchrotron ranges from 8 kD to 600 kD, within the range of most protein samples (Hammel & Tainer, 2021). Importantly, oligomeric states (Korasick & Tanner, 2018; Rambo & Tainer, 2013b), protein flexibility (Brosey & Tainer, 2019; Rambo & Tainer, 2011; Reyes, Schwartz, Tainer, & Rambo, 2014), and shape (Svergun, Petoukhov, & Koch, 2001) can be calculated from SAXS data. In our experience, the stoichiometry is particularly important as it is often not recognized. Our structural genomics study of 50 proteins, mostly from *Pyrococcus furiosus* showed that over half of the proteins studied formed homo-oligomers (Hura et al., 2009). It is estimated that over half of the proteins in the PDB are multimerizing (Marsh & Teichmann, 2015). Although we focus primarily on protein models, SAXS data can be collected on protein, DNA, and RNA individually or as complexes and compared to relevant atomic models. The impact of mutations and small molecules can also be probed through SAXS (Brosey et al., 2016; Pratt et al., 2014).

SAXS data can effectively validate protein structure prediction since it contains structural information on the distribution of electron pair distances of proteins in solution (Putnam, Hammel, Hura, & Tainer, 2007; Tsutakawa, Hura, Frankel, Cooper, & Tainer, 2007). Analogous to X-ray crystallography, SAXS data are collected in reciprocal space, with the Intensity (I) of the scattered X-rays plotted as a function of the scattering angle q. The scattering angle is typically calculated as momentum transfer q to incorporate beamline parameters and X-ray wavelength. The units can be in $Å^{-1}$ or nm^{-1}. Like X-ray crystallography, SAXS data can be converted to real space through the Fourier transform X-ray scattering relationship between reciprocal and real space (Putnam et al., 2007; Svergun, 1992). In real space, SAXS data are represented as a histogram of relation proportion P of electron pairs at distance r, also known as P(r). D_{max} is the longest electron pair distance in the P(r). SAXS data can be compared to protein structures in real or reciprocal space and is also called SAXS curves or SAXS profiles. Molecular envelopes can be calculated ab initio from experimental SAXS curves (Grant, 2018; Svergun et al., 2001). Although useful, ab initio envelopes are limited by the difficulty to differentiate between closely related models, so we find comparing SAXS data to atomic models directly more quantitative (Hura et al., 2013).

Here, we focus on comparing atomic models to SAXS data after the data has been collected and analyzed. This comparison allows us to validate a protein structure prediction and/or alter the prediction to better fit the solution data. A single SAXS curve for each condition has been generated from available SAXS data. Recent reviews provide detailed protocols for SAXS sample preparation, data collection, and analysis (Chinnam et al., 2022; Tully, Tarbouriech, Rambo, & Hutin, 2021). At our SIBYLS beamline at the Advance Light Source synchrotron, SAXS data can be collected in two modes: in a sample cell in high throughput (HT) SAXS (30 μL of a 0.5–2 mg/mL) or via Size-Exclusion Chromatography-coupled (SEC)-Multi-Angle Light Scattering (MALS)-SAXS (50–100 μL 5–20 mg/mL). HT-SAXS provides the best signal-to-noise. To compare SAXS to the atomic model, the sample ideally must be well behaved, have no aggregation, and be stoichiometrically monodisperse. SEC-MALS-SAXS will separate heterogeneity and provide a monodisperse sample for difficult samples but dilute the sample >4-fold (Hammel, Rashid, et al., 2021; Zhou et al., 2019). Making it a relatively inexpensive and available structural method, protein sample(s) can be mailed in at the SIBYLS beamline; the SAXS data are collected by experienced staff, and the SAXS data are made available to users.

The basic process for SAXS validation of protein structure predictions is to obtain atomic models from a publicly accessible server, predict the SAXS curves from the atomic models, compare these predicted SAXS curves to experimental SAXS data, and, if necessary, modify the models (e.g., change oligomerization state, change the protein conformation) to improve fit to the SAXS data (see Fig. 1A).

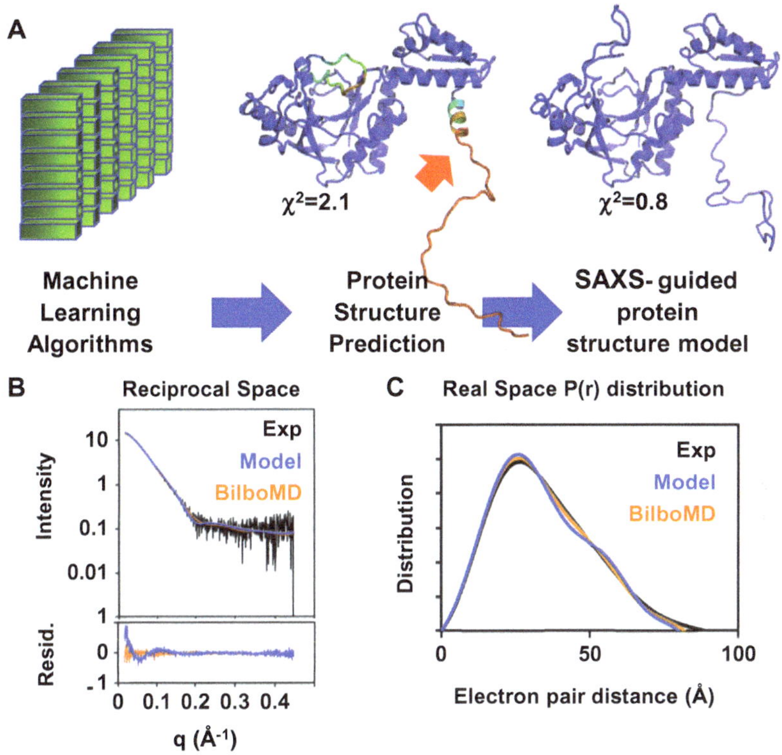

Fig. 1 From predictions to solution structures. (A) Schematic showing overall strategy. Shown are representative protein predictions from AlphaFold2 and modification of that model using BILBOMD. The AlphaFold2 prediction is colored from blue to red based on confidence. Note that BILBOMD improved the fit to the experimental SAXS data and suggests that a low confidence helix (see red arrow) does not occur significantly in solution. (B and C) Corresponding SAXS data in reciprocal and real space, respectively for experimental, AlphaFold2 protein structure prediction, and BILBOMD model.

2. Obtaining a protein structure prediction

2.1 Overview

Currently, of the many servers in protein structure prediction available, we are most familiar with RosettaFold (Baek, DiMaio, et al., 2021) created by David Baker's group, and AlphaFold2 (Jumper et al., 2021a) by the Deepmind Technologies company and recommend these programs here. However, accurate structure predictions can be obtained from other servers as well. We recommend experimental SAXS validations for any atomic model, regardless of source including crystallographic models (Chinnam et al., 2022).

In terms of accuracy of your protein structure prediction, it is our experience and documented in CASP14 that the accuracy of individual domains for most protein predictions and, in particular, confidence for their topology are high (Kryshtafovych et al., 2021). However, caution should be taken for domain-domain orientation, for proteins with low sequence homology and for regions that are predicted to be disordered. For the latter, likely due to their lack of representation in the PDB training database for the machine learning algorithms, there is currently failure for regions predicted to be disordered and these regions are predicted simply as random coil. Although SAXS-based atomic analyses can be informative with these type of proteins (Brosey et al., 2013; Brosey & Tainer, 2019; Hammel, Rashid, et al., 2021; Hammel, Rosenberg, et al., 2021; Hammel & Tainer, 2021; Tsutakawa et al., 2011, 2015; Wilamowski et al., 2021), care should be taken in conclusions as these regions adopt multiple conformations. It is also unclear how loop regions involved in domain swaps will be predicted, but they can play critical mechanistic roles as found for p62 in the TFIIH core 7-subunit complex that extends from the XPD ATPase site to its DNA binding groove (Yan et al., 2019). Another cautionary note for protein structure predictions is the possible presence of conformational changes, induced by small molecules or intrinsic to the protein domain. SAXS later on in the chapter can assess conformational flexibility and guide selection of alternative conformations.

Structures for homo- and heteromeric complexes can also be predicted using these protein structure prediction machine learning algorithms but have not yet been assessed by CASP (Baek, Anishchenko, Park, Humphreys, & Baker, 2021; Evans et al., 2021) (See Note 1). Predictions of complexes can be tested against experimental SAXS data similar to single chain models. For nucleic acid complexes, RNA prediction servers will also be helpful for this category of predictions and that SAXS will be of use for validation and tweaking the atomic RNA models to be more consistent with solution data.

2.2 Protein structure prediction servers

Follow the instructions for each server to obtain the prediction. Be sure that the sequence exactly matches the SAXS sample, including tags. Tags can also be added to already predicted models in the databases in modeling programs such as Modeler within CHIMERA, as previously described (Schneidman-Duhovny & Hammel, 2018). Follow the instructions for each site. These programs will typically output multiple models. All models can be compared to the SAXS data. As mentioned earlier, potential pitfalls are for proteins with little or no sequence homology to other proteins, as sometimes found for viral or bacterial proteins.

2.2.1 https://robetta.bakerlab.org/ This site hosts the RosettaFold protein structure prediction server. Homo-oligomeric and heteromeric complexes can also be predicted as described in RosettaFold's frequently asked questions (FAQ) (Baek, Anishchenko, et al., 2021).

2.2.2 https://alphafold.ebi.ac.uk/ In collaboration with EMBL-European Bioinformatics Institute (EMBL-EBI), this is a database of AlphaFold2 protein structure predictions for 20 model organisms including *Homo sapiens, Escherichia coli, Saccharomyces cerevisiae, Schizosaccharomyces pombe, Mus musculus, Rattus norvegicus, Caenorhabditis elegans, Danio rerio*, and *Drosophila melanogaster*. Alternatively, these models can be obtained on https://www.uniprot.org/ under individual protein entries. Besides the predicted structure in PDB format, the database also provides a per-residue confidence score (pLDDT) for each residue between 0 and 100. The regions below 50 pLDDT may be unstructured in isolation. Consequently, these values can be used to guide conformational sampling, as described in section 5. We note that the Swiss-Model also adds in structure predictions for complexes (Waterhouse et al., 2018).

2.2.3 ColabFold. AlphaFold2 can be run on this installation on a Google Colaboratory site. There are limitations in runtime, so large proteins may not be accurately predicted. AlphaFold-Multimer can be run at this site for atomic models of complexes, with the same time limit restrictions (Evans et al., 2021). https://colab.research.google.com/github/sokrypton/ColabFold/blob/main/AlphaFold2.ipynb.

2.3 Pitfalls to look out for

Atomic models should have secondary and tertiary structure in their predictions for ordered regions. If the prediction server has stopped in the middle because of a time limit given for the prediction (not uncommon occurrence for large proteins or complexes), these predicted-to-be-ordered regions will look garbled. To assess if these regions are predicted to be disordered or ordered, there are a number of highly accurate disorder prediction servers publicly available including PONDR and DisProt (Hatos et al., 2020; Piovesan et al., 2017; Quaglia et al., 2022; Xue, Dunbrack, Williams, Dunker, & Uversky, 2010) that predict disordered regions within the sequence. If the protein is homo-oligomerizing, there may be regions that overlap within the oligomer and that will need to be adjusted.

3. Prediction of SAXS curve from an atomic model

3.1 Overview

To validate structure predictions, we compare experimental SAXS data to the SAXS curve calculated from the structure prediction. Multiple programs can calculate a SAXS curve from atomic models of protein, DNA, and/or RNA (Forster et al., 2008; Schneidman-Duhovny et al., 2010, 2012, 2013, 2016; Svergun et al., 1995). We recommend FoXS, that uses the Debye formula to explicitly compute all inter-atomic distances including a predicted hydration layer and taking into account the displaced solvent (Schneidman-Duhovny et al., 2010) (Fig. 2). When input with experimental SAXS data, there are additionally two constants, c1 and c2, that adjust the total excluded volume and the density of the hydration layer, respectively, to better fit the experimental data (see Note 2 for potential issues with c1 and c2). The FoXS server is readily available and can be conveniently input with multiple atomic models uploaded as a compressed folder. A recent Methods paper gives a detailed review (Schneidman-Duhovny & Hammel, 2018).

3.2 Server for Prediction of SAXS curve from an atomic model

The server is located at https://modbase.compbio.ucsf.edu/foxs/.

3.3 Inputs for FoXS server

Use default parameters.

3.3.1 Atomic model(s) input for FoXS. The sequence should match the SAXS sample. Tags and missing sequence could be added to the atomic model by creating a homology model based on the protein structure prediction with Modeler implemented in Chimera. Atomic model(s) can be uploaded as a single structure or multiple structures as a zip-compressed file.

Debye Formula for calculating SAXS curve from atomic model

$I(q) = \sum_{i=1}^{N}\sum_{j=1}^{N} f_i(q)f_j(q)\frac{\sin(qd_{jn})}{qd_{jn}}$, where I is intensity as a function of q is a function of scattering angle θ and X-ray wavelength λ; q = 4π(sin (θ/2))/ λ; q = 4π(sin (θ/2))/ λ, d is the distance between atoms i and j, and N is the number of atoms. The form factor takes into account the displaced solvent and the density of the hydration layer (c1 and c2), which are adjusted when experimental data is input.
$f_i(q)=f_v(q)-c_1f_s(q)+c_2s_if_w(q)$, where $f_v(q)$ is the atomic form factor *in vacuo*, $f_s(q)$ is the form factor of the displaced solvent, s_i is the fraction of solvent accessible surface of the atom *i*, and $f_w(q)$ is the water form factor.

ChiSquare for the differential calculation of two SAXS curves, taking into account experimental error

$\chi^2 = \sum_{i=1}^{N}\frac{(\frac{I_1(q_i)}{m_1}-\frac{I_2(q_i)}{m_2})^2}{(\frac{\sigma_1(q_i)}{m_1}+\frac{\sigma_2(q_i)}{m_2})^2}$, where I1 and I2 are the intensities of the two curves, m1 and m2 are the Intensity means and σ1 and σ2 are the errors for the two SAXS measurements to be compared. L is the number of data points. The ideal is 1. Inclusion of error can drop the value < 1, so the minimum number would be considered the best fit. This is a difference-based metric, with bias towards low q and how the curves are scaled.

Volatility Ratio

$V_R = \sum_{i=1}^{N} abs[\frac{R(q_i)-R(q_{i+1})}{(R(q_i)+R(q_{i+1}))/2}]$, where R is the ratio of the intensities at q_i when ratio is normalized so the average over the range is 1 and where the ratio is binned at a minimal frequency (Δq = π/d where q is a function of scattering angle θ and X-ray wavelength λ; q = 4π(sin (θ/2))/ λ. By assuming a maximum dimension d < 40 nm, the number of bins N is 25 over a q range, q < 0.2Å^{-1}. This is a ratio-based metric, with weight distributed through the curve.

Fig. 2 Equations used for comparison of SAXS curves in reciprocal space.

3.3.2 Experimental SAXS data input for FoXS. Only one SAXS profile can be uploaded. The SAXS profile should have three columns: q, I(q), and error, where q is the momentum transfer. q = 4πsin (θ/2)/λ in Å^{-1}, θ is the scattering angle, λ is the X-ray wavelength, and I is scattering intensity as a function of q. Typically, most beamlines provide the SAXS curve calculated with q and considering beamline-specific parameters, including X-ray wavelength and distance from detector. One potential pitfall is that the scattering angle should be in Å^{-1} and not in nm^{-1}. Our SIBYLS beamline provides the scattering data in Å^{-1}.

3.4 Output from the FoXS server

Graphs showing fit of atomic models to experimental SAXS data. Below the graphs is a table of atomic models, calculated χ^2, c1, and c2 constants, ranked by the goodness of fit χ^2. The predicted SAXS curve for the atomic model can be obtained by clicking on the link on the right.

3.5 Processing of reciprocal space SAXS curves predicted from structure predictions

The output fit file has columns 1 to 4 for scattering angle q, I_{exp}, $error_{exp}$, and I_{model}. To use the I_{model}, the experimental column (I_{exp}) need to be removed. One can use the Linux command "awk '{print $1,$4}' fox.fit >model.dat" where foxs.fit is the FoXS-generated .fit file, $1 and $4 refer to the column # from left to right, and output file model.dat. Note that some programs output the model Intensity in column 3, so it is advised to check the output fit file to double check. In a basic text editor that does not add any hidden text to the file, remove unnecessary headers and tail text. The new file called model.dat can be compared to experimental data by V_R implemented in the SAXS similarity analysis (see below).

3.6 Generation of real space SAXS curves predicted from structure predictions

To generate the experimental and model SAXS curves in real space, use GNOM in the ATSAS package (Manalastas-Cantos et al., 2021; Svergun, 1992), as previously described (Chinnam et al., 2022). For the model, the experimental data must be removed, such that q and I_{model} are in the first two columns, as shown for the model.dat file in the previous section. Dmax must be input into the GNOM program but may differ for experimental and predicted SAXS data. See Note 3, for issues in experimental data.

4. Comparison of experimental and predicted SAXS curves

4.1 Overview

SAXS curves can be compared in both reciprocal space and real space. Quantitative comparison of experimental and predicted SAXS curves is an ongoing area of development in SAXS analyses, and we expect further improvements in this area.

4.2 Experimental SAXS data

SAXS data can be collected readily from any synchrotron, as described in other methods reviews (Chinnam et al., 2022; Tully et al., 2021). Data should be of high quality. Our SIBYLS beamline provides users preliminary assessment on data quality. Experimental data pitfalls include uninterpretable data caused by aggregation, bubbles in the sample cell, poor buffer subtraction (likely culprit when data are cut off at low q), too low protein concentration, contribution of detergent micelles, or inter-particle repulsion. We show examples of these pitfalls at our beamline web page. (https://bl1231.als.lbl.gov/htsaxs/statistics) (Dyer et al., 2014). In contrast, stoichiometric and conformational flexibility are amenable for the SAXS analysis programs using ensemble methods and can be dealt with using ensembles of atomic models. For example, in a study of SARS-CoV-2 replication machinery, we could detect multiple and distinct oligomerization states of Nsp7, Nsp8, and RNA (Wilamowski et al., 2021). Although mixtures do lower the resolution of what can be concluded from a SAXS curve, information can still provide valuable insights—in this example, we determined that the assembly of the viral replication/transcription complex was not linear and was RNA-dependent.

4.3 Comparison metrics (Fig. 2).

4.3.1 χ^2 in reciprocal space. The most common comparison metric and one that is implemented in FoXS is χ^2, a statistical test that quantitates the difference between the experimental and predicted data as a function of q, divided by the experimental noise. The lower the number, the better the model matches the experimental data. (See Note 4 for caveats in χ^2 applications in SAXS analysis). It is automatically generated in many of the SAXS modeling programs.

4.3.2 V_R in reciprocal space. A volatility ratio, originally developed for stock market analyses, uses a ratio to normalize the curves and spreads the weight of the comparison through the whole curve (Hura et al., 2013). V_R has been implemented in https://bl1231.als.lbl.gov/saxs-similarity/ and is described in detail in chapter "Visualizing and accessing correlated SAXS data sets with similarity maps and simple scattering web resources" by Murray et al., in this volume. As many as 100 SAXS curves (experimental or calculated) can be uploaded by drag-and-drop into the input box, and the web server automatically compares the similarity of each curve. As described on the web page, SAXS .dat files (reciprocal space) must be in plain text and simple file names are required. The SAXS .dat files must contain two columns, q (Å^{-1}) and Intensity. A third column containing the experimental errors is optional and can be used for χ^2 calculations on the SAXS-Similarity server. The default q range analyzed is 0.015–0.2 Å^{-1}, but the q range can be adjusted. A pairwise similarity plot is output with colors representing the Volatility ratio. Clicking on single cells shows the pairwise overlay and relative radius of gyration (Rg). An alternative view as a force plot is useful for visualizing curves grouped based on similarity.

4.3.3 SAXS data can also be compared in real space visually. Plots are normalized based on the area under the curve. Quantitative metrics have been tested (dos Reis, Aparicio, & Zhang, 2011), but none have been established. Nonetheless, it is useful to compare reciprocal and real space (see Note 5). In the real space curve, shoulders after the prominent peak indicate separated domains (beads on a string with varying distances).

4.4 Cautionary note

A cautionary note is the possibility for SAXS to validate a completely inaccurate protein structure prediction. In CASP13, atomic models were distorted to fit to the SAXS data in reciprocal space, with completely incorrect protein topology and geometry (Ogorzalek et al., 2018). However, in comparison of real space curves, there was clear differences between experimental SAXS data and the structure prediction. Thus, it is good practice to consider SAXS-validated structure predictions as a testable atomic model, look at plots in both reciprocal and real space and to make sure that geometry makes sense.

5. Fitting of the protein structure prediction(s) to the experimental SAXS data

5.1 Overview

Flexibility is a common component in protein mechanisms, and so many proteins will adopt multiple conformations. Flexibility can be disordered regions, particularly at tails or shifts in domains relative to each other. Based on our experiences, it is also more likely than not that the protein structure prediction is in a different oligomeric state, conformational state, or requires an ensemble of structures consistent with what is occurring in solution. Several programs will generate alternative conformations or complexes and score the output structures on their agreement to input SAXS data. More detail on these programs can be found in a recent methods paper (Schneidman-Duhovny & Hammel, 2018). A limitation is that these servers for these programs only work on proteins; small molecules and nucleic acids cannot be readily input. However, advanced molecular dynamic users can adapt these programs as needed. It should also be noted that one program, AllosMod, can include glycosylation.

5.2 Software to adjust protein model conformations or stoichiometry

5.2.1 BILBOMD (https://bl1231.als.lbl.gov/bilbomd/) is a stand-alone web server that will generate a large population of conformally diverse models (Pelikan, Hura, & Hammel, 2009; Schneidman-Duhovny et al., 2016). It performs all the multistate modeling stages: conformational sampling, SAXS profile calculation, and multistate models enumeration. The conformational sampling is based on the minimal molecular dynamics (MD) simulation using CHARMM. SAXS profile calculation and enumeration of multi-state models use FoXS (Schneidman-Duhovny et al., 2010, 2013) and MultiFoXS (Schneidman-Duhovny et al., 2016) programs, respectively. The entire protocol is fully automated and does not require user interaction (Schneidman-Duhovny & Hammel, 2018). More generally, BILBOMD explores conformational space based on molecular dynamics by keeping domains as rigid bodies and allowing flexible regions to move. Bond distances are maintained, but the temperature is set high, and bond angles are allowed to change to generate the most conformationally diverse population efficiently. As required input into BILBOMD, each segment or polypeptide is uploaded separately, plus the experimental SAXS data (see Note 6 for potential pitfalls). A Rg range is input, typically 10–20 Å below and above the experimentally defined Rg. The program uses the minimal ensemble algorithm and will output an ensemble of conformations most consistent with the experimental SAXS data.

5.2.2 AllosMod-FoXS is a combination of AllosMod and FoXS servers (Weinkam, Pons, & Sali, 2012) and generates a population of conformation by altering non-bonded distances in the input pdb and compares the altered atomic model to the experimental SAXS data. It differs from BilboMD, where defined regions are kept rigid. Instead, AllosMod uses molecular dynamics of the entire protein to generate an energy landscape of conformations around the atomic model. Practically, most conformational changes are in regions that have few contacts (e.g., linkers, thin regions) more than in well-folded globular regions, providing a means to generate conformational changes within the domain region itself. AllosMod has different sampling options, which can be tested. The higher temperature levels offer the most exploration of conformational space, but which is best is based on the individual protein structures. The

AllosMod server is at the Sali lab website (https://modbase.compbio.ucsf.edu/allosmod-foxs/). The initial output is a list of models, ranked based on fit to the input experimental data and a link to a compressed folder with all models. An option for ensemble analysis by MultiFoXS (see below) is available on this output page. A potential pitfall is that the protein's geometry can become overly distorted; loss of secondary structure elements indicates that this has occurred.

5.2.3 MultiFoXS is built on the minimal ensemble algorithm in BILBOMD (Pelikan et al., 2009) and is effective when an ensemble of conformations or different assembly states (e.g., a mix of monomer, dimer) fit the scattering data (Schneidman-Duhovny et al., 2016). MultiFoXS uses a minimalist approach and identifies the minimum number of conformations needed to agree to the experimental SAXS data and their relative population in the sample. These conformations are meant to serve as representative members of a larger population and should not be over-interpreted as "the true conformations" in the solution. The easiest way to use MultiFoXS is to input the multiple structures as a zip-compressed file in FoXS and then select further MultiFoXS analysis in the FoXS output window. If FoXS outputs high c2 constants near 4 for single structures, then use of ensembles in MultiFoXS can reduce the c2 constant, indicating that the protein is adopting multiple conformations in solution. The Porod exponent, determined from the reciprocal space SAXS curve, can provide additional corroboration for the presence of multiple conformations and flexibility (Brosey & Tainer, 2019; Chinnam et al., 2022; Rambo & Tainer, 2011; Reyes et al., 2014).

5.2.4 FoXSDOCK is a valuable program for two-component systems (Schneidman-Duhovny et al., 2016). Two structures can be rotated and translated around each other. Models are scored based on the interface, on the agreement to the experimental SAXS data, and/or a composite of the two. FoXSDOCK is available at the Sali lab website (https://modbase.compbio.ucsf.edu/foxsdock/). FoXSDOCK does not consider sequence conservation, so localization of conserved residue at the interface is expected and can serve as an additional check on accuracy (see Note 1).

5.2.5 Chimera offers a manual structure adjustment for proteins with multiple domains or oligomers. For beginners, this method leads to a more intuitive understanding of domain placement and the consequence to the scattering curve. Chimera can be downloaded from UCSF (https://www.cgl.ucsf.edu/chimera/download.html). The SAXS calculator can be found in the Tools→Higher Order Structure→Small Angle X-Ray Profile.

6. Example: XRCC1 solution state

Here we provide an example of how to determine the solution state of the transient dimer of scaffold-like protein XRCC1 (Hammel, Rashid, et al., 2021) by integrating the AlphaFold2 model, crystal structure of XRCC1-BRCT2 domain, and SAXS fitting.

6.1 AlphaFold2 model of human XRCC1

Initially, we obtained the AlphaFold2 model of human XRCC1 monomer from the database (https://alphafold.ebi.ac.uk/) in PDB format. (Fig. 3A). We used the per-residue confidence scores (pLDDT) below 50 to suggest regions unstructured in isolation (Fig. 3A) and to guide conformational sampling implemented in BILBOMD (Pelikan et al., 2009; Schneidman-Duhovny & Hammel, 2018).

6.2 Conformational sampling of XRCC1 monomer by BILBOMD

The XRCC1 rigid and flexible regions that drive conformational sampling were assigned based on pLDDT value taken from AlphaFold2 initial model. While residues with pLDDT <50 are flexible, the residues with pLDDT >50 were keep rigid (Fig. 3A). The BILBOMD server include an app that allowed interactive selection of regions within the AlphaFold2 model that create a constraints file, "const.inp" for the control of the conformational sampling. In the web app, rigid bodies are displayed as circles with the circle size proportional to the number of residues. At the same time, the flexible regions are shown as lines connecting to the circles (Fig. 3B). BILBOMD outputs contain top-scoring multistate models for 1–4 states delivered by email. Additional output includes foxs_rg.out the file with the list of Rg and maximal dimension (Dmax) values for all generated models. In the XRCC1 monomer, the χ^2 value of the best-scoring one-state model is 22.3 (Fig. 3C), and the best multistate model (4-states) is 2.9. As can be validated by residual of the SAXS fit, none of the single-models (Fig. 3C) or multistate models satisfactory represent the solution state. This is most likely due to the presence of XRCC1 dimeric state in solution

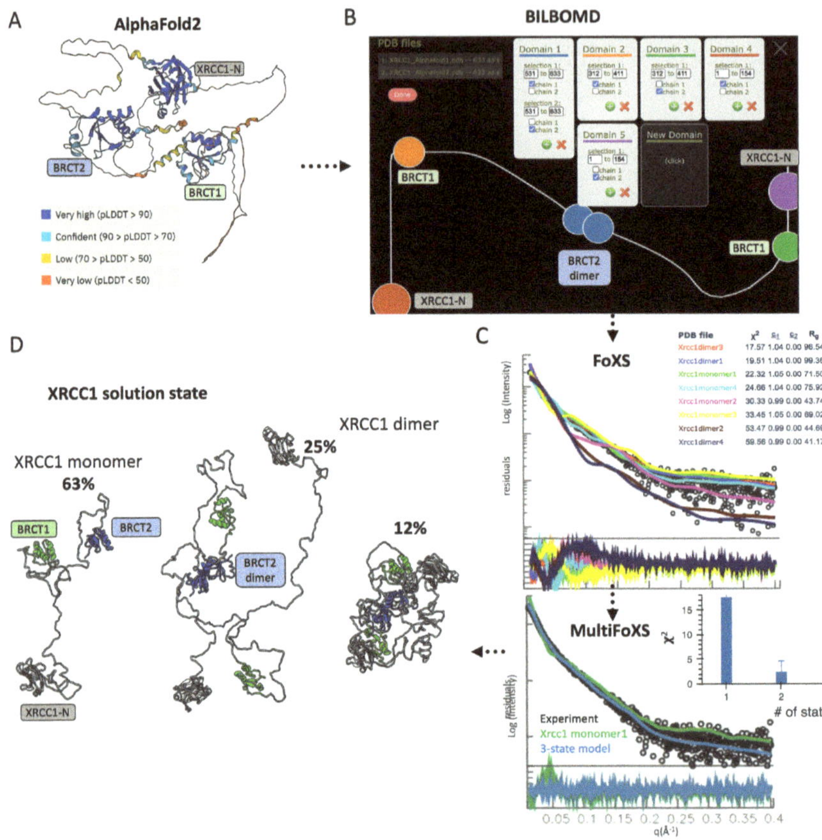

Fig. 3 XRCC1 solution state as identified by integrating AlphaFold2 protein structure prediction and SAXS modeling. (A) AlphaFold2 model colored by per-residue confidence score (pLDDT) values. (B) Building of restraining file that controls BILBOMD conformational sampling of XRCC1 dimer by selecting rigid domain regions based on pLDDT values. (C) FoXS web server output shows multiple SAXS fit, residual, χ^2 values, and Radius of gyration for four XRCC1 monomers and four dimer models derived from BILBOMD modeling. Bottom panel- MultiFoXS implementation in FoXS web server show χ^2 values for one-, two- and three state model. The plot show comparison of SAXS fit and fit-residual for the one- and three-state model. (D) Three state model is shown together with the percentage of each model used to fit the SAXS data shown in panel (C) Rigid body domains are highlighted.

(Hammel, Rashid, et al., 2021) therefore, we performed the subsequent conformational sampling of XRCC1-dimer.

6.3 Conformational sampling of XRCC1 dimer by BILBOMD

SEC-MALS shows that the XRCC1 coexists as a monomer and dimer in solution (Hammel, Rashid, et al., 2021). Thus, we perform conformational sampling of dimeric state using the BILBOMD approach as described in step 7.2. The initial XRCC1 dimer model was built by superimposing two AlphaFold2 models at its BRCT2 interface. In the XRCC1 dimer, we would like to maintain the dimer interface at the BRCT2 domains (the XRCC1-BRCT2 crystal structure PDB 3PC8, (Hammel, Rashid, et al., 2021)) as one rigid body during conformational sampling. As expected, the XRCC1 dimer single state or multistate model does not fit experimental SAXS data well (Fig. 3C). Therefore, we fit the SAXS data with a mixture of monomer and dimer models.

6.4 Determining XRCC1 monomer/dimer solution state by MultiFOXS

In the final step, we determined a multistate model for XRRC1 that represents the coexistence of monomeric and dimeric states in solution. As shown in Fig. 3D, the XRCC1 is a disordered scaffold-like protein and adopts a mixture of monomers and dimers through its BRCT1-BRCT1 domains interaction (Hammel, Rashid, et al., 2021). We initially compress (zip file) four monomer and four dimer models found in the BILBOMD outputs' four-state model from steps 6.2 and 6.3. We upload the zip file into the FoXS web server (Schneidman-Duhovny et al., 2010) and fit experimental SAXS data used in steps 6.2 and 6.3, respectively. Initially, we obtained fit for all eight models with the χ^2 values between 17.6 and 59.0 (Fig. 3D). By performing multistate model fit using the MultiFoXS option in the FoXS webserver (Schneidman-Duhovny et al., 2016), we obtain an excellent fit to the SAXS data with the χ^2 values of the three-state model 1.6 (Fig. 3C and D).

7. Notes

Note 1. As with any atomic model for complexes, interfaces should be assessed for high sequence conservation and charge/hydrophobic complementarity. The PISA server (https://www.ebi.ac.uk/pdbe/pisa/) provides an automated assessment for the latter (Krissinel & Henrick, 2007). Even in crystal structures, it is not unprecedented for interfaces to be mis-assigned, as we found for the repulsive interface between PCNA and sumo (Tsutakawa et al., 2015). It should be noted that SAXS.

Note 2. For the prediction of SAXS from atomic models, it is best to input the protein structure predictions with the experimental data, to improve the estimation of the hydration layer. The FoXS program uses two fitting constants, c1 and c2, to fit the experimental SAXS data. Note that a c2 constant close to 4, indicates overfitting to the experimental SAXS data. Often this occurs when the protein is adopting multiple conformations in solution and is reduced by analyzing against an ensemble of multiple structures (via MultiFoXS).

Note 3. In the GNOM analysis, aggregation of the SAXS sample may cause an artificial increase in Dmax. SEC-SAXS data should not have any aggregation; sometimes, there's been an accumulation of damaged protein on the sample cell window. Processing of the SEC-SAXS data with RAW can help minimize this type of aggregation (Hopkins, Gillilan, & Skou, 2017).

Note 4. As applied in SAXS, the χ^2 comparison metric does not evenly compare the reciprocal space curves from low to high q but is biased toward curves at the low q region. As SAXS is an exponentially decaying function with a 100 to 1000-fold difference in signal over the typical curve, the difference between curves of Intensity as a function of q (I_{exp}—I_{pred}) will be more significant at low q compared to high q, where a 1% difference would be 2–3 magnitudes smaller. How the curves are scaled will also bias χ^2. Another critical factor is that experimental noise or uncertainties of the data is in the denominator of χ^2, making noisier data appear to have a lower and better χ^2 fit (higher χ^2 scores) than better quality data with less noise. Thus, different models can be compared to one experimental data, but χ^2 fits to different experimental data to one model cannot be compared.

Note 5. In CASP13, computational scientists used experimental SAXS data to fit protein structure predictions (Ogorzalek et al., 2018). At that time, prediction scientists had the wrong topology for large proteins with no known structural templates, the most challenging targets. Their programs squashed the incorrect topology to fit the overall shape, and the reciprocal space fits were almost perfect. Yet, a comparison of these models in real space showed significant differences, showing that comparison in both reciprocal and real space is of value.

Note 6. For BilboMD, keep file names short and with no special characters. Input files may have to have unneeded headers and tails removed. Numbering is reset to the number one for the first residue in the file, so the residue number for the rigid domains may need to be adjusted accordingly. There is a 999 residue limit in the BILBOMD server.

8. Summary and conclusions

Machine learning algorithms have just started to solve protein structure prediction problems. Current limitations for homomeric and heteromeric complexes and complexes with nucleic acids and small molecules (ATP, drugs) will likely be addressed in the near future. Thus, it is important that SAXS can contribute and guide development of experimental models for conformations occurring in solution. Opening up the possibility for a SAXS-guided model being used as an AlphaFold2 template, Thomas Terwilliger has found that input of a crystallographic density-modified template into AlphaFold2 leads to a significantly improved output model that better fits the experimental density and shifts the register of a loop region (Terwilliger et al., 2022). Accurate solution structures will provide atomic models for testable hypotheses on substrate specificity, catalytic mechanisms, and regulation.

Although it is currently considered that an atomic structure cannot be determined from SAXS data alone, is this the point in history that we can consider SAXS validation of structure predictions as saying that an atomic structure can be determined by SAXS? With the exception of the highest resolution data, we could not derive atomic structures from crystallography, NMR, and cryo-EM data, without geometric constraints from knowledge of bond angles, bond length, secondary structures, etc. Indeed, low resolution crystallography and cryo-EM data is greatly aided by prior atomic models, either from other techniques and/or more recently, structure prediction algorithms. Given this use of structure predictions in other structural fields, can we now say that an atomic structure can be derived from SAXS data alone?

The potential application for accurate protein structure predictions goes beyond individual proteins with known functions. Multiple sequence alignments have been transformative in identifying critical residues but are dependent on sequence conservation. Although currently requiring deep sequence conservation, accurate protein structure predictions have the potential to provide a semi-orthogonal perspective for alignments. Proteins with only low sequence identity are challenging for sequence alignments, but if those critical residues co-localize in similar patterns, then better functional assignment and annotation can be achieved. High throughput structural techniques such as SAXS and SANS could validate predictions of proteins with shallow sequence depth. We have found co-localization of significant Evolutionary Tracing Scores mapped onto structures can pinpoint active sites and allosteric sites (Lees-Miller et al., 2021; Tsutakawa et al., 2021). This will be especially critical in the metagenomic sequence space that is growing exponentially every year (Chen et al., 2021). The Integrated microbial genomes and microbiomes (IMG) database holds

~70 billion genes, at the time of submission (https://img.jgi.doe.gov/cgi-bin/mer/main.cgi). Many viral and bacterial proteins currently have no or little sequence identity to other proteins, making accurate annotation difficult; anecdotally as much as 80% of a viral genome could not be annotated. As there are intensive efforts to mine the microbiome and viral metagenome for bioenergy, biomedicine, and environmental applications (Iglesias, Latorre-Perez, Stach, Porcar, & Pascual, 2020), opening up this area of genome science with accurate protein structure predictions, validated by experimental SAXS data, could accelerate finding solutions for global and health-related problems. It is also feasible that accurate and validated protein structure predictions for microbes or viruses associated with desired activity, could be run through metabolite docking programs to help identify proteins with that activity. The sensitivity of SAXS to induced domain movements onnotation in the metagenome through functional clustering of conserved sites.

Acknowledgments

We thank long-time collaborators and colleagues in the SAXS field, including Dmitry Svergun, Robert P. Rambo, Andrej Sali, Dina Schneidman-Duhovny, Jesse Hopkins, and Daniel Rosenberg for their many insights and contributions, including useful programs for SAXS data collection and analysis. Work is supported by NCI P01 CA092584 (to S.E.T., G.H., M.H., J.A.T.) and 1R01GM137021 (To S.E.T. and G.H.). The SIBYLS beamline's efforts are supported by DOE-BER IDAT under contract DE-AC02-05CH11231.

References

Andrec, M., Snyder, D. A., Zhou, Z., Young, J., Montelione, G. T., & Levy, R. M. (2007). A large data set comparison of protein structures determined by crystallography and NMR: Statistical test for structural differences and the effect of crystal packing. *Proteins*, *69*(3), 449–465. https://doi.org/10.1002/prot.21507.

Baek, M., Anishchenko, I., Park, H., Humphreys, I. R., & Baker, D. (2021). Protein oligomer modeling guided by predicted interchain contacts in CASP14. *Proteins*, *89*(12), 1824–1833. https://doi.org/10.1002/prot.26197.

Baek, M., DiMaio, F., Anishchenko, I., Dauparas, J., Ovchinnikov, S., Lee, G. R., et al. (2021). Accurate prediction of protein structures and interactions using a three-track neural network. *Science*, *373*(6557), 871–876. https://doi.org/10.1126/science.abj8754.

Brosey, C. A., Ho, C., Long, W. Z., Singh, S., Burnett, K., Hura, G. L., et al. (2016). Defining NADH-driven Allostery regulating apoptosis-inducing factor. *Structure*, *24*(12), 2067–2079. https://doi.org/10.1016/j.str.2016.09.012.

Brosey, C. A., & Tainer, J. A. (2019). Evolving SAXS versatility: Solution X-ray scattering for macromolecular architecture, functional landscapes, and integrative structural biology. *Current Opinion in Structural Biology*, *58*, 197–213. https://doi.org/10.1016/j.sbi.2019.04.004.

Brosey, C. A., Yan, C., Tsutakawa, S. E., Heller, W. T., Rambo, R. P., Tainer, J. A., et al. (2013). A new structural framework for integrating replication protein A into DNA processing machinery. *Nucleic Acids Research*, *41*(4), 2313–2327. https://doi.org/10.1093/nar/gks1332.

Chen, I. A., Chu, K., Palaniappan, K., Ratner, A., Huang, J., Huntemann, M., et al. (2021). The IMG/M data management and analysis system v.6.0: New tools and advanced capabilities. *Nucleic Acids Research, 49*(D1), D751–D763. https://doi.org/10.1093/nar/gkaa939.

Chinnam, N. B., Syed, A., Burnett, K. H., Hura, G. L., Tainer, J. A., & Tsutakawa, S. E. (2022). Universally accessible structural data on macromolecular conformation, assembly, and dynamics by small angle X-ray scattering for DNA repair insights. *Methods in Molecular Biology, 2444*, 43–68. https://doi.org/10.1007/978-1-0716-2063-2_4.

Classen, S., Hura, G. L., Holton, J. M., Rambo, R. P., Rodic, I., McGuire, P. J., et al. (2013). Implementation and performance of SIBYLS: A dual endstation small-angle X-ray scattering and macromolecular crystallography beamline at the advanced light source. *Journal of Applied Crystallography, 46*(Pt 1), 1–13. https://doi.org/10.1107/S0021889812048698.

Classen, S., Rodic, I., Holton, J., Hura, G. L., Hammel, M., & Tainer, J. A. (2010). Software for the high-throughput collection of SAXS data using an enhanced Blu-ice/DCS control system. *Journal of Synchrotron Radiation, 17*(6), 774–781. https://doi.org/10.1107/S0909049510028566.

dos Reis, M. A., Aparicio, R., & Zhang, Y. (2011). Improving protein template recognition by using small-angle x-ray scattering profiles. *Biophysical Journal, 101*(11), 2770–2781. https://doi.org/10.1016/j.bpj.2011.10.046.

Dyer, K. N., Hammel, M., Rambo, R. P., Tsutakawa, S. E., Rodic, I., Classen, S., et al. (2014). High-throughput SAXS for the characterization of biomolecules in solution: A practical approach. *Methods in Molecular Biology, 1091*, 245–258. https://doi.org/10.1007/978-1-62703-691-7_18.

Evans, R., O'Neill, M., Pritzel, A., Antropova, N., Senior, A., Green, T., et al. (2021). Protein complex prediction with AlphaFold-Multimer. *bioRxiv*. https://doi.org/10.1101/2021.10.04.463034. 2021.2010.2004.463034.

Forster, F., Webb, B., Krukenberg, K. A., Tsuruta, H., Agard, D. A., & Sali, A. (2008). Integration of small-angle X-ray scattering data into structural modeling of proteins and their assemblies. *Journal of Molecular Biology, 382*(4), 1089–1106. https://doi.org/10.1016/j.jmb.2008.07.074.

Grant, T. D. (2018). Ab initio electron density determination directly from solution scattering data. *Nature Methods, 15*(3), 191–193. https://doi.org/10.1038/nmeth.4581.

Hammel, M., Rashid, I., Sverzhinsky, A., Pourfarjam, Y., Tsai, M. S., Ellenberger, T., et al. (2021). An atypical BRCT-BRCT interaction with the XRCC1 scaffold protein compacts human DNA ligase IIIalpha within a flexible DNA repair complex. *Nucleic Acids Research, 49*(1), 306–321. https://doi.org/10.1093/nar/gkaa1188.

Hammel, M., Rosenberg, D. J., Bierma, J., Hura, G. L., Thapar, R., Lees-Miller, S. P., et al. (2021). Visualizing functional dynamicity in the DNA-dependent protein kinase holoenzyme DNA-PK complex by integrating SAXS with cryo-EM. *Progress in Biophysics and Molecular Biology, 163*, 74–86. https://doi.org/10.1016/j.pbiomolbio.2020.09.003.

Hammel, M., & Tainer, J. A. (2021). X-ray scattering reveals disordered linkers and dynamic interfaces in complexes and mechanisms for DNA double-strand break repair impacting cell and cancer biology. *Protein Science: A Publication of the Protein Society, 30*(9), 1735–1756. https://doi.org/10.1002/pro.4133.

Hatos, A., Hajdu-Soltesz, B., Monzon, A. M., Palopoli, N., Alvarez, L., Aykac-Fas, B., et al. (2020). DisProt: Intrinsic protein disorder annotation in 2020. *Nucleic Acids Research, 48*(D1), D269–D276. https://doi.org/10.1093/nar/gkz975.

Hopkins, J. B., Gillilan, R. E., & Skou, S. (2017). BioXTAS RAW: Improvements to a free open-source program for small-angle X-ray scattering data reduction and analysis. *Journal of Applied Crystallography, 50*(Pt 5), 1545–1553. https://doi.org/10.1107/S1600576717011438.

Hura, G. L., Budworth, H., Dyer, K. N., Rambo, R. P., Hammel, M., McMurray, C. T., et al. (2013). Comprehensive macromolecular conformations mapped by quantitative SAXS analyses. *Nature Methods*, *10*(6), 453–454. https://doi.org/10.1038/nmeth.2453.

Hura, G. L., Menon, A. L., Hammel, M., Rambo, R. P., Poole, F. L., 2nd, Tsutakawa, S. E., et al. (2009). Robust, high-throughput solution structural analyses by small angle X-ray scattering (SAXS). *Nature Methods*, *6*(8), 606–612. https://doi.org/10.1038/nmeth.1353.

Iglesias, A., Latorre-Perez, A., Stach, J. E. M., Porcar, M., & Pascual, J. (2020). Out of the abyss: Genome and metagenome mining reveals unexpected environmental distribution of Abyssomicins. *Frontiers in Microbiology*, *11*, 645. https://doi.org/10.3389/fmicb.2020.00645.

Jumper, J., Evans, R., Pritzel, A., Green, T., Figurnov, M., Ronneberger, O., et al. (2021a). Highly accurate protein structure prediction with AlphaFold. *Nature*, *596*(7873), 583–589. https://doi.org/10.1038/s41586-021-03819-2.

Jumper, J., Evans, R., Pritzel, A., Green, T., Figurnov, M., Ronneberger, O., et al. (2021b). Applying and improving AlphaFold at CASP14. *Proteins*, *89*(12), 1711–1721. https://doi.org/10.1002/prot.26257.

Jumper, J., & Hassabis, D. (2022). Protein structure predictions to atomic accuracy with AlphaFold. *Nature Methods*, *19*(1), 11–12. https://doi.org/10.1038/s41592-021-01362-6.

Korasick, D. A., & Tanner, J. J. (2018). Determination of protein oligomeric structure from small-angle X-ray scattering. *Protein Science: A Publication of the Protein Society*, *27*(4), 814–824. https://doi.org/10.1002/pro.3376.

Krissinel, E., & Henrick, K. (2007). Inference of macromolecular assemblies from crystalline state. *Journal of Molecular Biology*, *372*(3), 774–797. https://doi.org/10.1016/j.jmb.2007.05.022.

Kryshtafovych, A., Schwede, T., Topf, M., Fidelis, K., & Moult, J. (2021). Critical assessment of methods of protein structure prediction (CASP)-round XIV. *Proteins*, *89*(12), 1607–1617. https://doi.org/10.1002/prot.26237.

Lees-Miller, J. P., Cobban, A., Katsonis, P., Bacolla, A., Tsutakawa, S. E., Hammel, M., et al. (2021). Uncovering DNA-PKcs ancient phylogeny, unique sequence motifs and insights for human disease. *Progress in Biophysics and Molecular Biology*, *163*, 87–108. https://doi.org/10.1016/j.pbiomolbio.2020.09.010.

Manalastas-Cantos, K., Konarev, P. V., Hajizadeh, N. R., Kikhney, A. G., Petoukhov, M. V., Molodenskiy, D. S., et al. (2021). ATSAS 3.0: Expanded functionality and new tools for small-angle scattering data analysis. *Journal of Applied Crystallography*, *54*(1), 343–355. https://doi.org/10.1107/S1600576720013412.

Marks, D. S., Colwell, L. J., Sheridan, R., Hopf, T. A., Pagnani, A., Zecchina, R., et al. (2011). Protein 3D structure computed from evolutionary sequence variation. *PLoS One*, *6*(12), e28766. https://doi.org/10.1371/journal.pone.0028766.

Marsh, J. A., & Teichmann, S. A. (2015). Structure, dynamics, assembly, and evolution of protein complexes. *Annual Review of Biochemistry*, *84*, 551–575. https://doi.org/10.1146/annurev-biochem-060614-034142.

Morcos, F., Pagnani, A., Lunt, B., Bertolino, A., Marks, D. S., Sander, C., et al. (2011). Direct-coupling analysis of residue coevolution captures native contacts across many protein families. *Proceedings of the National Academy of Sciences of the United States of America*, *108*(49), E1293–E1301. https://doi.org/10.1073/pnas.1111471108.

Moult, J., Fidelis, K., Kryshtafovych, A., Schwede, T., & Tramontano, A. (2018). Critical assessment of methods of protein structure prediction (CASP)-round XII. *Proteins*, *86*(Suppl. 1), 7–15. https://doi.org/10.1002/prot.25415.

Ogorzalek, T. L., Hura, G. L., Belsom, A., Burnett, K. H., Kryshtafovych, A., Tainer, J. A., et al. (2018). Small angle X-ray scattering and cross-linking for data assisted protein structure prediction in CASP 12 with prospects for improved accuracy. *Proteins*, *86*, 202–214. https://doi.org/10.1002/prot.25452.

Pelikan, M., Hura, G. L., & Hammel, M. (2009). Structure and flexibility within proteins as identified through small angle X-ray scattering. *General Physiology and Biophysics*, *28*(2), 174–189.

Piovesan, D., Tabaro, F., Micetic, I., Necci, M., Quaglia, F., Oldfield, C. J., et al. (2017). DisProt 7.0: a major update of the database of disordered proteins. *Nucleic Acids Research*, *45*(D1), D219–D227. https://doi.org/10.1093/nar/gkw1056.

Pratt, A. J., Shin, D. S., Merz, G. E., Rambo, R. P., Lancaster, W. A., Dyer, K. N., et al. (2014). Aggregation propensities of superoxide dismutase G93 hotspot mutants mirror ALS clinical phenotypes. *Proceedings of the National Academy of Sciences of the United States of America*, *111*(43), E4568–E4576. https://doi.org/10.1073/pnas.1308531111.

Putnam, C. D., Hammel, M., Hura, G. L., & Tainer, J. A. (2007). X-ray solution scattering (SAXS) combined with crystallography and computation: Defining accurate macromolecular structures, conformations and assemblies in solution. *Quarterly Reviews of Biophysics*, *40*(3), 191–285. https://doi.org/10.1017/S0033583507004635.

Quaglia, F., Meszaros, B., Salladini, E., Hatos, A., Pancsa, R., Chemes, L. B., et al. (2022). DisProt in 2022: Improved quality and accessibility of protein intrinsic disorder annotation. *Nucleic Acids Research*, *50*(D1), D480–D487. https://doi.org/10.1093/nar/gkab1082.

Rambo, R. P., & Tainer, J. A. (2011). Characterizing flexible and intrinsically unstructured biological macromolecules by SAS using the Porod-Debye law. *Biopolymers*, *95*(8), 559–571. https://doi.org/10.1002/bip.21638.

Rambo, R. P., & Tainer, J. A. (2013a). Super-resolution in solution X-ray scattering and its applications to structural systems biology. *Annual Review of Biophysics*, *42*, 415–441. https://doi.org/10.1146/annurev-biophys-083012-130301.

Rambo, R. P., & Tainer, J. A. (2013b). Accurate assessment of mass, models and resolution by small-angle scattering. *Nature*, *496*(7446), 477–481. https://doi.org/10.1038/nature12070.

Reyes, F. E., Schwartz, C. R., Tainer, J. A., & Rambo, R. P. (2014). Methods for using new conceptual tools and parameters to assess RNA structure by small-angle X-ray scattering. *Methods in Enzymology*, *549*, 235–263. https://doi.org/10.1016/B978-0-12-801122-5.00011-8.

Schneidman-Duhovny, D., & Hammel, M. (2018). Modeling structure and dynamics of protein complexes with SAXS profiles. *Methods in Molecular Biology*, *1764*, 449–473. https://doi.org/10.1007/978-1-4939-7759-8_29.

Schneidman-Duhovny, D., Hammel, M., & Sali, A. (2010). FoXS: A web server for rapid computation and fitting of SAXS profiles. *Nucleic Acids Research*, *38*(Web Server issue), W540–W544. https://doi.org/10.1093/nar/gkq461.

Schneidman-Duhovny, D., Hammel, M., Tainer, J. A., & Sali, A. (2013). Accurate SAXS profile computation and its assessment by contrast variation experiments. *Biophysical Journal*, *105*(4), 962–974. https://doi.org/10.1016/j.bpj.2013.07.020.

Schneidman-Duhovny, D., Hammel, M., Tainer, J. A., & Sali, A. (2016). FoXS, FoXSDock and MultiFoXS: Single-state and multi-state structural modeling of proteins and their complexes based on SAXS profiles. *Nucleic Acids Research*, *44*(W1), W424–W429. https://doi.org/10.1093/nar/gkw389.

Schneidman-Duhovny, D., Kim, S. J., & Sali, A. (2012). Integrative structural modeling with small angle X-ray scattering profiles. *BMC Structural Biology*, *12*, 17. https://doi.org/10.1186/1472-6807-12-17.

Svergun, D. I. (1992). Determination of the regularization parameter in indirect-transform methods using perceptual criteria. *Journal of Applied Crystallography*, *25*, 495–503. https://doi.org/10.1107/S0021889892001663.

Svergun, D., Barberato, C., & Koch, M. H. J. (1995). CRYSOL—A program to evaluate X-ray solution scattering of biological macromolecules from atomic coordinates. *Journal of Applied Crystallography*, *28*(6), 768–773. https://doi.org/10.1107/S0021889895007047.

Svergun, D. I., Petoukhov, M. V., & Koch, M. H. (2001). Determination of domain structure of proteins from X-ray solution scattering. *Biophysical Journal, 80*(6), 2946–2953. https://doi.org/10.1016/S0006-3495(01)76260-1.

Terwilliger, T. C., Poon, B. K., Afonine, P. V., Schlicksup, C. J., Croll, T. I., Millán, C., et al. (2022). Improved AlphaFold modeling with implicit experimental information. *bioRxiv*. https://doi.org/10.1101/2022.01.07.475350. 2022.2001.2007.475350.

Tsutakawa, S. E., Bacolla, A., Katsonis, P., Bralic, A., Hamdan, S. M., Lichtarge, O., et al. (2021). Decoding Cancer variants of unknown significance for helicase-nuclease-RPA complexes orchestrating DNA repair during transcription and replication. *Frontiers in Molecular Biosciences, 8*, 791792. https://doi.org/10.3389/fmolb.2021.791792.

Tsutakawa, S. E., Hura, G. L., Frankel, K. A., Cooper, P. K., & Tainer, J. A. (2007). Structural analysis of flexible proteins in solution by small angle X-ray scattering combined with crystallography. *Journal of Structural Biology, 158*(2), 214–223. https://doi.org/10.1016/j.jsb.2006.09.008.

Tsutakawa, S. E., Van Wynsberghe, A. W., Freudenthal, B. D., Weinacht, C. P., Gakhar, L., Washington, M. T., et al. (2011). Solution X-ray scattering combined with computational modeling reveals multiple conformations of covalently bound ubiquitin on PCNA. *Proceedings of the National Academy of Sciences of the United States of America, 108*(43), 17672–17677. https://doi.org/10.1073/pnas.1110480108.

Tsutakawa, S. E., Yan, C., Xu, X., Weinacht, C. P., Freudenthal, B. D., Yang, K., et al. (2015). Structurally distinct ubiquitin- and sumo-modified PCNA: Implications for their distinct roles in the DNA damage response. *Structure, 23*(4), 724–733. https://doi.org/10.1016/j.str.2015.02.008.

Tully, M. D., Tarbouriech, N., Rambo, R. P., & Hutin, S. (2021). Analysis of SEC-SAXS data via EFA deconvolution and scatter. *Journal of Visualized Experiments, 167*, e61578. https://doi.org/10.3791/61578.

Tunyasuvunakool, K., Adler, J., Wu, Z., Green, T., Zielinski, M., Zidek, A., et al. (2021). Highly accurate protein structure prediction for the human proteome. *Nature, 596*(7873), 590–596. https://doi.org/10.1038/s41586-021-03828-1.

Varadi, M., Anyango, S., Deshpande, M., Nair, S., Natassia, C., Yordanova, G., et al. (2022). AlphaFold protein structure database: Massively expanding the structural coverage of protein-sequence space with high-accuracy models. *Nucleic Acids Research, 50*(D1), D439–D444. https://doi.org/10.1093/nar/gkab1061.

Waterhouse, A., Bertoni, M., Bienert, S., Studer, G., Tauriello, G., Gumienny, R., et al. (2018). SWISS-MODEL: Homology modelling of protein structures and complexes. *Nucleic Acids Research, 46*(W1), W296–W303. https://doi.org/10.1093/nar/gky427.

Weinkam, P., Pons, J., & Sali, A. (2012). Structure-based model of allostery predicts coupling between distant sites. *Proceedings of the National Academy of Sciences of the United States of America, 109*(13), 4875–4880. https://doi.org/10.1073/pnas.1116274109.

Wilamowski, M., Hammel, M., Leite, W., Zhang, Q., Kim, Y., Weiss, K. L., et al. (2021). Transient and stabilized complexes of Nsp7, Nsp8, and Nsp12 in SARS-CoV-2 replication. *Biophysical Journal, 120*(15), 3152–3165. https://doi.org/10.1016/j.bpj.2021.06.006.

Xue, B., Dunbrack, R. L., Williams, R. W., Dunker, A. K., & Uversky, V. N. (2010). PONDR-FIT: A meta-predictor of intrinsically disordered amino acids. *Biochimica et Biophysica Acta, 1804*(4), 996–1010. https://doi.org/10.1016/j.bbapap.2010.01.011.

Yan, C., Dodd, T., He, Y., Tainer, J. A., Tsutakawa, S. E., & Ivanov, I. (2019). Transcription preinitiation complex structure and dynamics provide insight into genetic diseases. *Nature Structural & Molecular Biology, 26*(6), 397–406. https://doi.org/10.1038/s41594-019-0220-3.

Zhou, Y., Millott, R., Kim, H. J., Peng, S., Edwards, R. A., Skene-Arnold, T., et al. (2019). Flexible tethering of ASPP proteins facilitates PP-1c catalysis. *Structure, 27*, 1485–1496.e4. https://doi.org/10.1016/j.str.2019.07.012.

CHAPTER THIRTEEN

Protein fibrillation from another small angle—SAXS data analysis of developing systems

Annette Eva Langkilde* and Bente Vestergaard
Department of Drug Design and Pharmacology, University of Copenhagen, Copenhagen, Denmark
*Corresponding author: e-mail address: annette.langkilde@sund.ku.dk; bente.vestergaard@sund.ku.dk

Contents

Abstract

During the fibrillation process amyloid proteins undergo structural changes at very different length and time scales. Small angle X-ray scattering (SAXS) is a method that is uniquely suitable for the structural analysis of this process. Careful measures must, however, be taken both in the sample preparation, data collection and data analysis

Methods in Enzymology, Volume 678
ISSN 0076-6879
https://doi.org/10.1016/bs.mie.2022.09.025

procedures to ensure proper data quality, coverage of the process and reliable interpretation. With this chapter, we provide many details about the data analysis of such developing systems. The recommendations are based on our own experience with analysis of data from several amyloid and amyloid-like proteins, with data decomposition being a central point in the procedure. We focus on two alternative approaches, one being a laborious, hands-on, iterative approach, the other being more automated, applying a chemometrics based software, developed for the purpose. Both methods can equally well be applied to other developing mixtures, but specific recommendations for amyloid samples are emphasized in this chapter.

1. Introduction

The formation of amyloid fibrils is indisputably associated with several grave, progressive diseases, notably major neurodegenerative disorders such as Alzheimer's and Parkinson's disease (Ankarcrona et al., 2016). The formation of amyloid fibrils is a multistep cascade of structural changes and is fascinating, but complex to study even in a simplified in vitro setting. It is common praxis to use a fibril-specific fluorescent dye, such as, e.g., thioflavin T (ThT) to follow the fibrillation process in vitro (Fodera et al., 2008; Gade Malmos et al., 2017), and by fitting to such fluorescence traces many amyloid and amyloid-like proteins have been suggested to follow a nucleation dependent polymerization process (Buell, 2019; Lee et al., 2007). The trace typically reveals a lag-phase, where no or little changes in the fluorescence signal are detectable. This is followed by an intermediate elongation phase, characterized by an exponential increase in the fluorescence signal, and finally the fluorescence signal levels at a plateau in the so-called steady state phase, completing the well-known S-shaped sigmoidal kinetics trace. The fluorescence trace conveniently visualizes the process, which also is useful when combining biophysical data from several methods. Evidently, the fluorescence trace does not reveal all the structural changes, that are taking place during the fibrillation reaction. Most of the intermediately formed structures are not detected by the dye and several fibril types and non-amyloid aggregates with diverging affinity for the fluorescent dye may co-exist in solution (Sidhu et al., 2018). It is, however, very useful to consider the fibrillation process in terms of an early, intermediate and a late phase, when describing the dominant structural changes, driving the fibrillation reaction: At early stages the native protein un- and refolds into non-native conformations, of which one or several are fibrillation prone. Various oligomeric structures form and re-dissolve in solution, at some point resulting in primary nucleation events. Such a nucleus primes the formation of fibril structures, polymerized by further addition of non-native protomers (fibril elongation), and secondary nucleation events subsequently significantly speed up the process, resulting in the exponential increase in fluorescence, characteristic for the intermediate elongation or growth phase. Gradually, the process eases off, eventually reaching a stage where the concentration of soluble material (protein not incorporated into a fibril structure), remains relatively steady.

This is, however, only a very general description, and it is important to consider a particular aspect of fibrillation: The process is chaotic, and each individual step is sensitive to the environment. At any timepoint during the fibrillation process the environment is changing, simply because the process is proceeding, with resulting changes in the volume fractions of each of the species formed. There are additional changes in the local environment, e.g., when particular ions incorporate into a fibril structure, and the growing surface area of the fibrils exhibits different physico-chemical properties than the soluble species (Kumari et al., 2021; Palmadottir et al., 2021). Also consider that each step is in principle reversible. The native structure, like any native protein structure, *breathes*, i.e., un- and re-folds to different extents, intermediate oligomeric structures may redissolve, rather than incorporate into a fibril, and protomers may leave the fibril structures and return to soluble states (Skamris et al., 2019). At any timepoint in the process each step takes place with different probability, and each species exists in different (at times negligible) proportions.

The net result is that each protein species at any given timepoint during the reaction adapts structurally to a changing environment. Each structural species thus co-exists with all other species, and cannot be isolated from the entire system, without disturbing both the system and the individual structures. That is, if you try to isolate, e.g., a volume of an intermediately formed oligomer from the solution, both the isolated protein and the remaining proteins will adapt to the changes in the environment. This means, that if we wish to study the fibrillation process and any of the intermediately formed structures, we must use a method that can be applied directly to mixtures. BioSAXS, small angle X-ray scattering of biomolecules in solution, is an ideal method for this. All components in a sample will scatter X-rays but by subtraction of the buffer scattering the isolated signal from the solvated proteins is obtained, free in solution, devoid of any effects from dehydration (e.g., TEM, fiber diffraction, solid state NMR), crystal packing (protein crystallography), surface adsorption (microscopy) or labeling (TEM, solid state NMR). Notably, all the macromolecular species that are present in the given sample, will contribute to the buffer-subtracted scattering data, weighted by their volume fractions. If the proteins exhibit pronounced attractive or repulsive effects, this will also influence the data. This is called the structure factor, whereas the scattering from the individual macromolecular species themselves constitutes the form factor (see (Svergun et al., 2013) for more detail). If the structure factors are negligible, BioSAXS data are additive, formally:

$$I_{total} = aI_A + bI_B + cI_C + \ldots nI_N$$

where I_{total} is the total scattering intensity curve measured, and nI_N denotes the scattering contribution to the total scattering curve from the individual Nth component, present in the volume fraction n, where $(a + b + c + \ldots n) = 1$.

In a simplified case of a fibrillation reaction only three species (a starting structure, an intermediate structure, and a fibril structure) contribute significantly to the total scattering pattern:

$$I_{total} \approx aI_{start} + bI_{intermediate} + cI_{fibril}$$

It is of interest to describe the intermediate structure both quantitatively and qualitatively, since intermediate oligomers, rather than the fibrils themselves, have been suggested to associate with cytotoxicity, and hence are also interesting drug targets (Wells et al., 2021). The intermediate species are also key to an understanding of fibril formation and elongation principles. Since intermediate species cannot be isolated without possibly influencing the individual structures, structural equilibria, and processes, we cannot measure the scattering curve from the pure species. Instead, we can isolate the scattering contribution from the intermediate oligomer from the total scattering pattern. In cases, where the starting structure is relatively stable, it is possible to measure I_{start}; likewise, in cases where only one fibril structure is formed, I_{fibril} can be measured. When collecting data at different timepoints during the fibrillation reaction, the evolving volume fractions, and the scattering curve from the intermediately formed species can be isolated by relatively simple procedures, which we have termed "manual decomposition" in Section 5, below.

In the perhaps more typical case where more than three species contribute significantly to the evolving scattering profiles, and/or where neither the starting structure nor the fibril structure at any point can be measured in an even relatively pure state, more advanced methods must be applied to decompose the data into the individual scattering components. With colleagues, we have developed the software COSMiCS for this procedure

(Herranz-Trillo et al., 2017; Sagar et al., 2021), a software which is based on chemometrics principles. This software can also be applied to decompose other types of evolving solutions (Bernado & Sagar, 2022), while our focus here is on the application for amyloid systems. We describe this approach in Section 6, below.

From our first experience with "manual decomposition" we revealed that a helical oligomer accumulated at intermediate timepoints during human insulin fibrillation, at concentrations correlating with the fibril elongation rate. We also published ab initio models of both the insulin oligomer and the repeating unit of insulin fibrils (Vestergaard et al., 2007). Our sample preparation and data collection strategy were simple, using batch fibrillation and manual loading of the samples into the SAXS capillary (see more details about data collection strategies in Vestergaard and Langkilde (2022). When later analyzing the fibrillation process of human α-synuclein (aSN) we used a plate reader for the fibrillation assay, and revealed how an annular structure accumulated during the fibrillation reaction, coinciding with increased permeabilization of anionic lipid vesicles (Giehm et al., 2011). In addition, we have shown how the oligomers interacts with a more physiologically relevant lipid model containing cholesterol and charge-neutral lipids (van Maarschalkerweerd et al., 2015b). This oligomer was stabilized by a low molecular weight compound, and decomposition of a SAXS data series revealed that the fibril elongation rate directly correlated with the concentration of the oligomer, strongly suggesting that this oligomer is a fibril building block (Nors Perdersen et al., 2015). In contrast, we did not observe accumulation of any oligomeric compounds during transthyretin (TTR) fibrillation (Groenning et al., 2015). Rather, when decomposing the SAXS data series from the fibrillation reaction, it was evident that a considerably unfolded monomer co-existed in solution with the early formed fibril species throughout the fibrillation reaction. Indeed, the formation of monomers from the native tetrameric form has been shown to be the rate limiting step in TTR fibrillation, forming the basis for the prevalent therapeutic approach against TTR related diseases (Bulawa et al., 2012).

In the above cases, it was possible to measure SAXS data from pure samples of both the start and end states of the fibrillation reaction, hence easing the decomposition process. When investigating a SAXS data series from the familial mutant aSN_{E46K}, neither the start nor the end states were homogeneous, and hence decomposition using the "manual approach" was not possible. Use of the chemometrics based software COSMiCS, however, enabled decomposition of this complex mixture (Herranz-Trillo et al., 2017). Surprisingly, the intermediately formed structure in this case was not comparable to the wildtype oligomer. Rather, a very large species with a low average mass density accumulated. Such a species could correspond to the liquid-liquid phase separated species, reported for many amyloid systems (Agarwal et al., 2021; Ray et al., 2020; Sawner et al., 2021; Xu et al., 2022).

Although this chapter primarily describes the analysis of SAXS data collected during the fibrillation process, with particular focus on intermediately formed structures, we mention that both we and colleagues have advocated the use of small angle scattering data in the elucidation of the final fibril structure as well. We analyzed the fibril structure of an amyloid-like peptide, a fragment of yeast prion protein Sup35p, revealing a ribbon-like, highly ordered, and hierarchical assembly (Langkilde et al., 2015). Combined with transmission electron microscopy, fiber diffraction data and knowledge from previously published high-resolution crystal structures (Nelson et al., 2005), SAXS data played a central role, revealing a consistently repeating distance between the individual protofilaments, forming ribbons with an increasing number of protofilaments, which aligned and eventually paired during fibril maturation (Langkilde et al., 2015). Small angle X-ray and neutron data likewise played a central role in the recent analysis of amyloid β42 fibrils, which in combination with solid state NMR data and Rosetta modeling resulted in a final atomistic model (Lattanzi et al., 2021). The structural information from SAXS is in its nature highly complementary to the high-resolution information obtained on the ordered fibril cores from, e.g., cryo-EM.

In the studies performed by us and mentioned here, we applied synchrotron radiation and ex situ fibrillation (i.e., fibrillation happens outside the SAXS sample cell), while colleagues have published studies from fibrillating glucagon (Oliveira et al., 2009) and FapC (Rasmussen et al., 2021), using in-house SAXS equipment and an in situ fibrillation scheme. While we apply actual data decomposition as analysis strategy, the SAXS data in the latter two studies are modeled with combinations of known analytical form factors, as we refer to further below (see Section 8).

In this chapter, we describe a step-by-step approach to analysis of the data from the developing, complex mixtures from fibrillating proteins. Please refer to our previous chapter (Vestergaard & Langkilde, 2022) for sample preparation advice and strategies for data collection. During the data analysis process, which often is a circular procedure, it is relevant to revisit this first chapter, where we also include descriptions of pitfalls and circumstances during the data collection process, potentially resulting in unforeseen or even unwanted features in the data.

2. General method notes

SAXS data should only be interpreted from samples of very high quality. In Vestergaard and Langkilde (2022), we stress the need to use highly purified and well-characterized samples for the data collection, which is an absolute prerequisite for this type of analysis. Please note that the most typical SAXS measurement is performed on monodisperse samples, and most of the data analysis methods available in the field build on the assumption that your samples meet the general recommendations for SAXS data collection and analysis (Graewert & Jeffries, 2017). A fibrillating solution is inherently polydisperse, which greatly complicates data analysis, but at the same time, SAXS data are uniquely suited for the structural investigation of these developing solutions, as we describe below. Here, we raise a flag, mentioning that the user must be aware of the limitations and challenges in this type of analysis and critically assess the relevance of available software for this type of measurements. In general, include all complementary biophysical data available, in support of your SAXS data analysis.

When publishing, follow the internationally agreed publication guidelines for the field (Jacques et al., 2012; Trewhella et al., 2017), providing ample details about the samples and fibrillation conditions used, data collection specifications and data analysis particularities. Also deposit data in a public archive. Full transparency will increase the impact of your results.

The only equipment needed for the data analysis is a standard computer, specialized software, and your data files. We provide recommendations for software below, but alternative software may be available from other authors. Please note that numerous software initiatives are running. With this chapter, we do not attempt to provide a review of these, for an overview, potentially consult http://smallangle.org.

3. Essential check of SAXS data from fibrillating systems

When collecting SAXS data at most of the large-scale facilities, and even on many lab instruments a series of advanced automated procedures will typically happen "behind the screen" before you see your SAXS data. This is highly advantageous for most types of data collection but given the special nature of fibril samples extra precautions must be taken to ensure correct data reduction procedures, and to check the basic quality of your dataset before moving along with the lengthy procedure of the decomposition analysis. For this reason, is it important that you do not only bring home the radially averaged data-files, but that you ensure that you can also access the 2D images from the SAXS data collection.

3.1 Non-exhaustive list of suitable software

Inspection of 2D images

- SASview (www.sasview.org)
- BioXTAS RAW (Hopkins et al., 2017; Nielsen et al., 2009)
- SAXSview (https://github.com/emblsaxs/saxsview)
- Fit2D (Hammersley, 1997)

Inspection of radially averaged files

- BioXTAS RAW (Hopkins et al., 2017, Nielsen et al., 2009)
- PRIMUS (Konarev et al., 2003)

3.2 Procedure

1. Does the beamline/instrument apply repeated exposures? If no, inspect the single 2D image from each sample exposure. If yes, select a 2D frame in the middle of the exposure series for inspection.
2. Carefully inspect the 2D file. The scattering profile must be isotropic (note 1). If this is not the case, the file must be discarded. The risk for non-isotropic scattering is higher, when the concentration of fibrils is higher and/or the fibrils have matured, both cases often leading to more viscous samples, hence pay extra attention at late timepoints.
3. Inspect radially averaged (and temporarily averaged if the data includes repeated exposures) buffer files. Plot all data from the buffers together. These files must superimpose perfectly. If this is not the case, you must pay extra attention. A few cases to watch out for are:
   a. Systematically increasing scattering intensity in the buffer files. This can be due to general insufficient cleaning of the sample cell during exposures and should have been noted during data collection. Most likely the entire data series must be discarded.
   b. Outliers. If a single buffer file does not superimpose, likely there has been temporary contamination in the SAXS exposure cell. Carefully inspect also the data from the immediately surrounding exposures of protein samples and consider discarding also these few files.
4. Inspect repeated exposures of protein samples (note 2). If all the repeated exposures superimpose, you can directly use the radially and temporarily

averaged file, provided by the automated software. If not, we distinguish the following cases:

a. There is a general trend of increased scattering intensity throughout the repeated frames. This is likely due to radiation induced aggregation, and the affected data must be discarded.

b. There is variation in the intensity, but not a general trend. This may simply be because the sample is non-homogeneous. At later timepoints, the sample consists of both non-fibrillated and fibrillated protein, and there may be some variation in the relative composition of the sample while it is flowing through the cell. In this case, it is important that all the repeated exposures are averaged, even if this is beyond the tolerance of the automated procedures for standard samples, or your data will not represent the total sample at the given timepoint.

5. Now plot all the remaining radially and temporarily averaged data from protein and buffer samples. Check that, as expected, all protein data has a higher scattering intensity, also at the highest scattering angles, compared to the buffer data. If not, consult (Nielsen et al., 2013) and only proceed the analysis after very careful consideration.

6. Average the buffer measurement before and after each protein measurement and use this file for background subtraction. Plot the background subtracted files together, this will be the basis of further analysis and decomposition.

3.3 Notes

1. The sample flows into the SAXS exposure cell via tubing, regardless of the beamline being equipped with a sample robot or not. In viscous samples there is a risk that the fibrils align in the tubing and exposure cell. Shorter tubing, manual loading or different types of pipetting robots (e.g., (Hura et al., 2009, Round et al., 2015)) may reduce but not abolish the risk. Sample alignment will result in non-isotropic scattering. Such non-isotropic scattering is not useful for this type of analysis and must be discarded. For an example of isotropic vs. non-isotropic scattering from fibril samples, see Fig. 2 in Langkilde et al. (2018).

2. Most beamlines, as well as lab instruments, use repeated exposures on the SAXS exposure cell, rather than one single exposure, for the sake of better signal statistics and to be able to check for sample stability during the data collection. At high-brilliance synchrotrons particular caution must be made to avoid radiation damage, and very short doses of X-rays are applied to the sample, while the sample is flowing through the sample chamber, resulting in the minimum average X-ray load on each fraction of the sample. This results in the collection of numerous datafiles from a single, flowing sample (Blanchet et al., 2015; Jeffries et al., 2015). Automated software pipelines are installed including procedures which can potentially discard data from the data series, if the data from the multiple repeated exposures change beyond a set tolerance, while the sample is flowed through the cell, since this normally signifies radiation damage. Such procedures are in general highly advantageous for the user, however, in the current situation, when collecting data from fibrillating protein, there may be exceptions to this, as described in the section.

4. Initial analysis of the background subtracted data

We advocate the decomposition of data series from mixed, developing solutions. But even before decomposing the data, there is a lot to learn from the data series, and indeed decomposition should not be attempted before careful calculation and inspection of all basic biophysical parameters from the data collected throughout the process, as described below.

4.1 Non-exhaustive list of suitable software

- BioXTAS RAW (Hopkins et al., 2017; Nielsen et al., 2009)
- PRIMUS (Konarev et al., 2003)

4.2 Procedure

1. Inspect the plotted background subtracted files together. You should be able to see a gradual development in the data over time ("time" being estimated either by clock after onset of fibrillation, or by advancement in ThT fluorescence signal, depending on the sample extraction approach). Note if there are outliers from the generally developing trends? If yes, check with your notes from the data collection whether anything happened during the experiment that may explain this observation? Also check whether you can observe crossover points in your data series, i.e., that numerous different data curves develop around particular points. For a two-phase system, one would expect a single crossover point. There may be more than one such point, at different stages of the fibrillation process. Make notes about any observations of crossover points.

2. Now inspect each of the datafiles at the lowest angles (also called the low-q area). Due to the relationship between real space and reciprocal (data) space, you will find information about the longest dimensions of the scattering particles at the lowest scattering angles (note 1). Check for inconsistencies at low-q. If there are inconsistencies, the data must be cropped at the lowest angle (note 2). The lowest angle data are crucial for the success of the analysis, and hence datapoints should only be omitted if necessary.
3. Perform an approximated Guinier analysis, resulting in a calculated radius of gyration (R_g) and extrapolated zero intensity (I(0)). Note that these estimates are only approximations, for two main reasons: firstly, these solutions are mixtures of several species with very different dimensions, and secondly, at later timepoints in the data series the dimensions of the fibril repeating unit is too large for the definition of a proper Guinier range (and you can only derive very rough estimates of a lower limit of R_g). The numbers are still useful, however, for the user's own evaluation of the data. If publishing your results, remember to note that a proper Guinier range is not covered, nor is the assumption of monodispersity fulfilled. Note down the R_g and I(0) values for each datafile. Plot the values against time (Note 3).
4. Calculate the pair distance distribution function (P(r)) for each datafile using a program for indirect Fourier transformation (e.g., GNOM (Svergun, 1992) or BIFT (Hansen, 2000)). Note that this is likewise difficult for mixtures of particles. The P(r) function will provide estimates of the maximum dimension of the particle(s) in solution (D_{max}), and alternative estimates of the R_g and I(0) values. When fibrils are present in the solution, the latter estimates are likely more plausible than those obtained by fitting to a compromised Guinier range. Plot the pair distance distribution functions. P(r) functions describing different shapes have different characteristics, hence this inspection may give you a first impression about the overall shape of the particles appearing in solution. Note the timepoint where the largest D_{max} value is reached for the first time. Also note how the distribution of particular distances gradually increase and decrease over time.
5. For very elongated, rod-like particles such as the fibrils, the scattering contribution from the longest axis can be separated from the scattering contribution from the cross section (Feigin & Svergun, 1987). This means that both the cross-sectional R_g and the cross-sectional zero angle scattering (and hence cross-sectional mass per unit length) can be

calculated (e.g., in PRIMUS). Note that this can be done accurately for monodisperse solutions of particles with an infinitely long longest axis, but an approximate cross-sectional analysis of the data collected from the fibril mixtures at later timepoints may still be informative, since the fibril scattering is highly dominant at the late stages of the fibrillation reaction. These calculated dimensions are useful by themselves, but also when later potentially selecting models for decomposition of the developing data or when calculating ab initio models (see Section 7). It is particularly interesting to observe if there is a development of these parameters during the so-called steady state, i.e., at late timepoints of the fibrillation reaction, where there is no observable development in the fibril-specific fluorescence signal (Langkilde et al., 2015; Rasmussen et al., 2021; Skamris et al., 2019).

6. In a similar manner the cross-sectional P(r) functions can be calculated for the later timepoints dominated by fibril species. Note that the commonly used method GNOM assumes monodisperse rod-like particles (Svergun, 1992), hence again the cross-sectional P(r) functions are only approximated. As in point 5 above, it is particularly relevant to investigate the steady state phase for any development. Such a development may be more pronounced in the cross-sectional analysis.
7. Plot all the calculated biophysical parameters together with the developing fluorescence signal against time. Note down any observations of coinciding changes in different biophysical parameters, changes during lag-phase or steady state phase, etc. Also compare such observations to the developing P(r) functions.
8. Note down the file number of your estimated "most pure" fibril data curve (note 4). You need this file number when applying a manual approach to data decomposition.

4.3 Notes

1. Protein fibrils are very long structures, often growing to μm lengths. The longest dimension that you can observe in a SAXS experiment depends on the angular range used for the data collection and depends on the overall settings at each individual beamline, but typically does not go beyond 1000 Å, although at some beamlines it may be possible to extend the data range to 10,000 Å. Even so, it is possible to analyze data from fibrillating solutions. The reason for this is, that fibrils can be regarded

as beads on a string, where the fibril twists forms repeating units. It is the overall dimensions of this repeating unit, which must be covered in the SAXS data range.

2. Note that there should not be general inconsistencies at low-q. Inconsistencies can be due to parasitic scattering or similar issues from the beam stop, but at synchrotron beamlines today such issues will be settled, even before you arrive at the beamline. Hence, if you observe general inconsistencies, you should consult the beamline staff. Note also, that in the decomposition process, all datafiles must have the same number of datapoints and the same angular range. There are objective criteria to estimate q_{min}. If the sample does not yet contain fibrils, you can include all points, that satisfy the Guinier approximation (Guinier et al., 1955). At later timepoints the Guinier approximation is no longer satisfied, and we recommend the inclusion of all data points in the low-q area, unless there are signs of parasitic scattering. For standard macromolecular modeling, q_{max} can be defined as the point which in a Porod representation does not adhere to linearity (i.e., the data drops with q^{-4} dependency). However, both intrinsically disordered proteins, liquid-liquid phase separated droplets and—perhaps—some fibrils do not have a well-defined boundary (in scattering terms) toward the solvent (Herranz-Trillo et al., 2017; Nielsen et al., 2013) and hence will not adhere to this principle. Another rule of thumb is to apply the rule that $q^{*}R_g \leq 7–8$, which however is difficult to estimate with the sometimes-poor approximations of R_g. Also, for the initial analysis of the number of components as well as the decomposition, information may be present by extending the data range otherwise used for modeling. Ultimately, we recommend that you test different data ranges during your decomposition efforts depending on the noise level in the data.
3. When plotting the biophysical parameters against time, use the time after onset of fibrillation. Do not simply use file-numbers as abscissa, since there may be different time intervals between different measurements.
4. The data curve representing the "most pure" fibril sample is not per definition the data curve measured at the latest timepoint. It is well known in the field that significant development of fibrils may take place during the steady state phase, despite the appearance of a relatively flat fluorescence trace. If the fibrils mature (change structurally to either different forms and/or multimers of the original fibrils), this part of the process is also a continuous process, non-separable from any of the earlier structural changes. It is, however, likely that the maturation of fibrils is much

more pronounced at later timepoints, than at earlier timepoints. Indeed, there are several cases where fibril maturation has been observed within much longer timeframes (days, weeks, months) than those applied during data collection (hours) and appearing close to stable within the timeframe investigated. Ideally, we would prefer to have a SAXS measurement from a sample which contains the first fibrils species formed during the process, since this would represent the dominant fibril species during the elongation process, and hence during the timepoints where we are attempting decomposition of the data. If it is evident from your basic analysis of the data that significant fibril maturation is happening also at earlier timepoints, manual decomposition is not possible.

5. Decomposition of data—Manual approach

It is possible to decompose data series of medium complexity without the use of COSMiCS (Herranz-Trillo et al., 2017; Langkilde et al., 2018; Sagar et al., 2021) (relevant use of COSMiCS is described in Section 6). Here, as it is more "hands-on" we call this approach manual. The manual approach is only possible if data from a solution with a pure starting point is available, or—alternatively—if a plausible molecular model of the starting point is available. We do not have significant experience with manual decomposition of developing solutions with more than three components and hence do not recommend this approach if analyzing more complex systems. For new users the learning outcome from a manual approach is likely considerable. The main steps of the manual decomposition are illustrated in Fig. 1.

5.1 Non-exhaustive list of suitable software

- PRIMUS (Konarev et al., 2003)
- OLIGOMER, special version (see point 9 below)

5.2 Procedure

1. Evaluate whether it is credible that you have collected data from a sample representing a pure starting point (i.e., the structural state representing the protein at the onset of fibrillation, and which is present in diminishing amounts during the fibrillation reaction. Note that this is not necessarily the same state as the native state). If *Yes*, proceed to point 3.

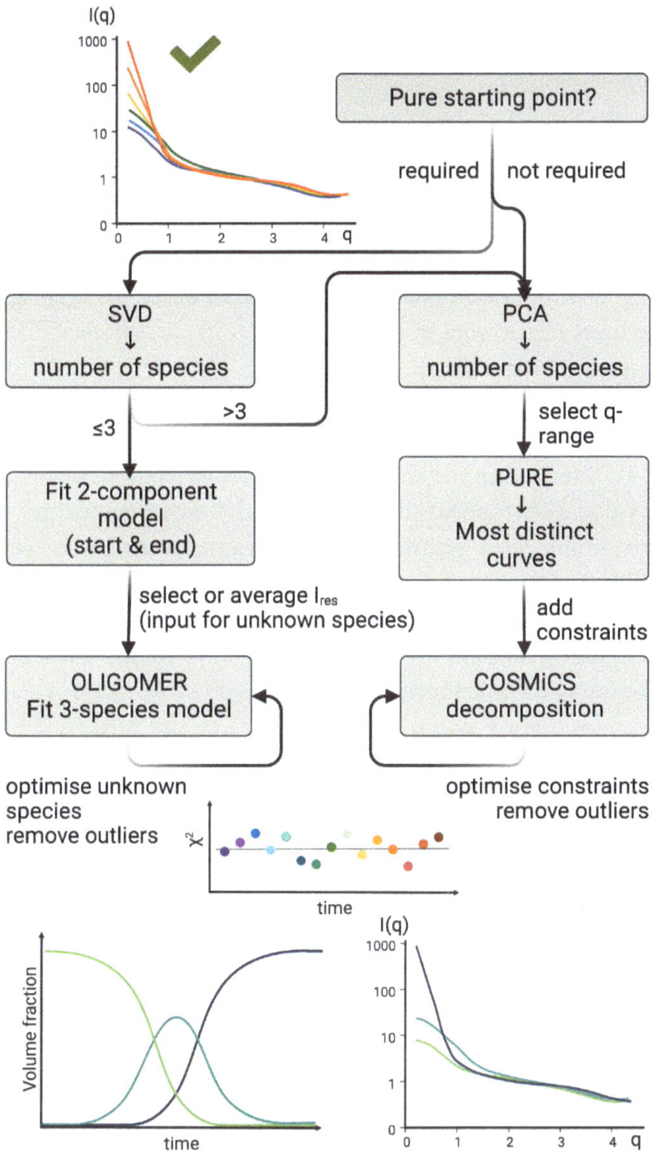

Fig. 1 Schematic view of the decomposition process. Assuming the data from the fibrillation process has been thoroughly checked a decision on manual (left side) or objective (right side) decomposition must be made. For the manual approach a known starting point (data from a pure and stable sample representing the onset of fibrillation or a corresponding validated model) is required and the manual approach is not recommended for more than three components. The objective automated process, on the other hand, requires at least twice the number of input data curves. The major steps in the two approaches are included in the flow diagram. Common to both is that the process is continued until a robust solution is reached with a good model fitting the total scattering at all time points, illustrated by a random distribution of χ^2 preferably close to 1. If this is not reached, it is necessary to return to earlier steps and redefine data ranges or input parameters. Once a solution is found the major result from both approaches is the developing volume fractions over time (bottom left) and the isolated scattering curves (bottom right) from the individual species. In the figure we have illustrated a solution for three species. *Created with BioRender.com.*

2. If you do not have data available from a pure starting point, evaluate whether you have a plausible molecular model of such a pure starting point, representing the onset as in point 1, above. If *Yes*, calculate a theoretical scattering curve from this molecular model, with the same number of datapoints and matching the range of your experimental data series. Scale the theoretical scattering curve based on the high-q range, to the level of the data from first timepoints. If *No*, use COSMiCS for the decomposition (Section 6).
3. After establishing that you have data (experimental or theoretical) representing your starting point: Plot the experimental data series in PRIMUS (emitting the theoretically calculated data curve). Apply singular value decomposition and inspect the eigen values and eigen vectors to establish an estimate of the minimal number of components contributing significantly to the scattering.
4. Now inspect your account from the initial general analysis of your data series, where you have noted down signs of potential outlier data curves and/or signs of development in the steady state phase. Remove such data curves one by one from the data series and rerun a singular value decomposition, to see how such individual data curves influence the estimate of the minimal number of components. No single data curve should change the number of contributing components, since the entire data series represents a continuously developing system. This means, that if any single data curve drastically changes the outcome of a singular value decomposition, there is reason to inspect the credibility of this data curve carefully.
5. It may also be relevant to test the outcome of a singular value decomposition with cropped data curves if there is a suspicion that high-angle data are noisy (e.g., if you have collected data from low concentration samples). Do not crop low-angle data at this point (evaluated in Section 4).
6. If you use a theoretical molecular model representing the pure starting point, now include the calculated data curve (obtained using CRYSOL (Franke et al., 2017), FoXS (Schneidman-Duhovny et al., 2013) or similar programs) in the data series and repeat the singular value decomposition. If your molecular model is adequate, the estimated number of components contributing to the data series should *not* change after including the theoretical data curve.
7. We recommend that you proceed with the manual decomposition approach if you have maximally three significantly contributing components in your solution. Alternatively, use COSMiCS (see below).

8. Now proceed to the actual data decomposition. You have defined the data curve representing the pure starting point (I_{start}), and the data file representing the "most pure" fibril species (I_{fibril}) (see Section 4, note 4). Use the program OLIGOMER (Konarev et al., 2003) or a similar program (Note 1). Your starting point for the fitting attempt is two data curves, representing the starting point (I_{start}) and fibril end point (I_{fibril}), respectively. If the singular value decomposition has suggested the presence of two species only, this should suffice, and you can proceed to point 11. We strongly recommend, however, to also test a three-species decomposition as in the following.

   If there is a third, unknown species (an intermediate structure, likely on-pathway to fibrillation) in solution, which contributes significantly to the scattering, you will not be able to fit the entire data series adequately with only two data curves (Note 1). Nevertheless, run OLIGOMER with only two input curves, expecting significant residual scattering (I_{res}), in accordance with the presence of a third species significantly contributing to the scattering. I_{res} is hence the residual scattering at each time point in the data series that cannot be accounted for by a linear combination of I_{start} and I_{fibril}. Plot the I_{res} curves from all time points. Your aim is to find the time frame of the fibrillation process, where the third species contributes most significantly to the total scattering, typically correlating with the timepoints where the 2-state fitting attempt has resulted in high χ^2 values. Here, you look for general trends in the I_{res} curves within this period. If you identify such general trends, you may scale and average the residual scattering curves from these timepoints, or you can simply use the I_{res} with the strongest scattering signal. This new curve ($I_{res\text{-}2\text{-}components}$) is your first approximation of a scattering curve, representing a third, unknown species, present in solution at intermediate timepoints (Note 2).
9. Rerun OLIGOMER with three input curves, I_{start}, $I_{res\text{-}2\text{-}components}$ and I_{fibril}. In this case, expect less significant residual scattering at each time point from the fitting procedure. Look for general trends in residual scattering curves which are significantly above noise level. Add such significant residual scattering from the fitting procedure to your first estimate of the scattering curve ($I_{res\text{-}3\text{-}components} = I_{res\text{-}2\text{-}components} + I_{res\text{-}new}$) (Note 3). Potentially scale and average your estimate from a selected time frame if general trends are observed. A special version of OLIGOMER, which automatically calculates the residual scattering at each timepoint and the second component in the input file, is available from the authors upon request.

10. Repeat point 9 until no residuals are above noise level, and χ^2 values are close to 1 (Note 4).
11. You may identify outliers in your data series at this point, if particularly strong residuals and high χ^2 values remain at single time points. Carefully inspect the corresponding files, also consulting your notes from the SAXS data collection and the results from calculating biophysical parameters (Section 4) and potentially remove the file from your data series, if you conclude that indeed the file is an outlier.

5.3 Notes

1. OLIGOMER calculates the volume fractions (a, b,..., n) providing the best possible fit of a linear combination of scattering curves (I_A, I_B,..., I_N) from a defined number (N) of components (A, B,..., N) to the experimental data curve (I_{exp}), or, formally

$$I_{exp} = A^* I_a + B^* I_b + \ldots + N^* I_n + I_{res}$$

where I_{res} is the residual scattering, not fitted by the linear combination. The fitting procedure minimizes χ_i^2, the discrepancy between the experimental and calculated curves at timepoint i, defined as

$$\chi_i^2 = [1/(K-1)] + \sum_{j=1}^{K} \{ [\mu I(q_j) - I_{exp}(q_j)] / \sigma(q_j) \}^2$$

with K data points in the experimental curve, μ being a scaling factor and $\sigma(q_j)$ the estimated standard error. The fits can thus be considered satisfactory if the residual scattering is indistinguishable from the noise level at all time points of the fibrillation reaction.

2. The first isolated I_{res}, after running OLIGOMER with only 2 input curves ($I_{res\text{-}2\text{-}components}$), is used as an approximation of the scattering from the unknown species, present at intermediate timepoints, but will be a relatively poor estimate of the component. This follows when considering that OLIGOMER will attempt the best possible fit to the experimental curve, using two components, which will have very different features in the scattering curve, being from a small and very elongated species, respectively. The fit to the experimental data will be off and the volume fractions of the two respective species will be poorly estimated, in particular at intermediate timepoints, since you cannot fit a series of experimental scattering files from solutions with different

volume fractions of three components, using only two scattering curves. $I_{res\text{-}2\text{-}components}$ is however used as a first approximation, thereby ensuring a model-free approach where the user has not made any prior assumptions about the structural features of the intermediately formed species.

3. Including $I_{res\text{-}2\text{-}components}$ as a third species when rerunning OLIGOMER with three (not two) components will already in the first attempt significantly improve the fits and estimate of the volume fractions of the start and fibril species. In this second OLIGOMER run, the residual scattering curves are now estimates of the structural features, which were not accounted for in $I_{res\text{-}2\text{-}components}$. The improved estimate of the scattering representing the third component will thus be ($I_{res\text{-}3\text{-}components} = I_{res\text{-}2\text{-}components} + I_{res\text{-}new}$).
4. Be aware that χ^2 is dependent on the estimated standard error (see note 1 in this section). If the standard errors are over- or underestimated the χ^2 values may differ from 1, but the residuals should be featureless and the χ^2 values stable along the timepoints.

6. Decomposition of developing data using COSMiCS

With colleagues we developed the software COSMiCS (*C*omplex *O*bjective *S*tructural analysis of *M*ult*i*-*C*omponent *S*ystems) (Herranz-Trillo et al., 2017; Langkilde et al., 2018; Sagar et al., 2021). COSMiCS applies a chemometrics approach, based on principles known as multivariate curve resolution using alternate least squares (MCR-ALS (Jaumot et al., 2005, Tauler, 1995). The power to derive information from complex data greatly improves when implying data from complementary sources, derived along the same reaction coordinate (e.g., data from different methods, measured from samples extracted at the same timepoint during the same process). In our application, we chose to use different representations of the same SAXS data files, essentially emphasizing different features of the data, as an alternative to complementary data. That is, we use the same experimental data set represented in absolute values (I(q)), in Holtzer representation ($I(q)\cdot q$) (Holtzer, 1955; Rambo & Tainer, 2013), in Kratky representation (($I(q)\cdot q^2$) (Glatter & Kratky, 1982) and in Porod representation (($I(q)\cdot q^4$) (Porod, 1951)), i.e., as four complementary data sets. In the different data sets the scattering intensity files are thus multiplied with q, lifted to increasing powers (0, 1, 2, 4). SAXS data contains information covering a wide range of dimensions (in real space), and when collecting data from a developing fibrillating solution the evolution of the scattering profile will be more pronounced in different q-ranges at different time frames of the process. In more qualitative terms, it is useful to look at an example from our own experience with fibrillation of aSN: The starting state is an intrinsically disordered protein, which is clearly distinguishable in a Kratky representation, with a very different profile than the data files at later time points, when more ordered oligomers and fibrils are formed. The Holtzer representation converges at higher q-values, irrespectable of the degree of flexibility (in contrast to the Porod and Kratky representations) and exhibits a maximum at lower q-values depending on the radius of gyration of the particles in solution, hence the position of this will vary, both when oligomers dominate the solution, and in particular when fibrils appear. The Porod representation shows a linear trend when there is a sharp boundary between the scatterer and the solvent, something which appears to not always be the case for protein fibrils (Herranz-Trillo et al., 2017; Nielsen et al., 2013) and hence expectedly data from solutions with large amounts of fibrils will be clearly distinguishable from data from solutions with less fibril content in this representation.

After considering this it is perhaps less of a surprise that the presentation of the same data set in four different formats can aid in an automated decomposition of the data. The method is also powerful for the isolation of scattering from transiently formed complexes in SAXS data from titration series (Sagar et al., 2021) and this, as well as the continued development of the program, is described in detail in Bernado and Sagar (2022). We have shown that COSMiCS provides a powerful means of decomposing data from complex developing amyloid systems, also when there is no measurement available from a stable structural starting point, or a pure end point (Herranz-Trillo et al., 2017). In contrast, pure scattering data from the starting point and a relatively pure fibril end point are requirements for a manual decomposition. Note, however, that a successful COSMiCS analysis is dependent on the availability of relatively many measurements, and in any case more, than what is necessary for a manual decomposition approach. This requirement must thus be considered when designing the data collection strategy (Vestergaard & Langkilde, 2022).

6.1 Non-exhaustive list of suitable software

- COSMiCS (https://github.com/BernadoLab/COSMiCS2021)

6.2 Procedure

1. Prepare the input data files (Note 1). The data files must have the structure (q, I(q), σ(q)) (momentum transfer, scattering intensity, standard error) and all files must have the same number of data points. Hence, if you have deleted data points in one of the files (cropped the data range at low-q or high-q) you must delete the same data points in all data files. The data files must be named with the same prefix and increasing file numbers and any headers in the files must have the same structure in all files (same number of lines). All files are placed in the same folder.
2. Perform the principal component analysis in COSMiCS. Select the expected number of significantly contributing components by inspecting the eigen values and eigen vectors plotted by COSMiCS.
3. Based on inspection of the plotted different presentations of the data (Absolute, Holtzer, Kratky, Porod) select the q-range to apply for each type of presentation, thereby focusing on the most pronounced features in each presentation.
4. Run the sub-program PURE (Windig et al., 2005), to select the most distinct SAXS curves within the data set (Note 2).
5. Select constraints for the procedure (Note 3). It is possible to apply the following constraints: non-negativity (intensities in scattering profiles and volume fractions have positive values); closure (mass conservation along the fibrillation process, i.e., it is expected that all the samples measured have the same total protein concentration); equality (can be applied if one of the files is known to derive from a monodisperse sample, e.g., if a stable starting point has been measured).
6. The user may also define optimization parameters, including the maximum number of iterations and the convergence criterion (implemented as a relative change of the standard deviation in the residuals after optimization).
7. You can now run the actual decomposition procedure in COSMiCS. The program will attempt decomposition on the data set on absolute scale (i.e., on a single representation of the data) and on all possible combinations of the different representation formats, i.e., in each case providing solutions that are mathematically satisfactory for the defined optimization parameters. COSMiCS has a graphical interface, and it is possible to plot the solutions for each subset of combinations.

The program will calculate the discrepancies of the fit to the individual experimental data curves (χ_i^2), and their average value for the fit to the overall dataset.

8. Inspect the solutions for each combination of data. Do the solutions appear physically meaningful? Are there generalities, repeatedly found for different combinations of data types? Are the χ_i^2 values acceptable? Are there individual fits (i.e., at a given time point) which has much higher χ_i^2 values than others? Are these time points perhaps associated with non-smooth areas in the concentration profile? Have you perhaps made notes, during your initial data analysis (Section 4) about inconsistencies at this time point? That is, in total, evaluate whether any data files (time points) are outliers. If yes, remove the file from the folder, and run COSMiCS again. COSMiCS will now in addition output the $\Delta\chi_i^2$, i.e., allow the user to compare the potential improvement in fit for each time point. If no improvement is observed, do not emit the time point.
9. Also evaluate your choice of constraints and optimization parameters, potentially modify the values and run COSMiCS again.
10. Carefully inspect the COSMiCS report, including your input parameters and the graphics illustrating the optimization.
11. Repeat points 7–10, until obtaining a robust solution.
12. Select the optimal combination of data representations. COSMiCS now calculates and assigns standard deviations to the decomposed scattering profiles (i.e., the scattering profiles of each species in solution) by the use of a Monte-Carlo error analysis (Svergun & Pedersen, 1994), and outputs the final result.

6.3 Notes

1. In future versions of COSMiCS, it will be possible to include data from complementary biophysical investigations alongside SAXS data. This will expectedly further improve the decomposition power of COSMiCS.
2. The most distinct data curves, selected when running the sub-program PURE (Windig et al., 2005), are used in the initial decomposition steps in the MCR-ALS approach. Do not consider these curves as actual estimates of scattering from individual components, that are actually present in the solutions. Rather, these selected curves are the scattering features, which, from a mathematical perspective, are most distinct, and hence

enables decomposition of the scattering features in the total data set. It is also possible for the user to select these initial "most distinct" data curves manually, but in general we recommend that the user applies PURE.

3. Consider, as mentioned above, that COSMiCS applies a strictly mathematical procedure, and hence non-intuitive solutions result, unless restraints provide the means to avoid them. That is, it is principally possible to decompose a given data set, resulting in (theoretical) scattering profiles and/or volume fractions with negative values, and this solution may give a mathematically satisfactory solution, although the result is physically meaningless. That is, the solution is ambiguous. Constraints reduce the ambiguity, but constraints should evidently be physically meaningful. Non-negativity is an obvious constraint to apply. Closure is meaningful in most cases but may be released if there is a risk that fibrils have precipitated in the sample vial, prior to collecting the SAXS data (although such data should be avoided if possible). Equality is a very powerful constraint, but should evidently be applied with great care, only in cases where it is fully plausible that one of the measurements represents one of the species present during the fibrillation reaction.

7. Further interpretation

The output from your decomposition analysis, assuming a robust solution properly fitting the experimental data, is a major result (Fig. 2). The individual scattering profiles of each component, present in the solution during the fibrillation reaction, will now allow you to continue the analysis, as if you had collected data from each isolated species, and you can perform fitting to known structures, ab initio modeling etc. The concentration profiles of the individual species that are appearing (or disappearing) during fibrillation is another main result and will provide you with knowledge about the underlying fibrillation principles, complementary to the type of information provided by for instance analysis of spectroscopic traces from fibrillation specific fluorescent probes. From a fluorescent trace, you get indirect indications of the presence of particular species, that is, the species that will bind the given fluorescent probe with varying affinity and specificity. Although very useful information about fibrillation kinetics has been derived by this method (for a recent example see (Rodriguez Camargo et al., 2021)), there is a risk, that the fluorescence signal does not adequately elucidate the presence of the species, investigated. A fluorescence probe may have affinity for some fibrils and lower or no affinity for other fibril types

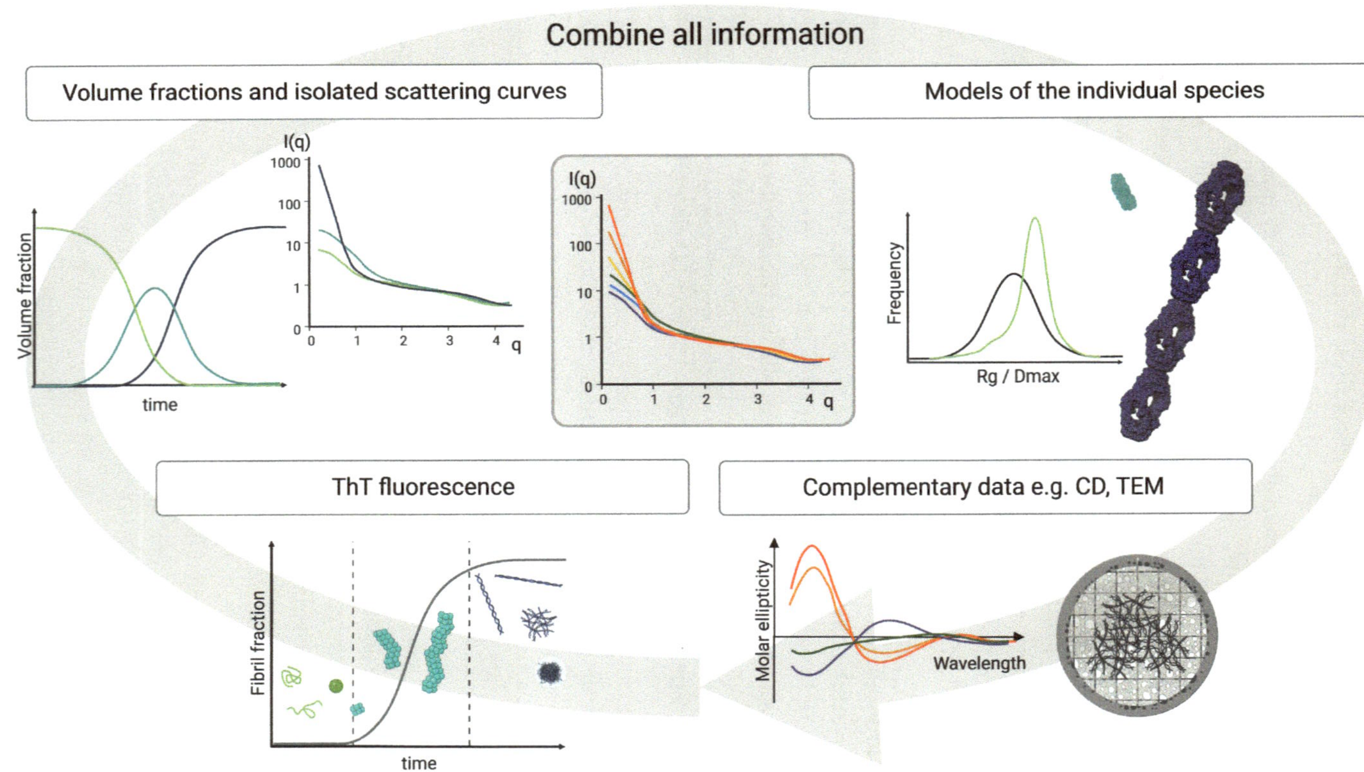

Fig. 2 See figure legend on opposite page.

and/or oligomeric species. In contrast, the decomposition analysis presented here, if correctly performed, will result in kinetics traces (concentration profiles) that account for all the dominant macromolecular species in solution. At least, it will consistently provide a physically meaningful and adequate fitting to all individual data points, using representative scattering profiles for the number of species, contributing to the total scattering data series, according to singular value decomposition and/or principal component analysis. On the other hand, an evident advantage of fitting to fluorescent traces is the means to fit to a very large number of kinetics traces collected at different experimental conditions, and in that sense the two methods are highly complementary and would likely be very beneficial to combine.

Below, we include suggestions for further analysis, after decomposing the data. Since principles in ab initio modeling and other structural modeling tools are very well described by others, we will not provide a step-by-step procedure here, but refer to the multiple protocols that can be found elsewhere. Rather, we provide a bulleted list of particularities and points of concern, when analyzing decomposed data from a fibrillation series.

- A simple, but very informative step is to calculate the first derivative of a fit to the concentration profile (volume fractions) of the fibril species. Compare this to the concentration profile of the second, intermediately formed species. If the two curves superimpose, this indicates that the second species is on-pathway to fibril formation (Nors Perdersen et al., 2015; Vestergaard et al., 2007)
- Compare your decomposed concentration profiles to the basic biophysical parameters (Section 4). Can you follow the development in, e.g., the pair

Fig. 2 The outcome of the main experiment. The SAXS data from the fibrillation process are central components in the structural analysis of the developing system (center of the figure). Once the decomposition is successful (upper left), the isolated curves can be used to obtain models of the species dominating the fibrillation process. This can e.g., be ensemble optimization of an intrinsically disordered starting state and ab-initio models of intermediate or fibril species (upper right). It is crucial to check that the decomposition and structural models make sense also in context of all available complementary data (exemplified here by fibrillation kinetics followed by fluorescence spectroscopy, circular dichroism data and transmission electron microscopy images (bottom)). With consistent results the experiment will give both structural and mechanistic insight in the fibrillation process. Created with BioRender.com with the addition of an ab initio fibril model based on scattering data from a previous study (Vestergaard et al., 2007), visualized with PyMOL (Schrödinger, 2020).

distance distribution functions, when comparing to the concentration profiles? Make sense of all your information in context of the new information from the decomposition.

- Compare your decomposed concentration profiles to all other available biophysical data. Does the appearance and disappearance of the intermediate and/or fibril species coincide with the appearance of other biophysical signals (see also Section 6, Note 1)?
- Perform a basic analysis of the individual decomposed scattering profiles, deriving all biophysical parameters from the individual species. This includes, e.g., Guinier analysis, pair distance distribution functions and Porod analysis or corresponding cross section analyses. In PRIMUS, the subroutine BODIES makes a fit of scattering curves to analytical form factors from standard geometrical shapes (Konarev et al., 2003). This can give you an initial indication about the overall dimensions of the individual components, supplementing the information you have from the pair distance distribution functions (note, however, that the length estimates of the fibrils are only for the repeating units of the fibrils, as the full dimension of the fibrils extend beyond the resolution limit of the SAXS experiment).
- The fibril species may also be fitted as cylinders with additional degrees of freedom, in terms of cross section polydispersity and cylinder flexibility using SASview (http://www.sasview.org/).
- You may also develop a structural model, fitting to your decomposed scattering profiles. For the starting point, consider the fact that many, if not most fibrillation reactions start by a (partial) un- or refolding of the native structure. This must evidently be considered, when evaluating the proper approach to structural modeling against any of the decomposed scattering curves. You can of course fit a theoretical scattering curve derived from a known, plausible atomic model to your SAXS curve using, e.g., the program CRYSOL (Manalastas-Cantos et al., 2021), but likely, the known structures from non-amyloidogenic conditions are not adequate in describing the protein.
- For partially or fully disordered proteins, several methods exist, for the description of the structural state. Disordered proteins should be considered as ensembles of protein structures and a method such as EOM (Bernado et al., 2007; Tria et al., 2015) uses genetic algorithms to select such ensembles from a pool of random structures, and is quite accessible in use, also for first-time users. We also refer to the protocols provided in

this and a previous volume, concerning extraction of additional biophysical parameters from single SAXS measurements on IDPs, in particular the solvent quality as well as the important integration of advanced computer simulations in the analysis of SAXS data from IDPs (see chapter "From dilute to concentrated solutions of intrinsically disordered proteins: Sample preparation and data collection" by Lenton et al., in this issue). Importantly, the inclusion of a proper solvent layer in the calculated scattering profile from IDP models is crucial for the relevant evaluation against SAXS data (Henriques et al., 2018). Finally, we may note, that we collaborated on the exploration of a significantly more elaborate and refined method, based on molecular dynamics simulations against experimental SAXS data, restrained and refined by metainference and Bayesian/Maximum Entropy, respectively, and using an implicit solvent model (Ahmed et al., 2021). Although this method undoubtedly provides more realistic structural information at the atomic level, such methods are not accessible for standard use at present.

- Low resolution ab initio models, using, e.g., the program DAMMIF (Franke & Svergun, 2009), from each of the individual components along the fibrillation pathway, may be relevant. Again, note that ab initio models should not be calculated if the structure is fully or partially disordered. Even if a model can be calculated, that will fit your curve, the model has no physical meaning. If applying DAMMIF, you need not select a starting volume for the model development. In our experience, however, a starting model may need to be provided by the user for the development of a model of the fibril structure. Here, the above mentioned BODIES analysis can assist the selection of a proper starting volume, used in combination with the older ab initio program DAMMIN (Svergun, 1999). Note also, that the fibril model obtained is not a model of the entire fibril structure, since we do not have data from the longest dimensions of fibrils. Fibrils often grow to the length of several μm, while we collect SAXS data, typically in the range 10–1000 Å. But most fibrils can be considered as beads on a string, with repeating units twisting around the central fibril axis, and this repeating fibril unit typically has dimensions within the boundaries of our SAXS data regime (below 1000 Å at the longest axis). A low-resolution ab initio model of the fibril repeating unit is a unique piece of structural information since most other methods available in the field will suffer from surface and/or dehydration effects.

8. Alternative approaches

We have described in detail how to prepare samples and collect data (Vestergaard & Langkilde, 2022). In this chapter we provide our suggestions of how to analyze the resulting data series, aiming at a decomposition of the data into scattering profiles, representing the individual species, present during the fibrillation reaction.

The ATSAS program suite now also includes an elegant method, where multiple SAXS data-curves from developing systems are fitted simultaneously, and ab initio models of the intermediate, unknown species are calculated directly in DAMMIX (Konarev & Svergun, 2018). The method has, among several examples, been applied with success to some of the datasets that we have referred to in the introduction, and the method is an appealing alternative analysis method, in systems where there is a well-described start- and endpoint, and with one unknown intermediate species.

A more classical approach is to analyze the data by selecting a set of predefined shapes, with known analytical form factors, and to fit a combination of such form factors to the experimental data. Examples would be to use a combination of cylinders (as a model for the fibril) and random coils (as a model for intrinsically disordered proteins), for which analytical form factors are known. We urge the reader to consult also publications from our colleagues, where this approach has resulted in insight into the fibrillation reaction of glucagon (Oliveira et al., 2009) and FapC (Rasmussen et al., 2021). Although at very different time-scales, similar fitting-principles were used on a stopped-flow time-resolved SAXS data-series capturing the initial (msec) structural response of an apomyoglobin mutant to a pH-jump (Ortore et al., 2011).

Important information can also be deduced from thorough analyses of average aggregation sizes and characteristics, as well as SAXS curve similarities. An example is the study of various mutants of superoxidase dismutase (Pratt et al., 2014), where SAXS data were collected at different timepoints in the fibril formation process of the various mutants in several different solution conditions.

Alternative methods have been developed by other groups for semi-automated decomposition of SAXS data series from developing mixtures, such as the program REGALS (Meisburger et al., 2021), based on principles comparable to COSMiCS. This method has also been applied with success to data from in-line SAXS/chromatography experiments, an important application, given the urgency of obtaining data from very pure and potentially unstable samples. A very recent method for decomposition of SAXS/chromatography data, based on evolving factor analysis, is described in (Konarev et al., 2022). We have also used decomposition of a particularly hard case of in-line size exclusion chromatography SAXS data (SEC-SAXS) from the native, intrinsically disordered protein α-synuclein. We showed that α-synuclein is prone to form specific di-tyrosine dimers during standard purification procedures and wanted to elucidate both the challenges, that we experienced during purifications and the effect of the dimer formation on the fibrillation kinetics. The monomer and dimer of the unfolded protein were both in natively disordered conformations (and hence with relatively featureless SAXS form factors) and co-eluted on the SEC column with only very modest separation, all in all causing the decomposition to be harder than usual. Initial estimates of the monomer and dimer SAXS curves, respectively, were generated with the aid of the chromatographic UV signal (UV_{280}) in a manual decomposition approach, and the two SAXS signals with only very few distinct scattering profile differences (but evidently different I(0), R_g and D_{max}) could hence be separated (van Maarschalkerweerd et al., 2015a).

As previously mentioned, the program COSMiCS, that we describe as a powerful tool for decomposition of SAXS (and other biophysical) data from protein fibrillation series, is also very useful for the analysis of, e.g., SAXS data from titration series resulting in the formation of transient protein complexes (Bernado & Sagar, 2022; Sagar, 2021). Hence, in conclusion, although the principles described in this chapter are with a focus on amyloid systems/protein fibrillation, all these latter cases exemplify that the same analysis principles can very well be applied to non-fibrillation data series, and hence in general to SAXS data from developing systems.

9. Summary and conclusions

Fibrillation processes and amyloid proteins are of interest both for their role in numerous fatal diseases, for their potential application as self-assembly nanomaterials, and on a basic scientific level for their intriguing structural properties and complexity. Well-prepared (Vestergaard & Langkilde, 2022), it is possible to collect SAXS data from such developing systems, enabling insight to the fibrillation mechanism without disturbing the fibrillation process. When successful, such an experiment results in unique structural information from transiently formed intermediate species as well as added information on the kinetic profile with concentration profiles of all the species, and the method is complementary to most other classic biophysical experiments. The analysis of such data demands special caution and particular approaches, which we describe here in detail, based on our own accumulated experience with amyloid and amyloid-like systems. We provide protocols for two alternative approaches to the central decomposition of the data series, one being a relatively simple, manual, iterative procedure, and the alternative applying a more automated approach using a chemometrics based principle. Ambiguity is inherent in SAXS data, thus carefully considering complementary information is important throughout the procedure, and we emphasize the importance of obtaining robust solutions before drawing conclusions. In the chapter, we aim to supply adequate warnings where the user should be particularly careful in the analysis, given the special nature of amyloid proteins. The approach of decomposition is, however, also more generally applicable to the analysis of SAXS data from other complex, developing mixtures of macromolecules.

Acknowledgments

The authors appreciate interest from and collaboration with many colleagues, both in the fields of protein amyloids and BioSAXS. In particular, we thank the group of Pau Bernado at CBS Montpellier, for the ongoing collaboration on the development and use of COSMiCS, and the entire Svergun group at EMBL-Hamburg, for the long-standing support, interest, and collaboration, as well as a significant amount of patience with our demanding ideas during beamtimes. The work has been performed, supported by funding from the Danish Council for Independent Research | Medical Sciences, the Lundbeck Foundation, the Novo Nordisk Foundation, the Carlsberg Foundation and DANSCATT. Last, but not least, we thank current and former members of the BioSAXS group at the University of Copenhagen for ideas, enthusiasm, and hard work.

References

Agarwal, A., et al. (2021). An intrinsically disordered pathological prion variant Y145Stop converts into self-seeding amyloids via liquid-liquid phase separation. *Proceedings of the National Academy of Sciences of the United States of America, 118*(45), e2100968118.

Ahmed, M. C., et al. (2021). Refinement of alpha-Synuclein ensembles against SAXS data: Comparison of force fields and methods. *Frontiers in Molecular Biosciences, 8*, 654333.

Ankarcrona, M., et al. (2016). Current and future treatment of amyloid diseases. *Journal of Internal Medicine, 280*(2), 177–202.

Bernado, P., et al. (2007). Structural characterization of flexible proteins using small-angle X-ray scattering. *Journal of the American Chemical Society, 129*(17), 5656–5664.

Blanchet, C. E., et al. (2015). Versatile sample environments and automation for biological solution X-ray scattering experiments at the P12 beamline (PETRA III, DESY). *Journal of Applied Crystallography, 48*(Pt 2), 431–443.

Buell, A. K. (2019). The growth of amyloid fibrils: Rates and mechanisms. *The Biochemical Journal, 476*(19), 2677–2703.

Bulawa, C. E., et al. (2012). Tafamidis, a potent and selective transthyretin kinetic stabilizer that inhibits the amyloid cascade. *Proceedings of the National Academy of Sciences of the United States of America, 109*(24), 9629–9634.

Feigin, L. A., & Svergun, D. I. (1987). *Structure analysis by small-angle X-ray and neutron scattering*. New York: Plenum Press.

Fodera, V., et al. (2008). Thioflavin T hydroxylation at basic pH and its effect on amyloid fibril detection. *The Journal of Physical Chemistry. B, 112*(47), 15174–15181.

Franke, D., & Svergun, D. I. (2009). DAMMIF, a program for rapid ab-initio shape determination in small-angle scattering. *Journal of Applied Crystallography, 42*(Pt 2), 342–346.

Franke, D., et al. (2017). ATSAS 2.8: A comprehensive data analysis suite for small-angle scattering from macromolecular solutions. *Journal of Applied Crystallography, 50*(Pt 4), 1212–1225.

Gade Malmos, K., et al. (2017). ThT 101: A primer on the use of thioflavin T to investigate amyloid formation. *Amyloid, 24*(1), 1–16.

Giehm, L., et al. (2011). Low-resolution structure of a vesicle disrupting α-synuclein oligomer that accumulates during fibrillation. *Proceedings of the National Academy of Sciences of the United States of America, 108*(8), 3246–3251.

Glatter, O., & Kratky, O. (1982). *Small angle X-ray scattering*. London: Academic.

Graewert, M. A., & Jeffries, C. M. (2017). Sample and buffer preparation for SAXS. *Advances in Experimental Medicine and Biology, 1009*, 11–30.

Groenning, M., et al. (2015). Considerably unfolded transthyretin monomers preceed and exchange with dynamically structured amyloid protofibrils. *Scientific Reports, 5*, 11443.

Guinier, A., et al. (1955). *Small-angle scattering of X-rays*. John Wiley & Sons, Inc.

Hammersley, A. P. (1997). *FIT2D: An introduction and overview. ESRF internal report, ESRF 97HA02T*. European Synchrotron Radiation Facility.

Hansen, S. L. (2000). Bayesian estimation of hyperparameters for indirect Fourier tranformation in small-angle scattering. *Journal of Applied Crystallography, 33*, 1415–1421.

Henriques, J., et al. (2018). On the calculation of SAXS profiles of folded and intrinsically disordered proteins from computer simulations. *Journal of Molecular Biology, 430*(16), 2521–2539.

Herranz-Trillo, F., et al. (2017). Structural analysis of multi-component amyloid systems by chemometric SAXS data decomposition. *Structure, 25*(1), 5–15.

Holtzer, A. (1955). Interpretation of the angular distribution of the light scattered by a polydisperse system of rods. *Journal of Polymer Science, 17*, 432–434.

Hopkins, J. B., et al. (2017). BioXTAS RAW: Improvements to a free open-source program for small-angle X-ray scattering data reduction and analysis. *Journal of Applied Crystallography, 50*(Pt 5), 1545–1553.

Hura, G. L., et al. (2009). Robust, high-throughput solution structural analyses by small angle X-ray scattering (SAXS). *Nature Methods, 6*(8), 606–612.

Jacques, D. A., et al. (2012). Publication guidelines for structural modelling of small-angle scattering data from biomolecules in solution. *Acta Crystallographica. Section D, Biological Crystallography, 68*(Pt 6), 620–626.

Jaumot, J., et al. (2005). A graphical user-friendly interface for MCR-ALS: A new tool for multivariate curve resolution in MATLAB. *Chemometrics and Intelligent Laboratory Systems, 76*(1), 101–110.

Jeffries, C. M., et al. (2015). Limiting radiation damage for high-brilliance biological solution scattering: Practical experience at the EMBL P12 beamline PETRAIII. *Journal of Synchrotron Radiation, 22*(Pt 2), 273–279.

Konarev, P. V., & Svergun, D. I. (2018). Direct shape determination of intermediates in evolving macromolecular solutions from small-angle scattering data. *IUCrJ, 5*(Pt 4), 402–409.

Konarev, P. V., et al. (2003). PRIMUS: A windows PC-based system for small-angle scattering data analysis. *Journal of Applied Crystallography, 36*, 1277–1282.

Konarev, P. V., et al. (2022). EFAMIX, a tool to decompose inline chromatography SAXS data from partially overlapping components. *Protein Science*, *31*(1), 269–282.
Kumari, P., et al. (2021). Structural insights into alpha-synuclein monomer-fibril interactions. *Proceedings of the National Academy of Sciences of the United States of America*, *118*(10), e2012171118.
Langkilde, A. E., et al. (2015). The architecture of amyloid-like peptide fibrils revealed by X-ray scattering, diffraction and electron microscopy. *Acta Crystallographica. Section D, Biological Crystallography*, *71*(Pt 4), 882–895.
Langkilde, A. E., et al. (2018). Noninvasive structural analysis of intermediate species during fibrillation: An application of small-angle X-Ray scattering. *Methods in Molecular Biology*, *1779*, 209–239.
Lattanzi, V., et al. (2021). Amyloid beta 42 fibril structure based on small-angle scattering. *Proceedings of the National Academy of Sciences of the United States of America*, *118*(48), e2112783118.
Lee, C.-C., et al. (2007). A three-stage kinetic model of amyloid fibrillation. *Biophysical Journal*, *92*(10), 3448–3458.
Manalastas-Cantos, K., et al. (2021). ATSAS 3.0: Expanded functionality and new tools for small-angle scattering data analysis. *Journal of Applied Crystallography*, *54*(Pt 1), 343–355.
Meisburger, S. P., et al. (2021). REGALS: A general method to deconvolve X-ray scattering data from evolving mixtures. *IUCrJ*, *8*(Pt 2), 225–237.
Nelson, R., et al. (2005). Structure of the cross-beta spine of amyloid-like fibrils. *Nature*, *435*(7043), 773–778.
Nielsen, S. S., et al. (2009). BioXTAS RAW, a software program for high-throughput automated small-angle X-ray scattering data reduction and preliminary analysis. *Journal of Applied Crystallography*, *42*, 959–964.
Nielsen, S. B., et al. (2013). Wildtype and A30P mutant alpha-synuclein form different fibril structures. *PLoS One*, *8*(7), e67713.
Nors Pedersen, M., et al. (2015). Direct correlation between ligand-induced alpha-Synuclein oligomers and amyloid-like fibril growth. *Scientific Reports*, *5*, 10422.
Oliveira, C. L. P., et al. (2009). A SAXS study of glucagon fibrillation. *Journal of Molecular Biology*, *387*(1), 147–161.
Ortore, M. G., et al. (2011). Time-resolved small-angle x-ray scattering study of the early stage of amyloid formation of an apomyoglobin mutant. *Physical Review. E, Statistical, Nonlinear, and Soft Matter Physics*, *84*(6 Pt 1), 061904.
Palmadottir, T., et al. (2021). Charge regulation during amyloid formation of alpha-Synuclein. *Journal of the American Chemical Society*, *143*(20), 7777–7791.
Porod, G. (1951). Die Röntgenkleinwinkelstreuung von dichtgepackten kolloiden Systemen. *Kolloid Zeitschrift*, *124*, 83–114.
Pratt, A. J., et al. (2014). Aggregation propensities of superoxide dismutase G93 hotspot mutants mirror ALS clinical phenotypes. *Proceedings of the National Academy of Sciences of the United States of America*, *111*(43), E4568–E4576.
Rambo, R. P., & Tainer, J. A. (2013). Accurate assessment of mass, models and resolution by small-angle scattering. *Nature*, *496*(7446), 477–481.
Rasmussen, H. O., et al. (2021). A multimethod approach for analyzing FapC fibrillation and determining mass per length. *Biophysical Journal*, *120*(11), 2262–2275.
Ray, S., et al. (2020). Alpha-Synuclein aggregation nucleates through liquid-liquid phase separation. *Nature Chemistry*, *12*(8), 705–716.
Rodriguez Camargo, D. C., et al. (2021). Surface-catalyzed secondary nucleation dominates the generation of toxic IAPP aggregates. *Frontiers in Molecular Biosciences*, *8*, 757425.

Round, A., et al. (2015). BioSAXS sample changer: A robotic sample changer for rapid and reliable high-throughput X-ray solution scattering experiments. *Acta Crystallographica. Section D, Biological Crystallography*, *71*(Pt 1), 67–75.
Sagar, A., et al. (2021). Structure and thermodynamics of transient protein-protein complexes by chemometric decomposition of SAXS datasets. *Structure*, *29*(9), 1074–1090.e4.
Sagar, A., & Bernadó, P. (2022). Disentangling polydisperse biomolecular systems by Chemometrics decomposition of SAS data. *Methods in Enzymology*, *677*, 531–555.
Sawner, A. S., et al. (2021). Modulating alpha-Synuclein liquid-liquid phase separation. *Biochemistry*, *60*(48), 3676–3696.
Schneidman-Duhovny, D., et al. (2013). Accurate SAXS profile computation and its assessment by contrast variation experiments. *Biophysical Journal*, *105*(4), 962–974.
Schrödinger, L. D. W. (2020). *The PyMOL molecular graphics system*. Schrödinger, LLC.
Sidhu, A., et al. (2018). Polymorph-specific distribution of binding sites determines thioflavin-T fluorescence intensity in alpha-synuclein fibrils. *Amyloid*, *25*(3), 189–196.
Skamris, T., et al. (2019). Early stage alpha-Synuclein amyloid fibrils are reservoirs of membrane-binding species. *Scientific Reports*, *9*(1), 1733.
Svergun, D. I. (1992). Determination of the regularization parameter in indirect-transform methods using perceptual criteria. *Journal of Applied Crystallography*, *25*, 495–503.
Svergun, D. I. (1999). Restoring low resolution structure of biological macromolecules from solution scattering using simulated annealing. *Biophysical Journal*, *76*(6), 2879–2886.
Svergun, D. I., & Pedersen, J. S. (1994). Propagating errors in small-angle scattering data treatment. *Journal of Applied Crystallography*, *27*, 241–248.
Svergun, D., et al. (2013). *Small angle X-Ray and neutron scattering from solutions of biological macromolecules*. Oxford University Press.
Tauler, R. (1995). Multivariate curve resolution applied to second order data. *Chemometrics and Intelligent Laboratory Systems*, *30*(1), 133–146.
Trewhella, J., et al. (2017). 2017 publication guidelines for structural modelling of small-angle scattering data from biomolecules in solution: An update. *Acta Crystallographica. Section D, Structural Biology*, *73*(Pt 9), 710–728.
Tria, G., et al. (2015). Advanced ensemble modelling of flexible macromolecules using X-ray solution scattering. *IUCrJ*, *2*(Pt 2), 207–217.
van Maarschalkerweerd, A., et al. (2015a). Formation of covalent di-tyrosine dimers in recombinant alpha-synuclein. *Intrinsically Disordered proteins*, *3*(1), e1071302.
van Maarschalkerweerd, A., et al. (2015b). Cholesterol facilitates interactions between alpha-synuclein oligomers and charge-neutral membranes. *FEBS Letters*, *589*(19 Pt B), 2661–2667.
Vestergaard, B., et al. (2007). A helical structural nucleus is the primary elongating unit of insulin amyloid fibrils. *PLoS Biology*, *5*(5), e134.
Vestergaard, B., & Langkilde, A. E. (2022). Protein fibrillation from another small angle—SAXS data analysis of developing systems. *Methods in Enzymology*, *677*, 291–321.
Wells, C., et al. (2021). The role of amyloid oligomers in neurodegenerative pathologies. *International Journal of Biological Macromolecules*, *181*, 582–604.
Windig, W., et al. (2005). A new approach for interactive self-modeling mixture analysis. *Chemometrics and Intelligent Laboratory Systems*, 77, 85–96.
Xu, B., et al. (2022). Manganese promotes alpha-synuclein amyloid aggregation through the induction of protein phase transition. *The Journal of Biological Chemistry*, *298*(1), 101469.

CHAPTER FOURTEEN

Visualizing and accessing correlated SAXS data sets with Similarity Maps and Simple Scattering web resources

Daniel T. Murray[a,†], David S. Shin[a,†], Scott Classen[a], Chris A. Brosey[b], and Greg L. Hura[a,c,*]

[a]Molecular Biophysics and Integrated Bioimaging Division, Lawrence Berkeley National Laboratory, Berkeley, CA, United States
[b]Department of Molecular and Cellular Oncology, The University of Texas MD Anderson Cancer Center, Houston, TX, United States
[c]Department of Chemistry and Biochemistry, University of California Santa Cruz, Santa Cruz, CA, United States
*Corresponding author: e-mail address: glhura@lbl.gov

Contents

[†] Contributed Equally

Methods in Enzymology, Volume 678
ISSN 0076-6879
https://doi.org/10.1016/bs.mie.2022.09.024

Abstract

Constructing a comprehensive understanding of macromolecular behavior from a set of correlated small angle scattering (SAS) data is aided by tools that analyze all scattering curves together. SAS experiments on biological systems can be performed on specimens that are more easily prepared, modified, and formatted relative to those of most other techniques. An X-ray SAS measurement (SAXS) can be performed in less than a milli-second in-line with treatment steps such as purification or exposure to modifiers. These capabilities are valuable since biological macromolecules (proteins, polynucleotides, lipids, and carbohydrates) change conformation or assembly under specific conditions that often define their biological role. Furthermore, mutation or post-translational modification change their behavior and provides an avenue to tailor their mechanics. Here, we describe tools to combine multiple correlated SAS measurements for analysis and review their application to biological systems. The SAXS Similarity Map (SSM) compares a set of scattering curves and quantifies the similarity between them for display as a color on a grid. Visualizing an entire correlated data set with SSMs helps identify patterns that reveal biological functions. The SSM analysis is available as a web-based tool at https://sibyls.als.lbl.gov/saxs-similarity/. To make data available and promote tool development, we have also deployed a repository of correlated SAS data sets called Simple Scattering (available at https://simplescattering.com). The correlated data sets used to demonstrate the SSM are available on the Simple Scattering website. We expect increased utilization of correlated SAS measurements to characterize the tightly controlled mechanistic properties of biological systems and fine-tune engineered macromolecules for nanotechnology-based applications.

1. Introduction

Biology fundamentally relies on precisely orchestrated chemical reactions. To coordinate the mechanism and timing that select which atomic bonds will be made or broken, macromolecular architectures have evolved to identify molecules and, through conformational changes, direct them into controlled environments. These environments concentrate reactants, promote specific rotamer orientations or orbital overlaps, and prevent reversion to initial states or unwanted side products. The macromolecules are also precisely tuned to function in context-specific ways based on solvent conditions and chemical inputs from other macromolecules. These functionalities remain unmatched by synthetic systems because the complexity and speed of macromolecular mechanisms are difficult to decipher experimentally or computationally. While recent advances in cryo-electron microscopy (cryo-EM) and continued innovation in macromolecular crystallography (MX) and nuclear magnetic resonance (NMR) provide atomic resolution information, these techniques host several shortcomings related to capturing the aforementioned macromolecular complexity of biological systems. SAS is well positioned to fill the gaps in our knowledge of macromolecular function, provide additional information that is unobtainable through other means, and enhance the value of every determined atomic resolution structure.

SAS analysis has proven to provide valuable, and often definitive, constraints on macromolecular conformational change and dynamics at the nanoscale. The number of constraints SAS generates on macromolecular structure has been estimated using a Shannon sampling formalism (Konarev & Svergun, 2015). Applying this formalism is necessary as SAS information content is not equivalent to the number of points in a SAS curve. The sampling of momentum transfer (q) as a function of scattered intensity usually far exceeds the number of points required to reconstruct the SAS curve. According to Shannon analysis, a single SAS measurement on a solution of macromolecules typically has 10–40 unique pieces of information (Konarev & Svergun, 2015). While alone, this information is insufficient to define the relative positions of each atom within a macromolecule and how they evolve over time, but when the information is used to constrain conformation based on a model generated by other methods, the resolution is high, can be generated in high-throughput, and, importantly, reflects the solution state. When multiple SAS measurements that probe multiple states are collected and analyzed together, the information content can be higher than the sum of Shannon channels and uniquely complementary to tools that resolve static structures.

The malleability of SAS measurements is an important asset during their application to providing insights on top of those gleaned from static structures. Due to low-throughput constraints needed to make measurements, cryo-EM, MX, and NMR often produce a set of atomic coordinates representing a single structure inhabiting a limited physiological context. Further, SAS profiles can be accurately calculated from an atomic model (Schneidman-Duhovny, Hammel, & Sali, 2010; Shin et al., 2009; Svergun, Barberato, & Koch, 1995). If a poor match is found between the measured and calculated profiles, the atomic model can be modified and a new SAS profile can be calculated and compared to generate new conformations that more closely match the measured data (Fig. 1). Therefore, a single atomic resolution structure, generated by a lower throughput technique, can be leveraged by SAS to define nanoscale changes that occur in other contexts. Integrating information from multiple contextualized SAS measurements will be central to applications of SAS over the next decade.

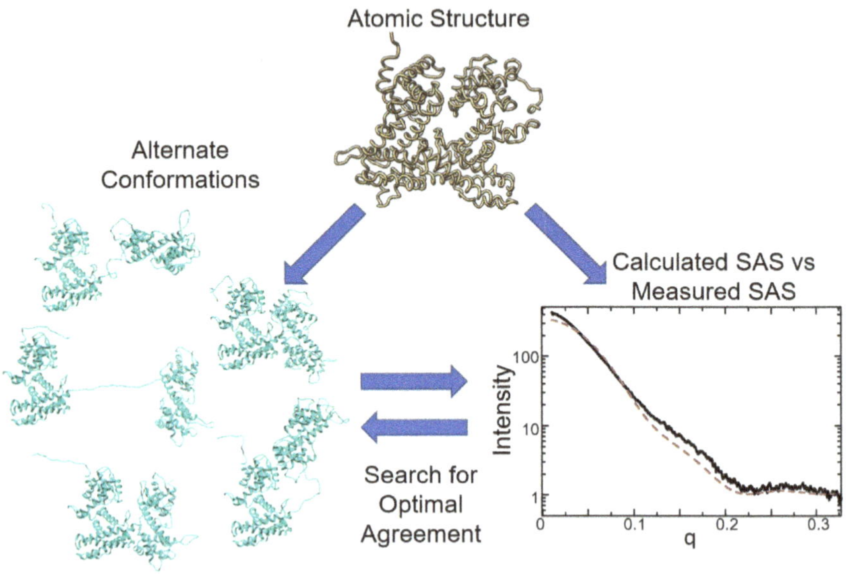

Fig. 1 Integrating SAS with atomic resolution measurements. A SAS signal can be accurately calculated from atomic coordinates determined either by a high-resolution technique or through structure prediction. Disagreement indicates a different solution state than that provided from the high-resolution model. Alternate conformations can be generated based on the atomic coordinates to search for models or ensembles of models that best fit the SAS data.

To facilitate the integration of information from multiple contextualized SAS measurements, the SAS community needs both access to correlated data sets and tools that analyze those data together. In this chapter, we describe our efforts to address both of these needs in two principle sections. In Section 2, we describe SAXS Similarity Maps (SSMs), which is a tool designed to visualize tens of SAXS measurements collected from a macromolecule or related macromolecules under a variety of contexts. We provide examples and also demonstrate the application of SSMs to derive insights in several cases. In Section 3, we describe a bespoke repository, called Simple Scattering, for organizing and holding this kind of data. Further, Simple Scattering can also be used to hold other types of data beyond those usefully analyzed by SSMs. Simple Scattering holds SAXS data generated during size-exclusion coupled SAXS (SEC-SAXS) and can also hold high-throughput (HT-SAXS), time-resolved, micro-fluidic-associated, and neutron-based SAS (SANS) data. These resources will be useful as the structural biology community seeks to leverage success in resolving static structures and produce dynamic and solution state perspectives.

SAXS is particularly well suited to collect structural information on multiple states or conditions of a macromolecule, can be conducted at a higher throughput than other structural techniques, and data acquisition times are often less than a second at synchrotron user facilities. (Yang et al., 2021). Many SAXS beamlines have automated sample loading schemes that allow hundreds of samples to be collected in only a few hours (Grant et al., 2011; Hura et al., 2009; Nielsen, Moller, & Gillilan, 2012; Round et al., 2015; Yang et al., 2021). Sample preparation is also straightforward, requiring microliter volumes of microgram quantities, and measurements can be performed in almost any biologically relevant condition. Additionally, sample preparation for SAXS data collection can also be automated with liquid handling robotics. The throughput and ease of sample preparation and data collection lends itself to conducting many experiments.

SAXS from many conditions that inform on macromolecules can be generated in several ways. Many complex macromolecular systems interact with multiple metabolites for their function. A common example are proteins that couple energy from ATP hydrolysis to other complex reactions (Krukenberg, Street, Lavery, & Agard, 2011; Rosenberg, Deindl, Sung, Nairn, & Kuriyan, 2005; Williams et al., 2011). Because these systems recognize both ATP and other proteins or metabolites and also need to release hydrolyzed ATP and the product of the reaction, they often cycle through a range of conformations. Multiple correlated SAXS measurements have been successfully used to identify which combinations of chemical signals activate different mechanisms. SAXS measurements can screen solution conditions or small molecules that trigger conformational changes. Using a mutation scanning approach on proteins, SAXS can be used to identify amino acid residues essential for facilitating substrate and partner binding and overall conformational change. Amino acids create specific and often nonintuitive interacting networks of hydrogen bonds that enable long distance communication between one region of a protein and another. Mutations are often critical to engineering macromolecules for specific structural and dynamic outcomes, which are readily identified by SAS.

Correlated SAXS data on macromolecules can also be obtained by collecting SAXS data during purification, exposure to light, or mixing. The flux of synchrotron X-ray sources and detector efficiencies ensures SAXS can be monitored as a function of time. The use of component mixing, light exposure, or temperature and pressure jumps can initiate reactions that induce conformational changes over time. For example, microfluidic technologies enable controlled mixing and modifications to a flow of macromolecules which can be examined at different points of the fluid flow (Malaby et al., 2015). Additionally, many macromolecular systems are transient in assembly and conformation. The equilibrium solution state of these transient systems is a mixture that can often be dissected during purification. SEC-SAXS provides access to transient structures from within a larger pool of mixed states and macromolecular species (Malaby et al., 2015; Mathew, Mirza, & Menhart, 2004).

SAS results do not always fit in a similar category as results from cryo-EM, NMR, and MX. At the Research Collaboratory for Structural Bioinformatics Protein Data Bank (RCSB PDB, https://www.rcsb.org/), each entry consists of the atomic coordinates of a single macromolecular conformation. A database with a similar purpose has been developed for SAS data, the Small Angle Scattering Biological Data Bank (SASBDB) (Kikhney, Borges, Molodenskiy, Jeffries, & Svergun, 2020). SASBDB is a valuable SAS community resource, as was its predecessor BioISIS (Hura et al., 2009). However, these databases have generally been created for a single type of SAS experimental configuration and result.

To compliment structural biology tools that provide perspectives on static structures, like the SASBDB, we have implemented Simple Scattering to store correlated data sets and deployed SSMs for the analysis of an important subset of these experiments. We have made the tools we describe in detail, below, accessible through a web browser. The SSMs are accessed at https://sibyls.als.lbl.gov/saxs-similarity/. The SAS data sets are stored at https://simplescattering.com/.

2. Visualization of correlated SAXS data

Integrating information from correlated SAXS data is not trivial. Ideally this integration would occur when atomic models have been determined from each independent measurement, as outlined in Fig. 1. However, as an averaging solution-based technique, many phenomena may be occurring at the same time. In complex macromolecular systems, for example, heterogeneity in conformation and assembly are common. Changes in solution conditions and the presence of substrates or products may push populations of conformations or assemblies toward one form or another. To diagnose these phenomena and extract meaningful biological insights, analysis of a data set as a whole is useful.

2.1 SAXS Similarity Map (SSM) test and demonstration

We have implemented a SAXS Similarity Map (SSM) or heat map approach to visualize correlated SAXS data sets (Hura et al., 2013). These maps are information rich and often provide mechanistic details. The order in which data are presented also often improves analysis. Before we present experimental examples, we will describe how this approach has been implemented and demonstrate its utility on calculated data.

An SSM displays data in a heat map form commonly employed with other types of data. SSMs are essentially symmetric matrices where each entry quantifies the difference between two SAXS curves. The score is represented as a color along a gradient from high to low. The central diagonal of the matrix compares a curve against itself (perfect agreement) and is blacked out to provide greater contrast. Every contributing curve is compared against all other curves in non-diagonal elements. Element 1,2 of the matrix compares the first curve against the second and 2,1 compares the second curve against the first. With most metrics developed to compare SAXS curves, these elements should have the same value.

Several metrics have been developed to compare SAXS curves. The most frequently used is χ^2 as defined in Eq. 1, where q_i is momentum transfer, I_1 and I_2 are the intensities, m_1 and m_2 the intensity means, and σ_1 and σ_2 the errors for separate SAXS measurements. We have developed our own metric called Volatility of Ratio (V_r) which is defined in Eq. 2, where R is the ratio of intensities at q_i (Hura et al., 2013). The mathematical measure of volatility is borrowed from the world of finance (Roll, 1984) and has been useful to describe how much variation there is over a stable trend. V_r is not without shortcomings. For instance, using V_r as a fitness measure for optimization in other applications is not as straightforward as with χ^2. Also, a $\chi^2 < 1$ describes an optimal fit, given the errors in the data, whereas V_r has not been implemented with the error of measurement considered. However, V_r has several valuable characteristics relative to χ^2 for comparison of SAXS data that prove its utility and are demonstrated below.

$$\chi^2 = \sum_{i=1}^{N} \frac{\left(\left(\frac{I_1(q_i)}{m_1}\right) - \left(\frac{I_2(q_i)}{m_2}\right)\right)^2}{\left(\left(\frac{\sigma_1(q_i)}{m_1}\right) + \left(\frac{\sigma_2(q_i)}{m_2}\right)\right)^2} \tag{1}$$

$$R(q) = \frac{I_1(q)}{I_2(q)}, V_r = \sum_{i=1}^{25} \left\| \frac{R(q_i) - R(q_{i+1})}{(R(q_i) + R(q_{i+1}))/2} \right\| \tag{2}$$

The first reason we favor V_r is that it operates on the ratio of two SAXS curves rather than the difference of two curves, as is the case with χ^2. SAXS data decay exponentially as a function of q. Therefore, differences in small q, which have exponentially higher values, can vastly outweigh differences in high q. The rate of exponential decay varies from structure to structure. However, with correlated SAXS measurements of the type typically analyzed in the SSMs—that is, a given macromolecular system in varying experimental conditions—the structures are often similar as is their rate of exponential decay. Thus, a ratio reduces the overwhelming impact of the exponential decay and provides more equal weighting to all regions of q. This can also be over-factored in that a q range of 0.01–0.2 $Å^{-1}$ covers a Bragg spacing of 628–31 Å while a q range of 0.2–0.4 $Å^{-1}$ covers 31–15.7 Å. The former range is frequently more important than the latter in terms of SAXS data. Therefore, our implementation of V_r by default only uses a q range of 0.01–0.2 $Å^{-1}$. A second reason we use V_r is that it bins the data to approximate Shannon sampling. Binning has the effect of increasing signal to noise since most SAXS profiles are over-sampled. The spacing between bins is linear in momentum transfer. Non-linear binning may have advantages that have yet to be explored.

We have calculated SAXS data from each step in a linear geometric morph of a designed protein cage (Lai, Cascio, & Yeates, 2012; Lai, Tsai, Sawaya, Asturias, & Yeates, 2013; Padilla, Colovos, & Yeates, 2001) starting from a compact state and gradually morphing to an expanded and open state (Fig. 2A and B). The cage resembles a closed triangular pyramid. The 11 curves calculated from models along the trajectory have been randomly ordered in an SSM (Fig. 2C, left). The first row in the map coincidently is the first step in the trajectory. Scanning the row of the curve that scored most similar (most red), the second step in the trajectory can be identified. By applying a simple clustering algorithm, the randomized curves align into a meaningful heatmap array that correlates with the closeness of atomic models along the morph trajectory (Fig. 2C, right). The clustering algorithm first finds the two curves with best agreement and pairs them at the start of the matrix list. It then finds the curves with the most disagreement relative to the pair with greatest agreement and places them at the end of the list. Finally, it works from both ends of the list finding the next nearest agreements and adds them to the list from both bottom and top.

The remarkable transformation of the randomized matrix to an ordered one shows that organizing SAXS data by their similarity correlates with organizing similar structures in solution. Additionally, using V_r to quantify

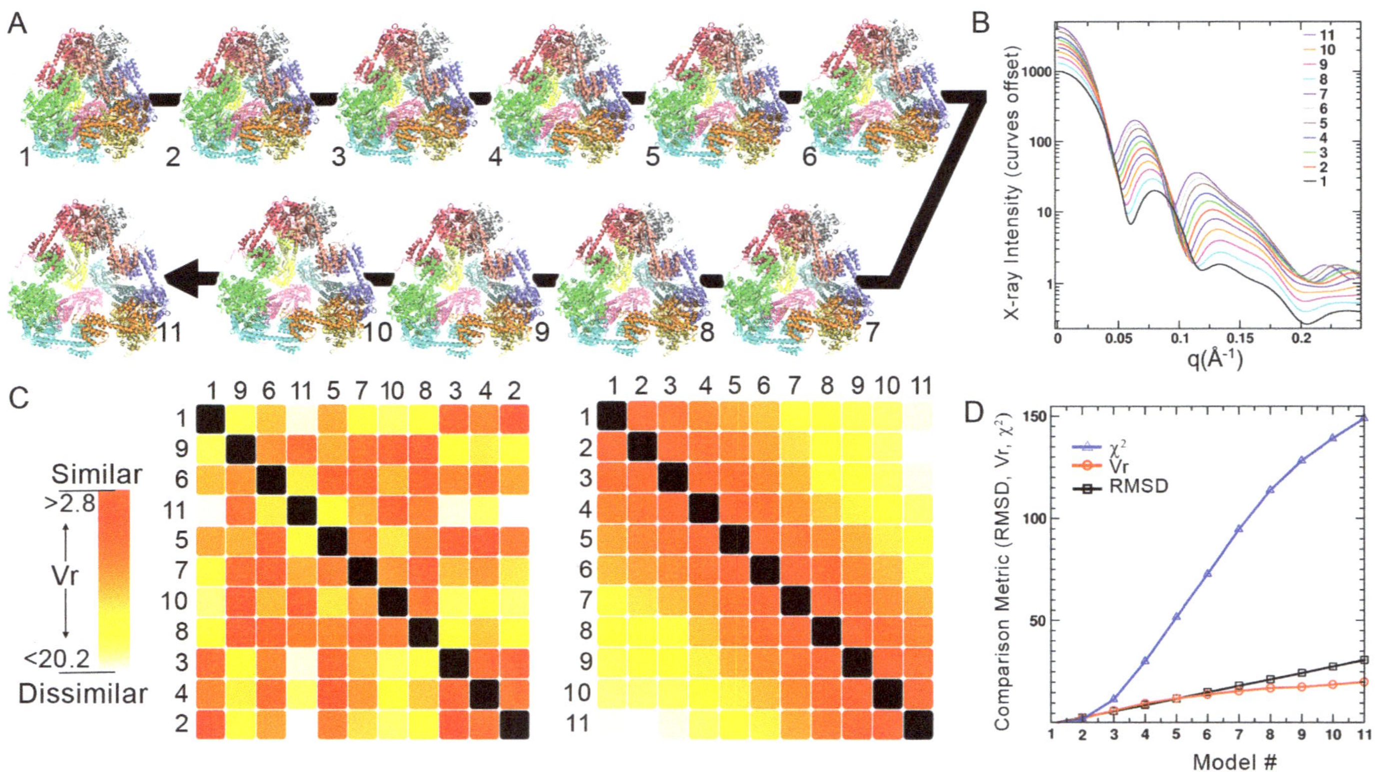

Fig. 2 See figure legend on next page.

SAXS, and therefore structural similarity, has advantages over χ^2. Fig. 2D plots the various comparison metrics (RMSD, V_r, χ^2) vs. models generated from the morph. V_r better captures the linearity of the morph RMSD relative to a χ^2 calculation on the same set of data. χ^2 changes exponentially and, as a result, loses its capacity to identify similarity with even modest changes in the Guinier region of experimental data since that region is weighted so heavily. χ^2 performs more as a binary metric; identifying data sets as either very similar or very different.

2.2 Data requirements and web-based application of SAXS similarity maps

As mentioned above, SSMs can be generated via a web-based interface at https://sibyls.als.lbl.gov/saxs-similarity/. The inputs are ASCI tabulated SAS data that take the form of three columns. The first column lists q, the second X-ray intensity, and the third error in measured intensity. The columns can be tab or space delimited.

Public web-based platforms are targets for hacking and security imperatives change frequently. Currently, pound ("#") symbol headers are tolerated. However, the closer the SAS files are to a simple three column format, the greater the chance they pass evolving security filtering process. Similarly, for security reasons, only files with ".dat" extensions are currently tolerated.

Fig. 2 Similarity of SAXS curves correlates directly to similarity in structure. (A) A designed tetrahedral protein cage was crystallized in a collapsed and compact state (conformation 1) relative to its idealized expanded design (conformation 11). A linear morph was created between the two (intervening conformations). (B) SAXS profiles were calculated from each step in the morph. The curves are offset in X-ray intensity for clarity. (C) These SAXS profiles were used to create a SAXS Similarity Map (SSM) where each cell is a comparison of two profiles (compared curves from each step in the morph are identified left and top of SSM). The similarity, numerically quantified by the metric V_r, is depicted as a color on a red to white gradient with the most similar curves (lowest V_r score) colored red and most dissimilar are white. On the SSM, (C, left) the curves were input in a random order. An automated clustering algorithm was applied to organize the SSM (C, right). (D) The root mean square deviation (RMSD) between the most compact structure (1) and later steps in the morph increases linearly (black squares), in agreement with the algorithm that created the morph. In contrast, the χ^2 agreement between SAXS curves exponentially increases (blue triangles). This extreme deviation complicates the interpretation of how different two structures are based on SAXS profile. The V_r agreement (red circles) follows the linear RMSD behavior more closely between morph steps and provides more correlated guidance on similar structures.

To be useful, more than two SAS profiles are required as the tool is purposed to compare different SAS profiles. Multiple SAS profiles can be selected and dropped into an active area in the web application or a folder can be dropped in the same way. The data do not need to have the same q spacing as an interpolation script is applied to all data sets. Errors in loading specific files are noted in a visual display. Once the simple file requirements are fulfilled and the files are dropped, an SSM is displayed.

The off-diagonal scores are used to adjust the range of the gradient. The two profiles that are most similar set one limit of the gradient while the profiles that score least similar set the other. The gradient can be adjusted with three options: red-yellow-white, blue-white-red and black-gray-white. The current implementation does not allow for further contrast adjustment. However, the SSM can be saved as an image and further manipulated by image processing software to highlight features that may not be visible by the default contrast settings.

The SSM can be manipulated on the web interface in several ways. The order of the files can be adjusted by dragging file names up or down on the list. Once a file is dragged to a different position, the entire SSM is reorganized as indicted with a visible effect. Tabs allow for the automated clustering or reversal of the order of data in the SSM.

More information about the difference between any two files is provided by clicking on the cell that compares the two. Once clicked, the two profiles are plotted below the SSM. The names of the two profiles are listed in the plot along with their calculated radii of gyration. The clicked cell in the SSM also lists the difference in radius of gyration, ΔRg, calculated from the two profiles. All ΔRg can be made visible by using a tab option.

An important capability is the adjustment of q range used for comparison. This capability makes possible the analysis of data with varying amounts of aggregation, other q dependent anomalies or when a specific region of q contains the most discrimination. For example, in drug screening, some drugs may be more insoluble than others. These may create large aggregates that solely affect the small q region. Adjusting the q range so that the small q region is ignored improves analysis. To adjust the q range, a cell must be clicked and, once the two profiles are plotted as described above, a q range selector is displayed for manipulation.

Four different metrics of comparison are available for the SSM, including χ^2. The default is V_r. As V_r has yet to consider relative signal to noise, the SSM may appear asymmetric for low signal data. V_r is ratio-dependent and low signal data will be near zero, thus affecting its scoring. For low signal data, the V_r score using Profile 1 divided by Profile 2 may not be the same as the score using Profile 2 divided by Profile 1. To adjust this asymmetry, the q range must be adjusted so that only the higher signal data is utilized. The matrix of numbers that quantifies the similarity can be viewed using a tab feature.

Tutorials, instructions, and test data are also available at the web site. Users of the web site that have trouble dragging and dropping their files are recommended to download the test data sets. Once downloaded, these data can be uploaded to the web site to generate an SSM, testing the web site's functionality. Once functionality is demonstrated, the test files can be inspected and mimicked in the data sets of interest.

2.3 Similarity Map of substrate binding

Many DNA repair proteins are significant in size and carefully orchestrate events to preserve genome integrity. Simply keeping DNA together is insufficient since several cellular malfunctions can occur with just one nucleotide base slip. Because of this complexity, DNA repair proteins are some of the most complicated macromolecular machines. The original motivating application for developing SSMs stemmed from mechanistic studies of a DNA repair protein MutS that binds multiple substrates and metabolites (Hura et al., 2013). MutS is greater than 100kDa and forms homo- or heterodimers. In most higher organisms, MutS is known as MSH2 and binds the homologous MSH3 or MSH6 to shift specificity for recognizing damaged DNA. MSH2/MSH3, also referred to as MutSβ (Lang et al., 2011), identifies unpaired loops or hairpins while MSH2/MSH6, also referred to as MutSα (Graham, Putnam, & Kolodner, 2018), identifies mismatches. Identification of damaged DNA is followed by a repair cascade with multiple binding partners that derive energy from an ATPase domain. Outstanding questions include how specificity is achieved, the mechanistic role of ATP hydrolysis, specific actions on DNA, and binding with partner proteins during the repair cascade. Knowledge of this system is key for cancer and antibody fields, where DNA mismatches are important for maintaining genomic integrity and generating new antibodies. Achieving a comprehensive understanding of MutS mechanisms with cryo-EM or MX alone has proven very challenging because many structures are necessary but have thus far been challenging to obtain. Correlated SAXS data sets were used to enhance overall understanding of these complex systems.

SAXS data was collected from a ten-member correlated SAXS data set: MutSβ either alone or in the presence of loop DNA with either ADP, ATP, ATPγS, or AMPpMp (Fig. 3). The latter two metabolites are non-hydrolysable analogues of ATP. The first row of the SSM compares the apo MutSβ against the other conditions. The most similar condition to apo MutSβ, along the first row, is in the presence of the substrate it is designed to recognize, a DNA loop. This is surprising as the protein surrounds the DNA in a clamp like manner and must involve a conformational change, though of an apparent lesser degree than the other conditions. For example, a greater dissimilarity in underlying conformation occurs in the presence of ADP and ATP.

Proceeding to the second row, where MutSβ with ADP is compared against all other conditions, a relative similarity is found with the solution

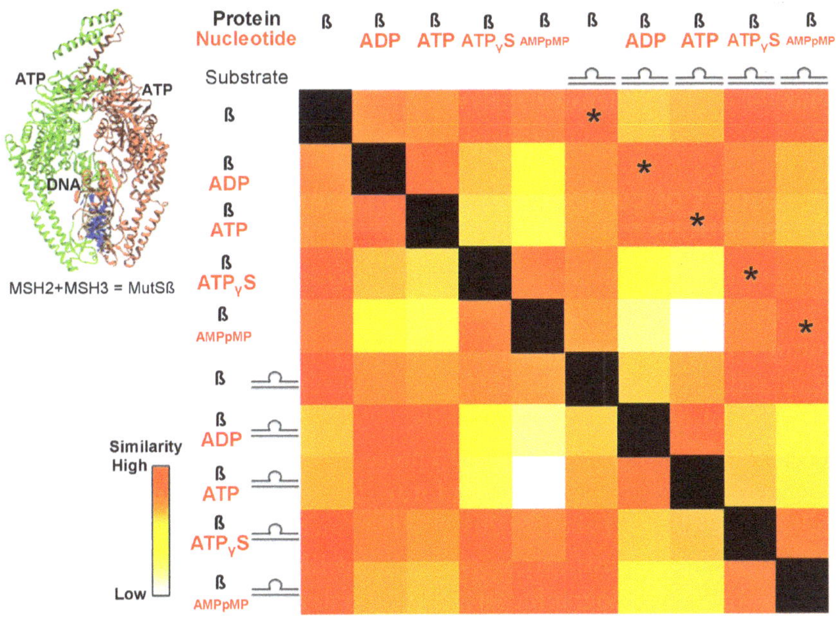

Fig. 3 Conformational changes of MutSβ due to substrate binding. MutSβ aids in repairing DNA loops and is powered by ATP hydrolysis. SAXS collected on MutSβ with and without DNA and with nucleotides ADP, ATP, and non-hydrolysable analogues AMPpMP and ATPγS were compared against one another in an SSM. The asterisked diagonal compares measurements with and without DNA in the presence of the same nucleotide. The indicated similarity across this diagonal highlights that, mechanistically, the nucleotide state determines the conformation of the protein, not DNA binding.

containing ATP and a relative dissimilarity is found with non-hydrolysable ATP analogues. This suggested, and was later corroborated by other methods, that MutSβ hydrolyzes ATP immediately, regardless of whether DNA is present (Hura et al., 2013). However, prior to hydrolysis, the ATP-bound state is in a different conformation that is more similar to the apo-state.

The above conclusions and a primary mechanistic insight about the system are further summarized in the diagonal of similarity, highlighted by the asterisks in Fig. 3. This diagonal compares ATP hydrolyzation states both on and off DNA. The SAXS curves from each ATP analogue, regardless of whether DNA is bound, are similar. Having verified that DNA is indeed bound using gel electrophoresis, the implication is that conformation is fixed by the ATP hydrolysis state and that its recognition process relies on finding DNA that is able to conform to this structure. This can be rationalized by understanding that DNA possessing a loop is more malleable than fully complementary DNA. SSM analysis requires ordering the conditions in a specific way and a careful examination for patterns. The benefit is that each measurement is represented and can be placed in the context of all other measurements for a comprehensive analysis.

2.4 Studies of conformation and assembly based on buffer conditions

Solution conditions influence macromolecular assembly and conformation. With designed proteins, understanding these factors can help improve or control the system (Lai et al., 2016). The protein cage system shown in Fig. 2 and Section 2.1, was designed, expressed and purified for further experimental analysis using correlated SAXS. A monomer in the dodecameric cage assembly is a fusion of a protein that forms trimers (the vertices of the pyramid) with a protein that forms dimers (the edges of the pyramid) (Padilla et al., 2001) (Lai et al., 2012, 2013). Successful formation of the pyramid was first demonstrated by cryo-stained EM and later by crystallography. In the crystallographic studies several crystal forms were found. However, none of the crystal structures matched the idealized pyramid. All of them were collapsed to varying extents, with compressed edges and a smaller internal cavity than expected from the initial design.

To determine whether salt or pH in the varying crystal forms affected the conformation of the cage we performed an extensive salt and pH screen. We modified the NaCl concentration using 10, 100, 300 and 500 mM.

We also analyzed each of these salt concentrations in single pH steps from pH of 4–11. This resulted in a total of 28 conditions. An SSM of these conditions is shown in Fig. 4.

The most obvious feature from the overall SSM is a grid like pattern. The vertical and horizontal strips of yellow and white occur at pH values of 10 and 11. Applying the clustering algorithm, all curves with pH of 10 and 11 cluster together (not shown).

Interpretation of SSMs is aided by including SAXS calculated from models expected in solution. In Fig. 4B we show an SSM of the 300 mM NaCl pH series including SAXS calculated from a model of the disassembled trimeric vertices of the cage and a crystal structure of the compact cage. At high pH the agreement to the model of the trimer is high, indicating that pH weakens the interaction between the dimer forming part of the fusion. Therefore, increasing pH can be used to disassemble the system. An SSM

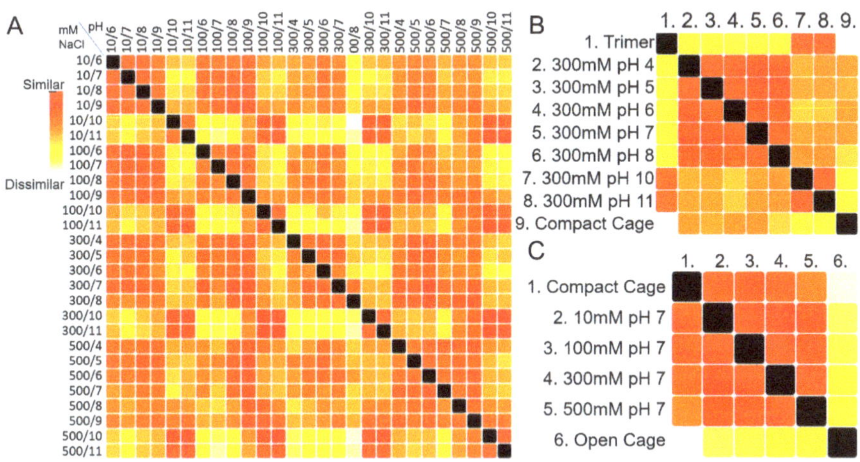

Fig. 4 Protein cage assembly can be controlled by NaCl concentration and pH. (A) SAXS data was collected on a dodecameric protein cage (atomic model shown in Fig. 2) as a function NaCl concentration and pH. These are identified at the top and left of the SSM using a mM NaCl/pH format. The striking grid like patterns in the SSM can be further understood looking at subsets of the data (B) pH can be used to control assembly. Looking at an SSM of a subset that varies pH but holds salt concentration constant (300 mM) shows that at the high pH values (10 and 11) the SAXS curves are most similar to that calculated from a trimeric disassembly product rather than the full cage. (C) Salt concentration modifies conformation of the cage. Comparing solutions that held a pH of 7 but varying salt concentration shows that the structures in solution are most similar to a compact state. However, increasing salt moves the overall structure toward the designed open state.

representing pH 7 with varying salt concentration (Fig. 4C) reveals that higher salt concentration increases the openness of the cage. Interpretation is aided by including theoretical SAXS curves calculated from the compact cage and the idealized open cage. A gradient in similarity toward the open conformation is revealed as salt concentration is increased. Changing the salt concentration is therefore a means to control cage expansion. While the designed conformation was not observed with this construct in any of the solution conditions tested, the results suggest that modifying the charges on the inside of the cage may help form the more open structure. With the current construct, salt and pH allow for significant control of assembly and conformation.

2.5 Structural impact of point mutations

Screening mutations in a protein has many applications in biology. In some cases, a single point mutation can inactivate or unfold a protein. Mutations can also affect affinity, allosteric ability to undergo structural rearrangement or multimerization. SAXS is sensitive to these types of changes.

The structural impact of mutations upon the protein Apoptosis-inducing factor (AIF) was assayed by SAXS (Brosey et al., 2016). A mitochondrial oxidoreductase, AIF facilitates metabolism by supporting import of OXPHOS components into mitochondria (Hangen et al., 2015). However, during DNA damage and PARP-1 hyperactivation, AIF is released from mitochondria and participates in a cell death pathway (Fatokun, Dawson, & Dawson, 2014; Morales et al., 2014). AIF functional regulation is believed to derive from architectural switching stimulated by binding of the metabolite NADH. NADH allosterically triggers AIF dimerization and release of a 50-residue surface loop (C-loop) from the C-terminal domain. Because the NADH binding site is distant from AIF's dimerization interface and C-loop (Fig. 5A), the "information" of NADH binding is communicated via intervening amino-acid cascades. Identifying amino acids along these cascades opens additional avenues for regulating AIF switching and functional responses. To define the amino acid cascade and controlling mechanisms of NADH responses, HT-SAXS was performed on a panel of AIF mutants with and without NADH.

An SSM helps characterize function-breaking mutations (Fig. 5B and C). In the absence of NADH, wild-type AIF is monomeric. However, mutants which remove the C-loop (ΔC-loop) or target residues S480, H454, W196, D485, and R529 resemble the dimeric NADH-bound wild-type state, pointing to their role in allosteric communication (Fig. 5B, *left*). In contrast,

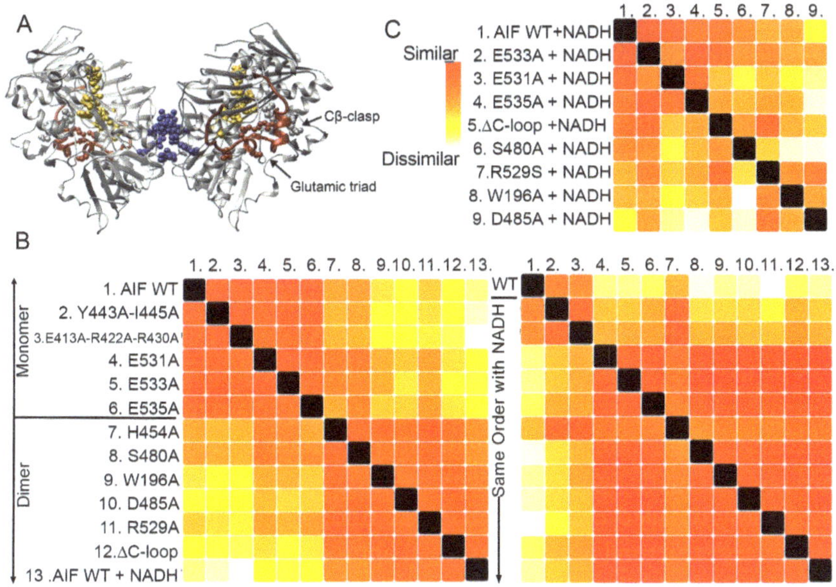

Fig. 5 SAXS uncovers allosteric communication between a metabolite binding site and dimer interface. (A) The protein AIF binds the metabolite NADH at its central active site (gold), triggering dimerization at a distal surface (blue) and release of a surface C-loop (red). Allosteric residues targeted for mutation are shown as spheres. Dimeric AIF supports OXPHOS biogenesis, a target of interest for cancer treatment. (B) SSM analysis of AIF mutants relative to wild-type (WT) monomeric AIF without NADH (left panel) distinguishes AIF mutants consistent with the AIF monomer (top) from AIF mutants that dimerize without NADH (bottom). Complementary SSM analysis of NADH-saturated AIF mutants (right panel) suggest high similarity to the wild-type AIF dimer except for Y443A-I445A and E413A-R422A-R430A (defective for dimerization) and H454A (defective for C-loop release). (C) Extended SSM analysis of NADH-saturated AIF mutants relative to wild-type AIF dimer conformationally partitions mutants of AIF's glutamic acid triad and Cβ-clasp.

mutations at the dimer interface, E413A-R422A-R430A and Y443A-I445A, prevent dimer formation, regardless of the presence of NADH (Fig. 5B, *right*). This result identified residues Y443 and I445 as novel contributors to the AIF dimerization interface. Notably, the H454A mutant exhibits intermediate similarity to wild-type AIF monomer and dimer. Complementary biochemistry experiments revealed that this mutant obligately dimerizes but does not release its C-loop in response to NADH (Brosey et al., 2016), uncovering its unique role in linking the active site and dimerization interface. Thus, SSM analysis also highlights mutants that present unusual and informative phenotypes.

To extract further detail, SSMs can be recalculated by systematic removal of outliers (here Y443A-I445A, E413A-R422A-R430A, and H454A). In Fig. 5C, NADH-saturated AIF mutants that successfully form dimers and release the C-loop can be compared directly to NADH-saturated wild-type AIF dimer to distinguish subtle impacts from individual mutations. As before, SSM analysis partitions different classes of allosteric mutants. The glutamic acid triad (E531, E533, E535) forms a network of exterior salt bridges securing the C-loop to AIF's surface, while residues of the Cβ-clasp (R529A, W196, D485) form a complex allosteric hydrogen bond network that is disrupted to release the C-loop upon NADH binding. Here, the SSM analysis for the wild-type dimer ranks collective similarity of the dimeric glutamic acid mutants higher than the dimeric Cβ-clasp mutants, suggesting a difference in conformation or monomer-dimer equilibria. As such, the SSM results frame and prioritize mutant classes for follow-up analyses, whether by detailed examination of the underlying SAXS curves or complementary structural and biochemical approaches.

2.6 Validating protein engineering designs with SSMs

A major motivator for establishing HT-SAXS is to confirm macromolecular designs from computationally driven macromolecular engineering. Recent machine learning computational programs have been successful in predicting fold based on sequence (Alexander et al., 2021; Baek et al., 2021; Jumper et al., 2021). Tackling the converse problem, desired structure to predicted sequence, has also been successful in several cases. Many engineered proteins express well, making them highly amenable to protein expression pipelines. The Baker lab and other macromolecular engineers have employed SAXS for experimental validation of their designs (Boyken et al., 2016; Chen et al., 2020; Langan et al., 2019).

In a foundational study on the simple alpha helix-turn-helix motif, the Baker lab has investigated the possible topologies that could be created (Brunette et al., 2015). Surprisingly many structures not yet observed in natural peptides seemed computationally feasible. Several of the topologies are shown in Fig. 6A. To test whether these computationally designed structures could be stabilized, sequences for over 40 topologies were expressed, purified, and verified by SAXS. Several were also verified from crystal structures.

An SSM was used to rapidly analyze the success of the computational designs (Fig. 6B). All experimental profiles are listed first followed by the

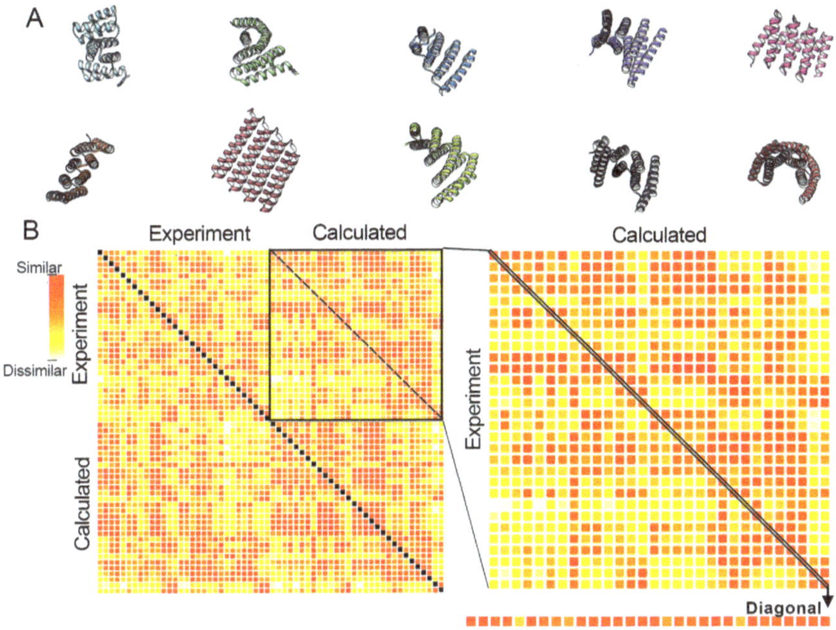

Fig. 6 Validating protein engineering principles with SSMs. Using a repeat helix-turn-helix design, over 80 protein topologies were computationally created. The designed topologies varied stacking, torsions, and angles between helices. Examples of topologies are shown in (A). Sequences from designed topologies were expressed, purified, and were measured by HT-SAXS. An SSM of 30 measured SAXS profiles and the corresponding calculated SAXS profiles is shown in (B). The SSM is ordered with all measured SAXS profiles in the first half followed by the calculated. The calculated profiles are in the same order (top to bottom) as the measured. Success could be assessed by analyzing the diagonal of the top right-hand quadrant of and SSM from all profiles, which is blown up on the right. The mostly red diagonal in this quadrant, which is further extracted in a linear form at the bottom, indicates success in many designs. Off-diagonal red squares indicate either poor SAXS discrimination between the topologies or sequences that favor other folds, as diagnosed by other quadrants.

calculated profiles in the same order as the experimental profiles. The SSM can be analyzed in quadrants; top left, top right, bottom left, and bottom right. The bottom right compares all the calculated scattering curves against one another. Given that constructs are approximately the same size and the overall shapes are mostly flat and elongated, the degree of variation in similarity in the SAXS curves confirms SAXS sensitivity to high resolution information. Some of the topologies create more unique SAXS profiles than others.

Experimental evidence for a match to the computational designed topologies can be observed in several ways from the SSM. The matching of patterns in the SSM of the top left quadrant (experiments against experiments) vs. the bottom right (computation against computation) is one perspective. A more direct approach is examining the top right quadrant (Fig. 6B, *blowup right*) where experimental data is compared to computation data. Since the topologies are listed in the same order, the agreement can be assessed by the degree to which there is a diagonal of similarity. The diagonal can be extracted to identify topologies that failed (Fig. 6B, *bottom*). These can be reviewed in detail for improvements in sequence or to help identify constraints on topology that have yet to be understood. The SSM therefore provides both a global perspective on the ability to form novel helix-turn-helix motifs and highlights specific instances for improvement in design.

3. Simple Scattering data set repository

The applications of SSMs described above is a necessary first step in a deeper analysis of those data sets. The drag and drop option of SSMs on a web browser make their application straightforward and potentially provide non-experts a tool to highlight outstanding conditions or constructs from a SAXS perspective. However, many other tools have been developed and could be applied in their analysis. Moreover, it is our sincere hope that new tools will continue to develop to extract further information from these rich data sets. To promote further analysis, we have deposited the data sets described above into the Simple Scattering web repository. The repository is available to store and access SAS data collected by others and we describe how SAS data can be deposited in Simple Scattering below.

Correlated SAS data can be generated in many ways, as described in the introduction. Besides the SSM examples described above that were collected by high-throughput means, Simple Scattering is also designed to hold correlated SAS data generated by other methods. Currently, Simple Scattering is also designed to hold SEC-SAS, time-resolved SAS (TR-SAS), and SAS from microfluidics. As a developing resource, to accommodate these various forms of data, data entry has significant flexibility with many optional fields while also accepting meta-data in various formats. This reduces the time to deposit a data set which is valuable for a technique that generates data quickly. The Simple Scattering client interface has been designed to allow automated deposition from particular types of correlated experiments directly from instruments, as described below. The repository was designed with a singular purpose to hold data rather than perform analysis on the data itself. From its inception Simple Scattering has also been designed with the goal of achieving FAIR (Findable, Accessible, Interoperable, and Reusable) principles of data management (Wilkinson et al., 2016).

Simple Scattering's purpose is distinct from the existing SAS repository SASBDB. The SASBDB (Kikhney et al., 2020) at https://www.sasbdb.org/ and its predecessor BioISIS (Hura et al., 2009) hold individual SAS curves that are analyzed to as full an extent as possible including atomic models derived from the SAS curves. The SASBDB and BioISIS resources are modeled after the PDB. As SASBDB and PDB entries are partially curated, database entry typically requires significant time. These databases were not specifically designed to store multiple data sets that characterize multiple correlated states. Simple Scattering, on the other hand, has been expressly designed to hold tens to hundreds of SAS curves from a correlated set of experiments collected from samples in varying solution conditions, related but slightly different constructs, or representing different sample preparation regimes (as exemplified in Section 2).

Simple Scattering provides several advantages over general purpose data repositories like Zenodo (https://zenodo.org/). As a niche site, such as SASBDB and the PDB, Simple Scattering provides a publicly available repository for researchers to discover large correlated SAS data sets. Simple Scattering provides control over how data is deposited, uniformity over how data may be displayed and the ability to further fine tune the site for future access by other tools. As noted above, we anticipate increased generation of correlated SAS data sets with a corresponding need for improved analysis. By building the database, we provide access and draw further attention to data set types that require the development of unique tools to complete their analysis.

Once deposition is complete an entry is paired with a unique identifier similar to those generated by the PDB and SASBDB. Simple Scattering identifiers start with the letters "XS" followed by six unique alphanumeric ASCII characters. A search tool is available to find an entry by the identifier among other key words described in proceeding sections.

3.1 Simple Scattering user storyboard

The current workflow for a user depositing correlated scattering data for public access has four main steps. (1) A data depositor sets up an account that is verified by the maintenance staff, (2) a single or correlated set of one-dimensional SAS data is uploaded by the depositor through a typical web-based graphical user interface or by an automated process from an instrument and is assigned a unique identifier, (3) the depositor adds descriptions and relevant meta-data, and (4) maintenance staff validate the deposition and release the entry for public access.

For the public, the latest entries may be found on the Simple Scattering "Dashboard" page, along with site statistics and blog posts. Entries for all available data are listed on the Simple Scattering "Open Datasets" pages. If the unique identifier code for a particular set of data is known, it can be appended to the site's URL after a slash (e.g., https://simplescattering.com/open_dataset/XSYY8DN2) and the page will be displayed. Alternatively, the code can be entered in the search bar atop each page to retrieve the entry. Data can also be identified by searching keywords associated with the deposition. The SAS data and any associated supplementary data can then be downloaded by clicking a link.

Data presented in Simple Scattering is intended as a supplement to a publication. In its current form, SAS data deposited in Simple Scattering is not extensively reviewed. Automated validation and analysis tools have yet to be developed and linked to the deposition process. Therefore, details associated with the Simple Scattering entry, such as data collection method and the file nomenclature used to delineate each SAS profile is the responsibility of the depositor. Ideally, the data set is associated with a reviewed publication and provides readers and reviewers access to the data described in the publication. To this end, depositors can update their deposition at any time so long as their Simple Scattering account is valid.

3.2 Deposition of data to Simple Scattering

Deposition of data into Simple Scattering can be approached in two ways. The first is by manual entry online through a web graphical user interface (Fig. 7A); the second is through an application programming interface (API) (Fig. 7B). The latter method lends itself to automation, which matches the pace of data collection at modern facilities wielding high-throughput configurations. The API is currently implemented at the SIBYLS beamline for its SEC-SAXS users. We will describe both processes below.

To manually deposit scattering data the user must register for an account. Registration requires the user's full name and email address, followed by the user's project leader, institute, and the country that the researcher belongs to. To deposit data, the account requires vetting of the user by the Simple Scattering system administration.

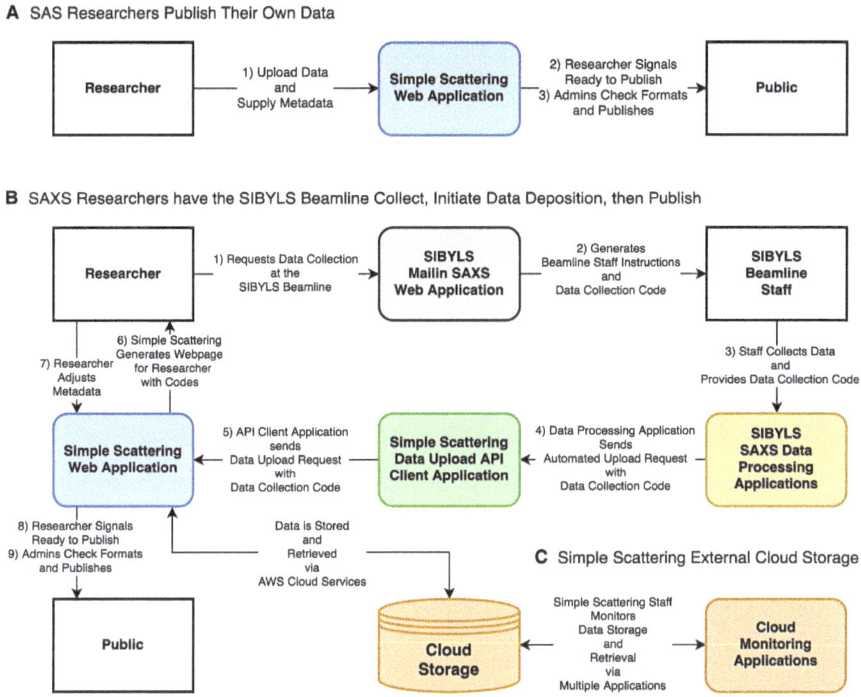

Fig. 7 Workflow for upload of data to Simple Scattering for SIBYLS beamline. (A) Upload and release of data from the Simple Scattering web application. (B) In a more automated pipeline, beamline users first request beamtime through our MailinSAXS site (https://bl1231.als.lbl.gov/htsaxs), which generates instructions and a unique data collection code. Beamline staff then collects data and converts image readouts to individual files that contain the scattering angle, intensities, and errors using in-house data processing applications. These applications can then initiate a data upload request to Simple Scattering on behalf of the user through a data upload API client application. Simple Scattering then generates its unique code for the data set and makes a request to store the data on Amazon Web Services. Upon success, Simple Scattering generates a draft web page for the beamline user, who then adds additional metadata and supplementary files. Upon publishing the data, the web page for the particular data set is made accessible to site visitors, and data can be retrieved from storage and sent to the visitor by clicking a link. (C) Staff monitors cloud storage through AWS cloud monitoring applications.

Once registered, the user uploads data sets by filling out a minimal set of required fields: Title of experiment, Data collection technique, Experimental description, File description. A user may also add additional metadata about the macromolecule, or in the case of complexes, macromolecules, including but not limited to: the sample type (protein, DNA, lipid, etc.), the sample name, the abbreviated name, and its polymer sequence (amino acid or nucleotide sequence) or chemical formula. Lastly, at minimum, a compressed (zip) file containing the data is uploaded. Other supporting files may also be uploaded.

The compressed file containing the data should consist of individual plain text files, one for each data collection experiment (if a single experiment was performed, the file should still be compressed). Each file within the compressed zip file should contain plain text, with three columns representing the momentum transfer, intensities and errors, in that order. Or alternatively, with two columns representing the momentum transfer and intensities. Headers or footers may be included in the individual data files provided each line is preceded by a hashtag or pound ("#") symbol. Each file within the compressed zip file should have the extension ".dat" or ".txt." To note, currently there is a memory limit of 20 MB for the compressed file. The raw data collection image files should not be included in the compressed file or any other files. These raw files could by uploaded to a service like Zendo (see above) and the unique identifier could be noted in the deposition.

In addition to the compressed file containing the data, researcher's may upload additional files. First, up to 100 individual experimental files contained within the compressed file may be uploaded. The purpose for this is so that visitors to the site may download one or a few files to inspect prior to downloading the larger data set. Second, researchers may add additional files that support the data set. This may include atomic coordinates generated by modeling or experimental methods such as MX, NMR, cryo-EM (.pdb or .cif files), or other data collected while running a size-exclusion column such as multi-angle light scattering (MALS) or quasi-elastic light scattering (QELS). Additionally, plain text or portable document format (.pdf) files with additional explanations or insights for the experiment may be included. MALS or QELS may be submitted in the form of portable document format (.pdf) files, or comma separated value (.csv) files. The Experimental description field is free text and serves a similar purpose as the methods and materials section in journals. Sample preparation information, the conditions tested, and the parameters that remained constant throughout the experiment are valuable for future analysis. Details of where and how the data was collected should also be included.

The file description field is intended to allow researchers to describe the file naming convention of their data files within the main compressed file. This does not preclude investigators adding pound tagged headers or footers inside the files themselves. If non-descriptive file names or headers/footers are used, the file description field will need to contain a file by file description of the SAS filenames. For a simple SEC-SAXS experiment, noting that the files are sequentially numbered according to the elution is sufficient. For more complex experiments, the investigators downloading data sets may need indicators that distinguish one file from another in a correlated data set. Often file base names provide a sufficient number of characters to be descriptive and are a useful way to distinguish one file from another. For example, in the case of the metabolite screen described in Section 2.2, the file named MutSBeta_DNA_ATP.dat is the SAXS profile collected from protein MutSβ mixed with hairpin DNA and ATP in contrast to the file named MutSBeta_ADP.dat which is the SAXS profile of MutSβ mixed with ADP. For the salt and pH screen conducted on the cage system described in Section 2.3, the files have names in the form of 10mM_pH7_PCTrip.dat. The mutation analysis (Section 2.4) contains the one-letter code mutations in the name.

Once all required information is input, the depositor can upload the data, and signal the Simple Scattering system administrators that a data set is either still in draft mode or ready for release to the public. System administrators can approve a data set for release or contact the user if there are issues with the submission. Depositors can continue to edit the information contained in their submission as necessary.

Simple Scattering also includes a private API that allows beamline scientists collecting data for their users to upload their data on their behalf to ease the deposition process. The API is implemented at the SIBYLS beamline at the Advanced Light Source to deposit SEC-SAXS data through an automated deposition system (Fig. 7B). When the SIBYLS beamline is operating, SEC-SAXS data are collected at a rate of approximately 20 samples per week. The deposition of an SEC-SAXS data set is accomplished by first using data processing software. This is followed by using a Python-based data upload API client application that sends requests to the currently private Simple Scattering API. The private API upload requires an email address and the location of the SAXS data files. The data is transferred off the local beamline file system to cloud storage, mediated by the Simple Scattering database. A unique identification code is generated for the data set and data availability is communicated to the user. At this stage the data remain private. However, the user is able to enter the other required information (meta data) described above. Once the user has filled out all required fields, the data can be submitted for approval by Simple Scattering system administration and publicly released.

3.3 Simple Scattering infrastructure

Simple Scattering is hosted on the Heroku cloud platform and offered to the public by the SIBYLS beamline. The frontend servers are provided via Heroku and as such the details of administering, maintaining and upgrading the physical frontend servers is left to them. By using an enterprise class cloud service provider such as Heroku, we are able to offload much of the nitty gritty details of server configuration, database maintenance, and service scalability to Heroku. If the demands on server capacity or database size/speed increase, Heroku provides the ability to seamlessly scale Simple Scattering should demand necessitate.

Simple Scattering is a Ruby on Rails application whose Object-Relational Mapping system managed by Active Record is coupled to a PostgreSQL relational database management system. Ruby on Rails is a pre-packaged server-side (backend) framework that utilizes Ruby to support the model-view-controller design pattern that is frequently used in web applications. It also includes the tooling for the HTML, CSS, and JavaScript, a high-level asynchronous programming language, required for client-side (frontend) programming to generate the web pages that outside users interact with. All metadata is stored in the PostgreSQL database provided as part of the Heroku service and is protected by both a local snapshot disaster recovery backup called "Postgres Continuous Protection" and a longer-term offsite backup solution called "PG Backups." All data files uploaded to Simple Scattering are stored in Amazon's Web Service (AWS) high durability, high availability S3 standard buckets which provide 99.999999999% durability and 99.99% availability over a given year (https://aws.amazon.com/s3/storage-classes/). We additionally monitor data flow to the cloud servers though Amazon's CloudWatch and CloudTrail services, and maintain separate backups of both the database and data. To connect to the API, a set of Python programs are used.

As the minimal core functionality of the Simple Scattering site has been completed, there are many avenues for improvements and new features. Future directions and features will be user focused, and we hope to include integrations with other beamlines. Suggestions include a public API for deposit, searching records and to download files. The ability to automate fetching data may allow other applications such as SAXS Similarity to access data from the unique Simple Scattering identification codes.

3.4 Current statistics from the Simple Scattering database

If Simple Scattering were to remain static with the current deposits, it is the largest public repository of correlated SAXS and SEC-SAXS data sets thus far. For developers of software tools, Simple Scattering is therefore a unique source of experimental data. As the site is just now being advertised to researchers, there are ~40 users. Thus far, over 180 SEC-SAXS data sets (each containing over 600 scattering profiles) have been deposited with nearly 40 released to the public. Three of the correlated SAXS data sets, reported above in Section 2, are also available. The Simple Scattering unique identifiers for the MutSβ, PCtrip, and AIF studies are XSYY8DN2, XSSQ87OY, and XSUTRZQL, respectively. Site statistics reveal that others from around the world are already accessing the sites, as anyone can download data without registration.

As noted above, there is a dashboard, https://simplescattering.com/dashboard, that shows the most recent uploads, the latest news and statistics about the data contained within the database. Currently, project titles, macromolecular names, and data set codes are searchable through the search field in the site's navbar. More advanced search features and page views will be available in the near future.

4. Summary and outlook

Macromolecular structural biology has made tremendous gains in recent years, including advances in cryo-EM and machine learning structure prediction such as AlphaFold2 (Jumper et al., 2021) and RoseTTAFold (Humphreys et al., 2021). These technologies, in conjunction with MX and NMR, are capable of providing near atomic resolution models of most soluble macromolecules. The remaining open challenges in characterizing macromolecules has shifted in light of these capabilities. One of the challenges receiving renewed focus is to derive an understanding of mechanisms given an atomic structure.

SAXS is well positioned to help address this challenge. Not only do SAXS data provide a means to confirm an atomic resolution model in solution (Fig. 1), but SAXS data can also be collected from macromolecules in hundreds of contexts. Above, we showed SAXS data collected from macromolecules as a function of multiple substrates (Section 2.3), pH and salt concentration (Section 2.3), and mutations (Section 2.5). When the data from multiple SAXS measurements are integrated together and displayed as an SSM, the conditions that maintain a structure or induce changes therein are readily apparent. In some cases, the SAXS calculated from atomic models are helpful as a reference. Conditions that deviate from this reference can be identified and further investigated. The available atomic model is modified to better agree with the SAXS data and provides insights into the factors necessary to change conformation or assembly. These approaches are powerful and can be applied to the engineering of designed macromolecules where the effect of many sequence modifications can to be tested for desired outcomes.

To provide access to correlated SAS data sets for further verification and analysis we have introduced the data depository Simple Scattering. Many of the data sets can be analyzed by SSMs or other algorithms designed for this purpose. SEC-SAXS has emerged as a powerful approach and analysis tools continue to develop and improve. We hope other synchrotron beamlines will also join in the effort in depositing SEC-SAXS data using the API. Providing access to many SEC-SAXS, HT-SAXS, SANS, microfluidic SAXS, and TR-SAXS data sets is expected to spur development of computational tools and drive sophisticated SAS analysis to advance our understanding of macromolecules and their mechanisms.

Acknowledgments

The acquisition of data at the SIBYLS beamline and the development of the web-based applications were largely funded by the Department of Energy, Office of Basic Energy Sciences, through the Integrated Diffraction Analysis Technologies (IDAT) program, supported by the DOE Office of Biological and Environmental Research under contract DE-AC02-05CH11231. Additional funding was provided by the National Cancer Institute (NCI) grants for Structural Biology of DNA Repair (SBDR) PO1 CA092584, CA220430, and the National Institute of Health project, ALS-ENABLE (P30 GM124169).

References

Alexander, L. T., Lepore, R., Kryshtafovych, A., Adamopoulos, A., Alahuhta, M., Arvin, A. M., et al. (2021). Target highlights in CASP14: Analysis of models by structure providers. *Proteins: Structure, Function, and Bioinformatics, 89*(12), 1647–1672. https://doi.org/10.1002/prot.26247.

Baek, M., DiMaio, F., Anishchenko, I., Dauparas, J., Ovchinnikov, S., Lee, G. R., et al. (2021). Accurate prediction of protein structures and interactions using a three-track neural network. *Science, 373*(6557), 871–876. https://doi.org/10.1126/science.abj8754.

Boyken, S. E., Chen, Z., Groves, B., Langan, R. A., Oberdorfer, G., Ford, A., et al. (2016). De novo design of protein homo-oligomers with modular hydrogen-bond network-mediated specificity. *Science, 352*(6286), 680–687. https://doi.org/10.1126/science.aad8865.

Brosey, C. A., Ho, C., Long, W. Z., Singh, S., Burnett, K., Hura, G. L., et al. (2016). Defining NADH-driven Allostery regulating apoptosis-inducing factor. *Structure, 24*(12), 2067–2079. https://doi.org/10.1016/j.str.2016.09.012.

Brunette, T. J., Parmeggiani, F., Huang, P. S., Bhabha, G., Ekiert, D. C., Tsutakawa, S. E., et al. (2015). Exploring the repeat protein universe through computational protein design. *Nature, 528*(7583), 580–584. https://doi.org/10.1038/nature16162.

Chen, Z., Kibler, R. D., Hunt, A., Busch, F., Pearl, J., Jia, M., et al. (2020). De novo design of protein logic gates. *Science, 368*(6486), 78–84. https://doi.org/10.1126/science.aay2790.

Fatokun, A. A., Dawson, V. L., & Dawson, T. M. (2014). Parthanatos: Mitochondrial-linked mechanisms and therapeutic opportunities. *British Journal of Pharmacology*, *171*(8), 2000–2016. https://doi.org/10.1111/bph.12416.

Graham, W. J. T., Putnam, C. D., & Kolodner, R. D. (2018). The properties of Msh 2-Msh6 ATP binding mutants suggest a signal amplification mechanism in DNA mismatch repair. *Journal of Biological Chemistry*, *293*(47), 18055–18070. https://doi.org/10.1074/jbc.RA118.005439.

Grant, T. D., Luft, J. R., Wolfley, J. R., Tsuruta, H., Martel, A., Montelione, G. T., et al. (2011). Small angle X-ray scattering as a complementary tool for high-throughput structural studies. *Biopolymers*, *95*(8), 517–530. https://doi.org/10.1002/bip.21630.

Hangen, E., Feraud, O., Lachkar, S., Mou, H., Doti, N., Fimia, G. M., et al. (2015). Interaction between AIF and CHCHD4 regulates respiratory chain biogenesis. *Molecular Cell*, *58*(6), 1001–1014. https://doi.org/10.1016/j.molcel.2015.04.020.

Humphreys, I. R., Pei, J. M., Baek, M., Krishnakumar, A., Anishchenko, I., Ovchinnikov, S., et al. (2021). Computed structures of core eukaryotic protein complexes. *Science*, *374*(6573), 1340. https://doi.org/10.1126/science.abm4805.

Hura, G. L., Budworth, H., Dyer, K. N., Rambo, R. P., Hammel, M., McMurray, C. T., et al. (2013). Comprehensive macromolecular conformations mapped by quantitative SAXS analyses. *Nature Methods*, *10*(6), 453–454. https://doi.org/10.1038/nmeth.2453.

Hura, G. L., Menon, A. L., Hammel, M., Rambo, R. P., Poole, F. L., Tsutakawa, S. E., et al. (2009). Robust, high-throughput solution structural analyses by small angle X-ray scattering (SAXS). *Nature Methods*, *6*(8), 606–612. https://doi.org/10.1038/nmeth.1353.

Jumper, J., Evans, R., Pritzel, A., Green, T., Figurnov, M., Ronneberger, O., et al. (2021). Highly accurate protein structure prediction with AlphaFold. *Nature*, *596*(7873), 583–589. https://doi.org/10.1038/s41586-021-03819-2.

Kikhney, A. G., Borges, C. R., Molodenskiy, D. S., Jeffries, C. M., & Svergun, D. I. (2020). SASBDB: Towards automatically curated and validated repository for biological scattering data. *Protein Science*, *29*(1), 66–75. https://doi.org/10.1002/pro.3731.

Konarev, P. V., & Svergun, D. I. (2015). A posteriori determination of the useful data range for small-angle scattering experiments on dilute monodisperse systems. *IUCrJ*, *2*(Pt 3), 352–360. https://doi.org/10.1107/S2052252515005163.

Krukenberg, K. A., Street, T. O., Lavery, L. A., & Agard, D. A. (2011). Conformational dynamics of the molecular chaperone Hsp90. *Quarterly Reviews of Biophysics*, *44*(2), 229–255. https://doi.org/10.1017/s0033583510000314.

Lai, Y. T., Cascio, D., & Yeates, T. O. (2012). Structure of a 16-nm cage designed by using protein oligomers. *Science*, *336*(6085), 1129. https://doi.org/10.1126/science.1219351.

Lai, Y. T., Hura, G. L., Dyer, K. N., Tang, H. Y., Tainer, J. A., & Yeates, T. O. (2016). Designing and defining dynamic protein cage nanoassemblies in solution. *Science Advances*, *2*(12), e1501855. https://doi.org/10.1126/sciadv.1501855.

Lai, Y. T., Tsai, K. L., Sawaya, M. R., Asturias, F. J., & Yeates, T. O. (2013). Structure and flexibility of nanoscale protein cages designed by symmetric self-assembly. *Journal of the American Chemical Society*, *135*(20), 7738–7743. https://doi.org/10.1021/ja402277f.

Lang, W. H., Coats, J. E., Majka, J., Hura, G. L., Lin, Y. Y., Rasnik, I., et al. (2011). Conformational trapping of mismatch recognition complex MSH2/MSH3 on repair-resistant DNA loops. *Proceedings of the National Academy of Sciences of the United States of America*, *108*(42), E837–E844. https://doi.org/10.1073/pnas.1105461108.

Langan, R. A., Boyken, S. E., Ng, A. H., Samson, J. A., Dods, G., Westbrook, A. M., et al. (2019). De novo design of bioactive protein switches. *Nature*, *572*(7768), 205–210. https://doi.org/10.1038/s41586-019-1432-8.

Malaby, A. W., Chakravarthy, S., Irving, T. C., Kathuria, S. V., Bilsel, O., & Lambright, D. G. (2015). Methods for analysis of size-exclusion chromatography-small-angle X-ray scattering and reconstruction of protein scattering. *Journal of Applied Crystallography*, *48*, 1102–1113. https://doi.org/10.1107/s1600576715010420.

Mathew, E., Mirza, A., & Menhart, N. (2004). Liquid-chromatography-coupled SAXS for accurate sizing of aggregating proteins. *Journal of Synchrotron Radiation, 11*, 314–318. https://doi.org/10.1107/s0909049504014086.

Morales, J., Li, L., Fattah, F. J., Dong, Y., Bey, E. A., Patel, M., et al. (2014). Review of poly (ADP-ribose) polymerase (PARP) mechanisms of action and rationale for targeting in cancer and other diseases. *Critical Reviews in Eukaryotic Gene Expression, 24*(1), 15–28. https://doi.org/10.1615/critreveukaryotgeneexpr.2013006875.

Nielsen, S. S., Moller, M., & Gillilan, R. E. (2012). High-throughput biological small-angle X-ray scattering with a robotically loaded capillary cell. *Journal of Applied Crystallography, 45*, 213–223. https://doi.org/10.1107/s0021889812000957.

Padilla, J. E., Colovos, C., & Yeates, T. O. (2001). Nanohedra: Using symmetry to design self assembling protein cages, layers, crystals, and filaments. *Proceedings of the National Academy of Sciences of the United States of America, 98*(5), 2217–2221. https://doi.org/10.1073/pnas.041614998.

Roll, R. (1984). A simple implicit measure of the effective bid-ask spread in an efficient market. *The Journal of Finance, 39*(4), 1127–1139. https://doi.org/10.1111/j.1540-6261.1984.tb03897.x.

Rosenberg, O. S., Deindl, S., Sung, R. J., Nairn, A. C., & Kuriyan, J. (2005). Structure of the autoinhibited kinase domain of CaMKII and SAXS analysis of the holoenzyme. *Cell, 123*(5), 849–860. https://doi.org/10.1016/j.cell.2005.10.029.

Round, A., Felisaz, F., Fodinger, L., Gobbo, A., Huet, J., Villard, C., et al. (2015). BioSAXS sample changer: A robotic sample changer for rapid and reliable high-throughput X-ray solution scattering experiments. *Acta Crystallographica Section D-Structural Biology, 71*, 67–75. https://doi.org/10.1107/s1399004714026959.

Schneidman-Duhovny, D., Hammel, M., & Sali, A. (2010). FoXS: A web server for rapid computation and fitting of SAXS profiles. *Nucleic Acids Research, 38*(Web Server issue), W540–W544. https://doi.org/10.1093/nar/gkq461.

Shin, D. S., DiDonato, M., Barondeau, D. P., Hura, G. L., Hitomi, C., Berglund, J. A., et al. (2009). Superoxide dismutase from the eukaryotic thermophile Alvinella pompejana: Structures, stability, mechanism, and insights into amyotrophic lateral sclerosis. *Journal of Molecular Biology, 385*(5), 1534–1555. https://doi.org/10.1016/j.jmb.2008.11.031.

Svergun, D., Barberato, C., & Koch, M. H. J. (1995). CRYSOL—A program to evaluate x-ray solution scattering of biological macromolecules from atomic coordinates. *Journal of Applied Crystallography, 28*, 768–773. https://doi.org/10.1107/S0021889895007047.

Wilkinson, M. D., Dumontier, M., Aalbersberg, I. J., Appleton, G., Axton, M., Baak, A., et al. (2016). The FAIR guiding principles for scientific data management and stewardship. *Scientific Data, 3*, 160018. https://doi.org/10.1038/sdata.2016.18.

Williams, G. J., Williams, R. S., Williams, J. S., Moncalian, G., Arvai, A. S., Limbo, O., et al. (2011). ABC ATPase signature helices in Rad50 link nucleotide state to Mre11 interface for DNA repair. *Nature Structural & Molecular Biology, 18*(4), 423–U454. https://doi.org/10.1038/nsmb.2038.

Yang, L., Lazo, E., Byrnes, J., Chodankar, S., Antonelli, S., & Rakitin, M. (2021). Tools for supporting solution scattering during the COVID-19 pandemic. *Journal of Synchrotron Radiation, 28*, 1237–1244. https://doi.org/10.1107/s160057752100521x.

FSC
www.fsc.org
MIX
Papier | Fördert
gute Waldnutzung
FSC® C083411

Kolibri 360 GmbH
Ferdinand-Jühlke-Straße 7
99095 Erfurt, Deutschland
produktsicherheit@kolibri360.de